D1547108

Ecological Studies, Vol. 138

Analysis and Synthesis

Ecological Studies

Volumes published since 1993 are listed at the end of this book.

Springer
New York
Berlin
Heidelberg
Barcelona
Hong Kong
London
Milan
Paris
Singapore
Tokyo

Eric S. Kasischke Brian J. Stocks
Editors

Fire, Climate Change, and Carbon Cycling in the Boreal Forest

With 95 illustrations, 22 in color

Springer

Eric S. Kasischke
Environmental Research Institute
of Michigan
Ann Arbor, MI 48105
USA
ekas@erim-int.com

Brian J. Stocks
Canadian Forest Service—Ontario Region
Sault Ste. Marie, Ontario P6A 5M7
Canada
bstocks@nrcan.gc.ca

Cover illustration: Photo: 1976 boreal crown fire in Northwestern Ontario, Canada. Photo: B.J. Stocks, Canadian Forest Service. Map: Distribution of large forest fires in Canada and Alaska, 1980–1994.

Library of Congress Cataloging-in-Publication Data
Fire, climate change, and carbon cycling in the Boreal Forest/edited by Eric S. Kasischke, Brian J. Stocks.
p. cm. — (Ecological studies; 138)
Based on papers from a workshop held in Fairbanks, Alaska, Sept. 11–14, 1995; with other contributed papers.
Includes bibliographical references and index.
ISBN 0-387-98890-4 (alk. paper)
1. Fire ecology Congresses. 2. Climatic changes Congresses. 3. Carbon cycle (Biogeochemistry) Congresses. 4. Taiga ecology Congresses. I. Kasischke, Eric S. II. Stocks, Brian J. III. Series: Ecological studies; v. 138.
QH545.F5F5735 2000
577.2—dc21 99-16556

Printed on acid-free paper.

Production coordinated by Princeton Editorial Associates, Inc., and managed by Francine McNeill; manufacturing supervised by Erica Bresler.
Typeset by Princeton Editorial Associates, Inc., Roosevelt, NJ, and Scottsdale, AZ.
Printed and bound by Maple-Vail Book Manufacturing Group, York, PA.
Printed in the United States of America.

9 8 7 6 5 4 3 2 1

ISBN 0-387-98890-4 Springer-Verlag New York Berlin Heidelberg SPIN 10732730

Preface

On September 11 to 14, 1995, a workshop sponsored by the U.S. Environmental Protection Agency's Office of Research and Development was convened in Fairbanks, Alaska. The overall goal of this workshop was to review the relationship between climate, fire, and carbon cycling in the North American boreal forest. The attendees included forty-three scientists from throughout Canada and the United States, with backgrounds and interests that were intentionally diverse, including forestry, forest and wetland ecology, paleoecology, biogeochemical cycling, atmospheric chemistry, remote sensing, fire behavior and prediction, and natural-resource management and policy. Although it would have been desirable to have an even broader range of participation, particularly from Russian scientists working in the Siberian boreal forests, time and budgetary constraints precluded this possibility.

An important aspect of the workshop was that the attendees not only represented academic and research institutes but also included representatives from federal, provincial, and state government agencies responsible for fire management and research in the North American boreal forest. Given this mixture of participants, a major focus of the workshop was on current fire policy and management practices in the North American boreal forest and the potential effects that climate change will have on these management activities.

The papers presented at the Fairbanks workshop provide the basis for many of the chapters in this book, but other sources were used as well. Several Russian colleagues agreed to contribute chapters on topics pertaining to the Siberian boreal forest region. Several chapters are based on recent research on the effects of fire on ecosystem processes in Alaskan boreal forests, as well as the use of satellite remote sensing systems to monitor the effects of fire in this region. Finally, the extensive research efforts conducted by the Canadian Forest Service on fire prediction and behavior modeling are the source of material presented in several chapters.

The book itself is divided into four sections that represent the different topic areas pertinent to understanding the role of fire and carbon cycling in the boreal forest. Although the workshop focused on the North American boreal forest, most of the principles and methods presented in the individual chapters are applicable to the boreal forest in general.

The Fairbanks workshop was supported by a grant from the U.S. Environmental Protection Agency to the Oak Ridge National Laboratory (ORNL) of the U.S. Department of Energy. We thank Sandra Davis of ORNL for her assistance in organizing and conducting the workshop. In addition, we thank Dr. Barbara Levinson of the EPA's Office of Research and Development for first suggesting the topic of this workshop and then arranging for its sponsorship by the EPA.

The material presented in the various chapters of this book is the result of years of research sponsored by a wide range of government agencies. We particularly acknowledge the support of the Canadian Forest Service, the U.S. Forest Service, the U.S. National Aeronautics and Space Administration, and the U.S. Environmental Protection Agency. The research discussed in this book has not been subjected to review by any of these agencies and therefore does not necessarily reflect their views, and no official endorsement should be inferred.

Ann Arbor, Michigan, USA — Eric S. Kasischke
Sault Ste. Marie, Ontario, Canada — Brian J. Stocks

Contents

Preface v
Contributors xi

1. **Introduction** 1
Eric S. Kasischke and Brian J. Stocks

Section I. Information Requirements and Fire Management and Policy Issues 7
Brian J. Stocks and Eric S. Kasischke

2. **Boreal Ecosystems in the Global Carbon Cycle** 19
Eric S. Kasischke

3. **Boreal Forest Fire Emissions and the Chemistry of the Atmosphere** 31
Joel S. Levine and Wesley R. Cofer III

4. **Eurasian Perspective of Fire: Dimension, Management, Policies, and Scientific Requirements** 49
Johann G. Goldammer and Brian J. Stocks

5. **Fire Management in the Boreal Forests of Canada** 66
Paul C. Ward and William Mawdsley

6. **Effects of Climate Change on Management and Policy: Mitigation Options in the North American Boreal Forest** 85
Perry Grissom, Martin E. Alexander, Brad Cella, Frank Cole, J. Thomas Kurth, Norman P. Malotte, David L. Martell, William Mawdsley, James Roessler, Robert Quillin, and Paul C. Ward

Section II. Processes Influencing Carbon Cycling in the North American Boreal Forest 103
Eric S. Kasischke

7. **Distribution of Forest Ecosystems and the Role of Fire in the North American Boreal Region** 111
Laura L. Bourgeau-Chavez, Martin E. Alexander, Brian J. Stocks, and Eric S. Kasischke

8. **Extent, Distribution, and Ecological Role of Fire in Russian Forests** 132
Anatoly Z. Shvidenko and Sten Nilsson

9. **Long-Term Perspectives on Fire-Climate-Vegetation Relationships in the North American Boreal Forest** 151
Ian D. Campbell and Michael D. Flannigan

10. **Controls on Patterns of Biomass Burning in Alaskan Boreal Forests** 173
Eric S. Kasischke, Katherine P. O'Neill, Nancy H.F. French, and Laura L. Bourgeau-Chavez

11. **Postfire Stimulation of Microbial Decomposition in Black Spruce (*Picea mariana* L.) Forest Soils: A Hypothesis** 197
Daniel D. Richter, Katherine P. O'Neill, and Eric S. Kasischke

12. **Influence of Fire on Long-Term Patterns of Forest Succession in Alaskan Boreal Forests** 214
Eric S. Kasischke, Nancy H.F. French, Katherine P. O'Neill, Daniel D. Richter, Laura L. Bourgeau-Chavez, and Peter A. Harrell

Section III. Spatial Data Sets for the Analysis of Carbon Dynamics in Boreal Forests 237
Eric S. Kasischke and Brian J. Stocks

13. **Carbon Storage in the Asian Boreal Forests of Russia** 239
Vladislav A. Alexeyev, Richard A. Birdsey, Victor D. Stakanov, and Ivan A. Korotkov

14. **Characteristics of Forest Ecozones in the North American Boreal Region** 258
Laura L. Bourgeau-Chavez, Eric S. Kasischke, James P. Mudd, and Nancy H.F. French

15. **Historical Fire Records in the North American Boreal Forest** 274
Peter J. Murphy, James P. Mudd, Brian J. Stocks, Eric S. Kasischke, Donald Barry, Martin E. Alexander, and Nancy H.F. French

16. **Fire and the Carbon Budget of Russian Forests** 289
Anatoly Z. Shvidenko and Sten Nilsson

17. **Using Visible and Near-Infrared Satellite Imagery to Monitor Boreal Forests** 312
Frank J. Ahern, Helmut Epp, Donald R. Cahoon, Jr., Nancy H.F. French, Eric S. Kasischke, and Jeffery L. Michalek

18. **Monitoring Boreal Forests by Using Imaging Radars** 331
Eric S. Kasischke, Laura L. Bourgeau-Chavez, Nancy H.F. French, and Peter A. Harrell

Section IV. Modeling of Fire and Ecosystem Processes and the Effects of Climate Change on Carbon Cycling in Boreal Forests 347
Eric S. Kasischke and Brian J. Stocks

19. **Modeling of Fire Occurrence in the Boreal Forest Region of Canada** 357
Kerry Anderson, David L. Martell, Michael D. Flannigan, and Dongmei Wang

20. **Climate Change and Forest Fire Activity in North American Boreal Forests** 368
Brian J. Stocks, Michael A. Fosberg, Michael B. Wotton, Timothy J. Lynham, and Kevin C. Ryan

21. **Carbon Release from Fires in the North American Boreal Forest** 377
Nancy H.F. French, Eric S. Kasischke, Brian J. Stocks, James P. Mudd, David L. Martell, and Bryan S. Lee

22. **Ecological Models of the Dynamics of Boreal Landscapes** 389
Herman H. Shugart, Donald F. Clark, and Amber J. Hill

23. **Using Satellite Data to Monitor Fire-Related Processes in Boreal Forests** 406
Eric S. Kasischke, Nancy H.F. French, Laura L. Bourgeau-Chavez, and Jeffery L. Michalek

Some see near end

24. Influences of Fire and Climate Change on Patterns of Carbon Emissions in Boreal Peatlands 423
Leslie A. Morrissey, Gerald P. Livingston, and Steven C. Zoltai

25. Effects of Climate Change and Fire on Carbon Storage in North American Boreal Forests 440
Eric S. Kasischke

Index 453

Contributors

Frank J. Ahern	Canada Center for Remote Sensing, Ottawa, Ontario K1A 0Y7, Canada
Martin E. Alexander	Canadian Forest Service, Edmonton, Alberta T6H 3S5, Canada
Vladislav A. Alexeyev	St. Petersburg Forestry Research Institute, St. Petersburg 194021, Russia
Kerry Anderson	Canadian Forest Service, Edmonton, Alberta T6H 3S5, Canada
Donald Barry	Bureau of Land Management, Alaska Fire Service, Fort Wainwright, AK 99703, USA
Richard A. Birdsey	USDA Forest Service, Radnor, PA 19087-4585, USA
Laura L. Bourgeau-Chavez	Environmental Research Institute of Michigan, Ann Arbor, MI 48105, USA

Donald R. Cahoon, Jr.	NASA Langley Research Center, Atmospheric Sciences Division, Hampton, VA 23681-0001, USA
Ian D. Campbell	Canadian Forest Service, Edmonton, Alberta T6H 3S5, Canada
Brad Cella	National Park Service, Alaska Regional Office, Anchorage, AK 99503, USA
Donald F. Clark	Department of Environmental Sciences, University of Virginia, Charlottesville, VA 22901, USA
Wesley R. Cofer III	NASA Langley Research Center, Atmospheric Sciences Division, Hampton, VA 23681-0001, USA
Frank Cole	Department of Natural Resources, Division of Forestry, Fairbanks, AK 99709, USA
Helmut Epp	NWT Centre for Remote Sensing, Department of Renewable Resources, Yellow Knife, North West Territories NTx1A3S8, Canada
Michael D. Flannigan	Canadian Forest Service, Edmonton, Alberta T6H 3S5, Canada
Michael A. Fosberg	IGBP/BAHC, Potsdam Institute for Climate Impact Research, D-14412 Potsdam, Germany
Nancy H.F. French	Environmental Research Institute of Michigan, Ann Arbor, MI 48105, USA
Johann G. Goldammer	Fire Ecology Research Group, The Global Fire Monitoring Center, Max Planck Institute for Chemistry, Freiburg University, D-79085 Freiburg, Germany
Perry Grissom	U.S. Fish and Wildlife Service, Yukon Flats NWR, Fairbanks, AK 99701, USA

Peter A. Harrell	Nicholas School of the Environment, Duke University, Durham, NC 27708-0328, USA
Amber J. Hill	Department of Environmental Sciences, University of Virginia, Charlottesville, VA 22901, USA
Eric S. Kasischke	Environmental Research Institute of Michigan, Ann Arbor, MI 48105, USA
Ivan A. Korotkov	V.N. Sukachev Forest Institute, Akademgorodok, Krasnoyarsk 660036, Russia
J. Thomas Kurth	Department of Natural Resources, Division of Forestry, Fairbanks, AK 99709, USA
Bryan S. Lee	Canadian Forest Service, Edmonton, Alberta T6H 3S5, Canada
Joel S. Levine	NASA Langley Research Center, Atmospheric Sciences Division, Hampton, VA 23681-0001, USA
Gerald P. Livingston	University of Vermont, School of Natural Resources, Burlington, VT 05405, USA
Timothy J. Lynham	Canadian Forest Service—Ontario Region, Sault Ste. Marie, Ontario P6A 5M7, Canada
Norman P. Malotte	Department of Natural Resources, Division of Forestry, Fairbanks, AK 99709, USA
David L. Martell	University of Toronto, Faculty of Forestry, Toronto, Ontario 5MS 3B3, Canada
William Mawdsley	Forest Management Division, Department of Resources, Wildlife and Economic Development, Fort Smith, North West Territories X0E 0P0, Canada

Jeffery L. Michalek — Environmental Research Institute of Michigan, Ann Arbor, MI 48105, USA

Leslie A. Morrissey — University of Vermont, School of Natural Resources, Burlington, VT 05405, USA

James P. Mudd — Environmental Research Institute of Michigan, Ann Arbor, MI 48105, USA

Peter J. Murphy — Canadian Forest Service—Ontario Region, Sault Ste. Marie, Ontario P6A 5M7 Canada

Sten Nilsson — International Institute for Applied Spatial Analysis, A-2361 Laxenburg, Austria

Katherine P. O'Neill — Nicholas School of the Environment, Duke University, Durham, NC 27708-0328, USA

Robert Quillin — Bureau of Land Management, Alaska Fire Service, Fort Wainwright, AK 99703, USA

Daniel D. Richter — Nicholas School of the Environment, Duke University, Durham, NC 27708-0328, USA

James Roessler — Bureau of Land Management, Alaska Fire Service, Fort Wainwright, AK 99703, USA

Kevin C. Ryan — USDA Forest Service, Intermountain Fire Sciences Laboratory, Missoula, MT 59802, USA

Herman H. Shugart — Department of Environmental Sciences, University of Virginia, Charlottesville, VA 22901, USA

Anatoly Z. Shvidenko — International Institute for Applied Spatial Analysis, A-2361 Laxenburg, Austria

Victor D. Stakanov	V.N. Sukachev Forest Institute, Akademgorodok, Krasnoyarsk 660036, Russia
Brian J. Stocks	Canadian Forest Service—Ontario Region, Sault Ste. Marie, Ontario P6A 5M7 Canada
Dongmei Wang	University of Toronto, Faculty of Forestry, Toronto, Ontario 5MS 3B3, Canada
Paul C. Ward	Aviation, Flood and Fire Management Branch, Ontario Ministry of Natural Resources, Sault Ste. Marie, Ontario P6A 6U5, Canada
Michael B.Wotton	Canadian Forest Service—Ontario Region, Sault Ste. Marie, Ontario P6A 5M7, Canada
Steven C. Zoltai (deceased)	Canadian Forest Service, Edmonton, Alberta T6H 3S5, Canada

1. Introduction

Eric S. Kasischke and Brian J. Stocks

The world's boreal forests play a critical role in a wide range of global change science issues. First, boreal forests are located in the region (e.g., high northern latitudes) where the current generation of general circulation models predicts there will be the most pronounced warming in response to the rising levels of atmospheric greenhouse gases (see Chapter 20). Second, because temperature is an important factor (if not the most important factor) controlling succession in boreal forests, the distribution of forest ecosystems in this region is likely to change dramatically in response to climate warming (see Chapter 22). Third, the boreal forest contains a significant portion of the world's timber resources, and changes to the growth patterns of these forests will affect the economies of many countries, particularly Canada and Russia. Fourth, fire is the major disturbance pattern in boreal forests, with its occurrence closely coupled to climate patterns; therefore changes in climate will result in changes to the fire regime as well (see Chapter 22). And finally, although boreal forests are presently one of the major terrestrial pools for carbon, shifts in the fire regime and ecosystem distribution in high northern latitudes associated with climate change are likely to result in significant increases in the atmospheric concentration of carbon dioxide and other greenhouse gases (Kasischke et al. 1995; Kasischke, this volume, Chapter 25).

A strong argument can be made that over the past quarter century, changes in climate and the fire regime have already started to cause a shift in the carbon dynamics in boreal forests (Kurz and Apps 1996). Figure 1.1[1] presents changes in

[1]Figure 1.1 will be found in the color insert.

average temperature over the past 30 years created through an analysis of surface weather records by Hansen and colleagues (1996). This figure shows that the most significant areas of warming coincide with the region occupied by the boreal forest. Given the close linkage between fire occurrence and climate (Flannigan and Harrington 1988; Stocks et al., this volume), there should be little surprise that over the past two decades there has been a significant increase (almost threefold) in the annual area burned in the North American boreal forest (Figure 1.2).

Changes in the fire regime and the vegetation and forest cover of the boreal forest will have far-reaching consequences within the policy and natural resource management communities. From the climate change policy perspective, it is essential to understand and quantify the role of the boreal forest as a net source or sink of atmospheric greenhouse gases, primarily carbon dioxide (Chapter 3). This information is essential for determining the total carbon budget for countries with significant areas of boreal forests, primarily Russia and Canada. In the light of the recent signing of the Kyoto Protocol of the United Nations Framework Convention on Climate Change, accurate information on the carbon budget of the boreal forest is required to develop policies designed to use boreal forests as a sequestra-

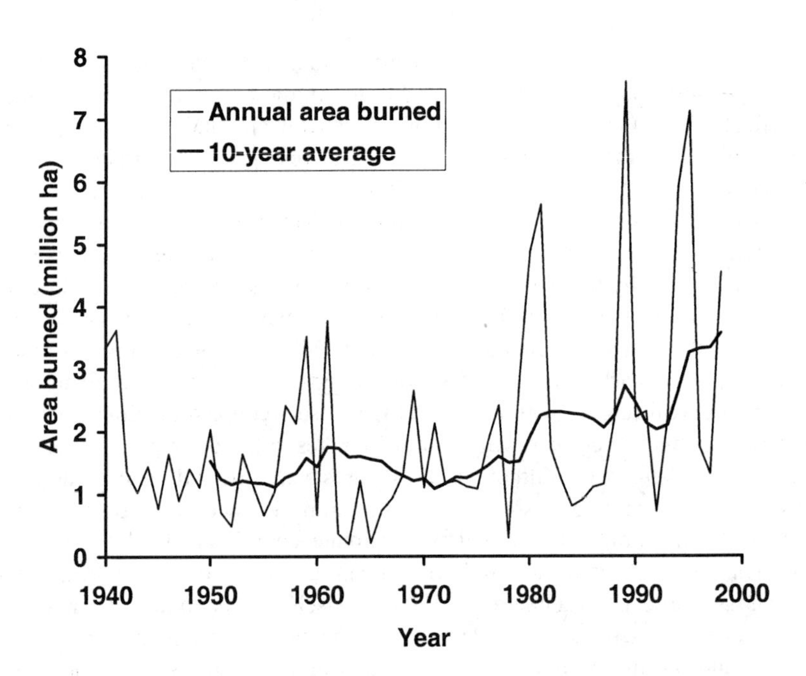

Figure 1.2. Annual area burned in the North American boreal forest region based on historical fire records.

tion mechanism to offset carbon released through the burning of fossil fuels. However, the value and practicality of implementing such carbon sequestration policies need to be weighed against their cost and alternative uses of the forested lands. These competing interests and the fact that the fundamental nature of the boreal forest is likely to be dramatically altered over the next half century present unique challenges to natural resource managers in this region (Chapters 4–6).

There are a significant number of scientific publications on the ecology of boreal forests (Larsen 1980, 1989; Van Cleve et al. 1986; Shugart et al. 1992; Paavilainen and Paivanen 1995), the role of fire in this ecosystem (Wein and MacLean 1983; Goldammer and Furyaev 1996), the potential effects of climate change on arctic ecosystems (Oechel et al. 1996), and the processes controlling net gas exchange and energy exchange in undisturbed boreal forests (Blanken et al. 1998, Goulden et al. 1998, McCaughey et al. 1997). Although most of these previous works address the role of climate and fire as a controlling factor of carbon cycling in boreal forest (particularly Wein and MacLean 1983 and Goldammer and Furyaev 1996), none does so in a completely integrated fashion. In addition, recent research has provided new insights into the importance of fire on ecosystem processes and carbon cycling.

The purpose of this book is to present an overview of our current understanding on the interrelationship between fire, climate, and carbon cycling in boreal forests. The chapters are organized into four sections. The chapters in the first section are intended to outline the types of information that scientists, natural resource managers, and policy makers require with respect to fire, its effects on the ecology of boreal forest and the atmosphere, and its influence on global climate change issues. In addition, chapters are presented that summarize the challenges faced by natural resource managers.

The chapters in the second section address the role that fire plays in the ecology of boreal forests. Many of the chapters in this section are based on recent research conducted in the boreal forests of interior Alaska. The chapters in the third section summarize different data sets that describe the distribution of fire and carbon throughout the boreal forest, as well as present overviews on using satellite imagery to monitor characteristics of the boreal forest. The chapters of the fourth section summarize approaches to modeling the occurrence of fire and ecosystem processes in boreal forests, carbon release, and how changes in climate will affect fire and carbon cycling in this biome.

The material presented in this book is intended to complement that published elsewhere. In particular, those interested in more detail on the role of fire in Eurasian boreal forest are referred to Goldammer and Furyaev (1996). Wein and MacLean (1983) present an excellent overview of the role of fire in boreal ecosystems, and Van Cleve and co-workers (1986) present what is probably the defining integrated set of research conducted over the past two decades on the ecology of a specific boreal forest ecosystem. Those interested in the development and application of quantitative models describing ecosystem processes in the boreal forest are referred to Shugart and associates (1992). Finally, those interested in contrasting the role of fire in boreal forests with other ecosystems should review Levine (1991, 1996a,b) and Crutzen and Goldammer (1993).

References

Blanken, P.D., T.A. Black, P.C. Yang, H.H. Neumann, Z. Nesic, R. Staebler, G. den Hartog, M.D. Novak, and X. Lee. 1998. Energy balance and canopy conductance of a boreal aspen forest: partitioning overstory and understory components. *J. Geophys. Res.* 102:28,915–28,927.

Crutzen, P.J., and J.G. Goldammer, eds. 1993. *Fire in the Environment: The Ecological, Atmospheric, and Climatic Importance of Vegetation Fires.* John Wiley & Sons, West Sussex, UK.

Flannigan, M.D., and J.D. Harrington. 1988. A study of the relation of meteorologic variables to monthly provincial area burned by wildfire in Canada. *J. Appl. Meteor.* 27:441–452.

Goldammer, J.G., and V.V. Furyaev, eds. 1996. *Fire in Ecosystems of Boreal Eurasia.* Kluwer Academic Publishers, Dordecht, The Netherlands.

Goulden, M.L., S.C. Wofsy, J.W. Harden, S.E. Trumbore, P.M. Crill, S.T. Gower, T. Fries, B.C. Daube, S-M. Fan, D.J. Sutton, A. Bazzaz, and J.W. Munger. 1998. Sensitivity of boreal forest carbon balance to thaw. *Science* 279:214–217.

Hansen, J., R. Ruedy, M. Sato, and R. Reynolds. 1996. Global surface air temperature in 1995: Return to pre-Pinatubo level. *Geophys. Res. Lett.* 23:1665–1668.

Kasischke, E.S., N.L. Christensen, Jr., and B.J. Stocks. 1995. Fire, global warming and the mass balance of carbon in boreal forests. *Ecol. Appl.* 5:437–451.

Kurz, W.A., and M.J. Apps. 1996. Retrospective assessment of carbon flows in Canadian boreal forests, pp. 173–182 in M.J. Apps and D.T. Price, eds. *Forest Management and the Global Carbon Cycle.* NATO ASI Series, subseries 1, vol. 40, Global Environmental Change. Springer-Verlag, Berlin.

Larsen, J.A. 1980. *The Boreal Ecosystem.* Academic Press, New York.

Larsen, J.A. 1989. *The Northern Forest Border in Canada and Alaska.* Ecological Studies 70. Springer-Verlag, New York.

Levine, J.S., ed. 1991. *Global Biomass Burning: Atmospheric, Climatic, and Biospheric Implications.* MIT Press, Cambridge, MA.

Levine, J.S., ed. 1996a. *Biomass Burning and Global Change. Vol. 1. Remote Sensing, Modeling and Inventory Development, and Biomass Burning in Africa.* MIT Press, Cambridge, MA.

Levine, J.S., ed. 1996b. *Biomass Burning and Global Change. Vol. 2. Biomass Burning in South America, Southeast Asia, and Temperate and Boreal Ecosystems, and the Oil Fires of Kuwait.* MIT Press, Cambridge, MA.

McCaughey, J.H., P.M. Lafleur, D.W. Joiner, P.A. Bartlett, A.M. Costello, D.E. Jelinski, and M.G. Ryan. 1997. Magnitudes and seasonal patterns of energy, water and carbon exchanges at a boreal young jack pine forest in the BOREAS northern study area. *J. Geophys. Res.* 102:28,997–29,007.

Oechel, W.C., T. Callaghan, T. Gilmanov, J.I. Holten, B. Maxwell, U. Molau, and B. Sveinbjornsson, eds. 1996. *Global Climage Change and Arctic Terrestrial Ecosystems.* Ecological Studies 124. Springer-Verlag, New York.

Paavilainen, E., and J. Paivanen. 1995. *Peatland Forestry.* Ecological Studies 111. Springer-Verlag, New York.

Shugart, H.H., R. Leemans, and G.B. Bonan, eds. 1992. *A Systems Analysis of the Global Boreal Forest.* Cambridge University Press, Cambridge, UK.

Van Cleve, K., F.S. Chapin III, P.W. Flanagan, L.A. Viereck, and C.T. Dyrness, eds. 1986. *Forest Ecosystems in the Alaskan Taiga.* Ecological Studies 57. Springer-Verlag, New York.

Wein, R.W., and D.A. MacLean, eds. 1983. *The Role of Fire in Northern Circumpolar Ecosystems.* John Wiley & Sons, New York.

I. INFORMATION REQUIREMENTS AND FIRE MANAGEMENT AND POLICY ISSUES

Brian J. Stocks and Eric S. Kasischke

Introduction

Between 1958 and 1992, the atmospheric concentration of CO_2 rose from 315 to 356 ppmv, a rate of 1.2 ppmv yr^{-1}. This was first documented through data collected at the Mauna Loa Observatory by Keeling (1982) and has since been documented in other locations throughout the world (Fig. I.1a). Analysis of the ratio of $^{13}C/^{12}C$ in air bubbles trapped within glacial ice cores shows that the concentration of atmospheric CO_2 has been increasing since at least 1750, when the concentration was 280 ppmv (Friedli et al. 1986).

There is no doubt that the atmospheric gases such as CO_2 create a "greenhouse" effect. While these gases transmit the shorter wavelength energy being emitted by the sun (which is absorbed and warms the earth's surface), they absorb the longer-wave infrared energy being emitted by the earth's surface. They are responsible for the warming of the earth's atmosphere. Without these greenhouse gases, the earth's air temperature would be at least 40°C cooler than it is today.

Because of its primary role in the radiative forcing of the atmosphere, throughout the latter part of this century there has been an ongoing debate as to the possible effects of the rise of the atmospheric concentration of CO_2 and other greenhouse gases. A scientific consensus is emerging that (1) the addition of CO_2 and other greenhouse gases is causing a rise in the average temperature in the atmosphere and (2) these rises will likely continue in the future with continuing

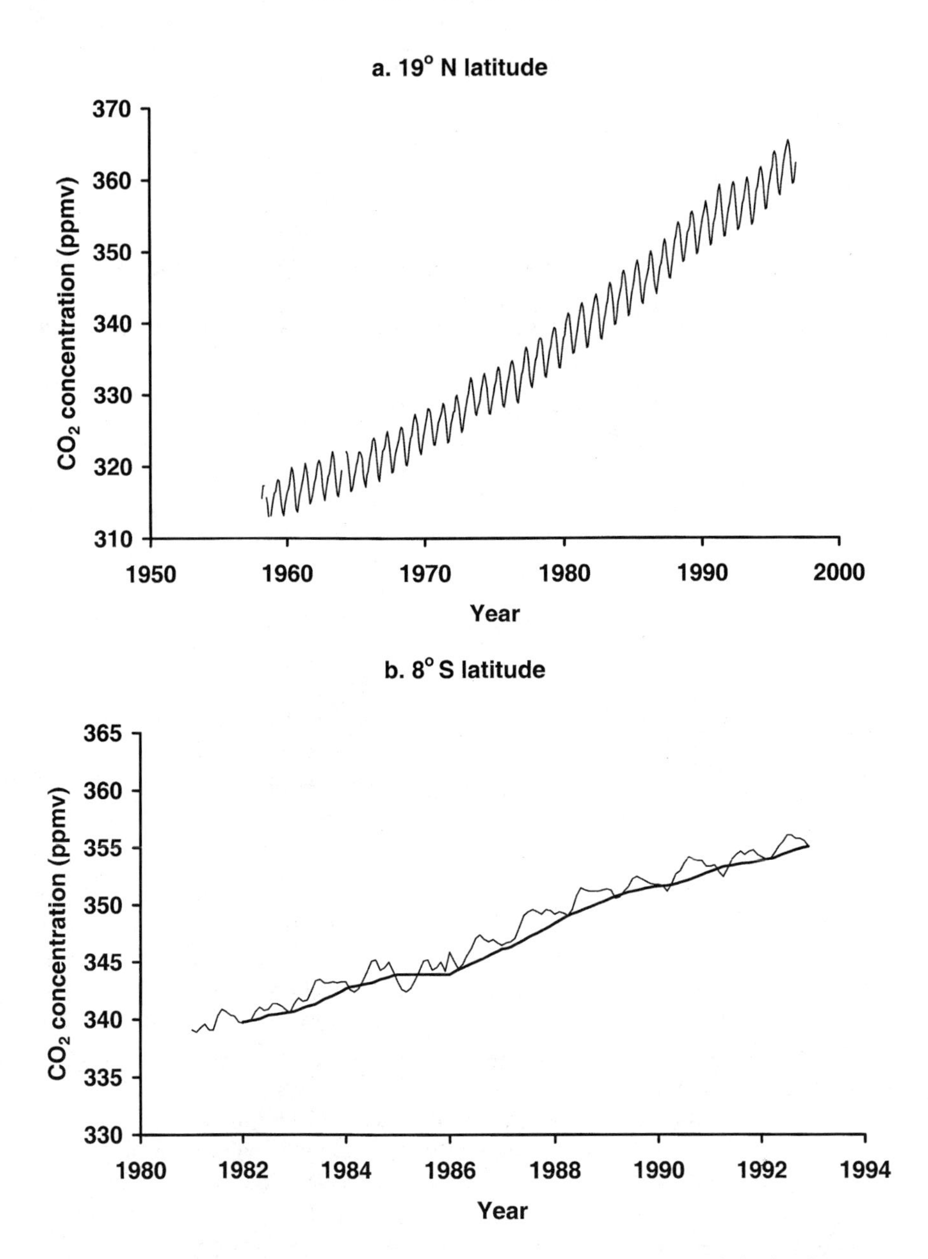

Figure I.1. Records of the atmospheric concentration of CO_2 at different latitudes. Data are for (a) 1958–1996 collected at the Mauna Loa Observatory, and from 1981 to 1992 at stations representing (b) tropical regions, (c) temperate regions, and (d) boreal regions. (Data were obtained from the Oak Ridge National Laboratory's Carbon Dioxide Information Analysis Center through the World Wide Web at http://cdiac.esd.ornl/home.html.)

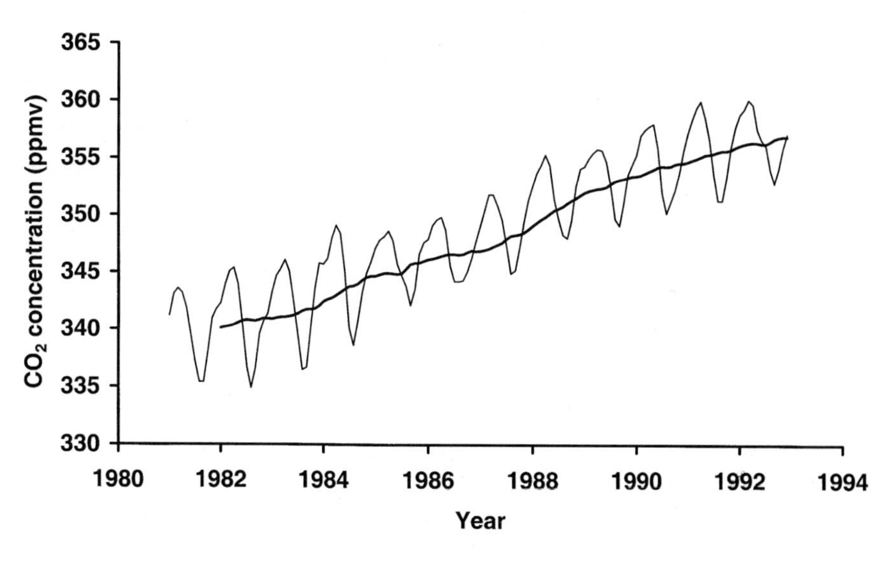

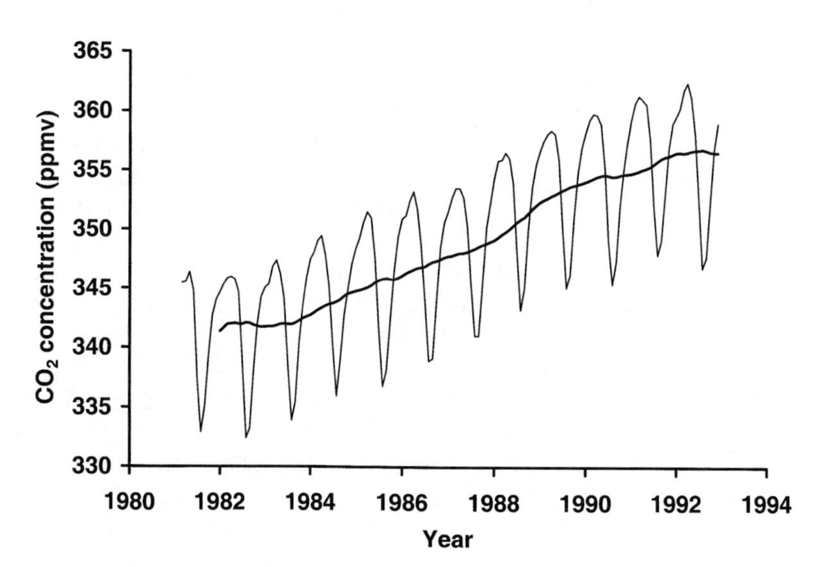

Figure I.1. (*continued*)

addition of greenhouse gases to the atmosphere (Houghton et al. 1990, 1996). What is even more certain in the minds of the scientific community is that the continual addition of greenhouse gases to the atmosphere will ultimately cause changes to atmospheric processes, which, in turn, will result in changes to the earth's biosphere and geosphere.

Great Carbon Debate

The majority of the increase in the atmospheric concentration of CO_2 over the past one and one-half centuries has been due to the combustion of fossil fuels. The burning of fossil fuels contributed 6.0 ± 0.5 Gt C to the atmosphere in 1990, up from 5.7 Gt C in 1987[1] (Watson et al. 1992). Between the beginning of the industrial revolution (ca. 1860) and 1987, Post (1993) estimates that enough carbon was added to the atmosphere to raise the concentration of CO_2 by 92 ppmv. Yet during the same time period, the atmospheric concentration of CO_2 rose by only 59 ppmv. To balance the atmospheric carbon budget requires a large net uptake of CO_2 by other components of the global carbon cycle. During the 1980s, the atmospheric concentration of CO_2 rose by an average of 1.4 ppmv yr^{-1}, or 3.3 Gt C yr^{-1}.

Determining the sources of this sink of atmospheric CO_2 is one of the major challenges facing the scientific community. There is currently 750 Gt C present in the atmosphere. In a single year, 25% of the entire pool of carbon present in the atmosphere is exchanged with the terrestrial biosphere and oceans. This means that the oceans and terrestrial biomes are annually releasing 30 times more CO_2 to the atmosphere than is released through the burning of fossil fuels. Although the rates of exchange with the atmosphere are roughly equal for the terrestrial and ocean sources, the mechanisms for exchange are different.

In the oceans, the atmospheric partial pressure of CO_2 and the equilibrium partial pressure of CO_2 in the surface waters drive the rate of exchange of carbon with the atmosphere. Thus, the rate of exchange of CO_2 between the atmosphere and oceans is largely dictated by physics (e.g., differences in the partial pressure of CO_2, the temperatures of the ocean and atmosphere, and local wind speeds). In the oceans, primary producers such as algae affect the carbon cycle by removing CO_2 from the water, thus lowering the partial pressure of CO_2 in the surface waters. The exchange of carbon between surface ocean waters and deep ocean waters is accomplished through deepwater upwelling and other ocean circulation, venting of CO_2 through the thermocline, and raining of detritus into the deeper ocean waters from dead plant and animal material from the surface waters.

By contrast, the direct exchange of CO_2 between the atmosphere and terrestrial biomes is through a combination of biological and physical processes. The biological processes include plant photosynthesis, plant respiration, and the decomposition of dead organic matter by fungi and microbes. The physical process is biomass burning during fire. All these processes, in turn, are controlled by the physical and chemical, characteristics of the environment.

---

[1]The global change community uses a number of different units to represent the amounts of carbon present in the world's oceans, biosphere, and atmosphere. 1 Gt = 1 gigatonne = 10^9 tonnes = 10^{15} g = 1 petagram = 1 Pg. Another common unit is the teragram: 1 teragram = 1 Tg = 10^{12} g. Note that 1 ppmv CO_2 in the global atmosphere equals 2.12 Gt C and 7.8 Gt CO_2.

Human activities more directly influence physical, chemical and biological processes in terrestrial ecosystems than they do in oceanic systems; thus, anthropogenic processes are more likely to perturb the carbon cycle in terrestrial biomes than in oceanic systems. For example, in addition to 6.0 Gt C yr^{-1} currently being added to the atmosphere from the burning of fossil fuels, land-clearing activities (primarily in the tropics) also add another 1.6 ± 0.7 Gt C yr^{-1} through the conversion of plant biomass to CO_2 (Houghton 1995).

Another example is the large amounts of carbon being added to the atmosphere through the burning of biomass from fires set by humans in savannas, about 1.0 Gt C yr^{-1} (Lacaux et al. 1993). Although carbon released through biomass burning in savannas has a relatively short life span in the atmosphere because of the rapid reaccumulation of biomass through plant regrowth, the addition of carbon-based gases to the atmosphere has profound influences on a wide range of atmospheric processes, particularly in the Southern Hemisphere (Crutzen and Carmichael 1993; Dickinson 1993).

Of the 7.6 Gt C yr^{-1} being added to the atmosphere through human activities, it is estimated that 2 Gt C yr^{-1} is being removed through increased absorption of CO_2 by the ocean (Watson et al. 1990, 1992). This leaves about 2.3 Gt C yr^{-1} unaccounted for, an amount commonly referred to as the "missing carbon sink." Arguments have been made that this sink must be located within the terrestrial biomes of the Northern Hemisphere (Tans et al. 1990; Ciais et al. 1995), based on observations that (1) the potential for oceans serving as this sink are limited by rates of diffusion between the ocean and atmosphere; (2) for the past several decades, there has been considerable deforestation in the tropical forests located in the Southern Hemisphere, making this region a net source for atmospheric CO_2, not at sink; (3) increased concentrations of atmospheric CO_2 should have a fertilization effect on plants, increasing the uptake of carbon on terrestrial biomes (Korner 1993); and (4) the largest portion of the earth's landmass is located in the Northern Hemisphere.

The source of the missing carbon sink remains a hotly debated subject. Based on an analysis of permanent sample plot data, Phillips and colleagues (1998) argue that tropical forests may, in fact, be a stronger sink of atmospheric carbon than previously thought. Based on an analysis of atmospheric CO_2 measurements collected at sites distributed throughout the world, Fan and co-workers (1998) use an inverse modeling approach to argue that the principal carbon sink lies within the North American temperate forest zone, with a strength of 1.7 Gt C yr^{-1}. This last study concludes that the boreal forest is a weaker carbon sink of 0.6 Gt C yr^{-1}.

One of the unique characteristics of gases present in the atmosphere is that although the rate of latitudinal mixing (e.g., mixing in a north-to-south direction) is relatively slow (e.g., on the order of several years to greater than a decade), mixing in an east-west direction occurs more quickly (on the order of several months). Because of this, atmospheric CO_2 measurements collected at a specific sampling station reflect processes occurring at the latitude where the data are collected. This phenomenon is illustrated in Figure I.1b to d (from data collected at three stations: one near the equator, one in the temperate region, and the third in the boreal region).

The intra- and interannual variations in the atmospheric concentration of CO_2 in Figure I.1 represent a number of processes occurring at different temporal and spatial scales, including (1) combustion of fossil fuels; (2) combustion of plant biomass from human activities and natural fires; (3) plant photosynthesis and respiration; (4) diffusion to and from the ocean surface; and (5) decomposition of dead organic material (heterotrophic respiration).

Comparison of the decadal increases in the atmospheric concentration of CO_2 at different latitudes shows a similar rise (between 1.35 and 1.40 ppmv yr^{-1}). The average atmospheric CO_2 concentrations in the temperate and boreal latitudes are 1.4 to 1.6 ppmv higher than at the equator because of the higher rates of fossil fuel burning in the industrialized nations of this region. Finally, the annual CO_2 fluctuations observed in the northern latitude data sets presented in Figure I.1 are due to seasonal variations in terrestrial ecosystem processes. Plant photosynthesis during the spring and summer lowers the level of atmospheric CO_2, and soil and plant respiration during the rest of the year result in an increased atmospheric CO_2 (Randerson et al. 1997; Field et al. 1998). The lower seasonal fluctuations occur in equatorial regions because of the lack of a distinct growing season.

Interannual variations in atmospheric CO_2 concentrations can also be detected in the data records, as illustrated in Figure I.2. This normalized plot of carbon exchange was created by subtracting the average CO_2 concentration of the previous 12 months from the monthly average and converting the result from ppmv CO_2 to Gt C. The data in Figure I.2 show that the average annual difference between the minimum and maximum levels of atmospheric carbon (the amplitude of the seasonal signature) is 30.2 Gt C. However, this value ranges between a low of 27.9 Gt C in 1987 and a high of 32.8 Gt C in 1985. The difference of 4.8 Gt C illustrates that shorter-term processes in the boreal forest region strongly influence atmosphere/land carbon exchanges. The challenge facing the boreal scientific community is determining how different terrestrial ecosystem processes contribute to variations in the atmospheric carbon budget represented in the plot in Figure I.2 at different time scales.

In addition to an increase in the concentration of atmospheric CO_2, the data records also show an increase in the amplitude (maximum minus minimum value) of the annual CO_2 oscillation (Fig. I.3). The magnitude of the rise in amplitude also varies as a function of latitude. At both the South Pole and Mauna Loa, Hawaii, the oscillation amplitude has risen by 0.025 ppmv CO_2 yr^{-1} since 1965. At Barrow, Alaska, the oscillation amplitude has risen by 0.158 ppmv CO_2 yr^{-1} since 1975. The cause of these rising amplitudes is subject to debate (Solomon and Cramer 1993). On the one hand, there is the argument that because CO_2 is a limiting factor in plant growth, increases in the atmospheric CO_2 will result in increases in photosynthesis without a corresponding increase in respiration and decomposition. In this line of reasoning, the increase in the oscillation amplitude is due entirely to uptake of CO_2 by plants; thus, the increasing amplitude supports the hypothesis of a terrestrial carbon sink. Others argue that the increasing ampli-

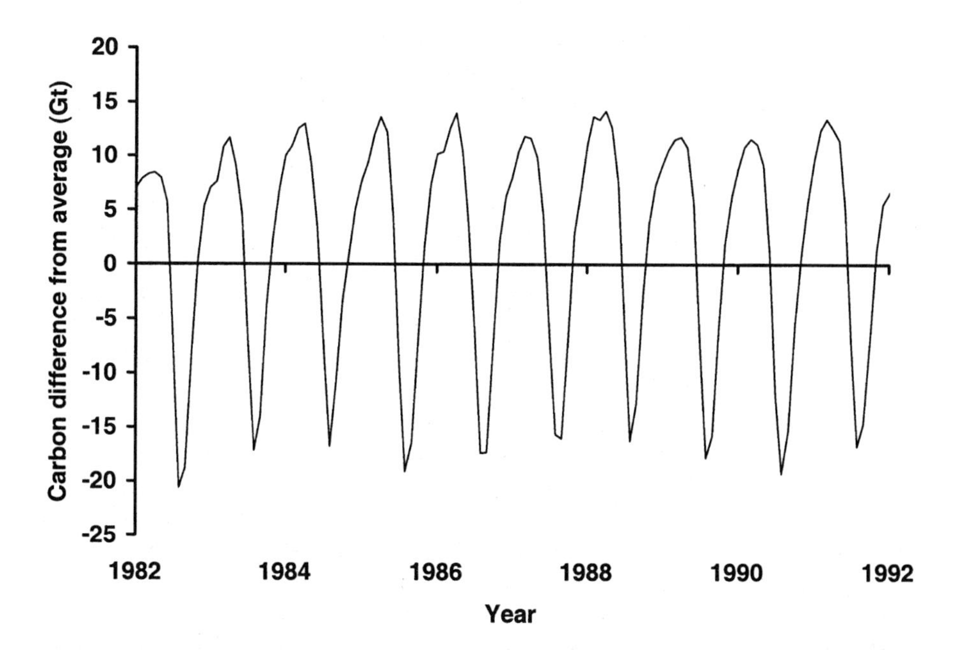

Figure I.2. Average seasonal variation in carbon exchange between the earth's surface and atmosphere in boreal regions, 1981–1992, based on normalizing the data presented in Figure I.1c.

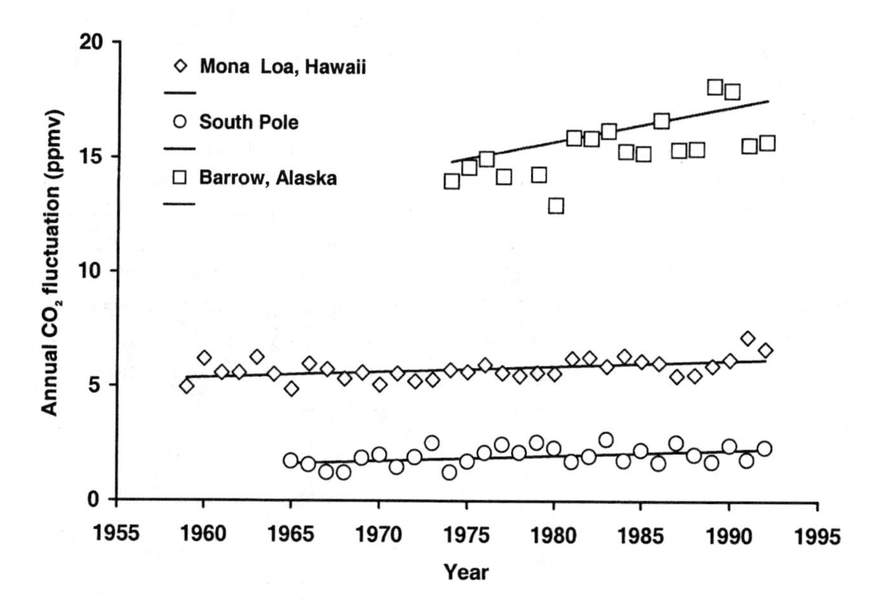

Figure I.3. Annual amplitude (maximum minus minimum) in the concentration of atmospheric CO_2 for three different latitudes based on data presented in Figure I.1.

tude may be due to increases in soil decomposition (which releases CO_2 to the atmosphere) because of a warming climate. This line of reasoning supports an argument that northern latitude terrestrial ecosystems are acting as a net source of atmospheric carbon. In either case, there are not sufficient data to prove or disprove either theory.

Finally, ecologists have long recognized the influence of climate on a wide range of ecosystem processes controlling plant distribution, ecosystem composition, and plant and forest succession. Holdridge (1947) showed that broad-scale vegetation formations could be defined in terms of bioclimatic variables: biotemperature, mean annual precipitation, and potential evapotranspiration. This classification scheme has allowed ecosystem modelers to estimate the effects of climate changes on vegetation distribution on a global basis (Emmanuel et al. 1985; Cramer and Leemans 1993). Modeling studies predict that if climate change does occur, the distribution of terrestrial ecosystems will be altered, with corresponding changes in the patterns of carbon storage in the terrestrial biome (Smith et al. 1992). Because cold temperatures are a limiting factor in ecosystem development in boreal regions and because all general circulation models (GCM) predict that the most significant changes in temperature will occur in high northern latitudes, there is little doubt that major changes in the distribution of ecosystems in this region will occur, as well as a wide range of processes within these ecosystems. For instance, Oechel and colleagues (1995) showed that over the past several decades, the gradual warming in the climate of the North Slope of Alaska has led to a drying out of the upper soil layers in the tundra ecosystems of this region. This drying has resulted in a corresponding increase in the rates of aerobic decomposition in the upper soil layers of these ecosystems, resulting in a net release of CO_2 to the atmosphere. More recently, Goulden and co-workers (1998) showed that recent increases in temperatures have resulted in increased soil respiration in a mature boreal forest in northern Manitoba.

In terms of studying long-term changes in land cover in boreal regions, one factor that has not been accounted for is the influence of fire. Flannigan and Harrington (1988), Flannigan and Van Wagner (1991), and Wotton and Flannigan (1993) studied the potential effects of changes in climate in fire regime in the Canadian boreal forest. The conclusion from these studies was that changes in climate projected for this region by the current generation of GCMs will not only result in longer fire seasons but also increase the probability of severe fire weather; thus, the fire frequency in boreal forests is also likely to increase due to climate warming. Scientists are just beginning to explore the linkages between climate change, fire, and the global carbon budget (Kasischke et al. 1995; Kasischke 1996).

Section Overview

In terms of the climate change and carbon debates, the important unanswered questions include: (1) What role does the boreal forest play in terms of adding and/or removing carbon from the atmosphere? (2) What form will greenhouse gas–induced changes to the atmosphere take at regional scales? (3) Over what time scales will these changes occur? (4) What options do we have for mitigating anthropogenic inputs of greenhouse gases to the atmosphere?

In the chapters of this section, we present an overview of the importance of these questions from a variety of different perspectives. In Chapter 2, Kasischke discusses the overall role of the boreal forest in the global carbon budget, including a comparison of the role of boreal forests versus temperate and tropical forests in the global carbon budget. In Chapter 3, Levine and Cofer present the importance of trace gases emitted from fires in boreal forests to important chemical processes and reactions that occur in the atmosphere. In this chapter, they point out that although the effects of biomass burning on atmospheric chemistry have received intensive study and attention in tropical and subtropical regions, the scientific community is only just beginning to focus attention on the same issues in high northern latitudes.

A key article of the Kyoto Protocol of the United Nations Framework Convention on Climate Changes requires the participating nations to create an accounting system to quantify the amounts of atmospheric carbon released from sources as well as absorbed by sinks, including both natural processes and human activities. Another article of this protocol allows Annex I countries (including the United States, Canada, and Russia) to trade carbon credits if more carbon is being absorbed by natural systems than is being released through human causes.

This latter article has led many nations to consider how they can amass carbon credits through adopting management strategies to reduce greenhouse gas emissions to the atmosphere from the various ecosystems that exist within their borders. Because they store large amounts of carbon, adoption of various management strategies for the boreal forest region is receiving a high degree of attention as a mechanism for creating carbon credits. Any of these strategies, however, will have significant costs associated with them. In Chapter 4, Goldammer and Stocks discuss current perspectives on fire management and policy in the countries that contain the Eurasian boreal forest, and in Chapter 5, Ward and Mawdsley present the Canadian perspective. Finally, in Chapter 6, Grissom and associates discuss the options and consequences of adopting different mitigation strategies for resource managers in the Alaskan boreal forest.

References

Ciais, P., P.P. Tans, M. Trolier, J.W.C. White, and R.J. Francey. 1995. A large Northern Hemisphere terrestrial CO_2 sink indicated by the 13C/12C ratio of atmospheric CO_2. *Science* 269:1098–1102.

Cramer, W.P., and R. Leemans. 1993. Assessing impacts of climate change on vegetation using climate classification systems, pp. 190–217 in A.M. Solomon and H.H. Shugart, eds. *Vegetation Dynamics and Global Change.* Chapman & Hall, New York.

Crutzen, P.J., and G.R. Carmichael. 1993. Modeling the influence of fires on atmospheric chemistry, pp. 89–105 in P.J. Crutzen and J. Goldammer, eds. *Fire in the Environment: The Ecological, Atmospheric, and Climatic Importance of Vegetation Fires.* John Wiley & Sons, New York.

Dickinson, R.E. 1993. Effects of fires on global radiation budget through aerosols and cloud properties, pp. 107–137 in P.J. Crutzen and J. Goldammer, eds. *Fire in the Environment: The Ecological, Atmospheric, and Climatic Importance of Vegetation Fires*. John Wiley & Sons, New York.

Emanuel, W.R., H.H. Shugart, and M.P. Stevenson. 1985. Climatic change and the broad-scale distribution of terrestrial ecosystem complexes. *Clim. Change* 7:29–43.

Fan, S., M. Gloor, J. Mahlman, S. Pacala, J. Sarmiento, T. Takahashi, and P. Tans. 1998. A large terrestrial carbon sink in North America implied by atmospheric and oceanic carbon dioxide data and models. *Science* 282:442–446.

Field, C.B., M.J. Behrenfeld, J.T. Randerson, and P. Falkowski. 1998. Primary production of the biosphere: integrating terrestrial and oceanic components. *Science* 281:237–240.

Flannigan, M.D., and J.D. Harrington. 1988. A study of the relation of meteorologic variables to monthly provincial area burned by wildfire in Canada. *J. Appl. Meteorol.* 27:441–452.

Flannigan, M.D., and C.E. Van Wagner. 1991. Climate change and wildfire in Canada. *Can. J. For. Res.* 21:61–72.

Freidli, H., H. Loester, H. Oeschger, U. Siegenthaler, and B. Stauffer. 1986. Ice core record of $^{13}C^{12}C$ ratio of atmospheric CO_2 in the past two centuries. *Nature* 324:237–238.

Goulden, M.L., S.C. Wofsy, J.W. Harden, S.E. Trumbore, P.M. Crill, S.T. Gower, T. Fries, B.C. Daube, S-M. Fan, D.J. Sutton, A. Bazzaz, and J.W. Munger. 1998. Sensitivity of boreal forest carbon balance to thaw. *Science* 279:214–217.

Holdridge, L.R. 1947. Determination of world plant formations from simple climatic data. *Science* 105:367–368.

Houghton, J.T., L.G. Meiro Filho, B.A. Callander, N. Harris, A. Kattenberg, and K. Maskell, eds. 1996. *The Science of Climate Change.* Contribution of Working Group I to the Second Assessment Report on the Intergovernmental Panel on Climate Change. Cambridge University Press, Cambridge, U.K.

Houghton, J.T., G.J. Jenkins, and J.J. Ephraums, eds. 1990. *Climate Change—The IPCC Scientific Assessment.* Cambridge University Press, Cambridge, U.K.

Houghton, R.A. 1995. Land-use change and the carbon cycle. *Global Change Biol.* 1:275–287.

Kasischke, E.S. 1996. Fire, climate change, and carbon cycling in Alaskan boreal forests, pp. 827–833 in J.S. Levine, ed. *Biomass Burning and Climate Change. Vol. 2. Biomass Burning in South America, Southeast Asia, and Temperate and Boreal Ecosystems, and the Oil Fires of Kuwait.* MIT Press, Cambridge, MA.

Kasischke, E.S., N.L. Christensen, Jr., and B.J. Stocks. 1995. Fire, global warming and the mass balance of carbon in boreal forests. *Ecol. Appl.* 5:437–451.

Keeling, C.D., R.B. Bascow, and T.P. Whorf. 1982. Measurements of the concentration of carbon dioxide at Mauna Loa Observatory, Hawaii, pp. 377–385 in W.C. Clark, ed. *Carbon Dioxide Review.* Oxford University Press, New York.

Korner, C. 1993. CO_2 fertilization: the great uncertainty in future vegetation development, pp. 53–70 in A.M. Solomon and H.H. Shugart, eds. *Vegetation Dynamics and Global Change.* Chapman & Hall, New York.

Lacaux, J.P., H. Cachier, and R. Delmas. 1993. Biomass buring in Africa: an overview of its impact on atmospheric chemistry, pp. 159–191 in P.J. Crutzen and J. Goldammer, eds. *Fire in the Environment: The Ecological, Atmospheric, and Climatic Importance of Vegetation Fires.* John Wiley & Sons, New York.

Oechel, W.C., G.L. Vourlitis, S.J. Hastings, and S.A. Bochkarev. 1995. Changes in arctic CO_2 flux over two decades: effects of climate change at Barrow, Alaska. *Ecol. Appl.* 5:846–855.

Philips, O.L., Y. Malhi, N. Higuchi, W.F. Laurance, P.V. Nunez, R.M. Vasquez, S.G. Laurance, L.V. Ferreira, M. Stern, S. Brown, and J. Grace. 1998. Changes in carbon balance of tropical forests: evidence from long-term plots. *Science* 279:439–442.

Post, W.M. 1993. Uncertainties in the terrestrial carbon cycle, pp. 116–132 in A.M. Solomon and H.H. Shugart, eds. *Vegetation Dynamics and Global Change.* Chapman & Hall, New York.

Randerson, J.T., M.V. Thompson, T.J. Conway, I.Y. Fung, and C.B. Field. 1997. The contribution of terrestrial sources and sinks to trends in seasonal cycles of atmospheric carbon dioxide. *Global Biogeochem. Cycles* 11:535–560.

Smith, T.M., R. Leemans, and H.H. Shugart. 1992. Sensitivity of terrestrial carbon storage to CO_2-induced climate change: comparison of four scenarios based on general circulation models. *Clim. Change* 21:367–384.

Solomon, A.M., and W. Cramer. 1993. Biospheric implications of global environmental change, pp. 25–52 in A.M. Solomon and H.H. Shugart, eds. *Vegetation Dynamics and Global Change.* Chapman & Hall, New York.

Tans, P.P., I.Y. Fung, and T. Takahashi. 1990. Observational constraints on the global atmospheric CO_2 budget. *Science* 247:1431–1438.

Watson, R.T., H. Rodhe, H. Oeschger, and U. Siegenthaler. 1990. Greenhouse gases and aerosols, pp. 5–40 in Houghton, J.T., G.J. Jenkins, and J.J. Ephraums, eds. *Climate Change—The IPCC Scientific Assessment.* Cambridge University Press, Cambridge, UK.

Watson, R.T., L.G. Meira Filho, E. Sanhueza, and A. Janetos. 1992. Greenhouse gases: sources and sinks, pp. 29–46 in J.T. Houghton, B.A. Callender, and S.K. Varney, eds. *Climate Change 1992—The Supplementary Report to the IPCC Scientific Assessment.* Cambridge University Press, Cambridge, UK.

Wotton, B.M., and M.D. Flannigan. 1993. Length of the fire season in a changing climate. *For. Chron.* 69:187–192.

2. Boreal Ecosystems in the Global Carbon Cycle

Eric S. Kasischke

Boreal Ecosystems and the Carbon Cycle

The terrestrial ecosystems found in the boreal region cover a little less than 17% of the earth's land surface, yet they contain more than 30% of all carbon present in the terrestrial biome (Table 2.1). For the purposes of this discussion, we divide the ecosystems found in this region into three broad categories: boreal forests, peatlands interspersed throughout the boreal forest, and tundra. Although there is room for debate as to the exact definition of these categories, they are used here for descriptive purposes only and are based on the criteria developed by Apps and colleagues (1993) to estimate the amount of carbon present in boreal forests and tundra. The percentages of total area and total terrestrial carbon were derived by using the estimates of Smith and co-workers (1993).

Table 2.2 compares the amounts of carbon stored in the three major forest complexes found on the earth's surface: boreal forests, temperate forests, and tropical forests. Although temperate and tropical forests combined cover almost twice the land area as the boreal forest, the boreal forest contains 20% more carbon than the other two forest types combined. It is estimated by Houghton (1996) and Kauppi (1996) that since 1850, both tropical and temperature forests have lost up to 100 Pg carbon due to human activities (e.g., forest clearing and forest conversion), an amount that is ten times that lost from boreal forests from deforestation during this time period (Houghton 1996). Although tropical forests are currently acting as a net source of atmospheric carbon, the strength of this

Table 2.1. Summary of Carbon Present in Boreal Ecosystems

Biome	Area ($\times 10^6$ ha)	Plant biomass carbon[a] (Pg = 10^{15} g)	Soil carbon[b] (Pg)
Boreal forest[c,d]	1,249	78	227
Peatlands[e]	260	<<1	397
Tundra[c]	890	<<1	180
Total	2,399	78	804

[a]Includes plant biomass and plant detritus.
[b]Includes forest soils and peat.
[c]Apps et al. (1993).
[d]Alexyev and Birdsey (1998).
[e]Zoltai and Martikainen (1996).

source may be shrinking rapidly because of reforestation in this region (Houghton 1996). Because of reforestation activities, it is thought that temperate forests are currently a net sink of atmospheric carbon (Kauppi 1996; Fan et al. 1998). Apps and associates (1993) estimate that vegetation in boreal ecosystems is currently a net carbon sink of 0.54 Gt C yr^{-1}. The soils of boreal forests and peatlands are currently a net carbon sink of 0.70 Gt C yr^{-1}, whereas tundra ecosystems are currently a net carbon source of 0.17 Gt C yr^{-1}.

During the early part of the next century, tropical and temperate forests are likely to remain net sinks of atmospheric carbon because of reforestation, whereas boreal forests may become a net source of carbon. The shift of boreal forests from a net sink to a net source of atmospheric carbon will likely originate from two sources, both anthropogenic in their origin. First, deforestation activities in the boreal forest are likely to increase, particularly in Russia. Second, climate change will cause an increase in the disturbances (fire and insects and pathogens) in this region, which, in turn, will cause a net release of carbon. This latter topic is discussed in more detail later in this chapter.

Geographically, boreal ecosystems are found on land surfaces above 50° north latitude, with the exception of central Siberia, where the boreal forest extends as far south as 45° north latitude (Fig. 2.1). The primary factors regulating the accumulation of carbon in this region include the seasonal patterns of solar il-

Table 2.2. Comparison of Carbon Storage in Boreal, Temperate, and Tropical Forests

Biome	Area ($\times 10^6$ ha)	Soil carbon (Pg)	Plant biomass carbon (Pg)	Total carbon (Pg)
Boreal forest[a]	1,509	625	78	703
Tropical forest[b]	1,756	216	159	375
Temperate forest[c]	1,040	100	21	121

[a]Area data from Apps et al. (1993); carbon data from Table 2.1.
[b]Area data from Brown (1996a); carbon data from Brown et al. (1993).
[c]Area and carbon data from Heath et al. (1993).

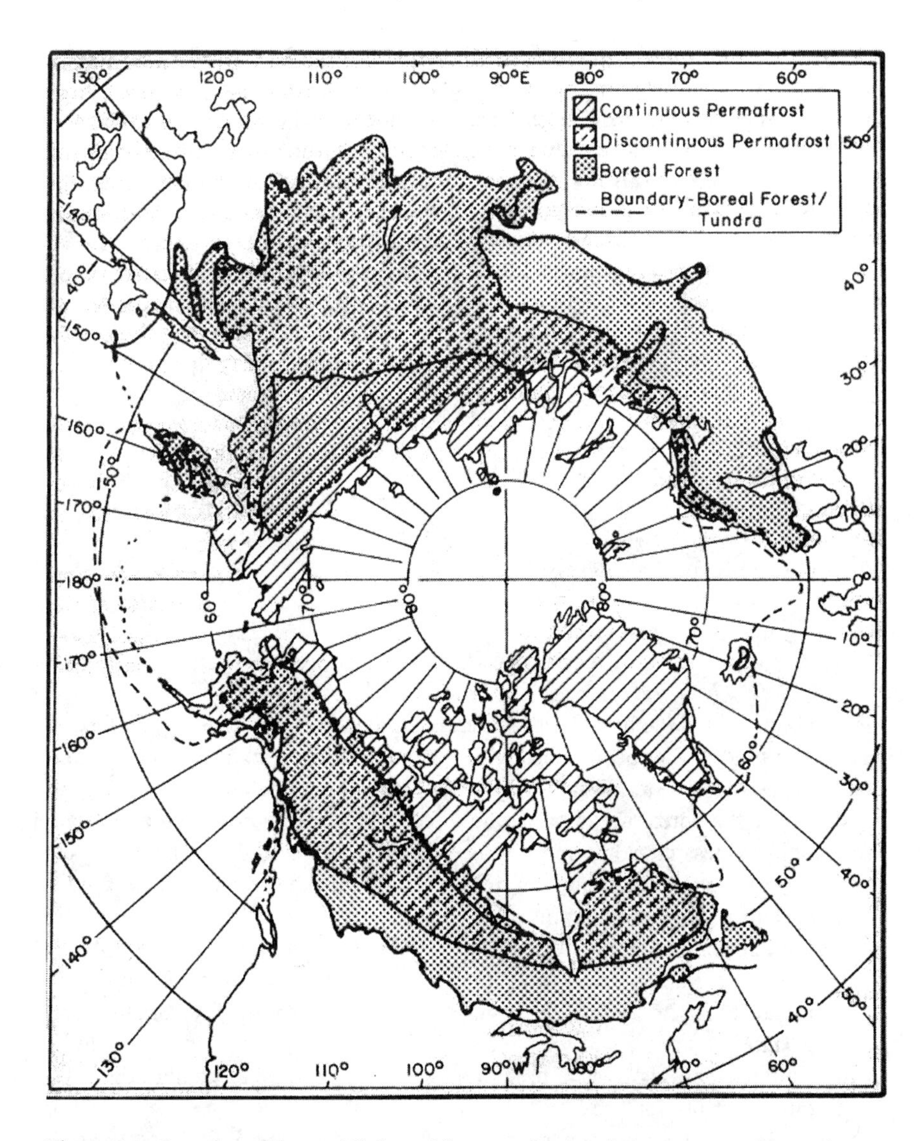

Figure 2.1. Location of the world's boreal forest region in relation to areas with continuous and discontinuous permafrost. (From Van Cleve et al. 1986.)

lumination, temperature, and precipitation. As with all terrestrial biomes, the source of inputs of carbon into boreal ecosystems is photosynthesis. The growing season for plants in the boreal region is much shorter than other regions because of patterns of solar illumination and temperature. Typically, the frost-free period in the boreal region is less than 90 days. However, because of the extremely long periods of daylight (24 hours at the summer solstice in regions above the Arctic Circle, or 66° 33′ north latitude), daytime temperatures can be warm in many boreal regions during the peak of the growing season. Because of this, rates of daily net primary production can be as high in boreal forests as in ecosystems found in temperate and tropical regions. Total annual net primary productivity is much lower, however, due to the shorter growing season.

A characteristic that sets boreal ecosystems apart from other biomes in terms of the carbon cycle is the large amounts of carbon stored in the soil compartment. This large mass of carbon is primarily the result of the extremely cold annual temperatures found throughout this region. These cold temperatures lead to the formation of continuous or discontinuous permafrost throughout most of the boreal region. The cold ground conditions significantly reduce the rates of decomposition and result in the deep, undecomposed organic soils found throughout this region. In addition, the presence of permafrost also severely impedes drainage of water from the land surface. Thus, even though annual precipitation in many boreal regions is low (≤300 mm), many of the soils in the ecosystems of the region remain saturated throughout the year. These saturated soils further impede decomposition and are a primary factor in the development of many of the peatlands found in this region (Paavilainen and Paivanen 1995).

Fire in the Boreal Region

Fire is ubiquitous in the boreal region, being extremely common in boreal forests and rare in tundra. Fire occurs throughout the boreal forest owing to a combination of factors related to climate, vegetation, and sources of ignition. Outside of coastal areas, the boreal region as a whole receives low amounts of precipitation. Precipitation of less than 300 mm yr^{-1} is typical for boreal forests, whereas levels of 150 mm yr^{-1} or less are common in areas along the Arctic coast, where tundra is found. During the summer months, daylight air temperatures are generally higher than 20°C, and temperatures greater than 30°C are not uncommon. Even though the poor drainage caused by permafrost leads to wet soil conditions in many boreal forests, during the warmer summer months the upper organic soil horizons (litter and fibric soil) can become desiccated. With higher temperatures in the summer, periods of low precipitation lead to low relative humidity and conditions favorable for fire.

Much of the understory vegetation and tree species present in the boreal region are fire adapted and readily ignite and burn during dry periods. In regions with close-crowned forests, many tree species (e.g., spruce and pine) retain their dead branches, which provides highly flammable fuel during fires (Cayford and McRae 1983; Viereck 1983). In addition, tree mortality from disease outbreaks provides significant fuel loads for biomass burning in many boreal regions (Furayev et al. 1983; Holling 1992). In open-crown forests (taiga) and forest-tundra, the ericaceous shrub layer provides the basis for the spread of fires over large regions. There are three reasons why this shrub layer promotes fires (Auclair 1983). First, ericaceous plants have a high percentage of ether extractions, which results in higher combustibility and heats of combustion than nonericaceous plants. Second, in open-crown forests, these shrub layers are nearly continuous, providing the fuel matrix required for fires to spread over large areas. And third, the dead litter from the shrubs adds considerably to the fuel matrix in the moss/lichen/litter layer, which dries out quickly during periods of low relative humidity and further promotes the spread of fire. Most fires in boreal regions are started by lightning strikes; the passage of weather fronts through a region can start hundreds of small fires in a region.

It has been estimated that between 5 and 12 million ha of boreal forest burns annually (Stocks 1991; Kasischke et al. 1993; Conard and Ivanova 1998). Although fire occurrence and location records are available from throughout the North American boreal forest (Alaska Fire Service 1995; Stocks et al. 1996; Murphy et al., this volume), similar comprehensive data for the Eurasian boreal forest are not (Korovin 1996; Shvidenko and Nilsson, this volume, Chapter 8). Based on recent analysis of satellite data collected over the Siberian boreal forest (Cahoon et al. 1994, 1996; Kasischke et al. 1999), it is known that the patterns of fire in the boreal forests of North America and Eurasia are similar.

Figure 1.2 presents a plot of annual area burned in the North American boreal forest for the period of 1940–1998. The records from which this plot was derived are discussed in detail in Chapter 15. It is thought that fire detection capabilities and recording protocols have improved significantly over the past two decades. A considerable number of fires probably were not reported prior to 1970; therefore, the annual area burned summarized for this time period in Figure 1.2 is low.

During the time period between 1940 and 1998, at least 115.0×10^6 ha of forest was burned in the North American boreal forest. Analyses of the fire records by Kasischke and associates (1995) and Stocks and co-workers (1996) show the fire return intervals vary between 30 and more than 200 years for the different regions in the North American boreal forest. On a decadal scale, it appears that the frequency of fire increased significantly during the 1980s and 1990s compared with previous decades, although an argument can be made that part of this increase is due to improved detection and reporting of fires.

Two characteristics of boreal forest fires are worth noting. First, the distribution of fires on an annual basis is not uniform, with more fire activity occurring during drier years (Flannigan and Harrington 1988). Since 1970, greater than 50% of the total area burned in the North American boreal forest occurred in the six most severe fire years. Second, most of the area burned in the North American boreal forest occurs in large fire events. Analysis of fire statistics shows that more than 98% of the total area burned occurs in fires larger than 2,000 ha in size. Fires greater than 100,000 ha in size are common in boreal forests, with fires greater than 1 million ha occurring frequently during severe fire years (see, e.g., Cahoon et al. 1994).

Fires, in combination with variations in soil and physiography, result in the mosaic of forest types with different stand ages found throughout the boreal forest. Figure 2.2[1] shows the location of large fires (> 400 ha in size) in the state of

[1]Figure 2.2 will be found in the color insert.

Alaska for the years 1954–1997. The location of these fires is largely restricted to the boreal forest region in Alaska (e.g., the area south of the Brooks Range and north of the Alaskan Range). When combined with other spatial information such as soil type, elevation, and forest cover, the information on fire location can be used to (1) determine the fire frequency in specific physiographical provinces within the state; (2) provide the basis for estimating stand age distribution; and (3) estimate how much carbon is released during individual fires (Kasischke et al. 1995).

Finally, although most studies on fire in boreal regions have been carried out in forest ecosystems, fires in tundra and peatlands have also been documented. A review of tundra fires by Wein (1976) revealed that most reported tundra fires were small in their area extent (< 50 ha), but numerous fires in the 100- to 2,000-ha range were reported, and several tundra fires more than 10,000 ha in extent were documented. Tundra and peatland fires occur during periods of extremely low rainfall and generally only consume the surface vegetation and upper layers of organic matter. However, fires in tundra and peatlands can smolder for weeks and months and eventually burn deeply into the organic soil layer (Wein 1983). Thus, the potential exists for increases in fires in these biomes if global warming results in warmer and drier conditions throughout the boreal region (Chapter 20).

Fire and Carbon Cycling in the Boreal Forest

Although it is believed that on a global scale boreal forests presently serve as a net sink of atmospheric carbon, on localized scales they act alternately as atmospheric carbon sources and sinks due to the influence of fire. The role of fire in carbon cycling in the boreal forest is a strong one, and its effects are both direct and indirect.

Stand replacement fires are the norm in most boreal regions, resulting in the death and partial burning (oxidation) of most overstory trees, complete burning of most understory vegetation, and partial burning of the mosses, lichens, litter, and organic soil present in the forest floor. In fires with a long smoldering phase, the organic soil can be burned all the way to the top of the mineral soil. Thus, a fire can result in a virtually instantaneous release of large amounts of carbon in the form of greenhouse gases (between 10 and 200 t C ha^{-1}) to the atmosphere (Chapter 10).

From an indirect standpoint, fire has multiple effects on carbon storage in the boreal forest. First, the frequency of fire has a direct influence on the stand age distribution in a region. The shorter the fire return interval, the larger the number of younger stands that are present. Because the level of biomass in both the living vegetation and ground layer (mosses, lichens, litter, and organic soil) generally increases as a stand matures, areas with shorter fire return intervals will have less carbon than areas with longer fire return intervals.

Fires have a strong influence on the thermal and moisture regime of the boreal forest, especially in areas underlined by permafrost. Soils tend to become warmer and drier immediately after a fire, and these changes influence patterns of plant and tree growth and soil respiration. Both of these processes have a strong influence on carbon cycling in the boreal region. Studies have shown that in the ground-layer there is actually a net loss of carbon during the first several years after a fire. This loss occurs because the ground layer temperature increases because of increases in direct solar insolation, reduction of surface albedo, and the loss of moss, litter, and organic soil, which serve as an insulating layer (Viereck 1983). The 5–10°C rise in forest floor temperature after fire results in a corresponding increase in the rates of decomposition, to the point at which the forest floor is acting as a net source of atmospheric CO_2. Van Cleve and colleagues (1983) showed that warming the ground layer of a black spruce forest in interior Alaska by 9°C during the growing season for 3 years resulted in 20% loss of ground-layer biomass. Studies have shown that the forest floor temperatures of boreal forests underlain by permafrost remain warmer than in unburned forests for at least 20–30 years after the fire. Rates of soil decomposition in Alaskan boreal forests remain elevated for several years after a fire (Chapter 11). Whether this pattern is present in other boreal forest regions needs to be determined. What is known is that at some point after a fire (usually 10–20 years) (1) the mosses and lichens (which produce much of the organic matter found in the ground layer) become reestablished; (2) the ground-layer temperature begins to decline; (3) soil moisture begins to increase; and (4) decomposition rates decrease to the point at which the ground layer becomes a net sink of atmospheric CO_2.

A sensitivity analysis is presented in Chapter 25 that integrates these various direct and indirect effects of fire and climate change on carbon storage in the boreal forest.

Fire Management Policy and Modification of Carbon Release in Boreal Forests

One of the key challenges today facing government policy makers is to decide what actions should and can be implemented to modify the rising levels of CO_2 in the earth's atmosphere (Houghton et al. 1990, 1996). The choices available to policy makers include (1) reducing the rate of burning of fossil fuels; (2) reducing the rate of emissions resulting from other human activities (e.g., large-scale deforestation in the tropics and other regions); and (3) implementing management practices that promote the sequestration of carbon within natural biomes (Apps and Prince 1996).

One option suggested by some researchers is to increase fire suppression efforts in the boreal forests (Dixon and Krankina 1993; Krankina 1993; Brown 1996b; Shvidenko et al. 1996, Guggenheim 1997). These management practices could reduce atmospheric CO_2 in two ways: (1) they could reduce the amount of CO_2 released into the atmosphere directly from fires; and (2) they could promote a longer-term sequestration of carbon through an increase of the fire return interval.

Although this option may appear to be attractive, in reality it may not be practical to implement and, in fact, may result in a more rapid release of carbon to the atmosphere over the longer term. Those advocating an increase in fire suppression activities in the boreal forests must face the fiscal realities being addressed by countries in the boreal region (Stocks et al. 1996). In Canada and the United States, forestry agencies are now facing reductions in operational funds because of decreasing federal, state, and provincial budgets. In Siberia, the shifting of the Russian government from a central to a market economy has resulted in severe budget reductions in all government agencies. Thus, even if it was decided that fire suppression in the boreal forest was a viable option for carbon emissions reduction, budgetary realities may preclude the implementation of this option.

Another reality that must be faced is that fire management agencies in boreal regions practice modified suppression of fires. In most areas, fires are only monitored and suppression activities initiated when there are direct threats to life, property, or features with a strong cultural significance.

Finally, in many instances, even when the decision is made to actively suppress a fire, they have grown to a size of which suppression is not possible. Because the boreal forest is so vast, fires are often tens and hundreds of kilometers from the nearest fire-fighting base. By the time sufficient manpower and equipment reach the fire, it has grown to such a sufficiently large size that suppression is difficult or impossible. The problem becomes even more acute during severe fire seasons, when there is often more fire activity than resources available to fight it.

A classic example of this problem is a forest fire that took place near Tok, Alaska, in the summer of 1990. This fire was started by a lightning strike within 2 km of the regional fire-fighting center. Because of the extremely dry fuel conditions as a result of a prolonged draught, the fire expanded rapidly, even though active suppression was pursued because of the fire's location near a major community. Despite 4 days of intensive suppression activity at a cost of more than $25 million, the only factor that prevented the fire from burning the entire community of Tok was a shift in the wind direction.

Even if the resources were available for fire suppression in boreal forests, policy makers should consider whether these efforts would be successful over the longer term. An argument can be made that the term *fire suppression* is really a misnomer. In reality, what is being practiced today in many forest regions is not fire suppression but fire delay. In areas where fire is a natural ecological process, suppression of fire usually only results in a buildup of fuels to a point that when a fire starts, suppression becomes impossible. A case in point is the series of fires that occurred in the greater Yellowstone National Park region in the United States during the summer of 1988 (Christensen et al. 1989).

Another factor worth serious consideration in planning fire suppression activities in the future is the effect of climate warming on fuel loads in forest ecosystems. One theory that has been advanced by forest ecologists is that increased air temperatures will increase the potential evapotranspiration demand of forest ecosystems to the point at which moisture stress will result in higher rates of tree mortality in many forest ecosystems, either directly or through increased infestations by insects (Holling 1992). This increased mortality would substantially increase fuel loads available for burning in boreal regions and may result in more intense burning of biomass in both the above- and below-ground layers than would have occurred under a natural fire regime.

In summary, there are arguments both for and against fire suppression as a policy option for reducing CO_2 emissions from boreal forests. Before such policies can be considered, several issues need further clarification, including

1. What are the actual costs in increasing the level of fire management activities in boreal forests?
2. Is fire suppression in boreal forests really practical?
3. How is the fire regime in boreal forests likely to change in a warming climate?
4. What are the effects of a changing climate on forest mortality? How does this mortality affect fuel loads and fire frequencies?

Summary

The ecosystems of the boreal region are currently playing and will continue to play a key role in global change research. Boreal forests and peatlands represent the largest terrestrial reservoir of atmospheric carbon. They occupy a region where significant warming has occurred over the past 30 years and where the current generation of climate models predict that the greatest warming will occur over the next century. There is little doubt that the amounts of carbon presently stored in these ecosystems will change dramatically along with the changing climate. There is also little doubt that if climate warming does occur, the frequency of fire will increase in the boreal forest as well. The increased fire activity will serve as a catalyst to the wide range of ecosystem processes controlling the storage of carbon in boreal forests. Developing a better understanding of how fire shapes the boreal landscape is key to understanding future carbon balances in this region.

Equally important to understanding the scientific questions and issues are the human dimensions of fire in the boreal forest zone. The boreal forest contains many natural resources that are valued by human society. It contains vast tracts of commercial timber of significant economic value to the countries of this region. The boreal forest is home to much wildlife, including many endangered species. The changing of forest cover in this region due to fire and climate change will undoubtedly result in major shifts in wildlife populations, which, in turn, will affect the humans who value wildlife as a renewable resource or for their aesthetics.

References

Alaska Fire Service. 1995. *1995 Fire Statistics and Season Summary.* Bureau of Land Management, Fairbanks, AK.

Alexeyev, V.A., and R.A. Birdsey. 1998. *Carbon Storage in Forests and Peatlands of Russia.* General Technical Report NE-244. U.S. Department of Agriculture, Forest Service, Northeastern Forest Experiment Station, Radnor, PA.

Apps, M.J., and D.T. Price, eds. 1996. *Forest Management and the Global Carbon Cycle.* NATO ASI Series, subseries 1, vol. 40, Global Environmental Change. Springer-Verlag, Berlin.

Apps, M.J., W.A. Kurz, R.J. Luxmoore, L.O. Nilsson, R.A. Sedjo, R. Schmidt, L.G. Simpson, and T.S. Vinson. 1993. Boreal forests and tundra. *Water Air Soil Pollut.* 70:39–53.

Auclair, A.N.D. 1983. The role of fire in lichen-dominated tundra and forest-tundra, pp. 235–253 in R.W. Wein and D.A. MacLean, eds. *The Role of Fire in Northern Circumpolar Ecosystems.* J. Wiley & Sons, New York.

Brown, S. 1996a. Tropical forests and the global carbon cycle: estimating state and change in biomass density, pp. 135–144 in M.J. Apps and D.T. Price, eds. *Forest Ecosystems and Forest Management and the Global Carbon Cycle.* NATO ASI Series I, Global Environmental Change. Springer-Verlag, Berlin.

Brown, S.A. 1996b. Management of forests for mitigation of greenhouse gas emissions, pp. 773–796 in R.T. Watson, M.C. Zinyowera, and R.H. Moss, eds. *Climate Change 1995—Impacts, Adaptations and Mitigation of Climate Change: Scientific-Technical Analysis.* Cambridge University Press, Cambridge, UK.

Brown, S., C.A.S. Hall, W. Knabe, J. Raich, M.C. Trexler, and P. Woomer. 1993. Tropical forests: their past, present, and potential future role in the terrestrial carbon budget. *Water Air Soil Pollut.* 70:71–94.

Cahoon, D.R., Jr., B.J. Stocks, J.S. Levine, W.R. Cofer III, and J.M. Pierson. 1994. Satellite analysis of the severe 1987 forest fires in northern China and southeastern Siberia. *J. Geophys. Res.* 99:18,627–18,638.

Cahoon, D.R., Jr., B.J. Stocks, J.S. Levine, W.R. Cofer III, and J.A. Barber. 1996. Monitoring the 1992 forest fires in the boreal ecosystem using NOAA AVHRR satellite imagery, pp. 795–802 in J.S. Levine, ed. *Biomass Burning and Climate Change. Vol. 2. Biomass Burning in South America, Southeast Asia, and Temperate and Boreal Ecosystems, and the Oil Fires of Kuwait.* MIT Press, Cambridge, MA.

Cayford, J.H., and D.J. McRae. 1983. The ecological role of fire in jack pines, pp. 183–200 in R.W. Wein and D.A. MacLean, eds. *The Role of Fire in Northern Circumpolar Ecosystems.* John Wiley & Sons, New York.

Christensen, N.L., J.K. Agee, P.F. Broussard, J. Hughes, D.H. Knight, G.W. Minshall, J.M. Peek, S.J. Pyne, F.J. Swanson, J.W. Thomas, S. Wells, S.E. Williams, and H.A. Wright. 1989. Interpreting the Yellowstone fires of 1988. *Bioscience* 39:678–685.

Conard, S.G., and G.A. Ivanova. 1998. Wildfire in Russian boreal forests—potential impacts of fire regime characteristics on emissions and global carbon balance estimates. *Environ. Pollut.* 98:305–313.

Dixon, R.K., and O.N. Krankina. 1993. Forest fires in Russia: carbon dioxide emissions to the atmosphere. *Can. J. For. Res.* 23:700–705.

Fan, S., M. Gloor, J. Mahlman, S. Pacala, J. Sarmiento, T. Takahashi, and P. Tans. 1998. A large terrestrial carbon sink in North America implied by atmospheric and oceanic carbon dioxide data and models. *Science* 282:442–446.

Flannigan, M.D., and J.D. Harrington. 1988. A study of the relation of meteorologic variables to monthly provincial area burned by wildfire in Canada. *J. Appl. Meteor.* 27:441–452.

Furyaev, V.V., R.W. Wein, and D.A. MacLean. 1983. Fire influences in *Abies*-dominated forests, pp. 221–232 in W. Wein and D.A. MacLean, eds. *The Role of Fire in Northern Circumpolar Ecosystems.* John Wiley & Sons, New York.

Guggenheim, D.E. 1997. Management of forest fires to maximize carbon sequestration in temperate and boreal forests. *World Resources Rev.* 9:46–47.

Heath, L.S., P. Kaupi, P. Burschel, H.D. Gregor, R. Guderian, G.H. Kohlmaier, S. Lorenz, D. Overdieck, F. Scholz, H. Thomasius, and M. Weber. 1993. Carbon budget of the temperature forest zone. *Water Air Soil Pollut.* 70:55–69.

Holling, C.S. 1992. The role of forest insects in structuring the boreal landscape, pp. 126–143 in H.H. Shugart, R. Leemans, and G.B. Bonan, eds. *A Systems Analysis of the Global Boreal Forest.* University Press, Cambridge, UK.

Houghton, J.T., L.G. Meiro Filho, B.A. Callander, N. Harris, A. Kattenberg, and K. Maskell, eds. 1996. *The Science of Climate Change.* Contribution of Working Group I to the Second Assessment Report on the Intergovernmental Panel on Climate Change. Cambridge University Press, Cambridge, UK.

Houghton, J.T., G.J. Jenkins, and J.J. Ephraums, eds. 1990. *Climate Change—The IPCC Scientific Assessment.* Cambridge University Press, Cambridge, UK.

Houghton, R.A. 1996. Land-use change and the temporal record, pp. 117–134 in M.J. Apps and D.T. Price, eds. *Forest Management and the Global Carbon Cycle.* NATO ASI Series I, Global Environmental Change. Springer-Verlag, Berlin.

Kasischke, E.S., N.H.F. French, P. Harrell, N.L. Christensen, Jr., S.L. Ustin, and D. Barry. 1993. Monitoring of wildfires in boreal forests using large area AVHRR NDVI composite data. *Remote Sens. Environ.* 44:61–71.

Kasischke, E.S., N.H.F. French, L.L. Bourgeau-Chavez, and N.L. Christensen, Jr. 1995. Estimating release of carbon from 1990 and 1991 forest fires in Alaska. *J. Geophys. Res.* 100:2941–2951.

Kasischke, E.S., K. Bergen, R. Fennimore, F. Sotelo, G. Stephens, A. Janetos, and H.H. Shugart. 1999. Satellite imagery gives clear picture of Russia's boreal forest fire. *EOS—Trans. Am. Geophys. Union* 80:141,147.

Kauppi, P.E. 1996. Carbon budget of temperate zone forests during 1851–2050, pp. 191–198 in M.J. Apps and D.T. Price, eds. *Forest Management and the Global Carbon Cycle.* NATO ASI Series, subseries 1, vol. 40, Global Environmental Change. Springer-Verlag, Berlin.

Korovin, G.N. 1996. Analysis of distribution of forest fires in Russia, pp.112–128 in J.G. Goldammer and V.V. Furyaev, eds. *Fire in Ecosystems of Boreal Eurasia.* Kluwer Academic Publishers, Dordrecht, The Netherlands.

Krankina, O.N. 1993. *Forest Fires in the Former Soviet Union: Past, Present and Future Greenhouse Gas Contributions to the Atmosphere.* EPA/600R-93/084. U.S. Environmental Protection Agency, Washington, DC.

Paavilainen, E., and J. Paivanen. 1995. *Peatland Forestry.* Ecological Studies 111. Springer-Verlag, New York.

Shugart, H.H. 1993. Global change, pp. 3–21 in A.M. Solomon and H.H. Shugart, eds. *Vegetation Dynamics and Global Change.* Chapman & Hall, New York.

Shvidenko, A., S. Nilsson, and V. Roshkov. 1995. *Possibilities for Increased Carbon Sequestration Through Improved Protection of Russian Forests.* IIASA Report WP-95-86. International Institute for Applied Systems Analysis, Laxenburg, Austria.

Smith, T.M., W.P. Cramer, R.K. Dixon, R.P. Neilson, and A.M. Solomon. 1993. The global terrestrial carbon cycle. *Water Air Soil Pollut.* 70:19–37.

Stocks, B.J. 1991. The extent and impact of forest fires in northern circumpolar countries, pp. 197–202 in J.S. Levine, ed. *Global Biomass Burning: Atmospheric, Climatic and Biospheric Implications.* MIT Press, Cambridge, MA.

Stocks, B.J., B.S. Lee, and D.L. Martell. 1996. Some potential carbon budget implications of fire management in the boreal forest, pp. 89–96 in M.J. Apps and D.T. Price, eds. *Forest Management and the Global Carbon Cycle.* NATO ASI Series, subseries 1, vol. 40, Global Environmental Change. Springer-Verlag, Berlin.

Van Cleve, K., L. Oliver, R. Schlentner, L.A. Viereck, and C.T. Dyrness. 1983. Productivity and nutrient cycling in taiga forest ecosystems. *Can. J. For. Res.* 13:747–766.

Van Cleve, K., F.S. Chapin, III, P.W. Flanagan, L.A. Viereck, and C.T. Dyrness, eds. 1986. *Forest Ecosystems in the Alaskan Taiga.* Ecological Studies 57. Springer-Verlag, New York.

Viereck, L.A. 1983. The effects of fire in black spruce ecosystems of Alaska and northern Canada, pp. 201–220 in R.W. Wein and D.A. MacLean, eds. *The Role of Fire in Northern Circumpolar Ecosystems.* John Wiley & Sons, Chichester, UK.

Wein, R.W. 1976. Frequencies and characteristics of tundra fires. *Arctic* 29:213–222.

Wein, R.W. 1983. Fire behavior and ecological effects in organic terrain, pp. 81–95 in R.W. Wein and D.A. MacLean, eds. *The Role of Fire in Northern Circumpolar Ecosystems.* John Wiley & Sons, New York.

Zoltai, S.C., and P.J. Martikainen. 1996. The role of forested peatlands in the global carbon cycle, pp. 47–58 in M.J. Apps and D.T. Price, eds. *Forest Ecosystems, Forest Management and the Global Carbon Cycle.* NATO ASI Series, vol. I40. Springer-Verlag, Heidelberg.

3. Boreal Forest Fire Emissions and the Chemistry of the Atmosphere

Joel S. Levine and Wesley R. Cofer III

Introduction

Over the 4.5-billion-year history of our planet, the composition and chemistry of the atmosphere, as well as the climate of our planet, have been affected by the production of atmospheric gases within the biosphere. These gases are produced by a variety of biological processes, including photosynthesis, respiration, decomposition, nitrification, denitrification, and methanogenesis. Recent research has identified another biospheric process that has both instantaneous and longer-term effects on the production of atmospheric gases—this process is biomass burning. The topic of biomass burning has been the subject of no less than six treatises over the past six years (Levine 1991; Crutzen and Goldammer 1993, Goldammer and Furyaev 1996; Levine 1996a,b; van Wilgen et al. 1997) and continues to be an active area of multidisciplinary research.

Biomass burning is the burning of the world's vegetation, including forests, savannas, and agricultural lands, for land clearing and land-use change. The immediate effect of burning is the production and release into the atmosphere of gases and particulates resulting from the combustion of biomass matter. The instantaneous combustion products of vegetation burning include CO_2, CO, CH_4, nonmethane hydrocarbons (NMHC), nitric oxide (NO), and methyl chloride (CH_3Cl). If the burned vegetation is not regrown, then the liberated CO_2 remains in the atmosphere. If the burned ecosystem undergoes regrowth, as do the savannas, the CO_2 is eventually removed from the atmosphere via photosynthesis and is incorporated in the new vegetation growth. However, gaseous emissions other than CO_2 remain in the atmosphere and are not reincorporated in the biosphere. Longer-term effects of biomass burning include enhanced biogenic soil emissions of NO and N_2O following burning. Gases produced by biomass burning that are environmentally significant include CO_2 and CH_4, greenhouse gases that affect global climate. Combustion particulates affect the global radiation budget and hence also affect climate. CO, CH_4, NMHCs, and NO are all chemically active gases and lead to the photochemical production of ozone (O_3) in the troposphere. Methyl chloride is also produced during biomass burning. Methyl chloride is a source of chlorine to the atmosphere, which leads to the chemical destruction of stratospheric O_3. A single chlorine atom can destroy 100,000 molecules of O_3 in the stratosphere. Biomass burning is also an important global source of atmospheric bromine in the form of methyl bromine (CH_3Br) (Mano and Andreae 1994). Bromine leads to the chemical destruction of O_3 in the stratosphere and is about 40 times more efficient in the destruction of stratospheric O_3 than is chlorine on a molecule-for-molecule basis.

In addition to gaseous emissions from biomass burning, smoke aerosols also influence global climate and atmospheric chemistry (Radke et al. 1991; Ward et al. 1991). Smoke aerosols can interact directly with solar radiation (Kaufman 1987; Lenoble 1991), act as condensation nuclei (Rogers et al. 1991; Penner et al. 1992), or form substrates for heterogeneous atmospheric chemical reactions (Andreae 1991). Complex biogeochemical interactions between biomass burning and ecosystems often involve an intricate exchange of aerosol-dispersed nutrients over large distances (Andreae 1991; Lacaux et al. 1991).

Estimating Emissions from Biomass Burning

Basic Approach

Knowledge of the geographic and temporal distribution of burning is critical to assess the emissions of gases and particulates from biomass burning into the atmosphere. More than a decade ago, Seiler and Crutzen (1980) showed that to a first approximation, the total amount of biomass (M) burned in a particular ecosystem expressed in grams dry matter (dm) per year can be given by the following equation:

$$M = A \times B \times a \times b \quad (3.1)$$

where A is the total land area burned annually ($m^2\ yr^{-1}$), B is the average organic matter per unit area in the individual ecosystem ($g\ dm\ m^{-2}$), a is the fraction of the aboveground biomass relative to the total average biomass B, and b is the fraction of aboveground biomass consumed during biomass burning. The parameters B, a, and b are determined by field measurements of biomass burning in diverse ecosystems. Determination of these parameters requires measurements before, during, and after burning. Once the total amount of biomass *M* burned is known, the total mass of carbon released to the atmosphere is typically estimated by assuming carbon makes up 45% of biomass by weight.

Patterns of Fire and Biomass Burning in Savannas, Tropical Forests, and Boreal Forests

Biomass burning primarily occurs in four different biomes of the world: tropical forests, savannas, boreal forests, and temperate forests. Although fires occur frequently in portions of the temperate forests of the American West, they are largely absent from other regions and overall make a relatively small contribution to greenhouse gas emissions. This section focuses attention on comparing and contrasting the characteristics of fire and biomass burning fires in the other three biomes.

In terms of quantifying the amounts of carbon released into the atmosphere, according to Eq. (3.1) three pieces of information are important: (1) how much total area burns per year; (2) how much biomass or fuel is present; and (3) what fraction of the fuel present is actually consumed during fire. The range of these parameters for the different biomes is summarized in Table 3.1 and discussed below.

It is generally believed that the vast majority of global biomass burning is human initiated and that it has significantly increased over the past few decades (Levine 1991). This may be true in tropical forests and savannas, but it is not true in boreal forests. South America and Africa contain the largest portion of the world's savanna biome, and humans ignite fires in these grasslands to suppress tree growth and to promote a more vigorous regrowth of grasses for grazing of livestock. Most savannas burn at least once every 2 years, resulting in 650 million ha of savannas burning every year (Hao et al. 1990). In tropical forests, biomass burning again is associated with human activities. In this biome, fire is used by humans to aid in the clearing of land, either to create agricultural fields or pasture. Over the past several decades, about 12 million ha yr^{-1} of tropical forests has been cleared and exposed to biomass burning.

By contrast, most fires in the boreal forest result through natural processes (i.e., ignition from lightning). Early estimates of the amounts of biomass burning in the

Table 3.1. Summary of Biomass Burning Parameters for Savannas, Tropical Forests, and Boreal Forests

Biome	Annual area burned ($\times 10^6$ ha)	Biomass density (t ha^{-1})	Fraction of biomass consumed
Savannas	650	6	0.8–1.0
Tropical forests	12	50–330	0.20–0.25
Boreal forest	8		
Aboveground vegetation		40–60	0.20–0.30
Organic soil		20–500	0.1–0.90

boreal forest were relatively low, 1.5 million ha yr^{-1} (Seiler and Crutzen 1980). However, a more complete compilation of fire records increased this estimate to 5–8 million ha yr^{-1} during the 1990s (Korovin 1996; Shvidenko and Nilsson, this volume, Chapter 8).

There is some skepticism about the amount of fire occurring in the Russian boreal forest based on official government statistics, reported to be about 1.5 million ha yr^{-1} (Korovin 1996). This skepticism has been fueled by analyses of satellite imagery collected over Russia during the late 1980s. One of the largest fires ever measured occurred in the boreal forests of the Heilongjiang Province of northeastern China in May 1987. In less than 4 weeks, more than 1.3 million ha of boreal forest was burned (Levine et al. 1991; Cahoon et al. 1994). At the same time, extensive fire activity occurred across the border in Russia, particularly in the area east of Lake Baikal between the Amur and Lena rivers (Cahoon et al. 1994). Estimates based on NOAA AVHRR imagery indicate that 14.446 million ha in China and Siberia was burned in 1987 (Cahoon et al. 1994). During a less severe fire year in 1992, Cahoon et al. (1996) estimated from satellite imagery that only 1.5 million ha burned in the Russian boreal forest.

Based on these observations, a strong argument can be made that the patterns of fire in the boreal forests of Russia and North America are similar and that the area burned in each region should be proportional. The fire records for North America show that during the 1980s, an average of 2.9 million ha yr^{-1} burned in North America. Because the Russian boreal forest covers an area twice as large as the North American boreal forest, it can be estimated that during the same time period, 8.1 million ha yr^{-1} burned in the boreal forest.

The density of the biomass fuels exposed to burning in the three different biomes is dramatically different. In savannas, the density of aboveground biomass is extremely low, about 6 t ha^{-1}. By contrast, the average aboveground biomass levels in tropical forests range between 50 and 330 t ha^{-1}. In the boreal forest, the average aboveground biomass densities are between 40 and 60 t ha^{-1}. The fuels available for burning in boreal forests differ from tropical forests and savannas, in that large amounts of organic soils are present in this biome. The average level of the fuels in these organic soils ranges between 20 and more than 500 t ha^{-1}.

The fraction of biomass consumed during fires varies greatly between the different biomes (Table 3.1). The burning efficiency in savannas is extremely high, with 0.8–1.0 of all aboveground vegetation being consumed during fires. The burning of the woody vegetation present in tropical and boreal forests is much less efficient, with the fraction of biomass consumption ranging between 0.2 and 0.3. Finally, the burning efficiency of the organic soils present in boreal forests is highly variable and dependent on the moisture conditions of the ground layer at the time of the fire. Studies in Alaskan boreal forests (Chapter 10) show that the fraction of biomass consumption ranges between 0.1 and 0.9.

Finally, an extremely important issue facing scientists trying to quantify the longer-term patterns in the atmospheric concentration of greenhouse gases relates to the overall net effect of biomass burning. Again, the answer to this question is dependent on the biome. It is thought that burning of savannas represents no net loss or gain of atmospheric carbon because of the rapid regrowth of the vegetation that is burned during the following growing season. An argument can be made that burning of tropical forests in boreal and tropical regions would not represent a net change over the time scale it takes forests that were killed by the fires to regenerate, on the time scale of several decades to several centuries. However, in tropical regions, because many forests are being converted to agricultural fields or pastures, their burning may represent a net loss of carbon to the atmosphere. Finally, the situation in terms of the organic soils in boreal forests is rather unique and often overlooked. Many of the deeper organic soil profiles are formed over time periods that span from several hundred to several thousand years. When these older soil profiles are burned, they represent a net loss of carbon to the atmosphere over shorter time horizons (< 200 years). Thus, the increase in fire severity and frequency projected for the boreal region is expected to result in a significant loss of stored carbon to the atmosphere (Kasischke et al. 1995).

Gaseous and Particulate Emissions During Biomass Burning

The carbon released during burning takes the form of several gaseous and particulate compounds, including CO_2, CO, CH_4, NMHCs, and particulate or elemental carbon. The ratio of any carbon compound to CO_2 produced in burning, or the ratio of a nitrogen compound to CO_2 produced in burning, may be determined by knowledge of the emission ratio (ER). The ER is the amount of any compound X produced during burning normalized with respect to the amount of CO_2 produced during burning. The ER is usually normalized with respect to CO_2 because CO_2 is the overwhelming carbon species produced during biomass burning and it is a relatively easy gas to measure. The ER is defined as

$$ER = \Delta X / \Delta CO_2 \tag{3.2}$$

where ΔX is the concentration of the species X produced by biomass burning, $\Delta X = X^* - X$, where X^* is the measured concentration in the biomass burn smoke, X is the background (out-of-plume) atmospheric concentration of the species, and ΔCO_2 is the concentration of CO_2 produced by biomass burning, $\Delta CO_2 = CO_2^* - CO_2$, where CO_2^* is the measured concentration in the biomass burn plume and CO_2 is the background (out-of-plume) atmospheric concentration of CO_2.

Emission factors (EF) are another method of evaluating emissions from biomass burning. EFs are usually expressed as the ratio of grams of species X produced during burning to the kilograms of fuel burned. Because much more information is needed to develop EFs than ERs (e.g., fuel consumption) they are not easily applied to fires of uncharacterized fuels. To remedy this, Ward and colleagues (1979) and Nelson (1992) have developed a technique referred to as the carbon mass balance (CMB) technique for calculating EFs. With CMB, total excess carbon (in-plume less background) is calculated from all measured excess of carbon products (i.e., CO_2 + CO + CH_4). Because at least 95% of the carbon released during burning is in the form of these three compounds, an estimate of fuel burned can be made on a mass of carbon/volume basis. This mass is multiplied by 2 because the carbon content of wood is approximately 50%. This technique has been applied to aircraft-obtained smoke plume measurements to determine EFs by Radke and co-workers (1988).

One of the important discoveries in biomass burning research over the past 5 years is that fires in diverse ecosystems are very different in the production of gaseous and particulate emissions. These emissions depend on the type of ecosystem burning, that is, the nature and distribution of the vegetation (i.e., forests or grass), the moisture content of the vegetation, and the nature, behavior, and characteristics of the fire (i.e., flaming or smoldering phase, fast or slow spreading). It is no longer correct to assume constant ERs for all fires and for the entire lifetime of a particular fire (Cofer et al. 1990, 1991).

The gaseous and particulate emissions produced during biomass burning are dependent on the nature of the biomass matter, the temperature of the fire, and the combustion efficiency of the fire, all of which are ecosystem dependent. In general, biomass is composed mostly of carbon (about 45% by weight) and hydrogen and oxygen (about 55% by weight), with trace amounts of nitrogen (0.3–3.8% by mass), sulfur (0.1–0.9%), phosphorus (0.01–0.3%), and potassium (0.5–3.4%) and still smaller amounts of chlorine and bromine. During complete combustion, the burning of biomass matter produces CO_2 and water vapor as the primary products, according to the reaction

$$CH_2O + O_2 \rightarrow CO_2 + H_2O$$

where CH_2O represents the average composition of biomass matter. Biomass burning can be considered as an abiotic equivalent of respiration/decomposition. Estimates of the release of carbon into the atmosphere from biomass burning for different ecosystems are summarized in Table 3.2.

The levels of biomass burning for temperate and boreal forests in Table 3.2 are based on the earlier research of Seiler and Crutzen (1980), which assumed only 1.5 million ha of boreal forest burned each year. If we assume that in fact 8 million

Table 3.2. Global Estimates of Annual Amounts of Biomass Burning and of the Resulting Release of Carbon into the Atmosphere[a]

Source of burning	Biomass burned (Tg dry matter yr^{-1})	Carbon released (Tg carbon yr^{-1})
Tropical forests	1,260	570
Temperate and boreal forests	280	130
Savannas	3,690	1,660
Agricultural waste	2,020	910
Fuel wood	1,430	640
Charcoal	21	30
World total	8,700	3,940

[a]Based on data from Andreae (1991).

hectares of boreal forest burn each year and that 20 t C ha^{-1} is released on average, then the value for temperate and boreal forests should be 267 Tg C yr^{-1}.

In the case of incomplete combustion in cooler and/or oxygen-deficient fires (i.e., the smoldering phase of burning), significantly more carbon is released in the forms of CO, CH_4, NMHCs, and various partially oxidized organic compounds, including aldehydes, alcohols, ketones, and organic acids (Lobert et al. 1991). Nitrogen is present in biomass mostly as amino groups (R-NH_2) in the amino acids of proteins. During combustion, the nitrogen is released by pyrolytic decomposition of the organic matter and partially or completely oxidized to various volatile nitrogen compounds, including molecular nitrogen (N_2), NO, nitrous oxide (N_2O), ammonia (NH_3), hydrogen cyanide (HCN), cyanogen (NCCN), organic nitriles [acetonitrile (CH_3CN), acrylonitrile (CH_2CHCN), propionitrile (CH_3CH_2CN)], and nitrates (Lobert et al. 1991). The sulfur in biomass is organically bound in the form of sulfur-containing amino acids in proteins. During burning, the sulfur is released mostly in the form of SO_2 and smaller amounts of carbonyl sulfide (COS) and nonvolatile sulfate ($SO_4^=$) (Lobert et al. 1991). About one-half of the sulfur in the biomass matter is left in the burn ash, whereas very little of the fuel nitrogen is left in the ash (Lobert et al. 1991).

By combining estimates for the global annual amounts of biomass burning given in Table 3.2 with information on the ERs for various compounds produced during burning, we can estimate the global emissions of various combustion products (both gaseous and particulate) and compare the source strength of these emissions to the production of these compounds by all sources, including biomass burning. Estimates for the global emissions of gases and particulates from biomass burning and from all sources are given in Table 3.3. Inspection of Table 3.3 indicates that biomass burning is a significant global source of many environmentally significant, radiatively active and chemically active species.

Emissions from Burning in the Boreal Forest: The Bor Forest Island Fire

An experimental high-intensity stand replacement fire was conducted in central Siberia on July 6, 1993. Science experiments designed around this fire were the first joint East-West fire research activities organized under the Fire Research Campaign Asia North and sponsored by the Siberian Branch of the Russian Academy of Sciences, the Russian Aerial Forest Fire Protection Service, the International Boreal Forest Research Association, the International Global Atmospheric Chemistry Project, part of the International Geosphere-Biosphere Program, and the Max Planck Institute for Chemistry. A research team from the Atmospheric Sciences Division of the NASA Langley Research Center participated in this experiment and obtained samples of gaseous and particulate emissions from the fire (Cofer et al. 1996). The Bor Forest Island site (60°45′ N, 89°25′ E) is about 600 km north of Krasnoyarsk, Siberia, and consisted of about 50 ha of 130-year-old, 20-meter-high, live and standing-dead Scots pine (*Pinus silvestris*),

Table 3.3. Comparison of Global Emissions from Biomass Burning with Emissions from All Sources (including Biomass Burning)[a]

Species	Biomass burning (Tg element yr^{-1})	All sources (Tg element yr^{-1})	Biomass burning (%)
Carbon dioxide			
Gross	3,500	8,700	40
Net	1,800	7,000	26
Carbon monoxide	350	1,100	32
Methane	38	380	10
Nonmethane hydrocarbons[b]	24	100	24
Nitric oxide	8.5	40	21
Ammonia	5.3	44	12
Sulfur gases	2.8	150	2
Methyl chloride	0.51	2.3	22
Hydrogen	19	75	25
Tropospheric ozone[c]	420	1,100	38
Total particulate matter	104	1,530	7
Particulate organic carbon	69	180	39
Elemental carbon (black soot)	19	<22	>86

[a]Based on data from Andreae (1991).
[b]Excluding isoprene and terpenes.
[c]Tropospheric ozone is not a direct product of biomass burning. The chemical precursors that lead to the production of tropospheric ozone include carbon monoxide, methane, nonmethane hydrocarbons, and nitric oxide, all of which are produced during biomass burning. The value for tropospheric ozone in the table presents the amount resulting from the photochemical production of the biomass burning precursor products.

with a forest floor covering composed essentially of lichen (*Cladonia* sp.) overlaying 5–7 cm of partially decomposed organic matter. Fuel loading was determined to be 34 t ha^{-1}. A high-intensity experimental fire was possible because the burn site was a small island of live trees surrounded by relatively wet marshlands—preventing fire control problems. The fire burned from about 1430 to 1700 local time, during which there were no significant natural winds. Due to the low wind conditions, the fire was ignited by hand torch along the perimeter, creating a convection fire. Once ignited, the fire generated its own wind field and produced a smoke column rising to about 5 km. The Bor Island fire can be divided into three distinct phases: (1) a flaming surface fire that consisted primarily of the combustion of litter, lichen, and duff, (2) an intense flaming fire involving explosive migrations of fire into needled tree crowns, and (3) a brief smoldering stage.

Grab samples were collected by using a Russian MI-8 helicopter fitted with the NASA Langley Research Center smoke sampling equipment. Particle-filtered samples were drawn (via a high-volume pump) through a nose-mounted probe, coupled to the high-volume pump with flexible hose. Smoke samples were fed into 10-L Tedlar gas sampling bags. Each bag constitutes one sample, which was composed of one helicopter pass through the smoke column. After sample collection, the gas was transferred into stainless steel grab sampling bottles, which were

Table 3.4. Emission Ratios (in %) Normalized with Respect to CO_2 for Bor Island Forest Fire[a]

	CO/CO_2	CH_4/CO_2	H_2/CO_2
Flaming 1 (4 samples)[b]	8.8 ± 2.7	0.5 ± 0.1	1.2 ± 0.2
Flaming 2 (5 samples)[c]	11.3 ± 2.7	0.4 ± 0.1	1.6 ± 0.1
Smoldering (4 samples)	33.5 ± 4.5	1.3 ± 0.2	2.2 ± 0.2

[a]Based on data from Cofer et al. (1996).
[b]Measurements represent flaming combustion before the ground fire erupted into the tree crowns and consisted of emissions produced by burning lichen, surface litter, and duff.
[c]Measurements represent flaming combustion both on the ground and in the tree crowns.

returned to our laboratory and analyzed for CO_2, CO, H_2, and CH_4 within 2 weeks of collection. Smoke sampling was conducted at altitudes as low as safety would permit, determined most often by fire intensity and smoke turbulence. Flight paths chosen during smoke plume/column sampling were based on visual keys such as smoke color, flame characteristics, apparent turbulence, and combustion stage. It is this ability to select and sample smoke from specific parts of a fire that is the real strength of helicopter sampling. For example, when the crowns of a section of trees were observed to explode into flames, we were able to chase and sample the resulting smoke. The helicopter was always moving forward during sampling at no less than 40 knots. At 40 knots, rotor downwash is well behind the sampling probe, thereby eliminating collection of smoke other than that targeted.

Mean ERs normalized with respect to CO_2, CO, H_2, and CH_4 determined for the Bor Forest Island fire are summarized in Table 3.4 and presented as mean EFs in Table 3.5. The average value presented in Table 3.5 was generated assuming that 90% of the biomass was consumed in a flaming fire and 10% in a smoldering fire. Also, an average carbon content of 50% was assumed for both aboveground and ground-layer biomass.

Because of the relative amount of inefficient smoldering fires that occur in boreal forests, calculations using the satellite-derived burn area and measured ERs of gases for boreal forest fires (Cofer et al. 1990) indicate that the Chinese and Siberian fires of the 1987 contributed about 20% of the total CO_2 produced by savanna burning, 36% of the total CO produced by savanna burning, and 69% of the total CH_4 produced by savanna burning (Cahoon et al. 1994). Because savanna burning represents the largest component of tropical burning in terms of the

Table 3.5. Emission Factors: Ratio of Mass of Combustion Gas to Mass of Biomass Fuel (in g kg^{-1}) for the Bor Island Forest Fire[a]

	CO_2	CO	CH_4	H_2
Crowning	1,500 ± 50	120 ± 30	2.0 ± 0.5	2.0 ± 0.2
Smoldering	1,100 ± 80	350 ± 45	8.0 ± 0.8	1.7 ± 0.4
Average	1,475 ± 40	180 ± 40	4.2 ± 0.8	0.8 ± 0.3

[a]Based on data from Cofer et al. (1996).

vegetation consumed by fire (Table 3.2), it is apparent that the atmospheric emissions from boreal forest burning are an important contributor to the global budgets for many species during the high fire years.

Role of Emissions from Biomass Burning in Atmospheric Chemical Reactions

Although much attention is being paid to the role of greenhouse gas emissions from biomass burning, these emissions also play a central role in a large number of important chemical reactions and processes. These processes are summarized in this section.

Chemistry of the Hydroxyl Radical in the Troposphere

The hydroxyl (OH) radical is the major chemical scavenger in the troposphere, and it controls the atmospheric lifetime of most gases in the troposphere. The atmospheric lifetime, t, of any gas, x_i, that reacts with the OH radical is given by the following expression:

$$t = 1/k\,[OH]$$

where k is the kinetic reaction rate for the reaction between OH and x_i and [OH] is the concentration of the OH radical (molecules cm^{-3}).

The concentration of the OH radical is controlled by the balance between its chemical production and destruction. The OH radical is formed by the reaction of excited atomic oxygen [$O(^1D)$] with water vapor:

$$O(^1D) + H_2O \rightarrow 2OH, \; k_1$$

Tropospheric $O(^1D)$ is produced by the photolysis of O_3. In the troposphere, this photolysis reaction occurs over a very narrow spectral interval, between 290 and 310 nm. The production of $O(^1D)$ decreases as the latitude increases because less incoming solar radiation for photolysis is available. The bulk of the water vapor in the atmosphere resides in the troposphere. The amount of H_2O in the atmosphere is controlled by the saturation vapor pressure, which decreases with decreasing atmospheric temperature (i.e., as altitude or latitude increases).

The OH radical is destroyed via its reactions with CO and CH_4, both important products of biomass burning:

$$OH + CO \rightarrow H + CO_2, \; k_2$$

and

$$OH + CH_4 \rightarrow CH_3 + H_2O, \; k_3$$

The concentration of the OH radical (molecules cm^{-3}) is determined by dividing the OH production term by its destruction terms:

$$[OH] = 2\, k_1\, [H_2O]/k_2\, [CO] + k_3\, [CH_4]$$

Production of Ozone in the Troposphere

In addition to controlling the chemical destruction of OH, the oxidation of CO and CH_4 by OH outlined above initiates the CO and CH_4 oxidation schemes, which leads to the photochemical production of O_3 in the troposphere in the following chemical reactions:

Carbon Monoxide Oxidation Chain

$$CO + OH \rightarrow CO_2 + H$$

$$H + O_2 + M \rightarrow HO_2 + M$$

$$HO_2 + NO \rightarrow NO_2 + OH$$

$$NO_2 + hv \rightarrow NO + O$$

$$O + O_2 + M \rightarrow O_3 + M$$

$$\text{Net reaction: } CO + 2O_2 \rightarrow CO_2 + O_3$$

where hv is ultraviolet radiation. Note that the key gases in the CO oxidation chain leading to the photochemical production of tropospheric O_3, CO, and NO are both produced by biomass burning.

Methane Oxidation Chain

$$CH_4 + OH \rightarrow CH_3 + H_2O$$

$$CH_3 + O_2 + M \rightarrow CH_3O_2 + M$$

$$CH_3O_2 + NO \rightarrow CH_3O + NO_2$$

$$CH_3O + O_2 \rightarrow CH_2O + HO_2$$

$$HO_2 + NO \rightarrow NO_2 + OH$$

$$2 \text{ x } (NO_2 + hv \rightarrow NO + O)$$

$$2 \text{ x } (O + O_2 + M \rightarrow O_3 + M)$$

$$\text{Net reaction: } CH_4 + 4O_2 \rightarrow CH_2O + H_2O + 2O_3$$

Note that the key gases in the CH_4 oxidation chain leading to the photochemical production of tropospheric O_3, CH_4, and NO are both produced by biomass burning.

Chemistry of Nitrogen Oxides in the Troposphere

In addition to being a key player in the CO and CH_4 oxidation chains leading to the chemical production of O_3 in the troposphere, NO also leads to the chemical production of nitric acid (HNO_3), the fastest growing component of acidic precipitation. NO is chemically transformed to nitrogen dioxide (NO_2) and then to NO via the following reactions:

$$NO + O_3 \rightarrow NO_2 + O_2$$

$$NO + HO_2 \rightarrow NO_2 + OH$$

$$NO + CH_3O_2 \rightarrow NO_2 + CH_3O$$

$$NO_2 + OH + M \rightarrow HNO_3 + M$$

Chemistry of the Stratosphere

Approximately 90% of the O_3 in the atmosphere is found in the stratosphere (at altitudes between 15 and 50 km), with about 10% in the troposphere (at altitudes between 0 and 15 km). Stratospheric O_3 is very important because it absorbs ultraviolet radiation (200–300 nm) from the sun and shields the surface from this biologically lethal radiation. Stratospheric O_3 is destroyed via a series of chemical reactions involving chlorine and bromine. Chlorine destroys stratospheric O_3 through the following catalytic cycle (Wayne 1991):

$$Cl + O_3 \rightarrow ClO + O_2$$

$$ClO + O \rightarrow Cl + O_2$$

$$\text{Net reaction: } O + O_3 \rightarrow 2O_2$$

Bromine destroys stratospheric O_3 via the following three reaction cycles (World Meteorological Organization 1995):

Cycle 1

$$BrO + ClO + h\nu \rightarrow Br + Cl + O_2$$

$$Br + O_3 \rightarrow BrO + O_2$$

$$Cl + O_3 \rightarrow ClO + O_2$$

$$\text{Net reaction: } 2O_3 \rightarrow 3O_2$$

Cycle 2

$$BrO + HO_2 \rightarrow HOBr + O_2$$

$$HOBr + h\nu \rightarrow OH + Br$$

$$Br + O_3 \rightarrow BrO + O_2$$

$$OH + O_3 \rightarrow HO_2 + O_2$$

Cycle 3

$$BrO + NO_2 + M \rightarrow BrONO_2 + M$$

$$BrONO_2 + M \rightarrow Br + NO_3$$

$$NO_3 + h\nu \rightarrow NO + O_2$$

$$Br + O_3 \rightarrow BrO + O_2$$

$$NO + O_3 \rightarrow NO_2 + O_2$$

There are two other channels for the reaction between BrO and ClO (Wayne 1991):

$$BrO + ClO \rightarrow Br + OClO$$

$$BrO + ClO \rightarrow BrCl + O_2$$

There is also a relevant reaction between BrO and itself (Wayne 1991):

$$BrO + BrO \rightarrow Br + Br + O_2$$

Biomass Burning and Atmospheric Nitrogen and Oxygen

Biomass burning is both an instantaneous source (combustion of biomass) and a long-term source (enhanced biogenic soil emissions via increased nitrification and denitrification in soil) of gases to the atmosphere. These gases affect the chemistry of the troposphere and stratosphere, as outlined in this chapter. In addition, biomass burning may also affect atmospheric concentrations and the biogeochemical cycling of nitrogen and oxygen, the two major constituents of the atmosphere. Through the process of nitrogen fixation, molecular nitrogen (N_2) is transformed to the surface in the form of "fixed" nitrogen [i.e., ammonium (NH_4^+) and nitrate (NO_3^-)]. Nitrogen fixation results from both natural processes (biological fixation in root modules in certain agricultural crops and atmospheric lightning) and human processes (the production of nitrogen fertilizer and high-temperature combustion). The world's use of industrially fixed nitrogen fertilizer has increased from about 3 Tg N/yr^{-1} in 1940 to about 75 Tg N/yr^{-1} in 1990 (Levine et al. 1996b). The "fixed" nitrogen in the forms of NH_4^+ and NO_3^- is returned to the atmosphere mainly in the form of N_2, with smaller amounts of N_2O, and still smaller amounts of NO by denitrification and in the form of NO by nitrification. Burning or "pyrodenitrification" may also be an important source of nitrogen, mostly in the form of N_2, from the biosphere to the atmosphere (Lobert et al. 1991; Levine et al. 1996b). The problem is that it is difficult to quantify the amount of N_2 released during burning (Lobert et al. 1991), however, biomass burning or pyrodenitrification may prove to be an important process in the recycling of nitrogen compounds from the biosphere to the atmosphere.

Burning affects the concentration of atmospheric oxygen in two ways. Carbon released during the burning of biomass combines with atmospheric oxygen to form CO_2. Hence, burning is a sink for atmospheric oxygen. In addition, biomass burning destroys the very source of atmospheric oxygen—its production in the biosphere via the process of photosynthesis in the world's forests.

Unique Aspects of Fire and Biomass Burning in the Boreal Forest

There are additional reasons why burning in the world's boreal forests is important in terms of the study of global climate change. First, the boreal forests are very susceptible to global warming. Small changes in the surface temperature can significantly influence the ice/snow/albedo feedback. Thus, infrared absorption processes by fire-produced greenhouse gases, as well as fire-induced changes in surface albedo and infrared emissivity, are more environmentally significant than in the tropics.

Second, in the world's boreal forests, global warming may result in warmer and drier conditions. This, in turn, may result in enhanced frequency of fire and the accompanying enhanced production of greenhouse gases that will amplify the greenhouse effect.

Third, fires in the boreal forests are perhaps the most energetic in nature. The average fuel consumption per unit area in the boreal forest is about 25–50 t ha^{-1}, which is about an order of magnitude greater than in the tropics. Large boreal forest fires typically spread very quickly, most often as crown fires. Large boreal forest fires release enough energy to generate convective smoke columns that routinely reach well into the troposphere and on occasion may directly penetrate across the tropopause. The tropopause is at a minimum height over the world's boreal forests. For example, a 1986 forest fire in northwestern Ontario (Red Lake) at times generated a convective smoke column that was 12–13 km high, penetrating the troposphere (Stocks and Flannigan 1987).

Fourth, the cold temperature of the troposphere over the world's boreal forests results in very low levels of tropospheric water vapor. The deficiency of tropospheric water vapor and the scarcity of incoming solar radiation over most of the year results in very low photochemical production of the OH radical over the boreal forests. The OH radical is the overwhelming chemical scavenger in the troposphere and controls the atmospheric lifetime of most tropospheric gases. The very low concentrations of the OH radical over the boreal forests will result in enhanced atmospheric lifetimes for most tropospheric gases, including the gases produced by fire combustion. Hence, gases produced by burning, such as CO, CH_4, and the oxides of nitrogen, will have enhanced lifetimes over the boreal forest, which in essence, strengthens their role as greenhouse gases.

And finally, measurements have shown that in addition to the instantaneous production of trace gases and particulates resulting from the combustion of biomass, burning also enhances the biogenic emissions of NO and N_2O from soil (Anderson et al. 1988; Levine et al. 1988, 1991, 1996b) and the biogenic emission of Co from soil (Zepp et al. 1996). It is believed that enhanced biogenic soil emissions of NO and N_2O are related to increased concentrations of NH_4^+ found in soil following burning. NH_4^+, a component of the burn ash, is the substrate in nitrification, which is the microbial process believed responsible for the production of NO and N_2O (Levine et al. 1988, 1991, 1996b). The postfire-enhanced biogenic soil emissions of NO and N_2O may be comparable with or even surpass the instantaneous production of these gases during biomass burning (Harris et al. 1996). In addition, the warming of soils occurring in boreal regions after fires leads to enhanced soil respiration and emissions of CO_2 (Chapter 11).

Conclusions

In addition to being a significant instantaneous global source of atmospheric gases and particulates, burning enhances the biogenic emissions of NO and N_2O from the world's soils. Biomass burning affects the reflectivity and emissivity of the earth's surface. Burning also affects the hydrological cycle by changing rates of land evaporation and water runoff. For the reasons outlined here, it appears that biomass burning may be a significant driver for global change. It seems appropriate to conclude this summary with a quotation from fire historian Stephen J. Pyne (1991): "We are uniquely fire creatures on a uniquely fire planet, and through fire the destiny of humans has bound itself to the destiny of the planet."

References

Anderson, I.C., J.S. Levine, M.A. Poth, and P.J. Riggan. 1988. Enhanced biogenic emissions of nitric oxide and nitrous oxide following surface biomass burning. *J. Geophys. Res.* 93:3893–3898.

Andreae, M.O. 1991. Biomass burning: its history, use, and distribution and its impact on environmental quality and global climate, pp. 3–21 in J.S. Levine, ed. *Global Biomass Burning: Atmospheric, Climatic, and Biospheric Implications.* MIT Press, Cambridge, MA.

Cahoon, D.R., B.J. Stocks, J.S. Levine, W.R. Cofer, and J.M. Pierson. 1994. Satellite analysis of the severe 1987 forest fires in northern China and southeastern Siberia. *J. Geophys. Res.* 99:18,627–18,638.

Cahoon, D.R., Jr., B.J. Stocks, J.S. Levine, W.R. Cofer III, and J.A. Barber. 1996. Monitoring the 1992 forest fires in the boreal ecosystem using NOAA AVHRR satellite imagery, pp. 795–802 in J.S. Levine, ed. *Biomass Burning and Climate Change. Vol. 2. Biomass Burning in South America, Southeast Asia, and Temperate and Boreal Ecosystems, and the Oil Fires of Kuwait.* MIT Press, Cambridge, MA.

Cofer, W.R., J.S. Levine, E.L. Winstead, and B.J. Stocks. 1990. Gaseous emissions from Canadian boreal forest fires. *Atmos. Environ. Part A* 24:1653–1659.

Cofer, W.R., J.S. Levine, E.L. Winstead, and B.J. Stocks. 1991. Trace gas and particulate emissions from biomass burning in temperate ecosystems, pp. 203–208 in J.S. Levine, ed. *Global Biomass Burning: Atmospheric, Climatic, and Biospheric Implications.* MIT Press, Cambridge, MA.

Cofer, W.R., E.L. Winstead, B.J. Stocks, L.W. Overbay, J.G. Goldammer, D.R. Cahoon, and J.S. Levine. 1996. Emissions from boreal forest fires: are the atmospheric impacts underestimated? pp. 834–839 in J.S. Levine, ed. *Biomass Burning and Climate Change. Vol. 2. Biomass Burning in South America, Southeast Asia, and Temperate and Boreal Ecosystems, and the Oil Fires of Kuwait.* MIT Press, Cambridge, MA.

Crutzen, P.J., and J.G. Goldammer, eds. 1993. *Fire in the Environment: The Ecological, Atmospheric, and Climatic Importance of Vegetation Fires.* John Wiley & Sons, Chicester, UK.

Goldammer, J.G., and V.V. Furyaev, eds. 1996. *Fire in Ecosystems of Boreal Eurasia.* Kluwer Academic Publishers, Dordrecht, The Netherlands.

Hao, W.M., M-H. Liu, and P.J. Crutzen. 1990. Estimates of annual and regional releases of CO_2 and other trace gases to the atmosphere from fire in the tropics, based on FAO statistics for the period 1975–1980, pp. 440–462 in J.G. Goldammer, ed. *Fire in Tropical Biota.* Ecological Studies 84. Springer-Verlag, New York.

Harris, G.W., F.G. Wienhold, and T. Zenker. 1996. Airborne observations of strong biogenic NOx emissions from the Namibian Savanna at the end of the dry season. *J. Geophys. Res.* 101:23,707–23,711.

Kasischke, E.S., N.L. Christensen, Jr., and B.J. Stocks. 1995. Fire, global warming and the mass balance of carbon in boreal forests. *Ecol. Appl.* 5:437–451.

Kaufman, Y.I. 1987. Satellite sensing of aerosol absorption. *J. Geophys. Res.* 92:4307–4317.

Korovin, G.N. 1996. Analysis of distribution of forest fires in Russia, pp. 112-128 in J.G. Goldammer and V.V. Furyaev, eds. *Fire in Ecosystems of Boreal Eurasia.* Kluwer Academic Publishers, Dordrecht, The Netherlands.

Lacaux, J.P., R.A. Delmas, B. Cros, B. Lefeivre, and M.O. Andreae. 1991. Influence of biomass burning emissions on precipitation chemistry in the equatorial forests, pp. 167–173 in J.S. Levine, ed. *Global Biomass Burning: Atmospheric, Climatic, and Biospheric Implications.* MIT Press, Cambridge, MA.

Lenoble, J. 1991. The particulate matter from biomass burning: a tutorial and critical review of its radiative impact, pp. 381–386 in J.S. Levine, ed. *Global Biomass Burning: Atmospheric, Climatic, and Biospheric Implications.* MIT Press, Cambridge, MA.

Levine, J.S., ed. 1991. *Global Biomass Burning: Atmospheric, Climatic, and Biospheric Implications.* MIT Press, Cambridge, MA.

Levine, J.S., ed. 1996a. *Biomass Burning and Global Change. Vol. 1. Remote Sensing, Modeling and Inventory Development, and Biomass Burning in Africa.* MIT Press. Cambridge, MA.

Levine, J.S., ed. 1996b. *Biomass Burning and Global Change. Vol. 2. Biomass Burning in South America, Southeast Asia, and Temperate and Boreal Ecosystems, and the Oil Fires of Kuwait.* MIT Press, Cambridge, MA.

Levine, J.S., W.R. Cofer, D.I. Sebacher, E.L. Winstead, S. Sebacher, and P.J. Boston. 1988. The effects of fire on bioenic soil emissions of nitric oxide and nitrous oxide. *Global Biogeochem. Cycles* 2:445–449.

Levine, J.S., W.R. Cofer, E.L. Winstead, R.P. Rhinehart, D.R. Cahoon, D.I. Sebacher, S. Sebacher, and B.J. Stocks. 1991. Biomass burning: combustion emissions, satellite imagery, and biogenic emissions, pp. 264–272 in J.S. Levine, ed. *Global Biomass Burning: Atmospheric, Climatic, and Biospheric Implications.* MIT Press, Cambridge, MA.

Levine, J.S., E.L. Winstead, D.A.B. Parsons, M.C. Scholes, R.J. Scholes, W.R. Cofer, D.R. Cahoon, and D.I. Sebacher. 1996a. Biogenic soil emissions of nitric oxide (NO) and nitrous oxide (N_2O) from savannas in South Africa: the impact of wetting and burning. *J. Geophys. Res.* 101:23,689–23,697.

Levine, J.S., W.R. Cofer, D.R. Cahoon, E.L. Winstead, D.I. Sebacher, M.C. Scholes, D.A.B. Parsons, and R.J. Scholes. 1996b. Biomass burning, biogenic soil emissions, and the global nitrogen budget, pp. 370–380 in J.S. Levine, ed. *Biomass Burning and Global Change. Vol. 1. Remote Sensing, Modeling and Inventory Development, and Biomass Burning in Africa.* MIT Press, Cambridge, MA.

Lobert, J.M., D.H. Scharffe, W.M. Hao, T.A. Kuhlbusch, R. Seuwen, P. Warneck, and P.J. Crutzen. 1991. Experimental evaluation of biomass burning emissions: nitrogen and carbon compounds, pp. 289–304 in J.S. Levine, ed. *Global Biomass Burning: Atmospheric, Climatic, and Biospheric Implications.* MIT Press, Cambridge, MA.

Mano, S., and M.O Andreae. 1994. Emission of methyl bromide from biomass burning. *Science* 263:1255–1257.

Nelson, R.M., Jr. 1992. *An Evaluation of the Carbon Mass Balance Technique for Estimating Emission Factors and Fuel Consumption in Forest Fires.* Research paper SE-231, Southeastern Forest Experimental Station, USDA Forest Service, Asheville, NC.

Penner, J.E., R.E. Dickinson, and C.A. O'Neil. 1992. Effects of aerosol from biomass burning on the global radiation budget. *Science* 256:1432–1433.

Pync, S.J. 1991. Sky of ash, earth of ash: a brief history of fire in the United States, pp. 504–511 in J.S. Levine, ed. *Global Biomass Burning: Atmospheric, Climatic, and Biospheric Implications.* MIT Press, Cambridge, MA.

Radke, L.F., D.A. Hegg, J.H. Lyons, C.A. Brock, P.V. Hobbs, R. Weiss, and R. Rasmussen. 1988. Airborne measurements on smokes from biomass burning, pp. 411–422 in P.V. Hobbs and M.P. McCormick, eds. *Aerosols and Climate.* A. Deepak Publishing, Hampton, VA.

Radke, L.F., D.A. Hegg, P.V. Hobbs, J.D. Nance, J.H. Lyons, K.K. Laursen, R.E. Weiss, P.J. Riggan, and D.E. Ward. 1991. Particulate and trace gas emissions from large biomass fires in North America, pp. 209–224 in J.S. Levine, ed. *Global Biomass Burning: Atmospheric, Climatic, and Biospheric Implications.* MIT Press, Cambridge, MA.

Rogers, F.R., J.G. Hudson, B. Zielinska, R.L. Tanner, J. Hallett, and J.G. Watson. 1991. Cloud condensation nuclei from biomass burning, pp. 431–438 in J.S. Levine, ed. *Global Biomass Burning: Atmospheric, Climatic, and Biospheric Implications.* MIT Press, Cambridge, MA.

Seiler, W., and P.J. Crutzen. 1980. Estimates of gross and net fluxes of carbon between the biosphere and atmosphere from biomass burning. *Clim. Change* 2:407–427.

Stocks, B.J., and M.D. Flannigan. 1987. Analysis of the behavior and associated weather for a 1986 northwestern Ontario wildfire: Red Lake #7, pp. 94–100 in *Proceedings of the Ninth Conference on Fire and Forest Meteorology,* April 21–24, 1987, San Diego. Society of American Foresters, Washington, D.C.

van Wilgen, B.W., M.O. Andreae, J.G. Goldammer, and J.A. Lindesay, eds. 1997. *Fire in Southern African Savannas: Ecological and Atmospheric Perspectives.* Witwatersrand University Press, Johannesburg, South Africa.

Ward, D.E., R.M. Nelson, and D. F. Adams. 1979. Forest fire smoke plume documentation. Paper 79–6.3 in *Proceedings of the 72nd Annual Meeting of the Air Pollution Control Association, June 21–29, 1979, Cincinnati, OH.*

Ward, D.E., A.W. Setzer, Y.J. Kaufman, and R.A. Rasmussen. 1991. Characteristics of smoke emissions from biomass fires of the Amazon region-BASE-A experiment, pp.

394–402 in J.S. Levine, ed. *Global Biomass Burning: Atmospheric, Climatic, and Biospheric Implications.* MIT Press, Cambridge, MA.

Wayne, R.P. 1991. *Chemistry of Atmospheres, 2nd ed.* Oxford University Press, Oxford, UK.

World Meteorological Organization. 1995. *Scientific Assessment of Ozone Depletion: 1994.* Global Ozone Research and Monitoring Project, Report 37, 10.18. Geneva, Switzerland.

Zepp, R.G., W.L. Miller, R.A. Burke, D.A.B. Parsons, and M.C. Scholes. 1996. Effects of moisture and burning on soil-atmosphere exchange of trace carbon gases in a southern African savanna. *J. Geophys. Res.* 101:23,699–23,706.

4. Eurasian Perspective of Fire: Dimension, Management, Policies, and Scientific Requirements

Johann G. Goldammer and Brian J. Stocks

Introduction

The world's boreal forests and other wooded land within the boreal zone cover 1,200 million ha, of which 920 million ha is closed forest. The latter number corresponds to 29% of the world's total forest area and to 73% of its coniferous forest area (ECE/FAO 1998). About 800 million ha of boreal forests with a total growing stock of about 95 billion m^3 is exploitable (41% and 45%, respectively, of the world total). The export value of forest products from boreal forests is about 47% of the world total (Kuusela 1990, 1992).

The vast majority of the boreal forests of Eurasia are located in Russia. The Federal Forest Service of Russia controls some 1,200 million ha of land and close to two-thirds of the world's boreal forests. The fact that the carbon stored in global boreal ecosystems corresponds to about 37% of the total terrestrial global carbon pool (plant biomass and soil carbon) suggests that it may play a critical role in the global climate system (e.g., as a potential sink or source of atmospheric carbon).

Among natural disturbances in the boreal zone, fire is the most important factor controlling forest age structure, species composition, and physiognomy, shaping landscape diversity, and influencing energy flows and biogeochemical cycles, particularly the global carbon cycle. Small and large fires of varying intensity have different effects on the ecosystem. High-intensity crown fires are most common in boreal North America, where they lead to the replacement of forest stands through succession. In average years, some 98% of all wildfires in Russia are surface fires (Odintsov 1999). These fires favor the selection of fire-tolerant tree species such as pines (*Pinus* spp.) and larches (*Larix* spp.) and may occur repeatedly within the life span of a forest stand without eliminating it. Extremely dry years (e.g., 1987 and 1998) allow the development of extended high-intensity stand replacement fires.

In recent years, wildfires have been essentially eliminated in boreal western Eurasia (Europe). The annual area affected by fire in Norway, Sweden, and Finland is less than 4,000 ha. Thus, the major impact of Eurasian fires is in the Russian Federation and other countries of the Commonwealth of Independent States. Statistics compiled by the Russian Aerial Fire Protection Service Avialesookhrana show that between 17,000 and 33,000 forest fires occur each year, affecting up to 2 to 3 million ha of forest and other land (Korovin 1996). Because fires are monitored (and controlled) only in certain protected areas, it is estimated that the actual area affected by fire in Eurasia's boreal vegetation is much higher (Chapter 8). For example, satellite-derived observations by Cahoon and colleagues (1994) indicate that during the 1987 fire season approximately 14.5 million ha was burned. In the same fire season, about 1.3 million ha of forests was affected by fire in the montane-boreal forests of northeast China, south of the Amur (Heilongjiang) River (Goldammer and Di 1990; Cahoon et al. 1991). Although highly episodic annually, it appears that, on average, areas burned in Eurasia are comparable with or greater than areas burned in boreal North America.

The predicted global warming over the next 30–50 years will be most evident in northern circumpolar regions. Fire may be the most important (widespread) driving force in changing the forest cover under climatic warming conditions. The prediction of increasing occurrence of extreme droughts in a 2 × CO_2 climate indicates that fire regimes will undergo considerable changes (Chapter 20). Increasing length of the fire season and possibly drought-induced mass outbreaks of insects will lead to a higher occurrence of large, high-intensity wildfires. Such fire scenarios may be restricted to a transition period until a new climate-vegetation-fire equilibrium is established.

Estimates of carbon stored in above- and below-ground live and dead plant biomass (without soil organic matter) in the global boreal forest area range between 66 and 98 Gt (Chapter 13). Additional large amounts of carbon are stored in boreal forest soils (200 Gt) and boreal peatlands (400 Gt) (Chapter 24). There is concern that changing fire regimes due to climate change will affect the balance of the boreal carbon pool and lead to the additional release of carbon into the atmosphere, thus acting as a temporary positive feedback loop to global warming. One of the major concerns is the possible impact of climate change on fire regimes on permafrost sites. Thawing of the active layer as a consequence of increased temperature and removal of vegetation cover by wildfires will lead to dramatic changes of ecosystems and release of radiatively active trace gases (Chapter 25).

Clearly, the future of fire in Eurasia is of critical global importance. In this chapter, the history of Eurasian fire is examined, and the current fire situation is explained in terms of fire statistics and policy consideration. Future management options are also discussed.

Eurasian Fire History

Throughout boreal Eurasia, charcoal in soils and lake sediments provides evidence of wildfires in forests since the glacial retreat and the application of fire in land-use systems since the Neolithic. Comprehensive historical studies on fire history of the Baltic region and Russia highlight the ecological and cultural significance of natural and human-caused fires (Heikinheimo 1915; Viro 1969; Zackrisson 1977; Clark and Richard 1996; Pyne 1997; Clark et al. 1998; Niklasson 1998).

Throughout the ages in Eurasia, fire has been an important tool for land clearing (conversion of boreal forest), silviculture (site preparation and improvement, species selection), and maintaining agricultural systems (e.g., hunting societies, swidden agriculture, and pastoralism) (Viro 1969; Pyne 1997). In addition to the natural fires, these old cultural practices brought a tremendous amount of fire into the boreal landscapes of Eurasia. In the early 20th century, fire use in the agricultural sector began to decrease because most of the deforestation had been accomplished for agriculture, and traditional use of fires (treatment of vegetation by free burning) was replaced by mechanized systems and the increasing use of fertilizers and pesticides. Despite the loss of traditional burning practices, humans are still the major source of wildland fires; only 15% of the recorded fires in the Russian Federation are caused by lightning (Korovin 1996).

Historic swidden agricultural and forestry systems with common methods and goals have been documented for the whole of temperate and boreal Europe. Swidden agriculture in Russia has been taken place since 946. In Germany slash-and-burn agroforestry systems are recorded back to the years 1290 and 1344 in the Odenwald and the Black Forest regions, respectively. Around the Baltic Basin (the transition zone between the temperate and boreal zones of western Europe), the burning systems in agriculture and forestry had a very common physiognomy and expanded greatly in the 18th and 19th century due to rapid demographic changes and increasing social problems (poverty, lack of land and food supply). Heikinheimo's studies (1915) reveal that shifting cultivation in Finland at that time was practiced over more than 4 million ha yr^{-1}. By the beginning of this century, some 50–75% of Finland's forest area had been exploited in this manner. In the eastern part of Finland, shifting cultivation was practiced longer and more intensively than anywhere else in the country.

With the introduction of more efficient land-use technologies at the beginning of the Industrial Revolution (including fertilizers and pesticides), slash-and-burn agriculture decreased but was still practiced until the middle of the 20th century. Observations by Heikeinheimo (1915) around 1900 in a number of areas of Russia reveal that at that time more than 40% of the cultivated land was on recently burned terrain (for more information on Russia's cultural fire history, see Pyne 1997).

Slash-and-burn agriculture was less common in Latvia and Lithuania because 85% of the forests were government owned and managed for timber production. Even so, the use of fire is well documented. In Lithuania, forests were cut and the woody material was transported to open fields and burned. In Latvia, woody material was burned in rows; a report from 1795 reveals that heap burning was preferred on poor soils. In Estonia, the wood of alder was preferred material for burning on the fields, where ash was distributed as fertilizer.

With the end of the era of shifting cultivation in Finland in the early part of this century, methods derived from this practice began to find use in the regeneration of forests of low productivity. Burning of logging waste and the raw humus layer was recommended as a means of promoting the natural restocking of regeneration sites. Broadcast-seeding on snow in spring, with prescribed burning preceding it, found widespread use in the 1920s. Prescribed burning in those times amounted to about 8,000 ha yr^{-1} but declined to only a few hundred hectares a year in the 1930s.

Prescribed burning enjoyed a resurgence after World War II, with a peak of more than 30,000 ha being burned annually in the mid-1950s. Because of the presence of deep organic soils, prescribed burning was particularly useful in assisting with the regeneration of northern Finland's spruce stands to pines. However, this second wave of prescribed burning came to an end during the latter half of the 1960s, when it was replaced by mechanized site preparation. The area annually treated with prescribed fire fell to 500–1,000 ha yr^{-1} and has remained at that level up to the recent past.

With the disappearance of swidden agriculture in Russia, fire was completely banned from the forests. Large fire events in the early 20th century, along with the Soviet perception that fire was evil, resulted in a dogmatic fire exclusion policy that governed the situation in the Soviet Union and modern Russia.

Current Eurasian Fire: Country Overviews

Central Eastern Asia: Russian Federation and Other Countries of the Former Soviet Union

With the breakup of the former Soviet Union, the Russian Federation inherited most of the forest cover in this region. Thus, forest fires play only a limited role on the non-Russian territories of the former Soviet Union. No statistical data are available from Azerbaijan, Kyrgystan, Tajikistan, and Uzbekistan. Kazakhstan reports 12,753 ha burned annually between 1994 and 1996 and Turkmenistan 522 ha burned over the same period. Thus, the focus of this section is on the Russian Federation.

Official fire statistics from the Federal Forest Service of the Russian Federation are available for the decade 1988–1997. No final information on the severe fire situation in the Russian Far East during the months August–October 1998 is yet available. In the past 10 years, the number of forest fires fluctuated from 17,600 to nearly 33,000 annually. The highest fire impact was recorded in 1996, when 32,833 fires burned an area of 1.3 million ha. However, it must be remembered that these numbers cover only fires that are monitored and controlled on protected forest and pasturelands. Satellite-derived observations suggest that the real figures on areas affected by fire in the territory of the Russian Federation are much higher (e.g., the figure of 14 million ha for the year 1987 [Cahoon et al. 1994] or 6 million ha for the year 1998 [Kasischke et al. 1999]).

Forestry practices in boreal Russia involve the use of heavy machinery and large-scale clear-cuts, leading to the alteration of fuel complexes. Many clear-cut areas reportedly are not regenerating forests but are degrading into grass steppes, which may become subject to short-return interval fires.

Additional fire hazards are created in forests damaged by industrial emissions (severe damages in the Russian Federation affect about 9 million ha). Radioactive contamination on an area of about 7 million ha creates special problems because fire can redistribute radionuclides (Dusha-Gudym 1996, 1999). Disturbance by insects (e.g., the large-scale infestation of central Siberia by the siberian silkworm [*Dendrolimus superans sibiricus*] or the mass outbreaks of the spruce budworm [*Choristoneura fumiferana*] in North America) creates fire hazards that threaten sustainable forestry in both continents.

Central Southeastern Asia: People's Republic of China and Mongolia

People's Republic of China

In the northern part of the People's Republic of China, the temperate-boreal forests and steppes are the vegetation types most affected by fire. The mountain-boreal forest of the Daxinganling Mountain range, Heilongjiang Province, northeast China, is dominated by pines (*Pinus sylvestris* var. *mongolica*) and larches (*Larix gmelinii*), which are favored by the continental climate. About 70% of the annual precipitation of 350–500 mm occurs between May and August. March–May and September–October are the months with highest fire danger, creating a bimodal fire season. Forest ecosystem dynamics and species composition in these areas are characterized by regular natural and, more recently, human-caused fires (Goldammer and Di 1990). The forests in Inner Mongolia are dominated by larch, Mongolian oak, and birch. The larch is the original vegetation, which would be replaced by oak and birch or poplar (in wet areas) if fires occur frequently over a short time.

The largest fire in recent history occurred in the Daxinganling region during the exceptionally dry months of May–June 1987. This fire affected a total land area of 1.3 million ha (most of the area was forested) and exhibited an exceptionally high intensity and spread rate. The main fire front traveled 100 km within 5 hours and burned 400,000 ha of forests within 32 hours on May 7 and 8, 1987. More than 200 individuals died, 56,000 persons lost their homes, and 850,000 m^3 of processed wood was consumed. Many additional infrastructures (bridges, railroad tracks, electricity and telephone lines) were burned (Ende and Di 1990; Goldammer and Di 1990). Evaluation of long-term statistical data reveals that between 1950 and 1990, 4,137 persons were killed in forest fires in the People's Republic of China (Goldammer 1994).

Between 1966 and 1987, recurrent fires burned 5.6 million ha of forest- and nonforestland in Heilongjiang Province. Because this area has a forest cover of 5.26 million ha, the total fire area includes repeatedly burned-over areas. In 1966, 10.4% of the forested area was burned, and in 1987 17.4% of the forest area was affected by fire (Goldammer 1993). In the period 1966–1986, more than one-third of all fires were started by lightning. In 1997, less than 10% of fires were lightning caused, indicating an increasing human influence on fire ignitions in this region of China. In the decade 1986–1996, the annual burned area in Heilongjiang Province was 95,000 ha.

Mongolia

In Mongolia, fire is a major factor determining the spatial and temporal dynamics of forest and steppe ecosystems (Valendik et al. 1998; Wingard and Naidansuren 1998). Of 17.5 million ha of forestland (corresponding to 8.1% of the territory of Mongolia), 75% are coniferous (predominantly larch [*Larix sibirica*] and pine [*Pinus sylvestris*]) and deciduous forest (with extended occurrence of birch [*Betula platyphylla*]). More than 4 million ha is disturbed by either fire (95%) or logging (5%). Logged areas have increased dramatically over the past 20–25 years. Approximately 600,000 ha of cutover land has not yet recovered. Steppe covers approximately 40% of the Mongolian territory and serves as main pasture resource for some 30 million heads of livestock. It is assumed that most of today's steppe vegetation is on fire-degraded sites formerly covered by forest (E. Naidansuren, personal communication).

The highest forest fire danger is characteristic of low-mountain pine and larch stands growing on seasonally freezing soils in areas characterized by an extremely continental climate. Annual air temperature fluctuations can approach 90°C, with the summer maximum above 40°C. Annual precipitation ranges from 250 to 350 mm. In dry years, this value does not exceed 200 mm in forest regions. The majority of forest fires burn within the central and eastern parts of the forested area of Mongolia. This can be attributed to the predominance of highly fire-susceptible (flammable) pine and larch stands. Moreover, economic activity is much higher in this area. In Mongolian forests, fire seasons are usually discontinuous (i.e., they have two peaks of fire danger). One peak is observed during the long dry spring (from March to mid-June) and accounts for 80% of all fires. The other fire danger peak falls within a short period in autumn (September–October) and accounts for 5–8% of fires. In summer, fires occur very rarely (only 2–5% of the total).

In one of the most sparsely populated countries in the world, it is difficult to obtain accurate information on fire history and causes. First fire history studies conducted by Valendik and co-workers (1998) indicate that most forest fires result from steppe fires invading adjacent forest stands. Lightning fires are common in the mountain taiga belt because of increasing storm activity in late May and early June. Extreme fire seasons occur every 3 years in Mongolia. The mean fire interval varies from 9 to 22 years, depending on forest type, slope and aspect, and human ignition sources.

The first attempts to manage fire in Mongolia began in 1921 and remained limited to local town fire departments until the 1950s. Relatively accurate records of area burned are available after 1981. From these data, it appears that Mongolia is experiencing an increase in wildfires. From 1981 to 1995, forest and steppe fires burned an average of 1.74 million ha annually. In 1996, 1997, and 1998, the area affected by fire was 10.7, 12.4, and 3.9 million ha, respectively: a sixfold increase. In a typical season (1981–1995), some 140,000 ha of forest areas burned. However in 1996 and 1997, this figure increased to 2.5 million ha annually, corresponding to about 22% of the total land area affected by fire. In these 2 years alone, more forested area burned than was harvested over the past 65 years (Naidansuren 1996; Wingard and Naidansuren 1998).

Scandinavia: Finland, Norway, and Sweden

In Finland, forests cover 26 million ha, corresponding to 74% of the total land area. Most of the Finnish forests are privately owned (63%). The majority of the forest is in the boreal zone and is dominated by conifers (Scotch pine, Norway spruce) and birch. Only a small part of northern Finland is mountain area. More than 30% of the forestland is dried swamp and peatland. Roughly 60,000 ha of swamp is managed for peat production.

The summer season is relatively short (May–September) but may often include several warm and dry spells with high forest fire risk. Finnish forest fire statistics are available since 1868. The worst forest fire in this century was a 3,000-ha fire that burned in Lapland after originating in Russia (H. Frelander, personal communication). The low fire activity in Finland is due to a well-developed fire monitoring and suppression system.

In Norway, the oceanic climate controls the composition and fire hazard of forests. Boreal coniferous forests stretch from the east toward the Scandinavian mountain range and its alpine ecosystems. The coastal area has been classified as a nemoboreal zone characterized by temperate coastal forests. In the south, there are smaller areas in a nemoral zone that today are strongly influenced by human activity. Lightning is frequent in Norway, but high precipitation and humidity in the western and central parts of the country do not allow ignition by lightning. The highest frequency of natural fires occurs in the boreal forests of the country's eastern lowlands, southwestward to the divide, and in the most continental part of central Norway (Mysterud et al. 1998). The forested area burned between 1986 and 1996 was 564 ha yr^{-1}.

Most of the territory of Sweden lies within the boreal and hemiboreal zone, with most of the terrain covered by a fairly flammable vegetation of coniferous trees, ericaceous dwarf-shrubs, and mosses. During the mid-1970s, fire was not considered a serious problem, and the collection of fire statistics was abandoned temporarily in 1975 but resumed in 1992 (Granström 1998). The average area burned between 1992 and 1996 was 2,500 ha yr^{-1}.

Baltic and Central Eastern European Countries: Belarus, Estonia, Latvia, Lithuania, and Poland

The fire problem zones in the countries bordering the southern Baltic Sea (Estonia, Germany, Latvia, Lithuania, Poland) and Belarus are located outside the boreal zone. These countries are included in this overview because they represent the transition to the temperate forest zone (nemoboreal or sub-boreal forests). The forest cover in these countries is dominated by pine forests, which are favored by the continental climate and show some similarities to the boreal pine forests.

The Republic of Estonia is covered by 2 million ha of forest, and since 1949 an annual average of 215 fires affected a forested area of 210 ha. In the neighboring Republic of Lithuania, forests cover 1.8 million ha, and the average annually burned area was 202 ha during the period 1984–1996. Forests in Latvia cover 45% of the surface of the country and are located primarily in the western region of the country. The average area burned between 1991 and 1996 was 1,295 ha yr^{-1} (Gertners 1998). In Belarus, about 5,000 ha of forests is affected by wildfires annually, predominantly young to middle-aged pine stands. The main fire problem in Belarus arises on forestlands contaminated by radionuclides from the Chernobyl nuclear power plant accident in 1986. Finally, in Poland the average annual area burned between 1980 and 1996 was 5,170 ha. New fire problems arise in those forested regions that are affected by severe industrial pollution.

Socioeconomic Background and Environmental Impacts of Fire: Implications for Fire Management Strategies and Public Policies

In this section, selected examples are given on specific fire phenomena and fire problems in various countries of Eurasia. Problem descriptions were defined by government agencies or individual analysts but do not necessarily reflect official policies or programs.

Fires in Radioactively Contaminated Terrain: Belarus, Russia, and Khazakhstan

According to Dusha-Gudym (1996), many areas within Russia, Belarus, and Khazakhstan were contaminated by radionuclides between 1949 and 1993. Atmospheric nuclear weapons testing began in Kazakhstan in 1949, and in succeeding decades radioactive material was transported over much of this region, both directly and indirectly through reactivation of radionuclides through forest fires. Industrial accidents were common in the former USSR, resulting in radioactive waste leakage into rivers and widespread atmospheric transport of radionuclides to surrounding forests. The Chernobyl nuclear power plant explosion in Belarus in 1986 created radioactive fallout over large areas. In addition to devastating direct environmental effects, radionuclides were deposited in many forested areas, killing trees and creating a severe forest fire hazard. Subsequent widespread forest fires resulted in the redistribution of radioactive material over long distances. Recent investigations (Dusha-Dudym 1996) confirmed high levels of cesium, strontium, and plutonium in fuels burned in forest fires in these regions, with a portion of radioactive ash remaining on-site while the remainder released as smoke aerosols and transported in the atmosphere.

People's Republic of China

In the People's Republic of China, most wildfires are human caused: fires escaping from agriculture maintenance burning; camp fires set by hunters, mining operations, and collectors of nonwood forest products; and fires started alongside roads and railroads. Lightning is a frequent cause only at the end of the spring fire season in the northern forest region, especially along the border with Russia. As a result, the focus in China is on the prevention of human-caused fires.

In China, local governments are responsible for fire prevention and suppression, declaring fire season restrictions, establishing fire control organizations, constructing wildfire control facilities, approving the use of prescribed fire, and conducting public education. The issuing of permits, the use of inspection checkpoints, and the monitoring of railway operations are all designed to prevent human-caused fires.

The Great China Fire of 1987 resulted in vastly improved fire prevention, detection, and suppression programs being developed in China, particularly in the northern provinces of Heilongjiang and Inner Mongolia. Forest fire control offices were established by the provincial governments, and the number of forest police and professional fire brigades was increased. A lightning detection system has been established in the Daxinganling Mountain forest region. Aerial fire-fighting capabilities have been expanded, including the use of air tankers and helicopters. An Internet and computer-based national fire information system is in use, and fire images received from NOAA weather satellites can be transmitted from the national office to all provinces in China. A national forest fire danger rating system for China is also under development.

As an important long-term strategy, plans have been made for the construction of "greenbelts," acting as a fuel-break network integrated into reforestation operations throughout China. The goal of this fuel-break network is to limit the size of wildfire-affected areas to a maximum of 100 ha (Shu 1998). A network of fire breaks is also being established along China's borders to prevent the movement of fires between China and neighboring countries.

Mongolia

Except for evidence provided by tree ring records, no information is available on wildfires in Mongolia before 1921. The nomadic people and livestock were highly dependent on well-preserved grazing resources. Wildfire incidence increased significantly during the 1950s and 1960s with rapid socioeconomic development in

Mongolia, the result of an increase in population and the use of agricultural machinery and equipment, the construction of the first railway, and the increasing use of forest materials for construction materials, timber trade, and fuel supply. As a consequence of increasing fire occurrence, the government of Mongolia established an aerial fire protection service in many provinces that used both helicopters and smoke jumpers. Today, each Mongolian province has a local civil defense department responsible for wildfire suppression.

Immediately following the severe 1996 fires, Mongolia received assistance from international organizations to help local people recover from the losses. American advice and training on disaster fire management were provided (Shulman 1996, 1997), and Germany contributed an emergency fire aid project carried out in the northern and eastern parts of the country. Since then, the government has been working to find long-term solutions to the problem of fire management. In a first step, the parliament passed laws designed to organize and improve firefighting efforts at all levels.

In 1998, the German and Mongolian governments signed an agreement under which Germany will provide long- and short-term experts, support staff, training, and equipment (Wingard and Naidansuren 1998) to develop a fire management plan compatible with both the protected area goals and the responsibilities of the local communities in one of the regions most affected by the 1996 fires. Fire management units in the local communities will receive professional training and basic hand tools suitable for regional conditions, and local centers will provide the necessary infrastructure for fire prevention activities, management information, training exercises, dispatch, and field organization. This project supports Mongolia by strengthening local capacities to address effectively the issues of fire prevention, presuppression planning, and suppression. It will do this by helping to organize cooperation between protected areas and local and national administrations responsible for fire management; by establishing the necessary infrastructure, providing training both in-country and abroad; and by including all stakeholders in the planning and implementation of fire management activities. A socioeconomic study is under way that will provide the design of a community-based fire management approach (Ing 1998).

Russia

A significant number of forest fires in Russia occur as a result of human activities, with the local population accounting for 60–80% of the total number of fires. According to a report of the Federal Forest Fire Service of Russia, a series of measures was recently launched for the prevention of human-caused fires, including construction of fire break systems, special access roads, water sources, and recreation areas for the public. Information to help prevent fires and for nature protection is becoming more important in Russia, in order to involve various levels of society, especially young people.

In accordance with the need to strengthen forest fire protection, Russian forest legislation has also been changed. In 1997, a new Forest Code of the Russian Federation was approved and several decrees and regulations were adopted. Responsibilities for fire management and disaster management in emergency situations connected with forest fires and for the supervision of the state forest fund were given to the Federal Forest Fire Service of Russia. Protection of forests from fires is being implemented by the ground and aviation subdivisions of the fire service, creating a single system for prevention, detection, warning, and suppression of fires.

For raising the level of forest fire protection and gradual reduction of forest fire intensity and the size of direct and indirect damage from forest fires, Russia adopted the state program Forest Fire Protection for 1993 to 1998. The draft federal program Forest Fire Protection for 1999 to 2005 has also been prepared.

Research and development are under way to improve forest fire danger forecasting; to develop new technical means of fire detection and suppression, including airborne technologies; and to create advanced communication systems. A program to reduce forest fire risk is under way; it includes the planning of preventive prescribed burning in forests. The adoption of these special programs by the Russian government emphasizes the significance of the work for the environmental protection of the nation.

Constraints in implementing the program have been imposed by the continuing economic problems in Russia. In a report on the 1996 fire season in Russia, the unfavorable situation in central and eastern Siberia was explained as a consequence of lowering the level of forest fire protection by cutting down the number of fire-fighting crews and reduction of aircraft flying time (Davidenko 1997). Fire detection flight patrols were cut to 40% of the average during the 1980s. The number of smoke jumper and helicopter-based crews was reduced by 50% from the levels of the early 1990s, and aerial fire fighter capability was cut by 70%.

The 1998 fire season was extremely severe in the Far East of Russia, and a lack of fire-fighting resources was one of the major impediments to appropriate response to the forest fires in this region. According to the Aerial Forest Fire Protection Branch (Avialesookhrana) of the Federal Forest Service of Russia, the Far East area burned in 1998 was ten times higher than the average over the previous decade. Preliminary assessment of satellite imagery indicates that the forested area affected by fire during the 1998 fire season will exceed 6 million ha in the Russian Far East and 8 million ha in all Russian territories (Kasischke et al. 1999).

Fire Science Programs with Relevance to Regional Fire Management and Policy Development

International Geosphere-Biosphere Program

The international vegetation fire research community has organized itself through various mechanisms. The International Geosphere-Biosphere Program (IGBP) is the most interactive platform, through which several major international and interdisciplinary fire research programs have been designed and implemented.

One of the operational IGBP core projects is the International Global Atmospheric Chemistry (IGAC) Project. One major IGAC activity involves investigating the impact of biomass burning on the biosphere and atmosphere (Biomass Burning Experiment). Since 1990, several research campaigns have been conducted. For the boreal Asian region, the Fire Research Campaign Asia-North (FIRESCAN) began in 1992. FIRESCAN addresses the role of fire in boreal ecosystems and the consequences for the global atmosphere and climate (FIRESCAN Science Team 1996; Goldammer and Furyaev 1996).

Additional fire experiments will be conducted jointly with scientists collaborating in the IGBP Northern Eurasia Study (NES), which is a joint effort of scientists representing several IGBP core projects, the Biospheric Aspects of the Hydrological Cycle, IGAC, and Global Change and Terrestrial Ecosystems projects. The unifying theme of the IGBP NES is the terrestrial carbon cycle and its controlling factors, and the main objective is to determine how these will change under the rapidly changing environmental conditions projected under global change (Steffen and Shvidenko 1996). The IGBP NES will consist of an integrated set of experimental and observational studies on a number of scales, modeling and aggregation activities, and supporting databases and geographic information system (GIS) capabilities. The major elements are transects and network sites; a water, energy, and carbon flux study; and detailed studies of disturbance regimes.

The fire component of the IGBP NES will have four components: (1) fire manipulations at individual forest sites; (2) a series of campaigns based on aerial and space-borne research platforms; (3) the construction of a fire database, relating the frequency, extent, and intensity of fires to vegetation and climatic conditions for present and historic conditions; and (4) development of aggregated models of forest fire frequency and extent, responsive to global change variables.

International Boreal Forest Research Association

The International Boreal Forest Research Association (IBFRA) was founded in 1991 after a meeting of the International Panel on Boreal Forests in Arkhangelsk, Russia. The Fire Working Group (originally called the Stand Replacement Fire Working Group) was one of the first working groups created under the IBFRA, and to date it has been the most active. Following an organizational meeting in Siberia in 1992, the Fire Working Group has strongly promoted and facilitated cooperative international and multidisciplinary boreal forest fire research between Russia and western boreal countries of Europe and North America (FIRESCAN Science Team 1996). Several collaborative studies dealing with global change/fire issues, remote sensing, fire behavior, fire danger rating, fire history, and fire ecology and effects have been initiated. A major conference and field campaign was carried out in central Siberia in 1993 in cooperation with FIRESCAN, with follow-up research activities planned beyond the year 2000. The International Crown Fire Modelling Experiment (Ft. Providence, Northwest Territories, Canada) began in 1997–1998 with involvement of European and Russian scientists and continued in 1999. A major fire research program involving scientists from the United States, Poland, and Russia is under way at present in Poland and Russia. It is designed to investigate fires of various intensities in European and boreal Asian pine forest ecosystems.

Russian Fire Science Initiatives

In Russia, several facilities of the state research organization (the Academy of Sciences) and the universities conduct research in basic questions of fire ecology, fire behavior, and technology development for fire intelligence and management.

The Sukachev Institute of Forests of the Russian Academy of Sciences, Siberian Branch, Krasnoyarsk, is Russia's center of excellence in fire research. The main foci are in fire ecology, biogeochemistry (carbon cycling), fire history, fire and fuel mapping, prescribed burning, and the use of remote sensing in fire management and fire impact assessment. The institute was host to the first international fire conference and fire experiment in modern Russia (FIRESCAN Science Team 1996; Goldammer and Furyaev 1996). The institute also hosts the International Laboratory of Forest Fire Ecology of the International Forestry Institute (IFI), of which several non-Russian scientists are members. The IFI Headquarters in Moscow focuses on fire database management, remote sensing of fires, and the development of a GIS for forest fires (Korovin 1996).

State forest research institutes involved in fire research are located in Krasnoyarsk, Ivanteevka (Moscow region), and St. Petersburg. The focus of work of the Research Institute for Forest Fire Protection and Forestry Mechanization in Krasnoyarsk is mechanical equipment for fighting forest fires on the ground (Yakovlev 1992). The Forest Research Institute in St. Petersburg has specialized on space-borne detection of fires and particularly the development of airborne fire suppression technologies. The Far East Forestry Research Institute in Khabarovsk has a research focus on fire problems in the Far East of the Russian Federation. The Laboratory of Forest Pyrology of the Research Institute of Forest Chemistry, Ivanteevka, Moscow region, is the leading fire laboratory in investigating the effects of fire on radioactively contaminated terrain.

At the university level, international cooperative efforts in the area of forest fire behavior modeling were initiated in 1994. The Canadian government translated a comprehensive Russian monograph on *Mathematical Modelling of Forest Fires and New Methods of Fighting Them* by A. Grishin of the Centre on Reactive Media Mechanics and Ecology, Tomsk State University. An international conference, Mathematical and Physical Modelling of Forest Fire and Ecology Problems, was held in Tomsk in July 1995. Several North American fire modelers participated, and the conference proceedings were published (Grishin and Goldammer 1996).

Intraregional and International Fire Management and Policy Programs

Economic Commission for Europe

One of the main activities of the Economic Commission for Europe (ECE) in the field of forest fires is the periodic collection and publication of fire statistics from the member states by the UN-ECE Trade Division, Timber Section (e.g., ECE/FAO 1997). The FAO/ECE/ILO Team of Specialists on Forest Fire also operates under the UN-ECE Trade Division. The team's main task is to provide a critical link in communication and cooperation between fire scientists, managers, and policy makers. The main activities embrace (1) the production of International Forest Fire News (IFFN), (2) the organization of seminars, and (3) the promotion of synergistic collaboration between individuals and institutions.

An ECE-wide study, Legislation and Regulations Related to Forest Fire Prevention and Control, was conducted in the mid-1980s (Goldammer 1986). Starting with the first issue of FAO/ECE IFFN in 1988, a biannual publication of the Timber Section, UN-Trade Division, a steadily increasing communication process in international fire matters was initiated. The IFFN provides an international information platform on which advances in fire research, technology, and policy development are reported and disseminated. Currently, the printed version of IFFN is distributed to about 1,000 agencies, research laboratories, and individuals all over the world. Starting with its 19th issue (August 1998), the IFFN became available on the Internet homepage of the Global Fire Monitoring Center, and all past issues are included (see http://www.uni-freiburg.de/fireglobe).

Global Fire Monitoring Center

The state of fire science (fundamental fire research, fire ecology) in most vegetation types and the results of biogeochemical and atmospheric sciences research over the past decade provide sufficient knowledge for supporting decision making at fire policy and management levels. However, despite the progress achieved during the past decade, many countries lack the expertise for developing adequate measures in fire policies and management. The fire and smoke episode of 1997–1998 in Southeast Asia was a good example of existing fire information systems or fire management expertise that was used only to a limited extent. These circumstances led to confusion at national and international decision-making levels and explain the critical delay of international response to fire and smoke emergencies. The same refers to the public response to fires in the Russian Far East in 1998. It was generally believed that the magnitude of the fires was unprecedentedly high. Such misjudgments could be avoided if a global fire information system was available for the use of national and international agencies involved in land-use planning, disaster management, or other fire-related planning and decision-making tasks.

In June 1998, the Global Fire Monitoring Center (GFMC) was established in accordance with the objectives of the United Nations (UN) International Decade of Natural Disaster Reduction and the recommendations of scientific forums and projects of several UN agencies and other international programs. The GFMC is established at the Fire Ecology and Biomass Burning Research Group of the Max Planck Institute of Chemistry, Germany. Following the principles that were developed for a scientific global vegetation fire information system in the early 1990s, the GFMC documents and provides real-time or near-real-time information related to fire. This includes linkages with other national, regional, and international information systems.

Conclusions

The causes and impacts of fire in the boreal Eurasian region are multifaceted due to its rich cultural diversity and a broad range of socioeconomic and environmental conditions. Thus, it is impossible to draw any generalized conclusion or develop recommendations that would lead to unified responses or to generally valid fire management strategies and public policies.

However, the country cases described here reveal that appropriate public policy responses to wildfires in forest and other vegetation (e.g., steppe fires) in general consider people to be the most important fire source. Prevention has been recognized as most important in designing policies and fire management strategies.

The main fire-related problems have been identified as (1) areas affected by industrial pollution or radioactive contamination; (2) lack of fire management resources in sparsely populated boreal forests of central Asia; and (3) regions where increasing burning activities lead to severe degradation in mountain forests, steppe, and steppe-forest ecotones.

New challenges in fire management are arising from the decision to use prescribed burning in forest fire management in Russia and in nature conservation and landscape management in western Europe. Advanced technology developments are under way (e.g., space-borne remote sensing of fire and fire impacts).

International cooperation in the field of fire management is increasing (e.g., mutual assistance agreements for wildfire disaster situations, exchange of fire management expertise, and international fire research programs), both inside the region and in exchange with other regions.

Although the magnitude of the fire problem in the Baltic region is smaller than that in Mediterranean Europe or elsewhere in the world, new initiatives in international collaboration in fire management are being established.

Public policies at national and international levels increasingly build on synergistic efforts, especially within the European community and the ECE region.

At the political level, both internationally and nationally, the case for fire management as part of a sustainable forest management strategy has been advanced. However, implementation of truly effective fire management calls for further action.

References

Cahoon, D.R., J.S. Levine, W.R. Cofer III, J.E. Miller, P. Minnis, G.M. Tennille, T.W. Yip, B.J. Stocks, and P.W. Heck. 1991. The great Chinese fire of 1987: a view from space, pp. 61–66 in J.S. Levine, ed. *Global Biomass Burning: Atmospheric, Climatic and Biospheric Implications.* MIT Press, Cambridge, MA.

Cahoon, D.R., B.J.Stocks, J.S.Levine, W.R.Cofer, and J.M.Pierson. 1994. Satellite analysis of the severe 1987 forest fires in northern China and southeastern Siberia. *J. Geophys. Res.* 99:18,627–18,638.

Clark, J.S., and P.J.H. Richard. 1996. The role of paleofire in boreal and other cool-coniferous forests, pp. 65–89 in J.G. Goldammer and V.V. Furyaev, eds. *Fire in Ecosystems of Boreal Eurasia.* Kluwer Academic Publishers, Dordrecht, The Netherlands.

Clark, J.S., J. Lynch, B.J. Stocks, and J.G. Goldammer. 1998. Relationships between charcoal particles in air and sediments in west-central Siberia. *The Holocene* 8:19–29.

Davidenko, E. 1997. The 1996 fire season in Russia. *Int. For. Fire News* 16:27–30.
Dusha-Gudym, S.I. 1996. The effects of forest fires on the concentration and transport of radionuclides, pp. 476–480 in J.G. Goldammer and V.V. Furyaev, eds. *Fire in Ecosystems of Boreal Eurasia.* Kluwer Academic Publishers, Dordrecht, The Netherlands.
Dusha-Gudym, S.I. 1999. Fires in forest vegetation zones contaminated by radionuclides, in J.G. Goldammer and D. Odintsov, eds. *Forest, Fire and Global Change.* SPB Academic Publishers, The Hague (in preparation).
ECE/FAO (Economic Commission for Europe/Food and Agricultural Organization of the United Nations). 1997. *Forest Fire Statistics 1994–1996.* ECE/TIM/BULL/50/4. United Nations, New York.
ECE/FAO (Economic Commission for Europe/Food and Agricultural Organization of the United Nations). 1998. The *Forest Resources of the ECE Region (Europe, the USSR, North America).* ECE/FAO/27. United Nations, Geneva.
Ende, J., and X. Di. 1990. The forest conflagration of May 1987 in northeastern China, pp. 169–174 in J.G. Goldammer and M.J. Jenkins, eds. *Fire in Ecosystem Dynamics. Mediterranean and Northern Perspectives.* SPB Academic Publishers, The Hague.
FIRESCAN Science Team. 1996. Fire in ecosystems of boreal Eurasia: the Bor Forest Island fire experiment, fire research campaign Asia-North (FIRESCAN), pp. 848–873 in J.S. Levine, ed. *Biomass Burning and Climate Change. Vol. 2. Biomass Burning in South America, Southeast Asia, and Temperate and Boreal Ecosystems, and the Oil Fires of Kuwait.* MIT Press, Cambridge, MA.
Gertners, A. 1998. Forest and forest fire prevention system in Latvia. *Int. For. Fire News* 19:52–57.
Goldammer, J.G. 1986. Legislation and regulations related to forest fire prevention and control, pp. 217–225 in *ECE/FAO/ILO Seminar on Methods and Equipment for the Prevention of Forest Fires.* Instituto Nacional Conservación de la Naturaleza (ICONA), Madrid, Spain.
Goldammer, J.G. 1993. *Feuer in Waldökosystemen der Tropen und Subtropen.* Birkhäuser-Verlag, Basel, Switzerland.
Goldammer, J.G. 1994. International Decade for Natural Disaster Reduction (IDNDR). *Int. For. Fire News* 11:31–37.
Goldammer, J.G., and X.Y. Di. 1990. The role of fire in the montane-boreal coniferous forest of Daxinganling, northeast China: a preliminary model, pp. 175–184 in J.G. Goldammer and M.J. Jenkins, eds. *Fire in Ecosystem Dynamics. Mediterranean and Northern Perspectives.* SPB Academic Publishers, The Hague.
Goldammer, J.G., and V.V. Furyaev, eds. 1996. *Fire in Ecosystems of Boreal Eurasia.* Kluwer Academic Publishers, Dordrecht, The Netherlands.
Granström, A. 1998. Forest fire and fire management in Sweden. *Int. For. Fire News* 18:75–78.
Grishin, A.M., and J.G. Goldammer, eds. 1996. *Mathematical and Physical Modelling of Forest Fire and Ecology Problems. Phys. Combust. Explos.* 32 (in Russian).
Heikinheimo, O. 1915. Der Einfluβ der Brandwirtschaft auf die Wälder Finnlands. *Acta For. Fennica* 4:1–59.
Ing, S.K. 1998. *The Social Conditions of Fire. A Sociological Study of the Mongonmort and Batschireet Regions of Mongolia.* Integrated Fire Management Project, German Mongolian Technical Cooperation, Khan Khentii Protected Area, Mongolia. GTZ IFM Project/Freiburg University, Fire Ecology Research Group.
Kasischke, E.S., K. Bergen, R. Fennimore, F. Sotelo, G. Stephens, A. Janetos, and H.H. Shugart. 1999. Satellite imagery gives clear picture of Russia's boreal forest fire. *EOS—Trans. Am. Geophys. Union* 80:141,147.
Korovin, G.N. 1996. Analysis of the distribution of forest fires in Russia, pp. 112–128 in J.G. Goldammer and V.V. Furyaev, eds. *Fire in Ecosystems of Boreal Eurasia.* Kluwer Academic Publishers, Dordrecht, The Netherlands.

Kuusela, K. 1990. *The Dynamics of Boreal Coniferous Forests.* The Finnish National Fund for Research and Development (SITRA), Helsinki, Finland.

Kuusela, K. 1992. Boreal forestry in Finland: a fire ecology without fire. *Unasylva* 43:22–25.

Mysterud, I., I. Mysterud, and E. Bleken. 1998. Forest fires and environmental management in Norway. *Int. For. Fire News* 18:72–75.

Naidansuren, E. 1996. Mongolia fire update. *Int. For. Fire News* 15:35–36.

Niklasson, M. 1998. Dendroecological studies in forest and fire history. *Acta Universitatis Agriculturae Sueciae Silvestria* 52:1–32.

Odintsov, D.I. 1999. Russian forests and their protection, in J.G. Goldammer and D. Odintsov, eds. *Forest, Fire, and Global Change.* SPB Academic Publishers, The Hague (in preparation).

Pyne, S.J. 1997. *Vestal Fire. An Environmental History, Told through Fire, of Europe and Europe's Encounter with the World.* University of Washington Press, Seattle.

Shu, L. 1998. The study and planning of firebreaks in China. *Int. For. Fire News* 19:51.

Shulman, D. 1996. Wildfires in Mongolia 1996. *Int. For. Fire News* 15:30–35.

Shulman, D. 1997. Strengthening disaster response capability in Mongolia. Project accomplishment summary. *Int. For. Fire News* 16:20–22.

Steffen, W.L., and A.Z. Shvidenko, eds. 1996. *The IGBP Northern Eurasia Study: Prospectus for Integrated Global Change Research.* The International Geosphere-Biosphere Program, International Council of Scientific Unions (ICSU), IGBP Stockholm.

Valendik, E.N., G.A. Ivanova, Z.O. Chuluunbator, and J.G. Goldammer. 1998. Fire in forest ecosystems of Mongolia. *Int. For. Fire News* 19:58–63.

Viro, P.J. 1969. Prescribed burning in forestry. *Comm. Inst. For. Fenn.* 67 (7):1–49.

Wingard, J.R., and E. Naidansuren. 1998. The German/Mongolian Technical Cooperation GTZ Integrated Fire Management Project, Khan Khentii Protected Area, Mongolia. *Int. For. Fire News* 19:64–66.

Yakovlev, B.P. 1992. All-Union Research Institute for Forest Fire Protection, Russia. *Int. For. Fire News* 16:11–12.

Zackrisson, O. 1977. Influence of forest fires on the north Swedish boreal forest. *Oikos* 29:22–32.

5. Fire Management in the Boreal Forests of Canada

Paul C. Ward and William Mawdsley

Introduction

The boreal forest dominates the Canadian landscape, as it often dominates the world's view of Canada as a country of vast unpopulated forests. The boreal forest region (Rowe 1972) sweeps across the country in a giant curve, covering almost 530 million ha of forestlands in portions of the Yukon and Northwest Territories (NWT), British Columbia, the prairie provinces, Ontario, Quebec, New Brunswick, and Newfoundland, and Labrador (see Fig. 14.1). Nova Scotia and Prince Edward Island are the only provinces that do not have a true boreal forest component.

These vast forest areas of Canada, composed of fire-dependent and fire-adapted species and ecosystems, have evolved in the presence of a fire regime driven originally by natural sources of fire ignition, by cultural practices of aboriginal people, and more recently, by the intrusion and development of modern industrial society. These forests are a source of raw material for a very large, export-oriented forest industry, a venue for recreational and cultural activities, and a supplier of many tangible and intangible products, services, and values. In recognition of these values, forest protection efforts were established in the late 19th century and have steadily developed to a point at which Canadian forest fire management agencies are recognized among the world's leaders in this field.

In this chapter, we describe the impact of fire on Canada's boreal forests and the delivery of fire management by Canadian agencies. Case studies of the fire management programs in Ontario and the NWT illustrate fire policies and operational approaches among different agencies. These studies also describe the challenges facing fire managers across the country, including (1) increasing demands for protection in some regions; (2) demands for more ecologically sound forest management practices; (3) fiscal pressures; and (4) a fire regime that may be changing as a result of climate change and other influences (see Chapters 19 and 20). Finally, we discuss some potential approaches to meeting those challenges.

Fire Management Responsibility

Under the Canadian Constitution, forest fire management activities on public (or Crown) lands are the responsibility of the provincial governments, which have jurisdiction over public lands and natural resources within their borders. The federal government is responsible for those public lands directly managed at the federal level, including national parks, lands reserved for First Nations people, and the Yukon Territory. Although most land in the NWT is under federal jurisdiction, the responsibility for forest and fire management has been transferred to the territorial government. Because more than 90% of Canadian lands are in public ownership, each of these agencies is responsible for a very large land base.

Forest Fire Impacts

Over the past 10 years, an average of 9,500 fires has occurred annually in Canada, burning an average of 3.2 million ha yr^{-1} (Chapter 15). National fire statistics for the past 30 years show a steady upward trend in both numbers of fires and area burned, which is causing significant concern among both fire management agencies and their clients. However, drawing firm conclusions about the meaning or cause of these trends is confounded by inconsistencies in the data and reporting practices. In earlier years, fire losses in remote areas were often underreported, and the reporting criteria, methods, reliability, and accuracy of all fire reports are open to question. Meanwhile, recent statistics include areas burned by fires that were allowed to burn unsuppressed through the deliberate application of an agency's fire management policies. These factors make it difficult to identify definitive trends from year to year in area burned data.

Fire Management Delivery

Most Canadian agencies deliver forest fire management directly as a government service, generally as part of the department that manages lands, forests, and other natural resources. For example, in Quebec, the protection service is delivered by a cooperative "protection company," jointly funded by the provincial government and other major landowners and stakeholders.

Fire management programs provide varying levels of prevention, detection, presuppression activities, suppression operations, and prescribed fire delivery. The level and type of activity in each category vary with each agency's natural resource management policies, protection priorities, financial resources, and in particular the ecological and physiographical conditions of the forest itself.

All agencies deliver an organized detection program. Fixed-wing aircraft detection is most common, but towers are still used in mountainous regions and where fixed detection sites offer recognized benefits (e.g., in high-value areas, near communities).

Suppression strategies use a mix of air and ground attack. In much of Canada, an abundance of water bodies allows skimmer air tankers (most commonly the Canadair CL-215 and CL-415) and ground crews using power pumps and hose to be the backbone of the fire suppression system. Retardant (land-based) tankers are the preferred air attack tool in much of western Canada (British Columbia, Alberta, the Yukon, and NWT) where terrain and suitable water sources limit the use of skimmer aircraft. Hand tools are also commonly used where water is limited.

There are other differences in organization structure and delivery among the agencies, based on the nature of the forests and fire regimes that they deal with and, to an extent, on organizational philosophy. Some agencies, such as those in Ontario and British Columbia, have relatively large full-time fire management organizations, in which suppression crews, helicopters, and other resources are hired on a seasonal basis to deal with moderate levels of fire activity without short-term or emergency assistance. Additional resources are acquired on an emergency basis when the fire situation escalates. Other agencies, such as those in Alberta, hire a relatively limited number of seasonal fire fighters and helicopters initially and expand or contract their suppression organization on a short-term basis through the use of auxiliary fire fighters and other resources as required in response to the current or anticipated fire load.

Fire Management Issues

All Canadian forest fire management agencies face a similar array of challenges. They are expected to provide a basic level of public security by protecting people and property from the negative impacts of fire. They are to provide natural resource protection and to protect recreational and cultural resources from the threat of fire. And most agencies deliver prescribed burning programs designed to apply managed fire as a tool for forest regeneration, ecosystem management, hazard reduction, and other purposes.

Meeting these objectives is increasingly challenging. As a government service, fire management has been subject to the financial pressures facing all levels of government, with deficit reduction efforts causing staff reductions and erosion of traditional levels of fire management resources. These resources are declining at a time when demands for protection from parts of the client base are increasing. Concurrently, the demands for ecologically appropriate forest management practices and concerns about the long-term impact of highly effective fire suppression have led to suggestions that in some areas more fire is needed in the forest to promote biodiversity and long-term sustainability. These factors are overlain by the debate about the impact of climate change on the forests of Canada and whether those changes are being manifested even now in an escalating fire regime.

To examine further these issues, we present two case studies: (1) fire management in Ontario, a populous province in central Canada where industrial, recreational, and social issues in the forest abound; and (2) fire management in the NWT, where a small population base, traditional life-styles, and large geographic extent impose both similar and contrasting issues to those faced in Ontario.

Case Study 1: Forest Fire Management in Ontario

Ontario is the second largest Canadian province, with a total land area of just more than 106 million ha. Almost 90% of the area is Crown (public) land, and most of that area is administered by the provincial government.

Ontario's Forests

Ontario is divided into four forest regions (Rowe 1972). A deciduous forest once covered southern Ontario, an area now heavily urbanized and developed for agriculture. Remnants of the original deciduous forests and outliers of the Carolinian forest type are still present. Much of the land base here is privately owned.

Central Ontario is home to the Great Lakes–St. Lawrence Region, typified by stands of red and white pine (*Pinus resinosa, Pinus strobus*) and tolerant hardwoods growing on the southern reaches of the Canadian (or Precambrian) Shield.

Most of Ontario is dominated by the boreal forest type, with mixed-wood forests of white and black spruce (*Picea glauca, Picea mariana*), balsam fir (*Abies balsamea*), jack pine (*Pinus banksiana*), and northern hardwoods in the east, trending to dominance by conifers, primarily black spruce and jack pine, in the north and west. Finally, a naturally nonforested area is reached at the extreme north of the province, along the coasts of Hudson and James Bay, where tundra conditions and sparse tree cover are present.

Ontario's forests are a major economic resource and host to diverse recreational and aesthetic values. The economy of the boreal forest area in particular is dominated by the forest industry, and the forest is home to a thin and broadly dispersed population of mostly resource-dependent communities.

Most of Ontario's forest types are fire dependent. Until the 20th century, Ontario's forest landscape was dominantly shaped by fire, and forest fires are still a considerable influence on the forest estate.

Fire Impacts in Ontario

In the 10-year period from 1988 to 1997, an average of 1,762 fires occurred annually in Ontario, burning over an average of 276,000 ha yr^{-1}. The Fire Management Program of the Ontario Ministry of Natural Resources currently spends an average of about $85 million (Canadian) annually on fire management and fire suppression activities. About 55% of the fires are the result of human activities, but there is a strong geographic gradient. Human-caused fires predominate in the more populated areas of the eastern and southern parts of the province, whereas lightning is responsible for about two-thirds of all ignitions in the more northern and western regions.

Fire management is governed by the Forest Fires Prevention Act (Revised Statutes of Ontario 1990), which specifies rules governing the use of fire, prevention, and equipment requirements when working in forested areas, hazard abatement, and the powers of fire wardens and other officers under the act. The act also specifies the official "fire season," from April 1 to October 31; many provisions of the act only apply during the fire season.

The fire response system in Ontario has been increasingly centralized over the past decade. Daily fire operations are now directed from a provincial response center and two regional response centers (east and west), which oversee a network of about 30 field fire management headquarters and attack bases. Fire suppression operations are designed around early detection and aggressive initial attack with 166 three- and four-person "unit crews" supported by nine Canadair CL-215 and five Twin Otter air tankers.

Fire management policy in Ontario recognizes that the exclusion of fire from the forest is neither economically feasible nor ecologically desirable (Ontario Ministry of Natural Resources 1991). The major fire policy objectives are

- To prevent personal injury, value loss, and social disruption resulting from a forest fire.
- To promote understanding of the ecological role of fire and to use its beneficial effects in resource management.

These objectives are addressed by a fire management infrastructure that ensures that every forest fire in Ontario receives a response and that each response is governed by the

- Predicted behavior of the fire.
- Potential impact of the fire on persons, property, and values.
- Estimated cost of the response.

Fire policy is implemented through regional fire management strategies that delineate fire management zones (Fig. 5.1), and the level of fire protection to be delivered in those zones.

The Intensive Fire Management Zone comprises the area of primary settlement, the commercial forest, and major recreation areas. In this zone, uncontrolled fire is generally undesirable and is addressed through early detection and aggressive initial attack.

The Measured Fire Management Zone is a band along the northern limit of the Intensive Zone. This area contains wood supplies that may be accessible or desirable in the medium to long term, areas used for remote tourism and recreation and some settlements. In this zone, organized detection is provided, and initial attack is carried out on all fires; if the initial attack effort is unsuccessful, further

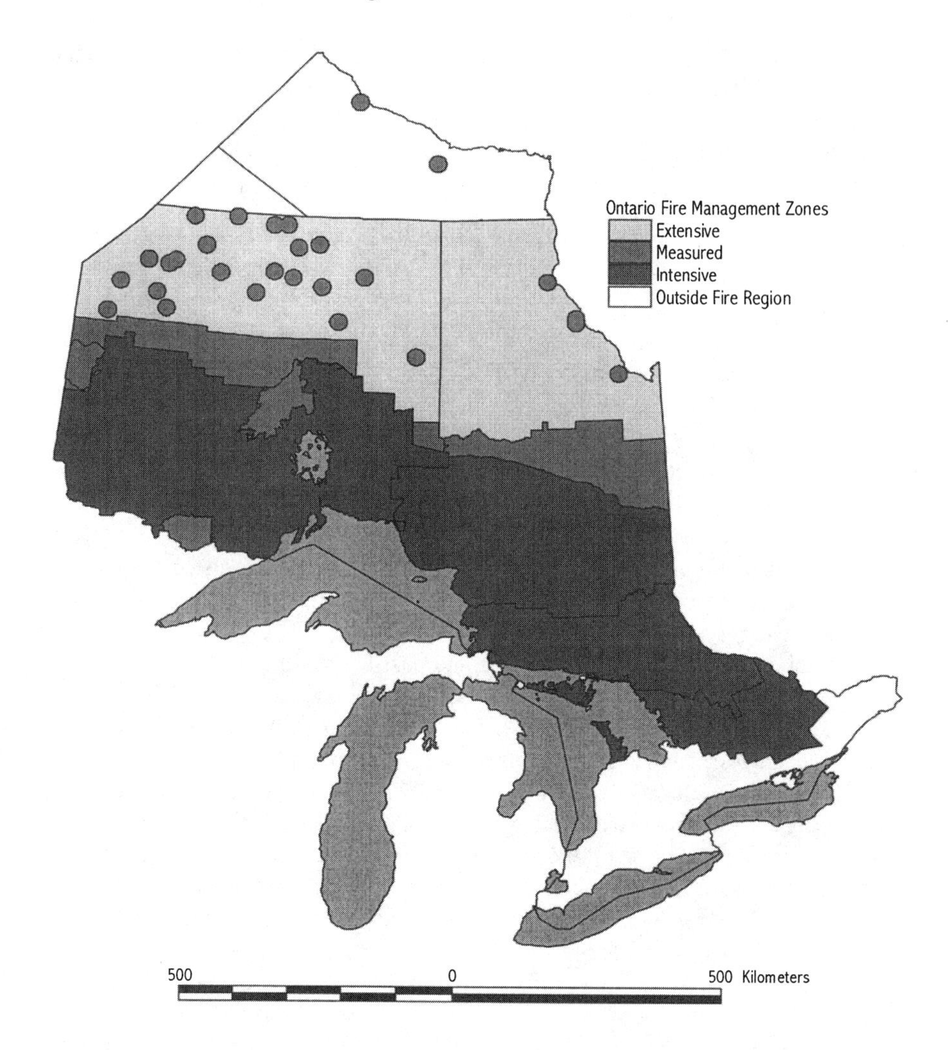

Figure 5.1. Fire management zones in Ontario, Canada.

action is dependent on an assessment of the values at risk and the cost of the response. On some fires, sustained suppression action will be maintained, or limited action on portions of the fire may be indicated, or suppression staff may be withdrawn from the fire and its activity monitored only.

The Extensive Fire Management Zone is located in the far north, an area mostly inaccessible by road and very sparsely populated and where fire-fighting operations are logistically difficult and expensive. Aboriginal and other communities in the zone are intensively protected, but suppression action is otherwise limited to the protection of identified values. Most fires are allowed to burn unsuppressed, maintaining the natural role of fire in the Extensive Zone. Smoke impacts though are a significant concern; remote communities in the north have suffered frequent evacuations in the past decade, caused generally not by the direct threat of fire to the community but by the incursion of smoke from distant large fires.

Ontario is currently involved in discussions and planning with residents of the far north to develop more refined fire management strategies tailored to better address the special needs of that remote part of the province.

Fire Management Effectiveness

When people speak about the impacts of forest fires, they tend to think of fire in singular terms—the impact of that fire on that specific part of the forest and its values. Lately, more interest has arisen in the overall impacts of fire, or the fire regime, on the whole forest, both in space and time. We also need to think about the impact of modern fire management practices on the forest landscape. Are we really effective at reducing the area burned by fires, and if so, what are the implications of reducing the "natural level" of fire in the forest, in terms of forest structure, successional patterns, and biodiversity? From a forest management perspective, what is the cumulative impact of all disturbance agents (natural and human-caused) on forest sustainability?

To gain some insight into the current and past fire regimes in Ontario and the effectiveness of modern fire suppression efforts, we compared recent fire occurrence data for the Extensive Fire Management Zone with similar data for the Intensive and Measured Zones combined (see Ward and Tithecott 1993 for a more detailed review of this analysis). It was broadly assumed that the Extensive Zone approximates a natural fire regime, where most fires were allowed to burn unsuppressed and where most of those fires are ignited by natural causes. In the Intensive/Measured Zone conversely, the majority of fires are human caused, and fire management resources intervene in that human-influenced fire regime and aggressively detect and attempt to suppress most fires.

We looked first at the range of fire sizes in each broad zone over a 15-year period from 1976 to 1990. In the Extensive Zone, we observed a large range of final fire sizes from small to very large (Fig. 5.2). These fires create a diverse mosaic of patch sizes on the landscape as the fire areas regenerate over time.

In the Intensive/Measured Zone, the effectiveness of current fire management efforts is reflected in the distribution of final sizes: the vast majority of fires are held to very small sizes (< 4 ha), with a few moderate sized fires and even fewer large fires (≥ 1,000 ha) (Fig. 5.2). The impact of this at the landscape scale is that fire is creating far fewer moderate- and large-sized disturbance patches than we observe in the Extensive Zone, where the fires are mostly unsuppressed.

We also looked at the reduction in area burned resulting from effective suppression operations. Using a variety of fire history studies, we coarsely estimated that the boreal and Great Lakes–St. Lawrence forest types that dominate Ontario's Intensive/Measured Fire Management Zone had a natural fire return interval of about 65 years, or an annual burn fraction of 0.015 (i.e., on average 1.5% of the land base would be burned annually).

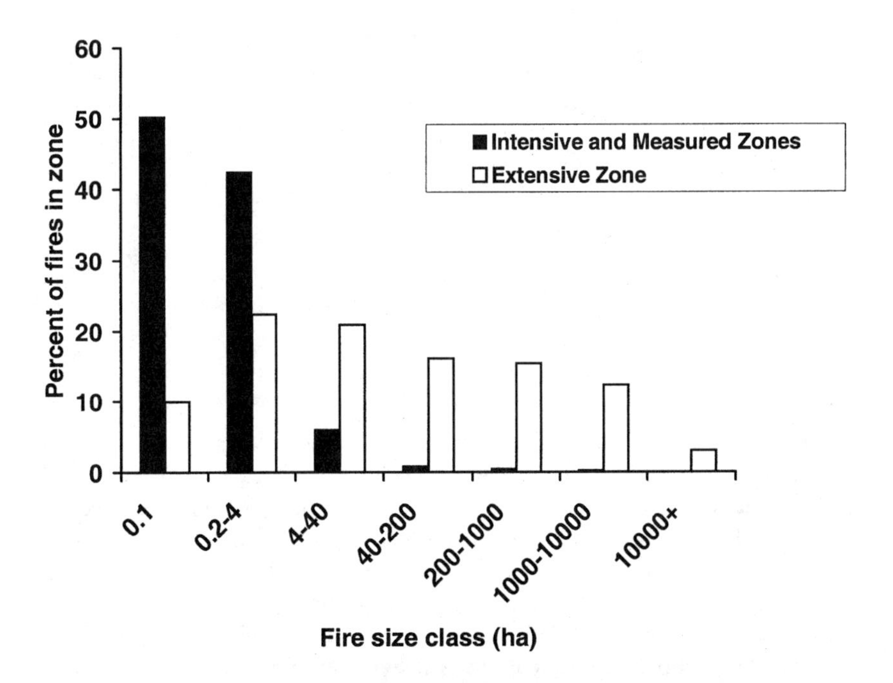

Figure 5.2. Distribution of plot of fires by size class in Ontario for 1976–1990. Data are presented for the Intensive and Measured Fire Management Zones (I+M) as well as for the Extensive Fire Management Zone (Ext).

The total area of the Intensive/Measured Zone is about 47 million ha, so with a 0.015 annual burn fraction, about 700,000 ha of the zone would have burned annually in the "presuppression fire regime." In reality, in the 15-year period from 1976 to 1990, on average only 80,000 ha was burned annually in the Intensive/ Measured Zone, an annual burn fraction of 0.00173 (or 0.173%), which translates into a fire return interval of roughly 580 years. These data suggest that modern fire management practices have reduced the area burned in the Intensive/ Measured Zone on Ontario by an order of magnitude when compared with the postulated presuppression fire regime.

This analysis has a number of interesting implications:

- Current fire management practices appear to have reduced the annual area burned in Ontario's commercial forest from 700,000 to 80,000 ha.
- Most fires are contained at small sizes, greatly reducing the relative proportion of moderate- and large-sized disturbances we observe when compared with the Extensive Fire Management Zone, where most fires are not suppressed.
- Fire management practices would seem to be leading to generally older forests and a reduction in the number of larger disturbance patches. Fire, of course, is only one of a number of natural and human disturbance agents in the forest, and the forest landscape is the net sum of those influences.

- Finally, as a sidelight to the global warming debate, if forest fires are a significant source of greenhouse gas emissions, effective fire management practices (based on the calculated 10-fold reduction in annual area burned) would seem to have significantly reduced the atmospheric impacts of fires in the Intensive/Measured Zone since the presuppression era.

Current Fire Management Issues

Despite this apparent success in its primary job—putting out fires—the Ontario forest fire management program faces a number of significant challenges as it heads toward the new century.

Increasing Demands for Protection

Forest resources in Ontario are in increasing demand. The areas exploited for commercial fiber supplies have expanded because of increased demand but also because new technologies have emerged to increase use of once-undesired species. The industry is concerned about the protection of its mature fiber supply and the protection of their regenerated areas and silvicultural investments.

Other values in the forest are also expanding. Demands for recreational opportunities and the expansion of settled areas and development into exurban areas are increasing protection requirements in both wildland-urban interface areas and more remote locations.

Aboriginal groups and other residents of remote areas in the Extensive Protection Zone have also raised issues about the need for more dedicated fire protection. Evacuations and community disruptions as a result of fire threats and smoke impacts are of major concern. These local residents are also concerned about the security of their local fiber resources, health impacts of smoke, and conflicts over how to assess the values at risk in and around their communities.

Ecological Concerns

Where once forest fire suppression was viewed almost universally as being worthwhile and valuable, there is now increasing concern about the long-term ecological impacts of highly effective suppression programs. Critics point to aging forests, potential fuel buildups, and unnatural age class and patch size distributions that result from the effective suppression of most fires.

Most forest management agencies are moving toward an ecosystem-based approach to management, seeking to look beyond individual trees and stands and instead managing the biodiversity and maintaining the sustainability of the whole forest. This landscape perspective calls for an examination of the role and impact of both the presence of fire in the forest and the impact of greatly reducing the influence of fire as an agent of disturbance and renewal that results from intensive fire suppression.

Issues around the role and potential use of fire (both wildfire and managed fire) are currently active on a number of fronts in Ontario: (1) in parks and wilderness management, such as in Quetico Provincial Park, where a fire management plan has been developed; and (2) in wildlife management, such as in Wabikimi Provincial Park, where landscape management for the conservation of woodland caribou habitat is a major issue. Prescribed burning has been used in other ecosystem restoration and rehabilitation efforts—red oak and yellow birch regeneration and the management of old-growth pine stands are just two examples. Stakeholders are considering the role of prescribed fire, prescribed natural fire, and reducing the level of suppression intensity in various areas as approaches to meeting a variety of ecological objectives.

The complexities involved in meeting these objectives while concurrently reducing program costs should not be underestimated. Choosing not to suppress certain fires does not automatically result in lower fire management costs.

Financial Pressures

Ontario's fire management program is dealing with the conflicting issues discussed above at a time when government efforts at lowering the deficit have reduced the program's financial resources and its flexibility to respond. Fire management is an expensive business; although fire agencies have made major efforts to maintain their technological advancement, fire suppression remains a strenuous, dirty, and labor-intensive effort.

Fire management places an added stress on government finances because of the variable nature of fire season severity. Ontario and most Canadian agencies use similar systems for funding fire operations: a planned base budget covers resource acquisition and readiness, but costs associated with actual suppression operations are paid for by an emergency fund allocated directly from the government's central treasury. The inability to predict in advance the yearly requirements for emergency funding causes unease within the government financial agencies; they are also uneasy about how effectively those funds are spent during a severe fire season.

Escalating Fire Regime

As if life were not difficult enough, Ontario and other Canadian fire agencies view with concern recent trends in fire occurrence and severity over the past several decades and, in particular, the increasing variability and severity of fire weather and climate during that period. Although several potential causes may be at work, climate change is a major concern. Other issues include increasing human activity in forested areas and the uncertain effect of older forests and possible large-scale fuel buildups resulting from a lengthy period of effective fire suppression.

Definitive evidence of the impact of these factors may not yet be universally accepted, but in Ontario the fire management program expects that it will continue to experience increasing variability in fire season severity. In general, the long-term trend is expected to be toward more difficult fire conditions, including the potential for longer active fire seasons and more severe burning conditions occurring on a frequent basis.

How will the program respond to these challenges? As a government agency, it is difficult to respond to long-term trends (e.g., climate change) that are still somewhat in the throes of scientific debate and whose evidence may still be obscured in the inherent variability of weather, climate, and human activities. But as a fire management agency, it is cognizant of the long-term implications of these changing conditions and of the need to define and describe the potential impacts to both the public and elected officials.

In the short to medium term, tangible actions can be taken:

- *Harder look at "level of protection":* There will be difficult decisions to be made about "how much protection is enough." It is now recognized that agencies will not be able to provide increased protection with declining budgets in a potentially more severe fire climate. There is a need for public debate on the value and role of forest fire protection and on how much society should be willing to spend on it.
- *More finely resolved zoning systems:* An important outcome of the above is the critical need for a better definition and evaluation of the values at risk and planning for a level of protection that balances the values with the cost of the protection effort. Resource managers, interest groups, and the public will have to deal with the difficult issue of what those values are—and how much the taxpayers can afford to spend on their protection—as well as what areas might benefit from an increased level of fire impact. Fire management decisions must be more integrated within the context of comprehensive resource management.

  The result will be a finer resolution in protection zones—fire managers will have to more closely examine the location of fires with respect to values at risk and protection priorities and the potential impact of each fire before determining the intensity of their response. It is not possible to maximize the protection of life, property, and values and the promotion of ecological objectives simultaneously on the same hectare of ground. Zones with different management objectives will likely be established, and fire agencies and the public will have to live with the risk management inherent in dealing with fire, which pays little heed to arbitrary boundaries.
- *Continued technology improvements:* Ontario and other agencies have responded to declining budgets and staffing levels by replacing labor with technology—improved suppression technologies such as air and ground-based foam application systems and lightweight equipment. They have also invested heavily in information and decision support systems to provide better situation assessment and analysis and to improve the ability to predict fire occurrence and behavior. Agencies must continue to improve their predictive tools and to better manage the risks involved in dealing with potentially more severe fire conditions with constrained resources.
- *More proactive management:* Fire and other resource managers will have to increase their efforts in looking at landscape level forest management. They will have to look at broader fuels management—the large-scale impact of harvesting patterns, access development, and fire disturbance patterns in an effort to ameliorate the potential for large fires. Fire managers need to provide more input to forest management planning in terms of access planning, and cut

layouts, shape, and continuity. The issue of using managed fire on a large scale for manipulating patch size and age class distribution across the landscape will need to be addressed.

The use of prescribed fire has declined in Ontario in recent years, as burning costs have increased and forest management funding has declined. Ecosystem management concerns though are reenergizing discussions about the use of prescribed burning and prescribed natural fire for manipulating the structure of the forest landscape.

- *Greater involvement of clients and stakeholders:* All the above actions suggest a greater role for clients, stakeholders and the public in fire management planning and delivery. The Ontario fire management program will likely see more external involvement in developing the level of protection strategies and setting of priorities. There will also be the need for more external involvement in the actual delivery or funding of fire management operations.

Case Study 2: Forest Fire Management in the NWT

The NWT extend across the top of the North American continent, encompassing the land area from 60° north latitude, the MacKenzie Mountains in the west, north and east to Greenland and the limits of the arctic archipelago. Its physiography includes portions of the western Cordillera, the interior plain, the Canadian (Precambrian) Shield, the mountains and fjords of Nunavut, and the polar deserts of the Arctic Islands.

Lands and waters fall under the jurisdiction, primarily, of the government of Canada, whereas forests and forest fire management have been the responsibility of the government of the NWT since April 1987. As land claims agreements are ratified with First Nations, large areas are being transferred to the collective ownership of the First Nations groups. At present, agreements with Inuit people (Inuvialuit and people of Nunavut) and the Dene of the Northern MacKenzie Valley (Gwich'in and Sahtu First Nations) have been ratified.

The forestland area of the NWT covers approximately 137 million ha, from the Yukon border east to the treeline. The treeline (the approximate northern limit of continuous forest) extends on a diagonal from approximately Hudson Bay in the very southeast to, and including, the MacKenzie River Delta in the northwest.

Delivery of forest fire management programs on public lands is the responsibility of the government of the NWT. The Establishment Acts for the First Nations claim settlements direct the government of NWT to continue to provide fire management services and programs on settlement land.

Forests of the NWT

The forested area of the NWT falls within the boreal forest region (Rowe 1972) and is within the Taiga Plains and West Taiga ecozones (Wiken 1986; see Fig. 14.1). The common tree species are white and black spruce (*Picea glauca, Picea mar-*

iana), trembling aspen (*Populus tremuloides*), balsam poplar (*Populus balsamifera*), jack pine (*Pinus banksiana*), and white birch (*Betula papyrifera*). On the mountain slopes in the west are concentrations of lodgepole pine (*Pinus contorta*) and subalpine fir (*Abies lasiocarpa*).

The area bounded by the Slave River in the east, the Liard River in the west, and north along the MacKenzie River is characterized by stands of mixed-wood, pure coniferous species and pure hardwood species on areas of favorable soils and climatic conditions, such as river drainages and well-drained upland soils. Much of the remainder of the area may be categorized as treed muskeg and wetlands, with associated vegetative cover (spruces, mosses, lichens, shrubs). East of the MacKenzie and Slave rivers, essentially to the treeline, is the Northwestern Transition Zone (Rowe 1972). Here, the forests are a mix of bog and muskeg, with large areas of open stands of smaller trees (primarily coniferous species) commonly with a ground covering of lichens. The forest productivity often reflects the unfavorable climatic conditions, poor soils, and forest fire history. West of the MacKenzie and Liard rivers is the alpine forest-tundra zone (Rowe 1972), where the forest cover is generally coniferous below the treeline, although there are concentrations of birch and poplar species in some locations.

The forests of the NWT have limited potential for industrial exploitation under present business practices. Concentrations of "merchantable" stands occur primarily along the major river drainages and productive areas of the southern reaches of the MacKenzie River, although forests with limited merchantable potential have been identified even in the far north. The forest area, however, is very important in the support of the traditional life-style of First Nations people. Hunting and gathering of big game, for family use or in an outfitting operation, and harvesting of valuable fur bearers are often the sole source of income. Outside of social welfare programs, this may be the only income available to many people. Managing the productivity and viability of forest ecosystems is thus of greater importance to northern resource managers than simply managing trees as wood products.

Forest Fire History in the NWT

Forest fires are recognized as "a significant and natural phenomenon in the forests of the Northwest Territories" (Government of NWT 1990). Alexander MacKenzie reported forest fires along the river on his voyage of exploration to the Arctic Ocean. First Nations people report that forest fires have been part of their people's existence as a benefactor, tool, and enemy.

Over the past 30 years, on average approximately 320 fires have occurred annually in the NWT, with an annual burned area of approximately 675,000 ha (see Fig. 17.8). The lowest number recorded in the past 20 years was 105 fires (1997), and the lowest area burned was 37,000 ha (1992). The highest number of fires in any year occurred in 1994, with 627 fires reported, and more than 3.0 million ha of forest land was affected. This was followed by 1995, where a lower number of fire reports (218) still accounted for almost 3.0 million ha.

The largest fire impact on record occurred in 1942, when a slash fire on the Alaska highway project was estimated to have burned approximately 30,000 square miles (in excess of 7.0 million ha) in British Columbia and NWT, with the greater portion in NWT (Janzen 1990). On an individual basis, the largest fire in recent times occurred in 1995, when one fire alone covered approximately 1.0 million ha. More the norm, in years of severe fire activity, a few fires account for large burn areas, often in one region of the NWT. For example, in 1994, a series of fires in the Caribou Range (Canadian Shield east of the Slave River and south of Great Slave Lake) burned more than 1.5 million ha. In 1986, a series of fires in the Inuvik region burned approximately 300,000 ha of the annual total of 321,000 ha across the NWT. As illustrated in the foregoing, the number of fires and the area burned are highly variable from year to year.

Forest Fire Environment in the NWT

The forests of the NWT are composed of fire-tolerant and fire-dependent species of trees and other flora. The most common fuel types in the north are spruce–lichen woodland and boreal spruce (Forestry Canada Fire Danger Group 1992). Naturally, stand age is a key determinant in the incidence, type, and effect of forest fires on the forest landscape. The most severe fire events occur over areas of mature forests where fuel loadings approach maximum natural levels.

Most forest fires in the NWT are caused by lightning, which occurs throughout the summer season. When significant lightning activity is coupled with dry forest fuels (as happens following periods of drought), fire numbers may be high.

The climate of the NWT is classified as part of the Subarctic Continental Zone, over most of the forested area (Environment Canada 1973). Average annual precipitation is approximately 350 mm, with slightly less than half falling as overwinter precipitation. Water deficits on a seasonal basis or over a period of years lead to drought conditions occurring locally or regionally. When the drought is severe and prolonged, either on a seasonal basis or over a number of years, a few fires may become very large. For example, the period leading up to and during the 1979–1981 fire seasons was a period of severe moisture deficits in the environment both over winter and summer. In this period, approximately 4 million ha burned. Again, during the 3-year period 1993–1995, 6.7 million ha was affected by forest fires. In each period, less than 10% of the fires accounted for most of the area burned.

In the NWT, the normal diurnal fluctuations of temperature and relative humidity may not occur. This phenomenon is most common north of the Arctic Circle but occurs in the southern NWT as well. During most of the summer fire season, the period of daylight varies from full sunlight 24 hours a day in the far north to short periods of twilight and long daylight periods (20 hours plus) in the south. Forest fires continue to burn throughout the "night" with little or no change in behavior if appropriate fuels and other behavior factors are also present. A fire in the Hay River area in 1981 burned 40–50 km between the 2200 MDT and 0400 MDT. In this case, the driving factor was an upper atmosphere ridge breakdown.

A statistical pattern relating fire numbers, area burned, and environmental patterns has not been determined for the NWT, except that, as our experience tells us, severe fire years occur during periods of drought and attendant severe fire behavior. During years of normal moisture balance, fires may be numerous, but suppression efforts may be successful, and individual and cumulative burn areas are significant but not noteworthy. In periods of drought, if fire occurrence is severe and timely, fire suppression demands may exceed the operational capabilities of the northern forest fire management organization.

Forest Fire Management Practices in the NWT: A Historical Perspective

Janzen's (1990) research of the history of fire management indicates that forest fire management practices in the NWT have mirrored southern North American activities in the period since European contact. In the NWT, forest fire management options have always been limited by the sheer size of the forested area. Total fire exclusion from the forest landscape has never been a viable alternative nor, when attempted, successfully practiced. Forest fire management policies, prior to those presently in effect, have included benign neglect and, most recently, defining areas or zones where fire suppression was stringently practiced and areas where limited or no fire suppression was contemplated. The last definition of zone boundaries enclosed an area of approximately 25 million ha in a fire action zone, with the remainder of the forest area defined as observation zone. In the fire action zone, fire management policy required initial attack and continued suppression action on all fires in an attempt to control each fire. In the observation zone, the norm has been no suppression effort.

Forest Fire Management in the NWT: 1990 to Present

In 1987, the government of the NWT assumed control over its forests and forest fire management. This event provided the opportunity for forest fire managers to redefine the fire management policies for the NWT to reflect the northern situation and the concerns and aspirations of the people of the north. In consultations with community and interest groups, several problem areas were identified, which led to a recrafting of the policy. Defining levels of fire protection on an area basis was considered an unacceptable practice. The people of the north wished to have control and influence over fire management decisions in their region of interest and to be consulted on fire management problems.

In 1990, the NWT Forest Fire Management Policy (Government of NWT 1990) came into effect. Key to the policy were the following principles:

- Fire is significant and natural to the northern forest regime.
- Forest fire management activities should include consideration of environmental, social, and economic factors.
- The first priority of forest fire management activities should be protection of human life and property.

- Forest fire management decisions should consider the value of the resources at risk.

Attached to the policy is a directive that provides, among other things, six criteria used by fire managers in determining forest fire response (Government of NWT 1990):

1. Values-at-risk, which are ranked in the order human life, property, renewable resource values, and cultural resource values.
2. Land and resource management objectives.
3. Availability of personnel and equipment.
4. Fire weather.
5. Fire risk in higher-valued areas.
6. Where property or resource values are being threatened, the relative value of the entity at risk.

The requirements of the policy and directive demand that communities and resource users be consulted on the location and existence of specific values and the relative importance of those values in their community interest area. Values identification has been an ongoing task of the organization, through local contacts and through specific consultation efforts especially in the winter season. For example, in the Sahtu region of the NWT, fire managers and interest groups have identified areas of priority for fire exclusion on area maps to be used for reference by fire managers at the time of fire occurrence.

The policy and directive recognize that complete exclusion of forest fires is neither practical nor feasible in the NWT. The policy specifically allows for wildfires to be left in an uncontrolled state and for fire actions to be limited to protection of specific values (e.g., fishing lodges, significant cultural values). Protection of specific values through fuels modification and hazard reductions, in advance of forest fire events, is also receiving more emphasis than was possible under previous policy guidance.

The policy requires interest groups to identify the values in the forest that may be affected by forest fires and the magnitude of the impact should that resource be destroyed by wildfire. For example, is the loss of an old-growth spruce forest critical in terms of a marten (*Martes americana*) trapping area, or will the postfire forest be more productive? Answering such questions has also placed more demands on the organization to evaluate the individual effects of forest fires on the environment.

Research is ongoing on the effects of fire on the environment, such as evaluation of the impact of forest fires on the wintering grounds of the Beverly and Qamanirjuaq caribou herds, east of Fort Smith (Beverly and Qamanirjuaq Caribou Management Board 1993). That research has indicated that there are areas where fire exclusion should be practiced to allow the habitat to recover and other areas where fires may not need to be controlled in terms of habitat requirements for caribou (*Rangifer tarandus*) wintering grounds.

The key to the success of the policy will be continued involvement of the owners of the forests (the people of the NWT) in predetermining resource values and in making timely decisions on complex fire events. The policy recognizes that forest fires will occur in the north. The direction to fire managers is that the people of the NWT will determine acceptable levels of forest fire impact, following which fire managers will take steps to implement those decisions.

Global Warming, Biomass Burning, the Carbon Cycle, and Management of Forest Fires in the NWT

Again, forest fires have been a factor in the boreal forests of the NWT since before European contact. Forest fires will continue to be a factor in the forest. From the perspective of fire managers in the NWT, the key questions are: Will the level of fire activity increase with global warming? What will global warming mean to the climate of the NWT? Although the answers to these questions are in dispute, in the meantime forest fire managers must deal with the existing fire environment. Fire management analyses and planning now consider the ecological impacts of forest fires on the northern environment and the potential for extensive disastrous fire events. Plans include methods of protecting communities from wildfire through proactive means as opposed to the traditional reactive means. Fire managers recognize that there will be, even in the present environment, cases in which where the fire event exceeds the resource capabilities at the location.

In terms of the issue of biomass burning, the forestlands of the NWT are a natural repository of materials that produce greenhouse gases of significant proportions. It is neither practical nor feasible to exclude forest fires from the northern landscape. Harvest of wood products, as a means of reducing the carbon base, would have limited effect, and it would not reduce the effect of fires on the north's wetlands. There will, then, be biomass burning.

Perhaps a continued natural fire regime will reduce the magnitude of burns as individual events, spreading a smaller impact over a longer time. Forest fire managers have observed, as expected, that once areas have been burned it is some time before a major fire event occurs again. The large burns of recent times have primarily consumed areas of old forest. Further, fire exclusion has generally proved to be a poor policy, as exclusion of fire without other activities in support is merely deferring the day of reckoning.

Fire managers in the north presently operate in a quasi-natural fire regime. Interference in natural processes has not proved wise or practical in the past. Would it be any more practical or desirable in the future?

Summary

The case studies presented here illustrate both the similarities and differences in fire management approaches and issues that would be observed across all Canadian fire management agencies. The intensity of efforts to suppress wildfires varies with the size of the province or territory's forested area, its accessibility, the level of commitment of its forest resource to the forest industry, and the demands and concerns of the people in the jurisdiction for other forest values: habitation, recreation, social, cultural, and aesthetic issues.

Where the forests are accessible and committed to a variety of human uses, agencies deliver intensive detection and suppression efforts. Where human settlement and demands are low, alternatives to full suppression efforts may be practiced, and a more natural fire regime is perpetuated. In general, fire management agencies deliver a level of fire protection commensurate with both the values at risk from fire and the cost of delivering the suppression response.

Canadian fire management agencies currently face a three-cornered challenge:

1. Meeting the demands from clients and stakeholders to maintain or increase the level of protection provided to critical wood supplies, settled areas, and recreational, cultural, and other human-centered values in the forest. Fire managers are discovering almost annually that the complex natural process that is a severe fire event can occur regardless of the level of expenditure and effort the agency brings to bear.
2. Addressing concurrent demands to restore natural ecological processes and to follow management practices that would emulate the natural disturbance patterns that shaped most of Canada's forests in the first place and sustained those forests over many centuries.
3. Having to address those foregoing concerns with constrained resources resulting from the continuing focus of most Canadian governments on expenditure and deficit reduction. (Those who believe that some forest conditions may benefit from an increased level of fire disturbance should understand that this is not a simple solution that would automatically result in reduced costs.)

Dealing with these three challenges is complicated by the growing concern that climate change, changes in human activity patterns, and changes in forest structure wrought by a century of increasingly intensive fire suppression may be collectively changing the fire regime with which Canadian fire managers must deal. Recent trends in increasing fire season severity and increasing variability in weather and fire conditions are causing significant unease in the fire management community as we approach the turn of the century. However, modifying management practices in response to those concerns will be difficult without a clearer understanding of future environmental patterns.

These potential changes have significant implications for the fire management policies and operational strategies that Canadian agencies have traditionally followed. Maintaining (or increasing) current levels of fire protection may be infeasible with static or declining levels of traditional fire management resources. Providing more selective levels of protection may be more appropriate ecologically but could, regardless, become an economic necessity, and Canadians may have to live with a greater general level of fire disturbance in their managed forests—unless public or stakeholder concerns convince funding agencies that those levels of fire loss are economically or socially unacceptable.

Fire managers will need to increase their consideration and understanding of fire impacts on a landscape scale and look at broader efforts in fuels and landscape management by better integrating fire management concerns in forest management planning. And they can expect to see greater involvement of client groups and stakeholders in planning and delivering fire management programs in the future.

References

Beverly and Qamanirjuaq Caribou Management Board. 1993. *A Fire Management Plan for the Forested Range of the Beverly and Qamanirjuaq Caribou Herds.* Technical Report 1, 2nd draft. Ontario Ministry of Natural Resources, Toronto.

Environment Canada. 1973. *The MacKenzie Basin.* Proceedings of the Intergovernmental Seminar Held at Inuvik, NWT, June 24–27, 1972. Environment Canada, Inland Waters Directorate, Winnipeg, Manitoba.

Forestry Canada Fire Danger Group. 1992. *Development and Structure of the Canadian Forest Fire Behaviour Prediction System.* Forestry Canada Information Report ST-X-3. Forestry Canada, Ottawa.

Government of the Northwest Territories. 1990. *Forest Fire Management Policy and Directive.* Government of the Northwest Territories, Yellowknife, NWT.

Janzen, S.S. 1990. *The Burning North: A History of Fire and Fire Protection in the Northwest Territories.* M.A. Thesis, University of Alberta, Edmonton.

Ontario Ministry of Natural Resources. 1991. *Forest Fire Management Policy for Ontario.* Aviation, Flood and Fire Management Branch Policy FM:1:01. Toronto.

Revised Statutes of Ontario. 1990. *Forest Fires Prevention Act.* Chapter F.24.

Rowe, J.S. 1972. *Forest Regions of Canada.* Department of the Environment, Canadian Forestry Service Publication 1300. Ottawa.

Ward, P.C., and A.G. Tithecott. 1993. *Impact of Fire Management on the Boreal Landscape of Ontario.* Ontario Ministry of Natural Resources, Aviation, Flood and Fire Management Branch Publication 305. Sault Ste. Marie, Ontario.

Wiken, E. 1986. *Terrestrial Ecozones of Canada.* Ecological Land Classification Series 19. Lands Directorate, Environment Canada, Ottawa.

6. Effects of Climate Change on Management and Policy: Mitigation Options in the North American Boreal Forest

Perry Grissom, Martin E. Alexander, Brad Cella, Frank Cole, J. Thomas Kurth, Norman P. Malotte, David L. Martell, William Mawdsley, James Roessler, Robert Quillin, and Paul C. Ward

Introduction

Over the next 50–100 years, the predicted doubling of the atmospheric carbon dioxide concentration is expected to increase summer temperatures up to 4–6°C at higher latitudes (Boer et al. 1992; Maxwell 1992; Ferguson 1995). In a $2 \times CO_2$ climate, summer precipitation is expected to be 105–160% of current, with more falling in storms and less as steady rain (Maxwell 1992; Ferguson 1995). This altered climate may include longer periods between rains (Wotton and Flannigan 1993).

As discussed in Chapter 20, with the predicted warming it is anticipated that fire incidence will increase, with more frequent major fire years. Increased fire activity will result not only from longer periods with favorable burning conditions but from increased frequency of fire ignition from lightning strikes over much of the boreal forest as a result of climate change (Price and Rind 1994a,b). The predicted climate warming and resultant increase in fire occurrence and severity would increase carbon emissions (Kasischke et al. 1995b).

As discussed in the chapters of Section II of this book, fire plays an important role in controlling many ecosystem processes in the boreal forest, burning an average of 5–12 million ha yr^{-1} (Kasischke et al. 1995a; Conard and Ivanova 1998). Because of fire, the boreal forest is a major source of greenhouse gases (Kurz and Apps 1995; French et al., this volume). It has been suggested that reducing fire incidence in the boreal forest represents a means to reduce greenhouse gas emissions and sequester carbon (Brown 1996; Izrael and Avdjushin 1997).

This suggested policy needs to be considered in terms of not only its implications for fire management but in terms of the effects of increased fire suppression on integrated management of the boreal forest ecosystem. Toward this end, this chapter discusses options available to fire and land managers in the North American boreal forest in regards to the predicted effects of climate change on the fire regime. These options include protection of current values or priorities in a more severe fire environment, as well as managing emissions from boreal forest fires that might contribute to global warming. The ecological consequences of mitigation of fire and other considerations are also discussed.

Options and Considerations for Fire Management in a 2 × CO_2 Climate

A reevaluation of fire management in the boreal forest is warranted, given the predicted change in climate and fire activity (Stocks et al. 1998; Stocks et al., this volume). Not only is there interest in changing fire management to mitigate carbon emissions, but climate change would affect existing protection priorities. To mitigate carbon emissions, fire managers can try to reduce fire occurrence or manage fires in ways that reduce their emissions. Although suppression of fires is often the first option that comes to mind, there is a variety of strategies and tactics available. In addition, the (U.S.) National Committee on Wildfire Disasters stresses that suppression alone cannot prevent forests from burning (Anonymous 1994).

Currently, managers over most of the North American boreal forest prioritize fire suppression (as well as preparedness activities) on values to be protected, including threats to human life and settlements, utility conveyances such as pipelines and electrical lines, cabins, commercial timber, and cultural sites (Taylor et al. 1985; Stocks 1993). This often results in fires being suppressed over large areas to protect relatively small, isolated sites. If the costs of protecting those values become excessive or if current tactics are no longer effective in protecting them in a new fire climate, managers are forced to reorder their priorities or change tactics or strategies. The result will likely be more individualized resource protection. Given the declined levels of public funding in both Canada and the United States, managers are being forced to do more with less funding. This requires closer cooperation and a coordination of goals and operations between a variety of governmental units.

If the policy emphasis shifts to mitigation of emissions from wildfires, specific stands with the potential for high emissions may be classified as resources at risk from fire. Management of high-emission stands would hinge on understanding the factors that influence combustion and emission as well as identifying and mapping high-emission areas. The expense to map such sites over the vast boreal forest could be high. Recent efforts by Ducks Unlimited and the Bureau of Land

Management to create an updated vegetation map in relatively small parts of interior Alaska cost about \$0.10 ha^{-1}.

Another option is to conduct controlled burns in potentially high-emission areas under favorable climatic and atmospheric conditions. Burning such areas at times when duff and organic soil moisture levels are high would reduce biomass consumption and therefore emissions.

Preparedness Activities

Preparedness involves fire management activities that take place before ignition of a wildfire, including fire prevention, fire detection, prepositioning of suppression resources, and fuels manipulation. Prevention efforts are intended to reduce the number of human-caused ignitions, mainly through education and enforcement activities. To help reduce emissions, education programs may be expanded to include information about the impacts of global warming. Prevention will likely play only a minor role, because it can affect only that area burned under human-ignited fires. Although humans cause 65% of fires in the North American boreal forest, those fires account for only about 15% of the area burned (Stocks 1991). Relative area burned from fires started by lightning strikes will likely increase if climate change increases the amount of lightning by 30% or more (Price and Rind 1994a). However, prevention may be more important than it is now because increased fire danger would facilitate ignition and increased fire intensity and severity would result in more damage.

Lightning detection systems and fire detection aircraft improve ability to find fires, the first step in any fire suppression system. However, much of the area burned in the boreal region is located in remote areas far from sites where detection aircraft are stationed; therefore, ignitions are not always found quickly. The delay and the slow response of distant suppression forces translate into reduced success at detecting and stopping the spread of fires while small. This results in larger fires that are costly and difficult to contain or extinguish. In major fire years, when numerous fires are producing large volumes of smoke, detection of new starts is made much more difficult. New technologies, such as rapid access to satellite observations, may provide a cost-effective approach for improving the rates of early detection of fires in remote boreal regions.

Manipulating fuel loads has the potential to dampen fire behavior and help protect public and fire fighter safety (IMRT 1994). Mechanical, chemical, and biological manipulation of fuels currently have limited use, restricted to easily accessible areas and to conditions producing minimal negative impacts on society. Controlling the timing and location of fire to achieve management objectives has greater potential for manipulating fuels (Stocks 1993). Prescribed, management-ignited fire or naturally ignited fire can be used to break up fuels or change stand structure and composition to reduce hazard and protect resources at risk (e.g., settled areas or high emission fuel types). Managers could allow potential high emission areas to burn under prescriptions (e.g., with high duff moisture content) that reduce the amount of biomass consumed (Andrasko et al. 1991).

Given the scale under consideration and existing staffing and fiscal constraints, burning with management-ignited fire may be too insignificant to affect the predicted severe fire regime and resultant emissions. Prescribed burning also has an inherent level of risk, and managers have generally been conservative in applying prescribed fire because of the potential for "unintended consequences" (USDA/USDI 1995). Heightened urgency because of global warming may require that more risks be taken, but the stakes would be higher because more severe conditions would result in greater effects from escaped fires. Careful prioritization and application could allow prescribed fire to be used to manage relatively small areas. A change in level of acceptable risk and a change in strategies would be required to make meaningful impacts on a large scale. One option might be to use numerous ignitions over large areas during mild years, so that duff/litter layer consumption is relatively low and continuous stands are broken up.

Suppression Activities

Fire suppression tactics and strategies are usually dictated by fire behavior, availability of suppression resources, potential threats to identified resources, and cost. Prediction of fire behavior will likely become more important in the future, given the predictions of increases in fire occurrence and severity.

To understand the impacts of using fire suppression to lessen greenhouse gas emissions, the fire control philosophy in place in Alaska and Canada until the 1980s can be studied. Before that time, fires were usually aggressively suppressed to minimize the area burned. That policy has largely been abandoned because of high costs, negative ecological impacts (Viereck 1983; Stocks 1993; Haggstrom 1994), and the fact that it was not always possible to effectively implement this policy. For example, some fires escaped initial attack efforts during the first day, then defied subsequent extended attack control efforts. In addition, fires were unofficially prioritized using values-at-risk when suppression resources became limiting during periods of high fire activity, resulting in some fires in remote areas never being actively suppressed (Pyne 1982).

Fire outbreaks usually occur on a regional scale, not on a continent-wide basis (Johnson 1992). When managers are able to predict fire outbreaks accurately, suppression forces are assigned to the potential trouble spots or high-priority areas. To accomplish rapid, massive movement of suppression resources requires timely accurate weather and fire behavior prediction, interagency cooperation, agreement on priorities, and adequate funding. In the future, if climate change increases fire danger, managers may be less willing or unable to dispatch their resources to areas outside their own jurisdiction, a practice that is common today. If predictions about fire outbreaks turn out to be wrong and their home unit gets several starts, the fire situation could quickly get out of hand.

Although the terminology and exact usage may vary, three general types of tactics are used for fire suppression: control, contain, and confine (see Pyne 1984). These tactics can be used alone or in combination on different parts of the fire. Control usually means direct attack at the fire edge, extinguishing the fire as soon as possible. Containment tactics use indirect attack involving construction or use of preexisting fuel breaks away from the actual fire front. This approach is used when fire intensity is too extreme for personnel to operate near the fire or where it may be unsafe to do so. Finally, confinement tactics attempt to keep a fire within a broader area, often using natural fuel breaks such as rivers or lakes. Sometimes the only suppression action on fires in a confinement strategy is to monitor the fire by aircraft to ensure that it has not burned beyond the intended area.

One of the predicted effects of climate warming is higher night-time temperatures (Loehle and LeBlanc 1996). Cooler temperatures at night result in lower relative humidity, which, in turn, lowers fire activity. Fire fighters often rely on this phenomenon to help suppress fires. Given warmer night-time temperatures, some adjustments in suppression operations will be required to continue to meet existing fire protection priorities.

To reduce area burned as much as possible, control tactics using more rapid initial attack are required to catch and stop fires while they are still small. Success requires rapid detection, mobilization, and attack, which, in turn, require higher levels of staffing and availability of aircraft. Because the boreal forest is sparsely settled, suppression forces are spread thin. Access to remote areas is slow and costly, making reliance on control tactics untenable for vast areas. Accurate prediction of areas where ignitions will likely occur helps mobilize adequate forces. Because of predicted increases in fire intensity, attack with hand crews alone will likely fail more often, requiring an increase in aerial application of water, foam, or retardant in the initial attack phase, which will greatly increase costs. In addition, strategies involving increased use of fire control tactics forego the benefits of having fuel breaks of less flammable fuel types that are created by large fires.

Managers of individual fires, governmental units, and regional dispatch centers must be cognizant of potential suppression resource sinks and allocate suppression resources prudently. Large fires that defy control efforts can tie up large numbers of personnel and equipment for long periods, greatly hindering overall ability to perform initial attack on new starts or to provide forces for other fires in the early stages of extended attack.

In the predicted increased fire scenario from climate change in the boreal region, containment tactics will likely be used for initial attack more frequently than at present, because increased fire intensity will prohibit direct attack more often. Containment is better suited to remote areas where fires may be too large for rapid control by the time suppression forces arrive. Fewer forces are usually required to implement this strategy than a control strategy, allowing managers to stretch their suppression resources. However, larger fires do reduce the options of managers. Some containment operations may be conducted solely by using aerial equipment. Aerial operations in remote areas may still be relatively slow and expensive because of logistical considerations, such as distance to travel for refueling.

Confinement tactics are currently used on many fires in remote areas. Although this strategy is least limiting to the area burned, it may become more common in a

$2 \times CO_2$ climate by default, dictated by more severe fire behavior and an increased fire load. Confinement tactics could be most helpful if fires were managed considering potential emissions from a site. Fires burning in low-emission fuel types might be confined and kept from reaching high-emission fuel types. Fires might be herded or fought with delaying tactics to keep them from reaching high-emission stands until late in the fire season or during wetter periods so that relatively little of the duff/litter layer burns and emissions are reduced.

Natural barriers help limit fire size and improve success and reduce cost of suppression operations by providing ready-made fuel breaks. Much of the boreal forest has vast expanses of highly flammable fuel types with few natural barriers (Taylor et al. 1985). If global warming lowers permafrost, many wetlands created by perched water tables may disappear (Kane et al. 1992), reducing the number of natural barriers even further. However, lower-flammability vegetation types created by fires would have increasing value in a $2 \times CO_2$ climate.

A flexible approach using all three types of tactics on the same fire might also be useful, either at the same time on different parts of the fire or at different times. For example, a fire might be managed first with a control strategy until it escapes initial attack, then confinement may become the best option if burning conditions are severe and if there is little chance of immediate containment. This would keep suppression resources available for initial attack in higher priority areas. When cooler or wetter weather appears for a few days or the fire approaches a major fuel break, such as a river, a containment strategy might then be used.

Fire managers almost always target suppression actions on the perimeter of fires; however, the smoldering phase of combustion behind the active flame front is sometimes a major source of emissions (Chapter 3). Increased attention might be focused on the interior of burns, especially if relatively small areas of deep organic soils are ignited. Treatment of large areas would be labor and cost prohibitive with existing methods.

Fire suppression organizations have adopted new technologies as they became available, including tracked vehicles, helicopters, fire-line explosives, fire-fighting foams, global positioning systems, thermal imaging, lightning detection systems, and computerized fire danger and behavior prediction systems. Improvements in these technologies would help tip the scale in favor of fire fighters, but the predicted worsening of fire behavior and fire activity could greatly outweigh those gains. New technologies, such as satellites that can detect new fires while small, would greatly aid managers.

Impacts and Considerations of Increased Fire Suppression

Agencies responsible for fire management in the North American boreal forest recognize that fire exclusion is not possible, nor is it economically or ecologically desirable (Pyne 1982; Stocks 1993). If more intensive or intrusive fire management to help reduce greenhouse gas emissions becomes a higher policy priority, fire management budgets will need to be increased. Although this would allow for more aggressive suppression, it would not guarantee success. If funding remains at current levels or continues to decline, long-term success is even less likely, especially given the current predictions for increased severity in the fire regime of the boreal forest.

Cost Considerations

Increasing suppression efforts will require a larger standing fire-fighting force. In addition, more aggressive suppression will entail increased operations and support costs. With elevated fire load and fire severity, those costs may be of a magnitude not yet encountered.

A recent fire suppression operation can be used as an example of the level of required support. The Big Lake (or Miller's Reach #2) fire occurred in June 1996 near Anchorage, Alaska. Suppression forces nearly contained this fire until a dry cold front passed through the area. The fire escaped initial attack and burned uncontrolled for 4 days. Because the fire occurred in the most heavily populated part of the state, additional suppression forces were quickly mobilized. During peak operations, the total suppression force (including resources from Canada and the lower 48 states) included more than 1,400 fire fighters, 75 engines, 20 water tenders, 66 bulldozers, 11 helicopters, and numerous retardant bombers.

During the time when the fire could not be controlled, suppression efforts were geared at evacuation of residents and protection of individual houses. When weather conditions became more favorable, the strategy was switched to control and stop the fire spread. The fire was eventually controlled 8 days after initial attack, after it had reached a size of just more than 15,000 ha. Suppression costs during peak operations were about \$1 million day^{-1}, and total fire suppression costs were more than \$10 million. More than 344 structures were burned by the fire.

This example shows that massive mobilizations and large expenditures do not always guarantee success.

Safety Considerations

Whether from decades of suppression (which has resulted in hazardous fuel situations) or from changing weather patterns, the area burned by wildfires has been increasing in the United States and Canada despite increasing fire management expenditures (Stocks 1991; Clark and Sampson 1995; Fig. 1.2). Worsening fire fuel conditions raise concern for the safety of the public and fire personnel. Fire suppression and prescribed burning are difficult hazardous jobs, and all operations have inherent risks. In the United States between 1910 and 1996, 699 wildland fire fighters died on the job. Fifty-eight percent of those (407 deaths) were directly caused by fire entrapment, including seven on prescribed burns. Other major causes of death included heart attacks, aircraft accidents, vehicle accidents, electrocution, and falling snags (NWCG 1997). Given the extreme fire behavior that fires in the boreal forest can exhibit, Alaskan and Canadian wildland fire fighters have suffered remarkably few deaths, especially resulting from entrapment by fire. However, the fires can exact a high human toll, such as the fires in northeastern China that killed more than 200 persons in 1987 (Cahoon et al. 1994).

Recently, more than 700 American fire fighters responded to a nationwide survey, and many identified inappropriate selection of strategies as a source of safety concerns (TriData 1996). One of the highest ranked problems was that fires were "fought in a more dangerous way than the values to be protected merited," which 43% of the respondents said happened often or very often. Thirty-six percent of respondents stated that fires were fought without adequate suppression resources, resulting in compromised fire fighter safety. Forces are often stretched thin and would be stretched farther by the predicted increase in fire activity associated with a 2 × CO_2 climate. More intense fires, an increased emphasis on suppressing fires to manage greenhouse gas emissions, and the creation of hazardous fuel situations will magnify those risks and may exact a higher toll in human fatalities and injuries.

Ecological Impacts

Successful suppression programs result in a significant lengthening of the fire return interval (Clark and Sampson 1995; Ward and Mawdsley, this volume). This increases the extent of stands containing flammable species (e.g., pine and spruce), causes stagnation of plant succession, and can affect the role of the ecosystem as a wildlife habitat (Wright and Heinselman 1973; Dyrness et al. 1986). However, given that not all fires can be successfully suppressed, larger and potentially more harmful fires are likely to occur. Attempting to replace lightning-caused fires with prescribed fires will also alter the landscape pattern and affect habitat diversity, because these fires tend to be relatively small and less severe compared with normal wildfires (Baker 1994).

Disruption of the fire regime perturbs the entire ecosystem, including hydrological, carbon, and nutrient cycles, landscape diversity, wildlife and plant species diversity, and species distributions and abundances (Clark and Sampson 1995; Murray 1996). The primary cause or threat of species extinction is habitat destruction and alteration, which is responsible for the listing of 73% of the threatened and endangered species in the United States (Wilson 1992). Imposing an unnatural fire regime will compound the effects of existing human activities. Fire is a critical influence in plant community dynamics, and it will play a major part in the ability of plant species and communities to adjust to a changing climate (Suffling 1995).

Other broad unforeseen effects may be felt in a ripple effect that propagates through the boreal forest and other ecosystems far away. Predicted climate change and a more aggressive fire suppression program would likely affect critical keystone species such as salmon that depend on hydrological and nutrient cycles of the boreal forest (Beamish 1995). The boreal forest of western Canada and interior Alaska has some of the largest runs of wild salmon in North America. Those fish are extremely valuable to subsistence users, recreational users, wildlife, and commercial fishers and their customers across the globe. Song bird abundance and diversity are correlated to primary productivity and plant community structural complexity (Spindler and Kessel 1980). Stagnation of forests and interrupting the disturbance regime will likely decrease suitable habitat for many species, particularly migrant insectivorous songbirds which comprise more than 90% of the bird populations in high-latitude forests (Diamond 1992; Kirk et al. 1996).

Forest health affects ecosystem function and can contribute to fire hazard. Forest disease and insect outbreaks are worsened by single age/species stands growing under stressed conditions, such as those created by successful fire exclusion (Clark and Sampson 1995). Individual plant stresses will also be increased by higher temperatures and higher temperature/precipitation ratios associated with the predicted $2 \times CO_2$ climate (Boer et al. 1992; Stocks 1993), which will make forest health more tenuous. In addition, the range of pest species will expand with warming temperatures (Loehle and LeBlanc 1996), leading to increased fuel loads and increased fire intensity.

Cultural Impacts

Obviously, the impacts to plant and wildlife communities from predicted climate changes would have major impacts on subsistence users. Loss of one food source will require more use of another. In western Alaska, rural residents eat about 300 kg per capita of meat and fish compared with 100 kg per capita in the Lower 48. If natural sources of meat are lost, replacing them with other food sources would cost between \$131 and \$218 million yr^{-1}, which is about half the income of those residents (Callaway 1995). About 57,600 persons live in the Northwest Territories of Canada, and 90% of them depend on wild game for subsistence. If that source were lost, replacement with purchased meats would cost \$37 million yr^{-1} (Gunn 1995).

Indigenous people have exhibited flexibility to continual change to the boreal forest, but the looming impacts of global warming may overwhelm their cultures (Langdon 1995). Norma Kassi of Old Crow, in western Canada, states (quoted in West 1995, p. 281):

> Our people are directly affected (by global climate change) . . . there are no compromises we can make. There are no changes we can make in these old ways. We cannot be compensated for any damage that might occur to our land, the birds, animals, water, fish, and thus our people as the result of another culture's indiscriminate activities or disregard for others. We have no alternatives to our way of life, this is the only one we know. Without this way of life, we will disappear.

Other nonaboriginal North Americans also gather cultural value from the boreal forest. Tourism, hunting, and fishing are growing industries in the boreal forest region, and many communities depend on visitors who come to recreate. In addition, many people never see the vast expanses of boreal forest but derive great pleasure from knowing that the wild places and wildlife simply exist (Wright 1974; Gunn 1995).

Discussion

Values of the boreal forest are many and varied and may be broken down into direct and indirect values derived from use as well as nonuse (UNCED 1991b). Direct-use values include indigenous consumption and commercial/industrial consumption, visual resources, recreation, and research and education. Indirect-use values include ecological services, watershed protection, protection of species and genetic diversity, soil formation and fertilization, socio-cultural values, and living environment for indigenous people. Nonuse values include existence values, option values, future values, others' use of resources, others' level of well-being, and the overall value of a healthy ecosystem. Costs of altering natural processes in the boreal forest to manage it as a carbon sink may eventually include management of new endangered species and welfare payments to subsistence users. Some potential costs are beyond financial realms, such as the loss of a way of life to native cultures and the loss of life among fire fighters exposed to more severe fire conditions with less available resources.

The potential coupling of climate warming with an altered fire regime presents serious challenges to resource managers and policy makers in the boreal forest. What exactly will the new fire climate be? How do managers identify what to manage for? Is it reasonable to manage for a previous vegetation formation and associated wildlife population that can no longer exist because of a climate changed by human influence? "The biophysical changes and philosophical questions posed by climate change will be enough to give anyone a headache" (Dennerlein 1995, p. 290). Cairns and Meganck (1994, p. 13) state that "global forestry, climate, and biodiversity agreements are less likely to be successful if they ignore social, ecological, political, and economic perspectives."

To a large extent, managers will be forced to address not only moral issues but also values that are deeply imbedded. How important is the boreal forest? How much should boreal forest function be impaired to help limit damage created by "unnatural" processes elsewhere?

Conflicting Goals and Policy

If managers look to policy, they often find not only guidance but conflicting goals as well. On the one hand, nations have agreed to promote and cooperate in the conservation and enhancement of natural sinks and reservoirs of greenhouse gases (INC-FCCC 1992). However, it is also recognized that management of forests must consider other environmental constraints and values. In addition, it is recognized that carbon sequestration is not the only solution for the problem of global warming.

The same entities responsible for greenhouse gas policy recommendations also place high value on biological diversity (UNCED 1991a). Nations have agreed to "promote the protection of ecosystems, natural habitats, and the maintenance of viable populations of species in natural surroundings" (UNEP 1992). Areas may only be used for carbon sinks where that strategy would be "ecologically viable, socially acceptable, and economically feasible in the broadest sense" (UNCED 1991b). These are complex, weighty, but vague considerations for local fire and land managers.

The (U.S.) President's Council on Sustainable Development also recommends policies to "enhance, restore, and sustain the health, productivity, and biodiversity of terrestrial and aquatic ecosystems" (PCSD 1995). This policy calls for reduced emissions of greenhouse gases, as well as for the use, conservation, and restoration of natural resources (land, air, water, and biodiversity) "in ways that help ensure long-term social, economic, and environmental benefits for ourselves and future generations" (goal 4). Canada's "Green Plan" (1990) calls for prudent management of resources and encourages "sensitive environmental decision making" (Marshall et al. 1993). The mission of Natural Resources Canada includes "sustainable development" of forests. More immediate to individual land and fire management agencies are their own policies, which vary because of differing missions (see Chapter 5).

Recognizing problems in American fire suppression programs, two groups recently came up with several recommendations, none of which included a more widespread, intensive fire control effort. Both the National Commission on Wildfire Disasters and the Interagency Management Review Team recommended more fuels management to reduce hazards and changes in suppression procedures to improve safety (Anonymous 1994; IMRT 1994). Those recommendations formed the basis of a recently adopted federal policy (USDA/USDI 1995), which emphasizes "the role of wildland fire as an essential ecological process."

Likelihood of Success

Implementing a policy of increased fire suppression to reduce greenhouse gas emissions involves issues related to increment trading and practicality. In this light, additional questions must be asked: How much should we attempt to intervene? How successful will we be? How much will be gained in helping prevent or reduce global warming?

From past experience, we know that the probability of success in excluding fire from the boreal forest for long periods of time is low (Taylor et al. 1985). It can be argued that fire exclusion is counterproductive (Stocks 1993), often leading to larger, more damaging, and more dangerous fires (Clark and Sampson 1995). We also know that ecological costs of fire exclusion are high (Viereck 1983).

Fire suppression operates on a "very narrow margin between success and failure" (Stocks 1993), and the predicted increase in fire activity caused by global warming would be magnified by the increasing proportion of fires that escape initial attack. "Despite public expectations, when the combination of excessive fuel build-up, topography, extreme weather conditions, multiple ignitions, and extreme fire behavior occurs, it is impossible to immediately suppress every wildland fire" (USDA/USDI 1995, p. 17). In the United States and Canada, severe fire seasons in recent years have seen larger areas burned, despite increasing suppression expenditures (Stocks 1993; Clark and Sampson 1995; Fig. 1.2).

The disturbance regime in the boreal forest is extremely variable (Cahoon et al. 1994; Kasischke et al. 1995a; Kurz and Apps 1995; Figs. 1.2 and 15.4). When a major fire year occurs, suppression resources are periodically overwhelmed. During these times, the area burned is largely determined by weather and fire behavior, because even though forces are available some fires are unmanageable.

Discussing fire behavior in spruce forests, Alexander and Cole (1994, in their Table 1) note that under extreme conditions "control is extremely difficult and all efforts at direct control are likely to fail. Direct attack is rarely possible given the fire's probable ferocity except immediately after ignition and should only be attempted with the utmost caution. Otherwise, any suppression action must be restricted to the flanks and tail of the fire. Indirect attack with aerial ignition . . . may be effective depending on the fire's forward rate of advance."

Under "super-critical" conditions, extreme fire behavior is certain, with behavior including "rapid spread rates, continuous crown fire development, medium to long-range spotting, firewhirls, massive convection columns, (and) great walls of flame." Suppression is "virtually impossible," and the only place for effective and safe control action is at the back and along the flanks until the fire stops its run.

In a 2 × CO_2 climate, fire season will likely be longer and more severe and major fire years would occur more often, so existing suppression strategies would fail more often (Stocks 1993). With extreme conditions, fires can burn through fuel types that normally have low flammability and burn across substantial barriers, such as rivers (Turner and Romme 1994). In addition, in the 2 × CO_2 climate, the effectiveness of creeks and lakes as barriers for confinement is lower (Kane et al. 1992).

In the relatively short term (5–30 years), we may be able to stop a majority of the fires (assuming massive increases in fire management budgets), but with the predicted, more severe climate, we almost surely will not. Attempted fire exclusion will cause bigger, more severe fires in the future, resulting in fire deferral. These deferred fires are likely to be larger and more intense, burning deeply to consume more of the duff and organic soils, and injecting smoke higher into the atmosphere, which is probably the worst possible scenario.

Conclusions

The predicted 2 × CO_2 climate will have profound effects on fire management in the boreal forest. Many only partially understood factors will influence the nature and magnitude of these changes. There will likely be a more severe fire regime over much of the boreal forest in the near future (<50 years) because predicted weather conditions will be more conducive to ignition and spread of fires. Beyond that, the current boreal forest region may see more frequent but smaller and more manageable fires because of the conversion of the vegetation to less flammable types.

The likelihood of success at significantly reducing fire occurrence in the boreal forest in order to use it as a carbon sink is low, especially given the predicted fire frequency and severity. Ultimately, weather, ignition sources, and fire behavior will largely dictate area burned in a 2 × CO_2 climate, and the future may hold more of the conflagrations that we have recently glimpsed (see Cahoon et al. 1994). If that occurs, managers would be forced to reprioritize fire protection efforts, increasing effort in high-value areas to maintain current levels of protection. That would mean that unless there were major increases in funding or significant changes in fire management, suppression effort would be reduced over vast areas of the boreal forest. Potential fire management adjustments to cope with a worsening fire situation include prescribed burning; more assertive prevention programs; improved detection, including increased reliance on remote sensing; less direct attack of fires and more indirect attack using natural barriers; and increased use of aerial resources.

A wide variety of strategies specifically for managing emissions is available, ranging from fire exclusion to burning potentially high-emission stands under conditions with relatively high fuel moisture levels that minimize consumption. Using a more aggressive fire management policy to reduce the area burned will have high economic, ecological, and cultural costs. Attempting to force ecosystems to respond to artificial objectives through "top-down, command-and-control management . . . usually results in unforeseen consequences for both natural ecosystems and human welfare" (Holling and Meffe 1996, p. 328). Living systems evolve with stresses and disturbances, and imposition of "unnatural" stresses results in "abnormal or pathological disintegration" (Regier 1993).

Research is needed to fully define potential costs, impacts, potential feedbacks, ways to effectively manage emissions, and methods of affixing value for increment trading. Managers may have to petition policy makers for their portion of the carbon budget. The issue quickly evolves into one of values and must be resolved by society.

The government's moral responsibility in the conservation of biodiversity is similar to that in public health and military defense. The preservation of species across generations is beyond the capacity of individuals or even powerful private institutions. Insofar as biodiversity is deemed an irreplaceable public resource, its protection should be bound into the legal canon (Wilson 1992).

Over the past two centuries, humans have greatly reduced emissions from wildland fires by altering vegetation and suppressing fires (Packham and Tapper 1996; Taylor and Sherman 1996). That increment gained by deferring wildland fires has been used up and surpassed by fossil fuel combustion and other anthropogenic sources of greenhouse gases. We have only recently begun to recognize in full the shortcomings of our fire suppression programs, including forest health problems, altered biotic communities, and an increase in hazardous wildland fuels (Clark and Sampson 1995). Stakes are high, and managers are faced with many unknowns, tough decisions, and possibly a grim future.

Acknowledgments

The authors thank Eric Kasischke (ERIM International) and Larry Vanderlinden (U.S. Fish and Wildlife Service) for their review of this chapter.

References

Alexander, M.E., and F.V. Cole. 1994. Predicting and interpreting fire intensities in Alaskan black spruce forests using the Canadian system of fire danger rating. Paper presented at Fire Working Group Technical Session and Poster Session at Society of American Foresters/Canadian Institute of Forestry Joint National Convention, Anchorage, AK, September 18–22, 1994.

Andrasko, K.J., D.R. Ahuja, S.M. Winnett, and D.A. Tirpak. 1991. Policy options for managing biomass burning to mitigate global climate change, pp. 444–456 in J.S. Levine, ed. *Global Biomass Burning—Atmospheric, Climatic, and Biospheric Implications.* MIT Press, Cambridge, MA.

Anonymous. 1994. Report on the National Commission on Wildfire Disasters. *American Forests* (Sept./Oct.):13–16.

Baker, W.L. 1994. Restoration of landscape structure altered by fire suppression. *Cons. Biol.* 8:763–769.

Beamish, R.J. 1995. Response of anadromous fish to climate change in the North Pacific, pp. 123–136 in D.L. Peterson and D.R. Johnson, eds. *Human Ecology and Climate Change—People and Resources in the Far North.* Taylor & Francis, Washington, DC.

Boer, G.J., N.A. McFarlane, and M. Lazare. 1992. Greenhouse gas-induced climate change simulated with the CCC second-generation general circulation model. *J. Clim.* 5:1045–1077.

Brown, S.J., ed. 1996. Management of forests for mitigation of greenhouse gas emissions, pp. 755–796 in R.T. Watson, M.C. Zinyowera, and R.H. Moss, eds. *Climate Change 1995—Impacts, Adaptations and Mitigation of Climate Change: Scientific-Technical Analyses.* Cambridge University Press, Cambridge, UK.

Cahoon, D.R., Jr., B.J. Stocks, J.S. Levine, W.R. Cofer III, and J.M. Pierson. 1994. Satellite analysis of the severe 1987 forest fires in northern China and southeastern Siberia. *J. Geophys. Res.* 99:18,627–18,638.

Cairns, M.A., and RA. Meganck. 1994. Carbon sequestration, biological diversity, and sustainable development: integrated forest management. *Environ. Manage.* 18:13–22.

Callaway, D. 1995. Resource use in rural Alaskan communities, pp. 155–168 in D.L. Peterson and D.R. Johnson, eds. *Human Ecology and Climate Change—People and Resources in the Far North.* Taylor & Francis, Washington, DC.

Clark, L.R., and R.N. Sampson. 1995. *Forest Ecosystem Health in the Inland West: A Science and Policy Reader.* Forest Policy Center, American Forests, Washington, DC.

Conard, S.G., and G. A. Ivanova. 1998. Wildfire in Russian boreal forests—potential impacts of fire regime characteristics on emissions and global carbon balance estimates. *Environ. Pollut.* 98:305–313.

Dennerlein, C. 1995. Preserving environmental values in parks and protected areas, pp. 289–298 in D.L. Peterson and D.R. Johnson, eds. *Human Ecology and Climate Change—People and Resources in the Far North.* Taylor & Francis, Washington, DC.

Diamond, A.W. 1992. Birds in the boreal forest, pp. 2–6 in D.H. Kuhnke, ed. *Proceedings, Workshop on Birds in the Boreal Forest,* March 10–12, 1992, Prince Albert, Saskatchewan, Canada. Forestry Canada NW Reg., N. Cent., Edmonton, AB, Canada.

Dyrness, C.T., L.A. Viereck, and K. Van Cleve. 1986. Fire in taiga communities of interior Alaska, pp. 74–86 in K. Van Cleve, F.S. Chapin III, P.W. Flanagan, L.A. Viereck, and C.T. Dyrness, eds. *Forest Ecosystems in the Alaskan Taiga.* Springer-Verlag, New York.

Ferguson, S.A. 1995. Potential climate change in northern North America, pp. 15–30 in D.L. Peterson and D.R. Johnson, eds. *Human Ecology and Climate Change—People and Resources in the Far North.* Taylor & Francis, Washington, DC.

Gunn, A. 1995. Responses of Arctic ungulates to climate change, pp. 89–104 in D.L. Peterson and D.R. Johnson, eds. *Human Ecology and Climate Change—People and Resources in the Far North.* Taylor & Francis, Washington, DC.

Haggstrom, D. 1994. The effects of fire and forest management policies on the boreal forest and wildlife of interior Alaska. *Wildfire* (Dec):31–38.

Holling, C.S., and G.K. Meffe. 1996. Command and control and the pathology of natural resource management. *Conserv. Biol.* 10:328–337.

IMRT. 1994. *Report of the Interagency Management Review Team: South Canyon Fire.* Boise Interagency Fire Center, Boise, ID.

INC-FCCC. 1992. *Report of the Intergovernmental Negotiating Committee for a Framework Convention on Climate Change on the Work of the Second Part of Its Fifth Session,* held in New York, April 30–May 9, 1992. A/AC.237/18 (Part II)/Add. 1. Intergovernmental Negotiating Committee for a Framework Convention on Climate Change, United Nations, New York.

Izrael, Y.A., and S.I. Avdjushin. 1997. *Russian Federation Climate Change Country Study. Vol. 4. Mitigation Analysis.* Russian Federal Service for Hydrometeorology and Environmental Monitoring, Moscow.

Johnson, E.A. 1992. *Fire and Vegetation Dynamics—Studies from the North American Boreal Forest.* Cambridge University Press, Cambridge, UK.

Kane, D.L., L.D. Hinzman, M. Woo, and K.R. Everett. 1992. Arctic hydrology and climate change, pp. 35–58 in F.S. Chapin III, R.L. Jefferies, J.F. Reynolds, G.R. Shaver, J. Svoboda, and E.W. Chu, eds. *Arctic Ecosystems in a Changing Climate: An Ecophysiological Perspective.* Academic Press, New York.

Kasischke, E.S., N.H.F. French, L.L. Bourgeau-Chavez, and N.L. Christensen, Jr. 1995a. Estimating release of carbon from 1990 and 1991 forest fires in Alaska. *J. Geophys. Res.* 100:2941–2951.

Kasischke, E.S., N.L. Christensen, Jr., and B.J. Stocks. 1995b. Fire, global warming and the mass balance of carbon in boreal forests. *Ecol. Appl.* 5:437–451.

Kirk, D.A., A.W. Diamond, K.A. Hobson, and A.R. Smith. 1996. Breeding bird communities of the western and northern Canadian boreal forest: relation to forest type. *Can. J. Zool.* 74:1749–1770.

Kurz, W.A., and M.J. Apps. 1995. An analysis of future carbon budgets of Canadian boreal forests. *Water Air Soil Pollut.* 82:321–331.

Langdon, S.J. 1995. Increments, ranges, and thresholds: human population responses to climate change in northern Alaska, pp. 139–154 in D.L. Peterson and D.R. Johnson, eds. *Human Ecology and Climate Change—People and Resources in the Far North.* Taylor & Francis, Washington, DC.

Loehle, C., and D. LeBlanc, D. 1996. Model-based assessments of climate change effects on forests: a critical review. *Ecol. Model.* 90:1–31.

Marshall, I.B., H. Hirvonen, and E. Wiken. 1993. National and regional scale measures of Canada's ecosystem health, pp. 117–129 in S. Woodley, J. Kay, and G. Francis, eds. *Ecological Integrity and the Management of Ecosystems.* St. Lucie Press, Delray Beach, FL.

Maxwell, B. 1992. Arctic climate: potential for change under global warming, pp. 11–34 in F.S. Chapin III, R.L. Jefferies, J.F. Reynolds, G.R. Shaver, J. Svoboda, and E.W. Chu, eds. *Arctic Ecosystems in a Changing Climate: An Ecophysiological Perspective.* Academic Press, New York.

Murray, M.P. 1996, Natural processes: wilderness management unrealized. *Nat. Areas J.* 16:55–61.

NWCG. 1997. Historical wildland firefighter fatalities, 1910–1996, 2nd ed. National Wildfire Coordinating Group, PMS-822, NFES 1849. National Interagency Fire Center, Boise, ID. Also available at http://fire.r9.fws.gov/fm/pms/docs/fat—pdf.

Packham, D., and N. Tapper. 1996. Climate change and biomass burning. Paper presented at the 13th Fire and Forest Meteorology Conference, Lorne, Australia.

PCSD. 1995. *Sustainable America: A New Consensus for the Future.* President's Council on Sustainable Development, Washington, DC, http://www.whitehouse.gov/WH/EOP/pcsd/tf-reports/amer-top.html.

Price, C., and D. Rind. 1994a. Possible implications of global climate change on global lightning distributions and frequencies. *J. Geophys. Res.* 99:10823–10831.

Price, C., and D. Rind. 1994b. The impact of a 2 × CO_2 climate on lightning-caused fires, *J. Clim.* 7:1484–1494.

Pyne, S.J. 1982. *Fire in America.* Princeton University Press, Princeton, NJ.

Pyne, S.J. 1984. *Introduction to Wildland Fire: Fire Management in the United States.* Wiley & Sons, New York.

Regier, H.A. 1993. The notion of natural and cultural integrity, pp. 3–18 in S. Woodley, J. Kay, and G. Francis, eds. *Ecological Integrity and the Management of Ecosystems.* St. Lucie Press, Delray Beach, FL.

Spindler, M.A., and B. Kessel. 1980. Avian populations and habitat use in interior Alaska taiga. *Syesis* 13:61–104.

Stocks, B.J. 1991. The extent and impact of forest fires in northern circumpolar countries, pp. 197–202 in J.S. Levine, ed. *Global Biomass Burning—Atmospheric, Climatic, and Biospheric Implications.* MIT Press, Cambridge, MA.

Stocks, B.J. 1993. Global warming and forest fires in Canada. *For. Chron.* 69:290–293.

Stocks, B.J., M.A. Fosberg, T.J. Lynham, L. Mearns, B.M. Wotton, Q. Yang, J-Z. Jin, K. Lawrence, G.R. Hartley, J.A. Mason, and D.W. McKenney. 1998. Climate change and forest fire potential in Russian and Canadian boreal forests. *Clim. Change* 38:1–13.

Suffling, R. 1995. Can disturbance determine vegetation distribution during climate warming? A boreal test. *J. Biogeog.* 22:501–508.

Taylor, D.L., F. Malotte, and D. Erskine. 1985. Cooperative fire planning for large areas: a federal, private, and state of Alaska example, pp. 206–214 in J.R. Lotau, B.M. Kilgore, W.C. Fischer, and R.W. Mutch, eds. *Proceedings, Symposium and Workshop on Wilderness Fire.* General Technical Report INT-182. USDA Forest Service, International Forest Range Experiment Station, Ogden, UT.

Taylor, S.W., and K.L. Sherman. 1996. *Biomass Consumption and Smoke Emissions from Contemporary and Prehistoric Wildland Fires in British Columbia.* Canadian Forest Service, Pacific Forestry Center, Victoria, BC, Canada.

TriData. 1996. *Wildland Firefighter Safety Awareness Study: Phase I—Identifying the Organizational Culture, Leadership, Human Factors, and Other Issues Impacting Firefighter Safety.* TriData Corp., Arlington, VA.

Turner, M.G., and W.H. Romme 1994. Landscape dynamics in crown fire ecosystems. *Landscape Ecol.* 9:59–77.

UNCED. 1991a. *Protection of the Atmosphere: Climate Change, Synthesis Report an Ongoing Processes and Recommendations.* A/CONF.151/PC/22. Preparatory Committee for the United Nations Conference on Environment and Development, 2nd session, 18 March–5 April 1991. United Nations, New York.

UNCED. 1991b. *Preparations for the United Nations Conference on Environment and Development on the Basis of General Assembly Resolution 44/228 and Taking into Account Other Relevant General Assembly Resolutions: Conservation and Development of Forests.* A/CONF.151/PC/64. Preparatory Committee for the United Nations Conference on Environment and Development, 3rd session, 12 August–4 September 1991. United Nations, New York.

UNEP. 1992. *Convention on Biological Diversity, Article 8.d.* Convention on Biological Diversity, United Nations Environment Program, United Nations, New York.

USDA/USDI. 1995. *Federal Wildland Fire Management: Policy and Program Review, Final Report.* U.S. Department of Agriculture and U.S. Department of Interior, Washington, DC.

Viereck, L.A. 1983. The effects of fire in black spruce ecosystems of Alaska and northern Canada, pp. 201–220 in R.W. Wein and D.A. MacLean, eds. *The Role of Fire in Northern Circumpolar Ecosystems.* J. Wiley & Sons, New York.

West, P.C. 1995. Global warming and conflict management: resident native peoples and protected areas, pp. 187–196 in D.L. Peterson and D.R. Johnson, eds. *Human Ecology*

and Climate Change—People and Resources in the Far North. Taylor & Francis, Washington, DC.
Wilson, E.O. 1992. *The Diversity of Life.* W.W. Norton, New York.
Wotton, B.M., and M.D. Flannigan. 1993. Length of the fire season in a changing climate. *For. Chron.* 69:187–192.
Wright, H.E., Jr. 1974. Landscape development, forest fires, and wilderness management. *Science* 186:487–495.
Wright, H.E., Jr., and M.L. Heinselman. 1973. The ecological role of fire in natural conifer forests of western and northern North America. *Quat. Res.* 3:319–328.

II. PROCESSES INFLUENCING CARBON CYCLING IN THE NORTH AMERICAN BOREAL FOREST

Eric S. Kasischke

Introduction

Scientists have long recognized the importance of fire as a major disturbance factor in the boreal forest (Wein and MacLean 1983), as well as the fact that changes in climate will significantly alter the patterns of forest cover in high northern latitudes (Smith et al. 1992). Although the central role fire plays in carbon cycling in the boreal forest has been recognized (Kasischke et al. 1995), many scientists conducting global-scale analyses on the effects of climate on vegetation and carbon cycling have yet to account for the effects of fire in their models (Smith and Shugart 1993; Raich and Potter 1995; McGuire et al. 1997; Randerson et al. 1997; Field et al. 1998).

For example, scientists developing models to exploit seasonal variations in the vegetation cover observed from satellite remote sensing systems tend to focus on models based on seasonal differences in photosynthetically active radiation or leaf area index (Running and Nemani 1988; Foley et al. 1997; Thompson et al. 1997). Although such models are used to study carbon cycling in other terrestrial biomes, they may not be appropriate for the boreal ecosystem for several reasons. First, most of the carbon in boreal forests resides in the soil layer (see Table 2.2). Although some researchers have developed models to link aboveground net primary productivity with soil respiration processes to study carbon balance issues (Thompson et al. 1997), this approach may not be appropriate for the boreal forest because (1) much of the litter entering the organic soil pool in boreal forest originates from moss, living biomass that is not directly viewed by satellite systems and therefore cannot be monitored by using a vegetation index; and (2) the coupling between aboveground vegetation condition and soil respiration used in these models may not be appropriate because the presence of permafrost results in a complex layered model for soil respiration that is often independent of air temperature and patterns of aboveground net primary productivity.

To account completely for the effects of fire on carbon cycling in the boreal forest requires an understanding of how the boreal forest functions as an integrated system. Although this may seem to be an obvious statement, this principle is often overlooked. It has been only recently that the broader scientific community studying terrestrial global change issues have turned their attention to the boreal forests, particularly those developing large-scale models of processes of controlling energy, water, and trace gas exchange between the atmosphere and land surface. Many of the approaches used in recent large-scale studies such as the Boreal Ecosystem-Atmosphere Study (BOREAS) (Sellers et al. 1997) were developed in ecosystems outside of the boreal region. These approaches have to be significantly altered to account for the major factors influencing ecosystem processes in the boreal region.

Fire has both direct and indirect roles in the exchange of carbon between the terrestrial surface and the atmosphere. Over the longer term, the indirect roles fire plays are perhaps more important than the direct roles. These indirect roles involve two aspects of the carbon budget: (1) the role fire plays in controlling the stand age distribution (Fig. 25.1); and (2) the role that fire plays in regulating the soil thermal and moisture regimes, which, in turn, control key metabolic processes that drive plant succession, photosynthesis, and soil microbial processes. Both of these controls are directly linked to variations in climate. Reducing the uncertainties in the rates of exchange of greenhouse gases as well as understanding how future climate change will influence the balance of carbon stored in the boreal forest requires a proper description of and accounting for the interactions between a variety of complex factors.

The focus of this introduction to Section II is on describing the boreal forest ecosystem in terms of those factors and processes controlling carbon cycling. Such a systems approach is not new, and similar approaches have been used to study carbon dynamics in the boreal forest (Kurz et al. 1991; Kurz and Apps 1993); however, much research is still needed to develop an integrated model of carbon cycling in the boreal forest. The intent of this introduction is not to discuss the factors and processes controlling carbon cycling in the boreal forest in depth, as these details are presented in the remaining chapters of this section and elsewhere (see, e.g., Bonan and Shugart 1989; Kasischke et al. 1995).

Major Carbon Reservoirs and Exchange Pathways in the Boreal Forest

Figure II.1 divides the boreal forest into eight major carbon reservoirs. Other studies have used different labels for boreal forest carbon reservoirs. For example,

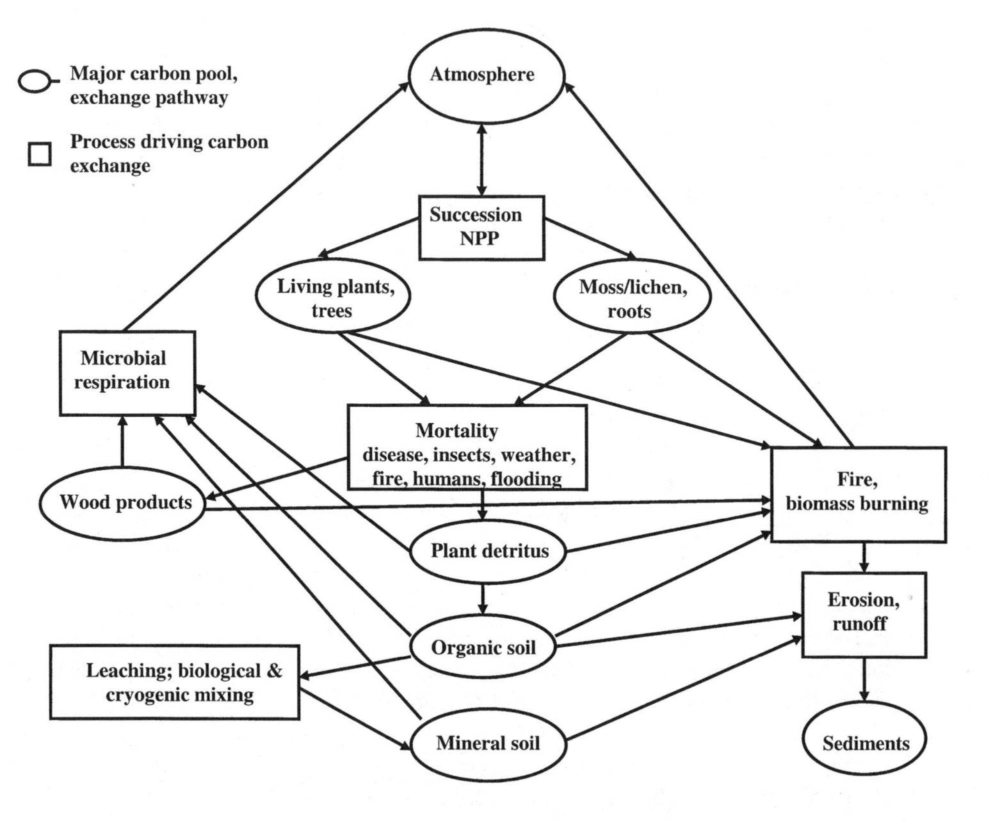

Figure II.1. Major carbon pools in the boreal forests and the processes controlling exchanges between these pools.

a common practice among Russian scientists is to divide the plant detritus pool into two categories: mortmass (standing-dead biomass) and litter (dead biomass lying on the ground surface) (see Chapters 8 and 16). One of the challenges in studying the boreal forest carbon budget lies in developing a common nomenclature to effectively compare and use data sets developed by different researchers.

The boreal forest has many characteristics that are unique when compared with other terrestrial biomes. Although other forest biomes have large amounts of carbon stored in their soils, boreal forests typically have deep layers of carbon-rich organic soils. Most of the carbon present in the boreal forest resides in three reservoirs: plant detritus, organic soil, and mineral soil. As noted previously, there are two categories of plant detritus in the boreal forest: litter and standing-dead trees. This latter category represents a major carbon reservoir because the boles of many coniferous tree species may remain erect for up to four or five decades after a fire due to low rates of decomposition. Finally, although the relative sizes of most of the major carbon pools in the boreal forest have been quantified (see Chapters 13 and 14), the amount of carbon entering lake/ocean/river sediments has not. In particular, charcoal deposits resulting from fires may represent an important long-term carbon sink (Kuhlbusch and Crutzen 1996).

The major processes and pathways controlling carbon exchange between reservoirs are also presented in the conceptual diagram in Figure II.1. Although the relative sizes of the major carbon pools are well understood at this time, the rates of exchange between the different carbon pools are not. To reduce the uncertainties in the rates of exchange of greenhouse gases between the boreal forest and the atmosphere, it is logical to focus on those exchange pathways and processes affecting the largest pools (e.g., plant detritus, organic soil, mineral soil). Given this perspective, it is easy to see that microbial respiration and fire play an important role in the boreal forest carbon budget because these are the only processes that result in transfer of carbon from plant detritus, organic soil, and mineral soil to the atmosphere.

When considering the processes and pathways controlling the exchange of carbon, the temporal dimension needs to be recognized. The net rate of carbon accumulation in the living biomass pool depends on variations in patterns of plant succession (which vary over yearly to decadal time scales) and variations in the rates of net primary productivity (which vary at daily to weekly to yearly time scales). However, accumulation of large amounts of carbon in the organic soil pool is a relatively slow process, occurring over time scales of centuries to millennia (Harden et al. 1997). This longer-term accumulation is controlled by (1) rates of input from dead plant material from moss/lichen and living biomass pools (which vary on the same time scale as net primary production); (2) rates of decomposition (which vary on time scales of weeks to years to decades to centuries, depending on the depth, composition, temperature, and moisture of the organic soils); (3) rates of output from runoff and cryogenic and biogenic mixing (which vary on decadal to century time scales); and (4) rates of disturbance of fire (which occur on decadal to century time scales).

Because fire is a nearly instantaneous process, it can cause large losses of carbon from the ground layer (20–>100 t C ha^{-1}) within hours or days (Chapter 10). Because deeper organic soils take centuries to millennia to accumulate, fires consuming large amounts of organic soil represent a net loss of carbon to the atmosphere, a fact overlooked in most terrestrial carbon analyses.

Factors Controlling Carbon Exchange in Boreal Forests

Figure II.2 illustrates the three major factors directly controlling the carbon exchange processes identified in Figure II.1. The thermal and hydrological regime of the soil layer has the strongest influence on the processes controlling carbon cycling in the boreal forest. Physiography, soil type, and climate also have strong direct controls on the carbon exchange processes but to a lesser extent than the ground thermal/hydrological regime.

Average annual temperatures of less than 0°C lead to the formation of permafrost in boreal forest ecosystems. The presence of frozen soil during the grow-

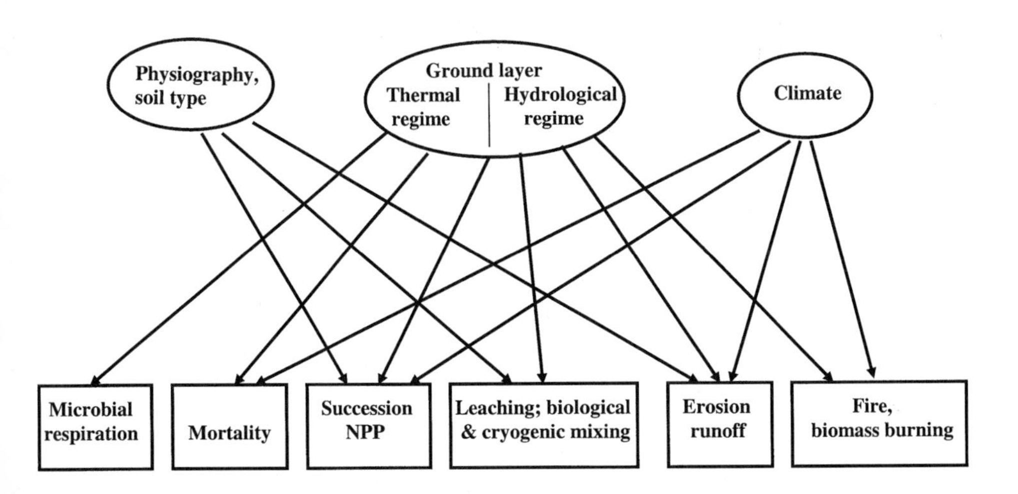

Figure II.2. Major direct influences on processes controlling carbon exchanges in boreal forests.

ing season not only limits the available rooting depth for plants but results in extremely wet or saturated soil conditions in many locations. These cold wet ground conditions have profound influences on metabolic processes that control microbial and plant respiration, plant mortality, the uptake of water and nutrients (which controls primary production and succession), and biological mixing in organic and mineral soils. They also exert strong controls on the physical processes of leaching, cryogenic mixing, erosion, runoff, and fire. A thorough review of the influence of these various factors on boreal forest processes is presented by Bonan and Shugart (1989).

In terms of carbon cycling in the boreal forest, not only are the direct controls illustrated in Figure II.2 important but so are indirect controls or feedback mechanisms, especially those related to fire. There are feedback mechanisms between the carbon reservoirs and the carbon exchange processes, as well as feedback mechanisms related to the effects of fire on the other controlling processes illustrated in Figure II.2.

Feedbacks between the different carbon reservoirs and the processes resulting in carbon exchange are illustrated in Figure II.3 and include changes in (1) plant composition during succession that influences rates of net primary productivity and direct carbon exchange with atmosphere; (2) plant composition and structure that influence energy and water exchange with the atmosphere through changes in (a) shading of the ground layer, (b) latent heat exchange through evapotranspira-

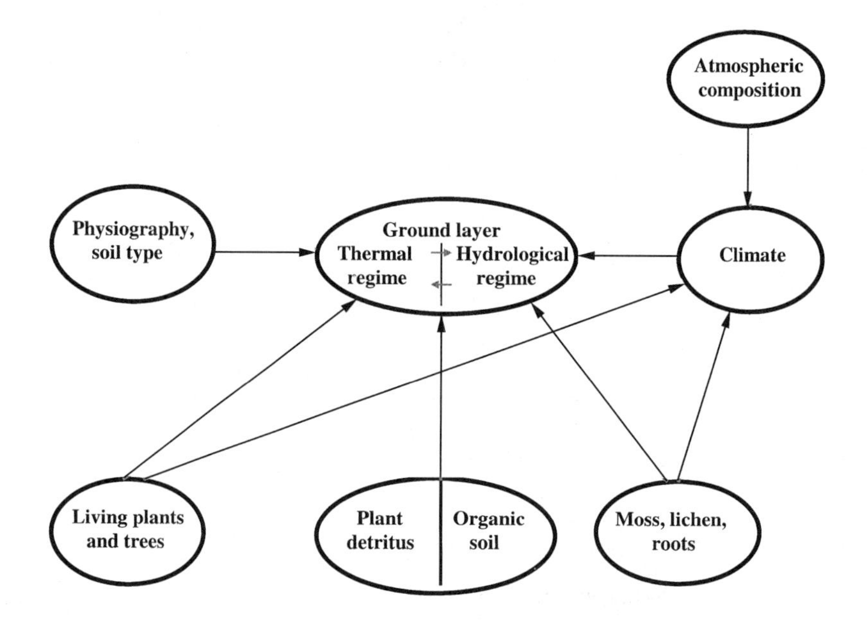

Figure II.3. Major indirect influences or feedbacks influencing carbon exchanges in boreal forests.

tion, and (c) sensible heat exchange through changes in the thermal conductivity of the vegetation layer; (3) plant structure and composition that directly influences patterns of fire and biomass burning; (4) moss community composition that results in changes of the C/N ratio of the litter, which directly influences rates of decomposition of plant material; (5) organic soil and plant detritus structure and composition that directly influence (a) rates of decomposition, (b) patterns of leaching and biological mixing, and (c) patterns of fire and biomass burning; and (6) organic soil, plant detritus structure and composition, and moss structure and composition that influence exchanges of energy and water between the land surface and atmosphere, including changes in the thermal conductivity and evaporation characteristics of the ground layer.

Figure II.3 diagrams the interactions between the different processes that control carbon exchange or the feedbacks between different carbon pools and these processes. This figure shows that the ground thermal/hydrological regime is strongly controlled by climate, site physiography (slope, aspect, and elevation), and soil type.

Of particular importance are the indirect effects that fire has on the ground thermal regime through its influence on characteristics of the different carbon pools. The direct release of atmospheric greenhouse gases from fire increases the capacity of the atmosphere to absorb energy being emitted by the earth's surface and thus has a direct influence on climate. Fire-induced changes in living mosses, lichens, and plants directly influence climate because of the role they play in the exchange of water and energy between the land surface and the atmosphere. Fire further influences the surface energy budget by (1) increasing the amount of solar energy reaching the earth's surface (through reducing the shadowing capacity of the plant canopy); (2) decreasing the albedo of the earth's surface; (3) depending on the severity of the fire, either increasing or decreasing the thermal emissivity of the land surface; and (4) through the consumption of organic soil, reducing the insulating capacity of the earth's surface. The net effect of fire is to significantly increase the ground temperature, reduce the amount of permafrost, and reduce soil moisture. In turn, these changes result in increases in heterotrophic respiration (Chapter 11) and strongly influence patterns of forest succession (Chapter 12).

Section Overview

The chapters of this section describe in more detail specific processes that are important in understanding the role of fire in carbon cycling in the boreal forest. In Chapter 7, Bourgeau-Chavez and associates describe the different boreal forest types found in North America and the role fire plays in their ecology. Shvidenko and Nilsson discuss the extent and role of fire in the Russian boreal forest in Chapter 8. This chapter is based largely on a review of the Russian scientific literature that is not widely distributed and therefore represents a unique assessment of research in this country. Another source of information on Russian forest fire research is Goldammer and Furyaev (1996). In Chapter 9, Campbell and Flannigan present an overview of using paleorecords derived from lake-bottom sediments to study historical patterns of fire in the North American boreal forests. Kasischke and co workers discuss factors controlling patterns of biomass burning and fire severity in Alaskan boreal forests in Chapter 10. These studies are important not only for estimation of direct carbon flux from fires (Chapter 21) but also for understanding the indirect effects of variations in fire severity. In Chapter 11, Richter and colleagues present the results of a study illustrating the influence of fire on soil respiration in three different Alaskan boreal forest types. Finally, in Chapter 12, Kasischke and associates discuss the effects of fire severity on the long-term patterns of forest succession in the Alaskan boreal forest.

References

Bonan, G.B., and H.H. Shugart. 1989. Environmental factors and ecological processes in boreal forests. *Annu. Rev. Ecol. Syst.* 20:1–28.

Field, C.B., M.J. Behrenfeld, J.T. Randerson, and P. Falkowski. 1998. Primary production of the biosphere: integrating terrestrial and oceanic components. *Science* 281:237–240.

Foley, J.A., I.C. Prentice, N. Ramankutty, S. Levis, D. Pollard, S. Sitch, and A. Haxeltine. 1997. An integrated biosphere model of land surface processes, terrestrial carbon balance, and vegetation dynamics. *Global Biogeochem. Cycles* 10:603–628.

Goldammer, J.G., and V.V. Furyaev, eds. 1996. *Fire in Ecosystems of Boreal Eurasia.* Kluwer Academic Publishers, Dordecht, The Netherlands.

Harden, J.W., K.P. O'Neill, S.E. Trumbore, H. Veldhuis, and B.J. Stocks. 1997. Moss and soil contributions to the annual net carbon flux of a maturing boreal forest. *J. Geophys. Res.* 102:28,805–28,816.

Kasischke, E.S., N.L. Christensen, Jr., and B.J. Stocks. 1995. Fire, Global warming and the mass balance of carbon in boreal forests. *Ecol. Appl.* 5:437–451.

Kuhlbusch, T.A.J., and P.J. Crutzen. 1996. Black carbon, the global carbon cycle, and atmospheric carbon dioxide, pp. 160–169 in J.L. Levine, ed. *Biomass Burning and Climate Change. Vol. 1. Remote Sensing, Modeling and Inventory Development, and Biomass Burning in Africa.* MIT Press, Cambridge, MA.

Kurz, W.A., and M.J. Apps. 1993. Contribution of northern forests to the global C cycle: Canada as a case study. *Water Air Soil Pollut.* 70:163–176.

Kurz, W.A., M.J. Apps, T.M. Webb, and P.J. McNamee. 1991. The contribution of biomass burning to the carbon budget of the Canadian forest sector: a conceptual model, pp. 339–344 in J.S. Levine, ed. *Global Biomass Burning: Atmospheric, Climatic and Biospheric Implications.* MIT Press, Cambridge, MA.

McGuire, A.D., J.M. Melillo, D.W. Kicklighter, Y. Pan, X. Xiao, J. Helfrich, B. Moore III, C.J. Vorosmarty, and A.L. Schloss. 1997. Equilibrium responses of global net primary production and carbon storage to doubled atmospheric carbon dioxide: sensitivity to changes in vegetation nitrogen concentration. *Global Biogeochem. Cycles* 11:173–189.

Raich, J.W., and C.S. Potter. 1995. Global patterns of carbon dioxide emissions from soils. *Global Biogeochem. Cycles* 9:23–36.

Randerson, J.T., M.V. Thompson, T.J. Conway, I.Y. Fung, and C.B. Field. 1997. The contribution of terrestrial sources and sinks to trends in seasonal cycles of atmospheric carbon dioxide. *Global Biogeochem. Cycles* 11:535–560.

Running, S.W., and R.R. Nemani. 1988. Relating seasonal patterns of the AVHRR vegetation index to simulated photosynthesis and transpiration of forests in different climates. *Remote Sens. Environ.* 24:347–367.

Sellers, P.J., F.G. Hall, R.D. Kelly, A. Black, D. Baldocchi, J. Berry, M. Ryan, K.J. Ranson, P.M. Crill, D.P. Lettenmaier, H. Margolis, J. Cihlar, J. Newcomer, D. Fitzjarrald, P.G. Jarvis, S.T. Gower, D. Halliwell, D. Williams, B. Goodison, D.E. Wickland, and F.E. Guertin. 1997. BOREAS in 1997: experiment overview, scientific results, and future directions. *J. Geophys. Res.* 102:28,731–28,769.

Smith, T.M., and H.H. Shugart. 1993. The transient response of carbon storage to a perturbed climate. *Nature* 361:563–566.

Smith, T.M., H.H. Shugart, G.B. Bonan, and J.B. Smith. 1992. Modeling the potential response of vegetation to global climate change. *Adv. Ecol. Res.* 22:93–116.

Thompson, M.V., J.T. Randerson, C.M. Malmstrom, and C.B. Field. 1997. Change in net primary production and heterotrophic respiration: how much is necessary to sustain the terrestrial carbon sink? *Global Biogeochem. Cycles* 10:711–726.

Wein, R.W., and D.A. MacLean, eds. 1983. *The Role of Fire in Northern Circumpolar Ecosystems.* John Wiley & Sons, New York.

7. Distribution of Forest Ecosystems and the Role of Fire in the North American Boreal Region

Laura L. Bourgeau-Chavez, Martin E. Alexander, Brian J. Stocks, and Eric S. Kasischke

Introduction

The boreal biome is characterized by several limiting factors that not only restrict species diversity and levels of primary productivity but also control specific processes in boreal ecosystems. These limiting factors include short growing seasons, low soil temperatures, extreme seasonal fluctuations in day length and temperature, continuous and discontinuous permafrost conditions, and the predominance of fire. This chapter synthesizes research results, field experience, and literature reviews to assess the influence of these abiotic factors on the boreal forests of North America. It provides information on the geology, ecology, soils, vegetation, and fire regimes of these forests.

Physical, Climatic, and Biological Setting

The boreal forest encompasses a circumpolar belt from approximately 47 to 70° north latitude comprising an area of 14.7 million km^2, 11% of the earth's land surface (Fig. 2.1). This biome lies to the north of the temperate forest and to the south of the vast arctic tundra. From Eurasia to North America, the boreal forest biome is dominated by coniferous tree species. Boreal forests of Eurasia include areas of Russia, China, Norway, Sweden, and Finland. In North America, the boreal forest covers an east-west region from Newfoundland to Alaska, with a northern limit that varies from 68° N latitude in the Brooks Range of Alaska to 58° N latitude on the western edge of Hudson Bay (LaRoi 1967). Thermal properties of large air masses influence the northern edge of the boreal forest. Humboldt observed in 1807 that certain isotherms parallel boundaries of the earth's terrestrial biomes. The boreal biome is bounded to the north by the July 13°C isotherm with deviations in montane and maritime regions (Larsen 1980). The southern limit is less distinct. In central and eastern Canada, the southern limit is the July 18°C isotherm. In areas of western Canada where precipitation is heavy, the southern border is north of the 18°C isotherm (Larsen 1980). Soil moisture appears to be a controlling factor in determining the southern border of the boreal zone (Hogg 1994), where a mosaic of coniferous and hardwood forest types occurs. The predominance of one forest type over the other is strongly influenced by the availability of soil moisture (Fosberg et al. 1996).

The physiognomy of the boreal forest biome is uniform from Eurasia to North America due to the dominance of conifers and similarity of their physical structure. From an aerial perspective, single-layered canopies of conifers underlain by low shrubs, herbaceous vegetation, and a carpet of mosses and lichens are common landscape features in Russia, Alaska, and Canada. However, closer observations reveal that a great degree of ecological diversity exists on both regional and local scales. The types of boreal forest that develop on a particular landscape are highly dependent on local climate, physiography, landform, soil, permafrost, and fire regime, as discussed in the following sections.

Geology and Soils

The bedrock of the North American boreal forest ranges from Precambrian granitic and gneissic rocks of the Canadian Shield in east and central Canada to Paleozoic and Cretaceous sedimentary rocks of southwest Hudson Bay, south central and western Canada, and Alaska (Elliot-Fisk 1988). Extensive glaciation of the North American boreal forest zone occurred as recently as the late Pleistocene (10,000 years ago). However, most of the Alaskan interior region was never covered by the continental glaciers of the Pleistocene. As a result, most soils of this region are not derived from glacial till and outwash material as is true in much of northern North America. Rather, most upland soils were formed from loessal deposits and, in floodplains, from alluvial deposits.

Canada's soils are primarily glacially derived and include Regosols, gleys, peats, brown earth soils, gray and brown wooded soils, and Lithosols (Larsen 1980). Podzol soils characterized by a bleached gray A_e horizon over a brown to reddish brown B horizon are frequently found in boreal forests. The distinct color difference found in podzols is due to the movement of iron, aluminum, and organic material from the A (bleached gray) to B horizon (brown). Movement of these minerals occurs because of the climate (low temperatures, high precipitation, and low evaporation), the chemical composition of the substrate, vegetation composition, and topography. Podzolization is particularly intense on acid sandy glacial deposits and weathered sandstone. These soil types are common on the Canadian Shield.

Climate and Permafrost

In general, the boreal forest biome is characterized by long cold winters and short, relatively cool summers, with low solar insolation compared with tropical and temperate regions. Local climatic conditions in North American boreal forests vary along gradients from north to south, and east to west. From north to south, the temperatures generally increase. The climate ranges from dry with extreme annual temperature variations in the west (interior Alaska) to a relatively warmer, wetter, maritime climate of eastern Canada (Bonan and Shugart 1989). Generally, from south to north there is a decline in prominence of hardwoods, an increase in conifers, and a decrease in tree heights and stand densities (Rowe and Scotter 1973). Lowland bogs and muskegs occur throughout the boreal region with little variability (Hare 1954).

Areas that are influenced by mountains or oceans have a greater climatic variation with more local differences than broad flat areas. Local climate is also strongly influenced by topography, slope, aspect, and edaphic factors (soil moisture and soil temperature), as well as the albedo, temperature, and emissivity of surface materials.

Although terminology varies from scholar to scholar (Hare 1954; Rowe 1972; Larsen 1980; Payette 1992), all describe north-to-south climatic subzones of boreal Canada as (1) boreal-tundra ecotone; (2) open boreal woodland; (3) main boreal forest; and (4) boreal-mixed forest ecotone. The boreal-tundra ecotone consists of patches of lichen woodland (open canopy forest) intermixed with tundra. Even at higher latitudes dominated by tundra, isolated individual trees sometimes survive, as do patches of forest in warmer protected sites such as valleys. Open boreal woodland (i.e., taiga) consists of open canopied forests dominated by conifers, primarily black spruce, and underlain by lichen. In North America, the main boreal forest is dominated by closed canopy coniferous forests of spruce, fir, and/or pine. Open forests occur in this zone on dryer sites. At the southern limit of the boreal forest, the coniferous forests of the boreal region gradually give way to prairies and temperate forests (boreal-mixed forest ecotone).

The north-to-south pattern of climatic subzones described above is not present in Alaska or the Yukon Territory due to the mountainous terrain. It is also not maintained in northern Ontario because of the marine deposits of Hudson Bay (Hare 1954) lowlands, where bogs, fens, and muskegs dominate.

Permafrost is the thermal condition of soil when the temperature is less than 0°C for more than 2 years (Brown 1970). In the North American boreal forest, permafrost conditions range from continuous in the open boreal forest to discontinuous in interior Alaska and western Canada to permafrost-free in much of eastern Canada (Fig. 2.1). In the discontinuous zone, the presence or absence of permafrost is determined by a variety of factors that contribute to soil temperature. Cold ground temperatures reduce decomposition rates, allowing buildup of organic material. Some locations, such as north-facing slopes, are predisposed to low ground temperatures simply because of low amounts of insolation. In most cases, north-facing slopes will have a well-developed permafrost layer, whereas south-facing slopes will be permafrost-free (Van Cleve et al. 1986). In areas where temperature conditions are marginal, such as east and west slopes and flat terrain, the presence of permafrost is dependent on other site factors. These include overstory and ground-cover vegetation, soil moisture, and microclimate. Low depressions, where temperature inversions create cold air temperatures, will usually have underlying permafrost. Poorly drained areas, slope bases, and hillside saddles will develop permafrost from the colder temperatures that result from wet ground conditions. Shading of the sun's radiation and protection from wind by trees can also create conditions favorable for permafrost development. The presence of permafrost itself causes cooling of the ground by perching water, creating a deep and well-developed permafrost layer at sites undisturbed for long periods of time (Bonan and Shugart 1989).

Coniferous tree species well adapted to fire dominate sites with permafrost; therefore, wildfire is common in permafrost areas. Disturbance of the organic layer by fire often causes permafrost soils, in particular discontinuous permafrost soils, to change. The removal of sun-shading tree canopies and insulating moss layers as well as the charring of the ground surface by fire results in changes in the ground-surface energy exchange regime (Brown 1983). Increased soil temperature results from reduced evapotranspiration and decreases in the surface albedo. Warmer soils result in permafrost melting and thus changes in hydrology (Brown 1983; Viereck 1983). Permafrost melting along with decreased transpiration rates (due to tree death) often causes an increase in ground moisture for several years after a fire (Dyrness and Norum 1983; Viereck 1983). However, in some cases soils have been observed to become drier postburn (Kershaw and Rouse 1971; Swanson 1996). The overall effects of fire on soil moisture are a function of fire severity, soil type, and the presence or absence of permafrost. In addition to changes in soil moisture, fire often results in a deepening of the active layer, nutrient mineralization, and increased soil microbial activity, which collectively create favorable growing conditions for the establishment of seedlings.

Major Tree Species and Factors Controlling Their Distribution

Although thousands of canopy tree species exist in tropical regions, and hundreds exist in temperate forests, fewer than 50 tree species survive in boreal regions because of the harsh climate and cold soils. Ten tree species dominate the North American boreal forest, and 14 dominate boreal Eurasia (Nikolov and Hemisaari 1992).

The predominant conifer species of North American boreal forests include black spruce (*Picea marianna* [Mill.] BSP.), white spruce (*Picea glauca*

[Moench] Voss.), balsam fir (*Abies balsamea* L.), jack pine (*Pinus banksiana* Lamb.), lodgepole pine (*Pinus contorta* L.), and alpine fir (*Abies lasiocarpa* [Hook.] Nutt.). The deciduous species include trembling or quaking aspen (*Populus tremuloides* Michx.), balsam poplar (*Populus balsamifera* L.), and white or paper birch (*Betula papyrifera* [Marsh.]). There is one deciduous conifer found in the North American boreal forest, tamarack (*Larix laricina* [DuRoil] K. Kock).

Black spruce, white spruce, and tamarack are found throughout the North American boreal forest, occurring from Alaska to Newfoundland. They all have a northern treeline reaching into northern Quebec, the northern Mackenzie River Valley in the Northwest Territories, and the southern foothills of the Brooks Range in Alaska.

The relative abundance of the coniferous species varies from region to region. Black and white spruce dominate in Alaska and central Canada. Balsam fir (nonexistent in boreal Alaska) and tamarack are predominant in central and eastern Canada. Lodgepole pine (primarily in British Columbia) and alpine fir are montane species, and jack pine is a continental species (all nonexistent in boreal Alaska). Alpine fir occurs on north- and east-facing mountain slopes and ridges up to timberline. It is found as far north as central Yukon Territory, extends southeast to central Oregon, and south to southern New Mexico and southern Arizona (Harlow et al. 1996). Jack pine extends from Quebec to Alberta and from the Mackenzie River in the north to its southern limit in Michigan.

All the deciduous species are distributed throughout the North American boreal forest and are typically pioneer species, invading recently disturbed landscapes.

Feather mosses, sphagnum moss, and lichens are prevalent ground covers, often forming continuous mats beneath the tree canopies. These mats are the primary humus builders; they tie up nutrients, insulate cold soils, and have a strong role in succession (Heinselman 1981a; Viereck et al. 1986).

The distribution of all tree species are influenced by mean annual air temperature, length of growing season, and annual potential evapotranspiration (Hare 1954). Species are restricted to regions that have climatic conditions they can tolerate and where they do not receive competition from better-adapted species.

Balsam fir dominates on well-drained sites in southeastern Canada. It is usually a dominant species in old-growth forests and occurs most frequently in the moister regions that have low fire frequencies, such as Newfoundland, New Brunswick, and the Laurentian Mountains of Quebec (Halliday and Brown 1943). In south central and southwestern Canada, well-drained sites are dominated by white spruce. Black spruce and tamarack dominate the wet and peaty soils of this region. Jack pine is an associate on dry sites except in the extreme east where its distribution ends (Rowe and Scotter 1973). In the east, frequent fires favor black spruce, paper birch, and jack pine because balsam fir is not well adapted to fire. In the west, sites recently disturbed by fire are inhabited by pioneer species such as aspen, jack pine, black spruce, and birch. White spruce is a late-successional species on upland sites in the west (Rowe and Scotter 1973). Northward, there is an increase in shallow cold soils, permafrost, peaty soils, and harsh climates that result in an increase in black spruce dominance.

The Cordillera region of northwestern Canada is dominated by black spruce, white spruce, lodgepole pine, and at higher altitudes, alpine fir. Lodgepole pine dominates recently burned sites. Alpine fir, like balsam fir, seems poorly adapted to invading recently burned sites. In the south, where hardwoods increase in abundance, the soils are typically deeper and more fertile. Here, black spruce loses its dominance, and the deciduous species of aspen and balsam poplar as well as white spruce predominate.

Several temperate tree species have their northern ranges in the boreal zone and are typically found scattered or in isolated patches. The temperate species sometimes found in the boreal zone include black cottonwood (*Populus trichocarpa*), yellow birch (*Betula alleghaniensis*), eastern white pine (*Pinus strobus*), red pine (*Pinus resinosa*), eastern hemlock (*Tsuga canadensis*), eastern white cedar (*Thuja occidentalis*), sugar maple (*Acer saccharum*), red spruce (*Picea rubens*), black ash (*Fraxinus nigra*), green ash (*Fraxinus pennsylvanica*), Manitoba (box elder) maple (*Acer negundo*), bur oak (*Quercus macrocarpa*), and white elm (*Ulmus americana*).

Landscape-Level Fire Characteristics in North American Boreal Forests

Fire is a natural ecosystem process in the boreal forest, and it has many vital functions (Wright and Heinselman 1973; Alexander and Euler 1981). In fire-dependent ecosystems, fire controls the distribution of plant species and communities and in turn regulates fuel accumulation at both the stand and landscape scales. Thus, the vegetation-type mosaic (i.e., stand age-class distribution and developmental stages) of a given region is, to a large extent, an expression of its fire regime. A fire regime refers to the kind of fire activity or pattern of fires that generally characterizes a given area. Some important elements of the fire regime include the fire cycle or fire interval, fire season, and the number, type, and intensity of fires (Merrill and Alexander 1987; Van Wagner and Methven 1980; Heinselman 1981b). Fire regimes in the North American boreal forest vary from short-interval crown fire/high-intensity surface fire regimes to very long-interval crown fire regimes. Regional fire climate, lightning incidence, physiography, and the balance between dry matter production and decay all determine the fire regime. As Van Wagner and Methven (1980) state: "For each vegetation type to be perpetuated by periodic fire, there will exist an optimum fire regime that best fulfills its ecological requirements."

The fire cycle constitutes the number of years required to burn over an area equal to the total area of interest (Van Wagner 1978). The random nature of fire starts means that some areas may not burn during a given fire cycle, whereas other areas may burn more than once. Natural fire cycles in the boreal forest have been determined to vary from about 40 years in interior Alaska (Yarie 1981) to about 500 years in southeastern Labrador (Foster 1983). As a result of attempted fire exclusion practices during the past 50–80 years, fire cycles in some regions of the North American boreal forest have lengthened considerably (Alexander 1980; Ward and Tithecott 1993), whereas others have maintained near-natural levels due to fire management policies (Stocks et al. 1996). Fire interval refers to the average number of years between the occurrence of fires at a given point on the landscape (Merrill and Alexander 1987). Mean fire intervals of 25–40 years have been recorded in some sections of the boreal forest (Carroll and Bliss 1982; Delisle and Dube 1983; Lynham and Stocks 1991).

Although fires exceeding 1 million ha have been recorded in the boreal forest (e.g., the 1950 Chinchaga River fire in British Columbia/Alberta, and the 1995 Horn Plateau fire, Northwest Territories), such occurrences are believed to be rare. Evidence from northwestern Ontario suggests that under natural conditions (i.e., prior to the implementation of fire control), some regions of the boreal forest experienced far more intermediate-sized fires (Ward and Tithecott 1993). This is contrary to the popular adage that 95% of the area burned is the result of 2–5% of the fires (this is a valid suppression era statistic). This is no doubt a reflection of differential burning potential in the fuel type mosaic.

The season or time of year during which a fire occurs is of significance in terms of plant phenology (Robbins and Meyer 1992) because soil and fuel moisture conditions and the likely degree of fuel consumption influence vegetative response following fire. Generally, the fire season in the North American boreal biome runs from April to September, with the number of fire starts and area burned peaking from June through August. This is largely due to lightning fire ignitions and a gradual seasonal drying trend in most years (Harrington 1982). Lightning fire occurrence increases significantly as one moves from the eastern to western regions of the boreal forest (Simard 1975; Stocks and Hartley 1979). Fires resulting from human causes are particularly prevalent in the spring following snow melt and prior to the flushing of ground vegetation and hardwood overstories. Forest floor consumption at this time of year is typically not high because of the recent saturation effects of melting snow. However, a combination of below-normal overwinter precipitation and low ground fuel moisture levels at the end of the previous season can lead to deep-burning ground fires in the spring (Lawson and Dalrymple 1996).

Although there are a number of descriptors of fire behavior (Albini 1993), fires are typically characterized by the fuels that are involved in the combustion process. Three broad types are commonly recognized: ground fire (Fig. 7.1a),[1] surface fire (Fig. 7.1b), and crown fire (Fig. 7.1d). Intermittent crown fires (Fig. 7.1c) reflect the transitional or developmental stage that takes place between pure surface fires below the crowning threshold and fully developed crown fires. A burned area generally exhibits all three types of fires, as well as unburned islands within the perimeter, which can constitute up to 10% of the total area. Patterns of burn depend on a host of factors including fuel moisture status, soil types, geology, alignment and extent of water bodies, and the existing fuel continuity.

For a fire to spread through a fuel complex, there must be sufficient fuel of appropriate size, arrangement, and dryness (Van Wagner 1983). One immediately

[1]Figure 7.1 will be found in the color insert.

evident impact of fire is the amount of fuel consumed, which in turn results in a dramatic reduction in the flammability of the forest. Relatively recent burns are commonly regarded by fire management practitioners as virtual barriers to fire spread.

The type of fire and its intensity have important ramifications in the management of unwanted wildfires that threaten life, property, and other values-at-risk (Alexander and Cole 1995). Ground fires and the vast majority of surface fires can be controlled by fire fighters using available equipment and resources such as hand tools, water delivery systems, earth-moving machinery (bulldozers), helicopters, and air tankers. However, despite advances in modern fire-fighting technology, crowning forest fires are nearly impossible to stop once initiated due to their high intensity levels. These fires continue to burn until the fuel source becomes exhausted and/or the winds subside. Both high-intensity surface and crown fires can generate airborne embers, which may be deposited up to 2 km down wind of the main advancing flame front. These embers, in turn, start new fires and thereby greatly complicate the task of containing or suppressing the entire fire.

Fire Behavior Characteristics of Boreal Fires

Frontal fire intensity, or energy release, is generally considered the most comprehensive quantitative measure of fire intensity and the best measure of a fire's resistance to control and is defined by Byram (1959) as

$$I = Hwr \tag{7.1}$$

where I is the fire intensity (kW m^{-1}), H is the net low heat of combustion (kJ m^{-2}), w is the amount of fuel consumed in the flaming or active combustion phase (kg m^{-2}), and r is the linear rate of fire spread (m sec^{-1}). Fire intensity defined in this manner is directly related to flame size and radiant heat. In contrast to fire intensity, fire severity refers to the combined effects of both flaming and smoldering combustion as manifested in various fire behavior characteristics (e.g., flame front dimensions including residence time, fire persistence, or total burn-out time). Byram's formula embraces the most fundamental features of a wildland fire from the standpoint of fire behavior; namely, that as a fire spreads, it consumes fuel and it emits heat energy. Fire intensity varies over an immense range, from about 10 kW m^{-1} (the threshold for fire spread by flaming combustion) to more than 100,000 kW m^{-1} for major conflagrations. Based on their physical fire behavior characteristics, boreal fires can generally be classified into three categories (Van Wagner 1983): (1) ground or subsurface fires in deep organic layers with frontal fire intensity levels less than 10 kW m^{-1}; (2) surface fires with intensities ranging between 100 and 6,000 kW m^{-1}; and (3) crown fires with intensities from 6,000 to more than 100,000 kW m^{-1}. Ground or subsurface fires spread very slowly (perhaps 1 m h^{-1} at most) through smoldering, with few or no visible flames (Wein 1983; Frandsen 1991). Surface fires in natural forest stands may spread very rapidly (seldom exceeding more than 6 m min^{-1}) and generate moderately high flames with intensities up to 5,000 kW m^{-1} or advance as a low, creeping flame front with intensities less than 500 kW m^{-1} (Norum 1982; Alexander and De Groot 1988; Stocks 1989; Alexander et al. 1991; Quintilio et al. 1991). Tree mortality varies widely depending on species and stand age. Crown fires spread as intense walls of flame extending from the ground surface up to two or three times the forest canopy height. Crown fires spread at rates of 1–6 km h^{-1} with intensities of at least 10,000–30,000 kW m^{-1} (Stocks 1987a,b; Stocks and Flannigan 1987, Alexander and Lanoville 1989; Stocks and Hartley 1995) but often approaching 100,000 kW m^{-1}. In general, surface fires must consume enough fuel (a function of fuel moisture levels) and generate sufficient intensity to involve the crown layer before crowning can begin. Convection, with some assistance from radiation is the principal means of vertical heat transfer from the surface to the crown layer (Byram 1959). Surface fires consume forest floor material and surface fuels such as shrubs and downed woody material. There is considerable overlap between intense surface fires and incipient crown fires in terms of intensity. Crowning generally results in complete defoliation of the canopy and widespread tree mortality. Incomplete crowning or partial consumption of the ladder or bridge fuels in a forest stand generally begins to occur at intensities of about 4,000 kW m^{-1} but depends on various crown fuel properties such as moisture content, canopy height, and density (Van Wagner 1977; Alexander 1998).

Fuel consumption can vary considerably both within and between boreal fires. In general, however, boreal crown fires consume 20–30 tons of fuel (oven dry weight) per hectare (Stocks 1991; Stocks and Kauffman 1997). On average, roughly two-thirds of the fuel consumed involves the forest floor (moss, litter, and humus), and the remaining one-third consumed involves crown fuels (needles and fine twigs).

Once crowning occurs, the fire has access to the ambient wind field that greatly accelerates the fire's spread rate until an equilibrium rate of spread is achieved. The surface and crown phases of the fire advance downwind as a linked dependent unit, with the crown fire dependent on the surface fire for sustained energy. Crown fires can be intermittent with trees torching individually or in clumps (reflecting the transition between surface fire and fully developed crown fires). They can also be active with solid flame development in the crowns (Van Wagner 1977). Active crown fires are the most common in Canada and Alaska. Intermittent crown fires generally spread at 5–10 m min^{-1}, whereas fully developed, active crown fires can attain spread rates greater than 100 m min^{-1} (Stocks and Kauffman 1997). Although the fast-spreading active crown fires that dominate the North American boreal landscape require sufficient fuel availability (low fuel moisture contents), they are primarily driven by strong winds and are aided by both short- and long-range spotting of fire brands ahead of the flame front.

The high fuel consumption rate in active boreal crown fires (an order of magnitude greater than that of grassland fires) typically continues for several hours a day during a burning period. Active crown fires result in the development of sustained, very-high-energy release rates that generate towering convection columns.

These columns generally reach heights of 10 to 12 km and directly enter the free troposphere.

In general terms, the requirements for extreme fire behavior in a forest environment (e.g., the occurrence of a "blow-up" or high-intensity crown fire run) are fairly well known given an ignition source, dry and plentiful fuels, strong winds and/or steep slopes, and an unstable atmosphere (Byram 1954). One very common misconception concerning the development of large high-intensity crowning forest fires is the notion that a prolonged or severe drought is a necessary prerequisite, but this is not necessarily the case. Stocks and Walker (1973) presented evidence for the propensity for crown fires to occur during the spring of the year throughout much of the North American boreal forest. The periodic establishment and breakdown of the atmospheric upper-level ridge that occurs throughout the fire season is especially conducive to the process of fuel drying, ignition by lightning, and rapid fire growth resulting from strong winds (Nimchuk 1983).

The Canadian Forest Fire Danger Rating System (Stocks et al. 1989) has two major subsystems, the Canadian Fire Weather Index system (Van Wagner 1987) and the Canadian Forest Fire Behavior Prediction (FBP) system (Forestry Canada Fire Danger Group 1992). Based on a series of empirical predictive equations developed through an extensive experimental burning and wildfire documentation program over many years, the FBP system predicts fire behavior in a number of forest types and is used throughout Canada and Alaska. Accurate prediction of spread rates, fuel consumption, and intensity levels gives fire managers using the FBP system the capability to anticipate fire behavior in many boreal fuel types, including jack pine, lodgepole pine, black spruce, aspen, balsam fir, and boreal mixed woods.

Fire Ecology of Major Boreal Tree Species

Historically, wildfire seems to have always dominated the boreal zone. The prevalence of wildfire has lead to adaptations to fire (e.g., serotinous cones) for many boreal tree species (Clark and Richard 1996). Of the ten dominant boreal tree species in North America, only balsam fir and alpine fir are not well adapted to fire. These two species are typically found in old-growth forests where fire frequency is low.

In studying functional adaptations of plants to fire, Rowe (1983) categorized boreal species based on strategies and means of reproduction. The five categories identified by Rowe (1983) include invaders, evaders, avoiders, resisters, or endurers. A species can belong to more than one category if they exhibit more than one type of adaptation.

Invaders are pioneer species inhabiting recently disturbed landscapes. They have wind-disseminated seeds whose viability is short-lived. Invaders include paper birch, white spruce, and trembling aspen.

Avoiders are shade-tolerant species that invade forests later in the successional chronosequence. From this perspective, white spruce is classified as an avoider, as is balsam fir.

Endurers are species that vegetatively reproduce after fires. They have shallow to deeply buried vegetative parts (roots and tree trunks) that often survive fire and are capable of sprouting. Endurers include trembling aspen and paper birch

Evaders are tree species whose seeds are both long-lived and are stored in places protected from fire (e.g., serotinous cones). Jack pine, lodgepole pine, and black spruce are all evaders.

Finally, resisters are those species that can survive low-intensity ground fires. Jack pine and lodgepole pine are resisters once their canopies are high enough above the ground to avoid crown fires.

Postfire patterns of succession on a specific site are closely linked to prefire vegetation patterns, neighboring vegetation, fire regimes, and permafrost conditions. Decomposition rates are often low in boreal forests and result in a buildup of organic matter and a tie up of nutrients in the ground layer. The depth of burn of the organic ground layers influences the type of regeneration that occurs on a given site. Light burns result in high root sprouting and other forms of vegetative reproduction. By contrast, deep burns kill roots but create favorable seedbeds. Thus, high levels of seedling development are common in deeply burned landscapes.

Fire plays a key role in nutrient cycling of boreal forests, both directly and indirectly. The burning of organic matter results in direct nutrient mineralization. The aftereffects of fire, including increased soil temperature and increased microbial activity, result in heightened nutrient release through organic matter decomposition for several years postburn. These direct and indirect processes promote favorable conditions for the establishment of seedlings and increased site productivity, especially in regions underlain by permafrost.

Different fire cycles result in different adaptations to fire (Rowe 1983). The shorter fire cycles typically found on drier sites favor endurers such as trembling aspen. Intermediate fire cycles favor resisters and evaders such as jack pine, lodgepole pine, and black spruce. Very long fire cycles occurring on moist sites favor avoiders such as balsam fir, alpine fir, and white spruce. In general, shade-intolerant pioneer species such as aspen, jack pine, and lodgepole pine are replaced in the absence of fire by shade-tolerant species such as black spruce, white spruce, balsam fir, or alpine fir (Rowe 1983). The first of these species, black spruce, is both an early and late successional species. Fire in black spruce will generally lead to regeneration of black spruce; however, in some regions such as the Cordillera it will often be replaced by lodgepole pine, and on dry sites in central Canada by jack pine. Instances have been found in Alaska where a severe fire that removes most of the organic soil allows aspen to invade black spruce forests (Chapter 12). Fire on a site dominated by jack pine will typically regenerate to jack pine, fire in lodgepole pine will regenerate to lodgepole pine and fire in aspen will regenerate to aspen. By contrast, fire in white spruce, balsam fir, or alpine fir will regenerate to willow, aspen, alder, jack pine, black spruce, lodgepole pine, or a combination of two or more of these depending on the seed source, climate, and site conditions.

The typical postfire successional patterns of various forest species are described below. Although a generalized discussion of successional trends is provided, the actual fire ecology and postfire regeneration for a particular site will vary depending on climate, physiography, landform, season of burn, age of preburn trees, neighboring ecosystems, postfire weather patterns, and other factors.

Aspen

Aspen is a short-lived, shade-intolerant, clonal species that frequently reproduces vegetatively after fire. Aspen seedlings can become established and grow in sites with exposed mineral soils (see Chapter 12, Fig. 12.5a). Despite a need for fire, aspen stands do not ignite easily and require specific site and climatic conditions for fires to occur (Wright and Bailey 1982). Aspen forests are most flammable in the spring and late summer to early fall during leaf-off conditions (Duchesne et al., in press). At these times, the surface vegetation is dormant and dry, resulting in higher flammability of the fuel matrix. In general, fires in younger aspen stands are of low intensity. Crowning fires do not occur unless there is sufficient dead fuel on the forest floor; therefore, older stands with higher fuel accumulations on the forest floor have higher-intensity fires. Fire frequencies of 3–15 years were common in aspen parklands of the west during presettlement times, resulting in shrublike cover in the understory (Duchesne et al., in press). In the absence of fire, aspen is replaced by shade-tolerant conifers or grasses and shrubs, depending on the seed source (Krebil 1972).

Due to its short life span, favorable adaptations to fire, and shade intolerance, pure stands of aspen are perpetuated only by higher fire frequencies. In the absence of fire or under low-intensity fires that do not kill trees growing in the understory, aspen is succeeded by different conifer species. Aspen is a pioneer species in a variety of forest types from white spruce to jack pine to balsam fir.

Jack Pine

Jack pine is a short-lived, shade-intolerant species that is succeeded by shade-tolerant species such as black spruce or white spruce in western Canada and balsam fir or white cedar in eastern Canada (Fowells 1965). Wildfire has a definite ecological role in the succession of jack pine forests. It has been suggested that jack pine would disappear from the boreal forest without fire (Cayford and McRae 1983), because in the absence of fire, organic soils deepen (resulting in cooler, wetter conditions that are not conducive to establishment of jack pine seedlings) and shade-tolerant species become established in the understory.

Regeneration of jack pine is greatest after high-intensity fires that consume most of the duff (litter and organic soil). Jack pine foliage is highly combustible, and trees are killed by crown fires at any age (Cayford and McRae 1983). In the southern part of its range, surface fires of moderate intensity will kill only portions of stands, resulting in patchy multiaged stands (Heinselman 1981a). Although jack pine seedlings produce cones within 3–5 years after a fire, they do not produce sufficient seed crops to ensure reproductive success until the stands reach 10 years of age. Germination of seeds is greatest on bare mineral soil.

The Canadian FBP system recognizes two jack pine fuel types: (1) mature jack pine characterized by pure but more open stands that have a closed crown, high canopy base, light litter levels, and deep organic layers with feathermoss cover; and (2) immature jack pine, which is a pure dense stand undergoing natural thinning with much dead standing stems and downed woody fuel, shallow organic soil layers, and little moss cover (Johnson 1992). Van Wagner (1977) hypothesized that young jack pine forests are more susceptible to crowning than mature stands, and subsequent field studies by Stocks (1987a, 1989) found this to be true.

Jack pine, lodgepole pine, and aspen typically have high crown bases and shed lower dead branches allowing survival in low-intensity surface fires. Crown base heights vary not only by species and tree age but also by site quality and stand density (Johnson 1992). In some mature jack pine and lodgepole pine forests of western Canada, low-intensity surface fires occur at 25-year intervals, and whole stands are not killed because the crown base heights are high enough to avoid crown fires and crown death

At some sites, jack pine forests will be invaded early in the successional chronosequence by other species (Heinselman 1973; Day and Woods 1977). At sites where fire does not remove all the organic soil, pine regeneration is generally low. A lower success in pine germination during the first 15 years postburn results in the development of a denser herb and shrub layer. The absence of a dense jack pine thicket allows other species—including black spruce or balsam fir, white spruce, and paper birch—to establish themselves in the understory for the next 30 years. If fire remains absent from the site over the following 30 years, these species will become canopy codominants with jack pine. After 170 years, jack pine will likely vanish from the canopy.

Lodgepole Pine

Lodgepole pine has a very similar fire ecology to that of jack pine. However, it occurs over a wider range of sites—from mesic to flooded to dry exposed sites—and the fire regimes are thus highly variable. The most flammable lodgepole pine stands have an abundant lichen ground cover. Crown fires in these forest types are usually of low intensity unless there is a large accumulation of woody debris on the ground. This leads to stand-replacing, high-intensity fires. Young, dense lodgepole pine stands and overmature stands with an understory of shade-tolerant conifers have a higher crown fire hazard than moderately dense to open mature stands (Lotan et al. 1984). The most common type of fire occurring in lodgepole pine stands is low-intensity surface fires. This type of fire is common when surface fuels are low and canopy bases are high. Under severe drought conditions, high-intensity crown fires occur and burn very large areas. These high-intensity fires result in mortality of all canopy and understory trees and produce conditions highly favorable to regeneration of lodgepole pine.

Frequent fires in western North America have led to a widespread abundance of lodgepole pine forests. Lodgepole pine trees have cones with viable seeds at an age of less than 10 years. Fires not only release these seeds but also often result in the bare mineral soils that are the best substrate for seed germination. Lodgepole pine seedlings, with their abundant and rapid germination and growth, outcompete seedlings of other species.

Young, even-aged stands of lodgepole pine will usually be replaced by more shade-tolerant species between years 50 and 200 postburn (Lotan et al. 1984). However, light periodic surface fires may eliminate the shade-tolerant understory species, and when the canopy becomes open enough lodgepole pine will regenerate in the understory.

Typically, there is an initial herb and shrub stage immediately after a fire, which is short-lived and quickly progresses to dominance of lodgepole pine seedlings and saplings. By the time the pine matures, an understory of spruce and alpine fir develops. Usually by this stage a stand-thinning fire or stand-replacing fire occurs, and thus the dominance of lodgepole pine is maintained.

Black Spruce

Black spruce, white spruce, alpine fir, and balsam fir all retain lower dead branches and have low crown bases. Therefore, crown fires are common in stands of these species. Most fires in the black spruce forest type are high-intensity crown or surface fires that kill overstory trees (Viereck 1983). Black spruce is especially well adapted to high fire frequencies. The combination of low precipitation, ericaceous (and flammable) shrub layers, and persistent lower branches result in large intense fires in these ecosystems. The ericaceous shrub layers act as a fuel ladder to the dead branches on the trees, which, in turn, act as a fuel ladder to the canopy. Because of this fuel matrix, crown fires are common in this forest type.

Black spruce stands are found on both upland and lowland sites. In the southern part of its range, it occurs primarily on lowland sites with deep wet organic soils, and its associates include balsam fir (in the east) paper birch, tamarack, and trembling aspen (Hare 1954).

The Canadian FBP system recognizes two types of black spruce forests in terms of fuel types: (1) spruce-lichen woodland types, which are open stands that grow on well-drained uplands with continuous mats of lichen ground cover and thin organic layers typically in the subarctic; and (2) boreal spruce types, which are defined as moderately well-stocked closed canopy black spruce stands with a carpet of feather or sphagnum mosses and deep organic layers on uplands or lowlands (Johnson 1992). The spruce-lichen woodlands are much more flammable than the boreal spruce type because the canopy is very open, allowing direct insolation and drying of the lichen forest floor. Fire tends to be more frequent in black spruce types of western Canada and Alaska, with frequency decreasing in eastern Canada due to a more humid climate.

Typically, some of the organic soil remains after a fire; however, severe drought conditions in late summer or very dry windy conditions may result in consumption of the entire organic layer and exposure of mineral soil. Consumption of the ground layers within a specific burn site can be highly variable, often creating a mosaic of deeply burned to unburned organic layers.

Black spruce trees produce cones as early as 10–15 years after fire with optimum seed production between 50 and 150 years. This allows the species to survive and regenerate under a wide range of fire frequencies.

The successional chronosequence found in Alaskan black spruce forests has been described by Van Cleve and Viereck (1981). Vegetative regeneration begins almost immediately in a burned black spruce boreal forest via sexual and asexual reproduction. Typically, during the spring following a fire, the charred landscape is invaded by mosses, liverworts, lichens, and herbaceous plants during the initial stage of postfire succession that lasts from 2 to 5 years, depending on the site and the burn severity (Van Cleve and Viereck 1981). During this stage, black spruce seedlings become established. In sites where severe fires have occurred and removed most of the organic soils, aspen and birch seedlings can also become established (see Chapter 12).

For moist black spruce-feathermoss forests, the next successional stage (lasting from 5 to 25 years) is dominated by shrubs that have resprouted. Black spruce trees begin to dominate the site between 25 and 50 years after a fire. At this time, an invasion of feathermosses (Pleurozium spp.) and sphagnum (Sphagnum spp.) occurs and the permafrost layer begins to rebuild to its prefire depth. Black spruce are slow growing on these sites due to cold, wet, shallow soils and low nutrient availability. During years 50 to 100 the stand thins to a lower stem density as the canopy closes. After 100 years, gaps begin to develop in the canopy, density decreases further, and the ecosystem begins to degrade. Without fire, the forest continues to decrease in density and further degrade.

For open-canopied black spruce-lichen forests, the initial stage of regeneration is dominated by mosses, lichens and sometimes grasses, with black spruce seedlings becoming established (Duchesne et al., in press). Between 10–60 years, the ground cover becomes dominated by fruticose lichens and low ericaceous shrubs (Black and Bliss 1978). By year 100, an open black spruce–canopy has formed.

Black spruce bogs are distributed throughout the North American boreal forest. Black spruce occurs in pure stands or in association with tamarack, eastern white cedar, or paper birch. Pure tamarack or eastern white cedar bogs also occur, but the black spruce dominated bogs are the most common type. Due to the wet conditions (high water table and high humidity) of bogs, they are not as susceptible to fire as other lowland and upland forests. Tamarack typically grows on wetter sites than black spruce, and the species is very resistant to fire. Very thick organic layers develop on bogs, sometimes greater than 1 m in depth. This moss layer generally has a high moisture content, which deters fire ignition. However, under severe drought conditions, bogs will burn. Drought conditions typically occur between July and September when the water table is at its lowest (Duchesne et al., in press). Fire in bogs can lead to increased paludification. When the transpiring canopy is removed, water tables rise and light levels increase, promoting establishment and growth of sphagnum mosses. Conditions are not conducive to germination of tree seedlings (Duchesne et al., in press), and thus a tundralike community may develop.

White Spruce

Upland mature white spruce forests accumulate deep organic layers that make them susceptible to fire. Crown fires are common due to the low canopy heights. Fires are most common in July and August due to lightning strikes (Heinselman 1981a). Fires are sometimes very large in this forest type, often covering thousands of hectares (Duchesne et al., in press).

Floodplain white spruce forests are less susceptible to fire because of higher moisture conditions in the ground layer. These sites typically have shrubs in the understory (Alnus and Salix spp.) that are not flammable during the growing season. These sites typically have fire cycles greater than 300 years (Rowe and Scotter 1973).

White spruce will sometimes regenerate postfire if fire occurs in late summer during seed dispersal. Seed regeneration is improved if the moss and lichen are burned during fire, exposing mineral soil. However, white spruce is often outcompeted by pioneer species that will dominate the site for several years. The sequence of regeneration in floodplain white spruce generally begins with mosses and herbs in the first 5 years followed by dominance of willow in years 6–30. Trembling aspen, white spruce, and paper birch typically establish during the first 5 years, and these deciduous species take over the canopy between years 31 and 45 and dominate for up to 150 years. In contrast, white spruce slowly develops in the understory and eventually dominates the canopy between 150 and 300 years.

Balsam Fir

Although balsam fir is not adapted to fire as well as other boreal tree species, fire does play a role in the ecology of forests where it is found. Balsam fir forests have a relatively lower fire hazard and thus a lower fire frequency than pines and spruces because of the more humid climates in the areas in which they dominate (Furyev et al. 1983). Fire occurs in balsam fir stands only after long droughts. During such times, crowning fires are common because balsam fir trees tend to retain their lower dead branches, which act as a fire ladder.

Fire hazard in balsam fir stands is greatly increased by insect defoliation such as spruce budworm outbreaks (Stocks 1987b). Defoliation kills trees and results in increased insolation and thus a drying of the surface layer. Spring fires have higher rates of spread in these dead forest types, and crowning is more likely than during summer fires. The lower summer fire hazard is due to the green understory that develops beneath insect-killed trees, which reduces the rate of spread. Spring fires rarely burn much of the organic layer because of high ground moisture.

Because balsam fir and alpine fir have low fire tolerance, postfire regeneration tends to lead to dominance of other tree species. Shade-tolerant fir only becomes a dominant canopy tree in old-growth stands in the absence of fire. A typical successional trend will be for aspen and/or paper birch to dominate the canopy immediately after fire, with white spruce or balsam fir developing an understory canopy that will eventually replace the hardwoods in dominance. Fire suppression and logging in eastern Canada have favored the establishment of pure balsam fir stands due to its shade-tolerant characteristic.

Summary

In this chapter, we have reviewed the major tree species found in the North American boreal forest and discussed those factors that control their distribution. We have paid particular attention to the role of fire in the ecology of these species and, in turn, how the characteristics of these species influence the fire regime. This understanding is particularly important when developing approaches to explore how fire influences carbon cycling at continental scales. Thus, this chapter contains background material that provides the foundation for studies presented in later chapters (Chapters 23 and 24).

Four of the chapters of this section explore additional aspects of the relationship between fire and ecosystem processes in the North American boreal forest. Chapter 9 presents an overview of paleoecology studies that show the long-term perspective of how fire has been a major factor in this region for at least 10,000 years. Chapter 10 explores how specific climatic and ecosystem factors influence patterns of biomass burning and direct release of carbon into the atmosphere from fires in Alaskan boreal forests. Chapter 11 presents results from a study that explores how fire influences soil carbon emissions from black spruce forests in Alaska. Finally, Chapter 12 explores the interactions between fire and climate on controlling longer-term patterns of forest succession in Alaskan boreal forests.

References

Albini, F.A. 1993. Dynamics and modelling of vegetation fires:observations, pp. 39–52 in P.J. Crutzen and J.G. Goldammer, eds. *Fire in the Environment: The Ecological, Atomspheric, and Climatic Importance of Vegetation Fires.* John Wiley & Sons, Chichester, UK.

Alexander, M.E. 1980. Forest fire history research in Ontario: a problem analysis, pp. 96–109 in *Proceedings of the Fire History Workshop.* General Technical Report RM-81. USDA Forest Service, Rocky Mountain Forest Range Experiment Station, Fort Collins, CO.

Alexander, M.E. 1998. Crown fire thresholds in exotic pine plantations of Australasia. PhD Thesis, Australian National University, Canberra, Australia.

Alexander, M.E., and F.V. Cole. 1995. Predicting and interpreting fire intensities in Alaskan black spruce forest using the Canadian system of fire danger rating, pp. 185–192 in

Managing Forests to Meet People's Needs: Proceedings of the 1994 Society of American Forestry/Canadian Institute of Forestry Convention. SAF Publication 95-02. Society of American Forestry, Bethesda, MD.

Alexander, M.E., and W.J. De Groot. 1988. *A Decision Aid for Characterizing Fire Behavior and Determining Fire Suppression Needs.* Canadian Forest Service, Northern Forestry Center, Edmonton, Alberta, Canada. Poster with text.

Alexander, M.E., and D.L. Euler. 1981. Ecological role of fire in the uncut boreal mixedwood forest, pp. 42–64 in *Proceedings Boreal Mixedwood Symposium.* COJFRC Symposium Proceedings O-P-9. Canadian Forest Service, Great Lakes Forest Research Center, Sault Ste. Marie, Ontario, Canada.

Alexander, M.E., and R.A. Lanoville. 1989. *Predicting Fire Behavior in the Black Spruce–Lichen Woodland Fuel Type of Western and Northern Canada.* Forestry Canada, Northern Forestry Center, Edmonton, Alberta, Canada, and Government of the Northwest Territories, Department of Renewable Resources, Fort Smith, Northwest Territories. Poster with text.

Alexander, M.E., B.J. Stocks, and B.D. Lawson. 1991. *Fire Behavior in Black Spruce-Lichen Woodland: The Porter Lake Project.* Information Report NOR-X-310. Forestry Canada, Northwest Region, Northern Forestry Center, Edmonton, Alberta, Canada.

Barney, R.J. 1969. *Interior Alaska Wildfires, 1956–1965.* USDA Forest Service, Pacific Northwest Forest and Range Experimental Station.

Black, R.A., and L.C. Bliss. 1978. Recovery sequence of Picea mariana–Vaccinium uliginosum forest after burning near Inuvik, Northwest Territories, Canada. *Can. J. Bot.* 56:2020–2030.

Bonan, G.B., and H.H. Shugart. 1989. Environmental factors and ecological processes in boreal forests. *Annu. Rev. Ecol. Syst.* 20:1–28.

Brown, R.J.E. 1970. *Permafrost in Canada—Its Influence on Northern Development.* University of Toronto Press, Toronto, Canada.

Brown, R.J.E. 1983. Effects of fire on the permafrost ground thermal regime, pp. 97–110 in R.W. Wein and D.A. MacLean, eds. *The Role of Fire in Northern Circumpolar Ecosystems.* John Wiley & Sons, New York.

Byram, G.M. 1954. *Atmospheric Conditions Related to Blowup Fires.* Station Paper 35. USDA Forest Service, Southeast Forest Experiment Station, Asheville, NC.

Byram, G.M. 1959. Forest fire behavior, pp. 90–123 in K.P. Davis, ed. *Forest Fire: Control and Use.* McGraw-Hill, New York.

Carroll, S.B., and L.C. Bliss. 1982. Jack pine-lichen woodland on sandy soils in northern Saskatchewan and northeastern Alberta. *Can. J. Bot.* 60:2270–2282.

Cayford, J.H., and D.J. McRae. 1983. The ecological role of fire in jack pine forests, pp. 183–195 in R.W. Wein and D.A. MacLean, eds. *The Role of Fire in Northern Circumpolar Ecosystems.* John Wiley & Sons, New York.

Clark, J.S., and P.J.H. Richard. 1996. The role of paleofire in boreal and other cool-coniferous forests in the boreal forest, pp. 65–89 in J.G. Goldammer and V.V. Furyaev, eds. *Fire in Ecosystems of Boreal Eurasia.* Kluwer Academic Publishers, Dordrecht, The Netherlands.

Day, R.J., and G.T. Woods. 1977. The Role of Wildfire in the Ecology of Jack and Red Pine Forests in Quetico Provincial Park. Ontario Ministry of Natural Resources, Quetico Provincial Park Fire Ecology Study Report 5.

Delisle, G.P., and D.E. Dube. 1983. *One and One-Half Centuries of Fire in Wood Buffalo National Park.* Forestry Report 28:12. Canadian Forest Service, Northern Forestry Research Center, Edmonton, Alberta, Canada.

Duchesne, L.C., A. Applejohn, L. Clark, and C. Mueller-Rowat. In press. *Fire in Northern Ecosystems.* USDA Forest Service Publication.

Dyrness, C.T., and R.A. Norum. 1983. The effects of experimental fires on black spruce forest floors in interior Alaska. *Can. J. For. Res.* 13:879–893.

Elliott-Fisk, D.L. 1988. The boreal forest, pp. 33–62 in M.G. Barbour and W.D. Billings, eds. *North American Terrestrial Vegetation.* Cambridge University Press, Cambridge, UK.

Fowells, H.A. 1965. *Silvics of Forest Trees of the United States.* Agriculture Handbook 271. U.S. Department of Agriculture, Washington, DC.

Forestry Canada Fire Danger Group. 1992. *Development and Structure of the Canadian Forest Fire Behavior Prediction System.* Information Report ST-X-3. Forestry Canada, Ottawa, Ontario, Canada.

Fosberg, M.A., B.J. Stocks, and T.J. Lynham. 1996. Risk analysis in strategic planning: fire and climate change in the boreal forest, pp. 495–505 in J.G. Goldammer and V.V. Furyaev, eds. *Fire in Ecosystems of Boreal Eurasia.* Kluwer Academic Publishers, Dordrecht, The Netherlands.

Foster, D.R. 1983. The history and pattern of fire in the boreal forest of southeastern Labrador. *Can. J. Bot.* 61:2459–2471.

Frandsen, W.H. 1991. Burning rate of smoldering peat. *Northwest Sci.* 65:166–172.

Furyaev, V.V., R.W. Wein, and D.A. MacLean. 1983. Fire influences in Abies-dominated forests, pp. 221–232 in R.W. Wein and D.A. MacLean, eds. *The Role of Fire in Northern Circumpolar Ecosystems.* J. Wiley and Sons, New York.

Halliday, W.E.D., and A.W.A. Brown. 1943. The distribution of some important forest trees in Canada. *Ecology* 24:353–373.

Hare, K.F. 1954. Climate and zonal divisions of the boreal forest formation in eastern Canada. *Geog. Rev.* 40:615–635.

Harlow, W.M., E.S. Harar, J.W. Hardin, and F.M. White. 1996. *Textbook of Dendrology,* 8th ed. McGraw-Hill, New York.

Harrington, J.B. 1982. *A Statistical Study of Area Burned by Wildfire in Canada, 1953–1980.* Information Report PI-X-16. Canadian Forest Service, Petawawa National Forest Institute, Chalk River, Ontario, Canada.

Heinselman, M.L. 1973. Fire in the virgin forests of Boundary Waters Canoe Area, Minnesota. *Quat. Res.* 3:329–382.

Heinselman, M.L. 1981a. Fire and succession in the conifer forest of northern North America, pp. 374–405 in D.C. West, H.H. Shugart, and D.B. Botkin, eds. *Forest Succession—Concepts and Application.* Springer-Verlag, New York.

Heinselman, M.L. 1981b. Fire intensity and frequency as factors in the distribution and structure of northern ecosystems, pp. 7–57 in H.A. Mooney, T.M. Bonnicksen, N.L. Christensen, J.E. Lotan, and W.A. Reiners, eds. *Fire Regimes and Ecosystem Properties.* General Technical Report WO-26. USDA Forest Service, Washington, DC.

Hogg, E.H. 1994. Climate and the southern limit of the western Canadian boreal forest. *Can. J. For. Res.* 24:1835–1845.

Johnson, E.A. 1992. *Fire and Vegetation Dynamics: Studies from the North American Boreal Forest.* Cambridge University Press, Cambridge, UK.

Kershaw, K.A., and W.R. Rouse. 1971. Studies on lichen-dominated systems. I. The water relations of Cladonia alpestris in spruce-lichen woodland in northern Ontario. *Can. J. Bot.* 49:1389–1399.

Krebil, R.G. 1972. *Mortality of Aspen on the Gros Ventre Elk Winter Range.* Research Paper INT-129. USDA Forest Service, Intermountain Forest and Range Experiment Station, Moscow, Idaho.

LaRoi, G.H. 1967. Ecological studies in the boreal spruce-fir forests of the North American taiga. I. Analysis of the vascular flora. *Ecol. Mon.* 37:229–253.

Larsen, J.A. 1980. *The Boreal Ecosystem.* Academic Press, New York.

Lawson, B.D., and G.N. Dalrymple. 1996. *Ground-Truthing the Drought Code: Field Verification of Over-winter Recharge of Forest Floor Moisture.* FRDA Report 268. Canadian Forest Service, Pacific Forestry Center and B.C. Ministry of Forestry, Research Branch, Victoria, British Columbia, Canada.

Lotan, J.E., J.K Brown, and L.F. Neuenschwander. 1984. Role of fire in lodgepole pine forests, pp. 135–152 in D.M. Baumgartner, R.G. Krebill, J.T. Arnott, and G.F. Weetman, eds. *Symposium Proceedings of Lodgepole Pine—The Species and Its Management.* Cooperative Extension, Washington State University, Spokane, WA.

Lynham, T.J., and B.J. Stocks. 1991. The natural fire regime in an unprotected section of the boreal forest in Canada, pp. 99–109 in *Proceedings, Tall Timbers Fire Ecology Conference.* Tall Timbers Research Station, Tallahassee, FL

Merrill, D.F., and M.E. Alexander, eds. 1987. *Glossary of Forest Fire Management Terms,* 4th ed. Publication NRCC 26516. Natural Resource Council of Canada, Canadian Committee on Forest Fire Management, Ottawa, Ontario, Canada.

Nikolov, N., and H. Helmisaari. 1992. Silvics of the circumpolar boreal forest tree species, pp. 13–84 in H.H. Shugart, R. Leemans, and G.B. Bonan, eds. *A Systems Analysis of the Global Boreal Forest.* University Press, Cambridge, UK.

Nimchuk, N. 1983. *Wildfire Behavior Associated with Upper Ridge Breakdown.* ENR Report T/50. Alberta Energy and Natural Resources, Forest Service, Edmonton, Alberta, Canada.

Norum, R.A. 1982. *Predicting Wildfire Behavior in Black Spruce Forests in Alaska.* Research Note PNW-401. USDA Forest Service, Pacific Northwest Forest Range Experiment Station, Portland, OR.

Payette, S. 1992. Fire as a controlling process in North American boreal forest, pp. 144–169 in H.H. Shugart, R. Leemans, and G.B. Bonan, eds. *A Systems Analysis of the Global Boreal Forest.* University Press, Cambridge, UK.

Quintilio, D., M.E. Alexander, and R.L. Ponto. 1991. *Spring Fires in a Semi-mature Trembling Aspen Stand in Central Alberta.* Information Report NOR-X-323. Forestry Canada, Northwest Region, Northern Forestry Center, Edmonton, Alberta, Canada.

Robbins, L.E., and R.L. Myers. 1992. *Seasonal Effects of Prescribed Burning in Florida: A Review.* Miscellaneous Publication 8. Tall Timbers Research Station, Tallahassee, FL.

Rowe, J.S. 1972. *Forest Regions of Canada.* Pub. 1300. Department of the Environment, Canadian Forestry Service, Ottawa, Canada.

Rowe, J.S. 1983. Concepts of fire effects on plant individuals and species, pp. 135–151 in R.W. Wein and D.A. MacLean, eds. *The Role of Fire in Northern Circumpolar Ecosystems.* John Wiley & Sons, New York.

Rowe, J.S., and G.W. Scotter. 1973. Fire in the boreal forest. *Quat. Res.* 3:444–464.

Simard, A.J. 1975. *Wildland Fire Occurrence in Canada.* Canadian Forest Service, Ottawa, Ontario, Canada. Map with text.

Stocks, B.J. 1987a. Fire behavior in immature jack pine. *Can. J. For. Res.* 17:80–86.

Stocks, B.J. 1987b. Fire potential in spruce budworm-damaged forests of Ontario. *For. Chron.* 63:8–14.

Stocks, B.J. 1989. Fire behavior in mature jack pine. *Can. J. For. Res.* 19:783–790.

Stocks, B.J. 1991. The extent and impact of forest fires in northern circumpolar countries, pp. 197–202 in J.S. Levine, ed. *Global Biomass Burning: Atmospheric, Climatic, and Biospheric Implications.* MIT Press, Cambridge, MA.

Stocks, B.J., and M.D. Flannigan. 1987. Analysis of the behavior and associated weather for a 1986 northwestern Ontario wildfire: Red Lake #7, pp. 94–100 in *Proceedings of the Ninth Conference on Fire and Forest Meteorology* (April 21–24, 1987, San Diego, CA). Society of American Forestry, Washington, DC.

Stocks, B.J., and G.R. Hartley. 1979. *Forest Fire Occurrence in Ontario.* Canadian Forest Service, Great Lakes Forest Research Center, Ottawa, Ontario, Canada. Map with text.

Stocks, B.J., and G.R. Hartley. 1995. *Fire Behavior in Three Jack Pine Fuel Complexes.* Northern Ontario Development Agreement, Sault Ste. Marie, Ontario, Canada. Poster with text.

Stocks, B.J., and J.B. Kauffman. 1997. Biomass consumption and behavior of wildland fires in boreal, temperate and tropical ecosystems: Parameters necessary to interpret historic fire regimes and future fire scenarios, pp. 169–188 in J.S. Clark, H. Cachier,

J.G. Goldammer, and B.J. Stocks, eds. *Sediment Records of Biomass Burning and Global Change.* NATO ASI Series, Subseries 1, Global Environmental Change, Vol. 51. Springer-Verlag, Berlin, Germany.

Stocks, B.J., and J.D. Walker. 1973. *Climatic Conditions before and during Four Significant Forest Fire Situations in Ontario.* Information Report O-X-187. Department of Environment, Forestry Service, Sault Ste. Marie, Ontario, Canada.

Stocks, B.J., B.D. Lawson, M.E. Alexander, C.E. Van Wagner, R.S. McAlpine, T.J. Lynham, and D.E. Dube. 1989. The Canadian Forest Fire Danger Rating System: an overview. *For. Chron.* 65:450–457.

Stocks, B.J., B.S. Lee, and D.L. Martell. 1996. Some potential carbon budget implications of fire management in the boreal forest, pp. 89–96 in M.J. Apps and D.T. Price, eds. *Forest Ecosystems, Forest Management and the Global Carbon Cycle.* NATO ASI Series, Subseries I, Vol. 40, Global Environmental Change. Springer-Verlag, Berlin.

Swanson, D.K. 1996. Susceptibility of permafrost soils to deep thaw after forest fires in interior Alaska, U.S.A., and some ecologic implications. *Arc. Alp. Res.* 28:217–227.

Van Cleve, K., and L. Viereck. 1981. Forest succession in relation to nutrient cycling in the boreal forest of Alaska, pp. 184–211 in D.C. West, H.H. Shugart, and D.B. Botkin, eds. *Forest Succession: Concepts and Application.* Springer-Verlag, New York.

Van Cleve, K., F.S. Chapin III, P.W. Flanagan, L.A. Viereck, and C.T. Dyrness, eds. 1986. *Forest Ecosystems in the Alaskan Taiga.* Springer-Verlag, New York.

Van Wagner, C.E. 1977. Conditions for the start and spread of crown fire. *Can. J. For. Res.* 7:23–34.

Van Wagner, C.E. 1978. Age-class distribution and the forest fire cycle. *Can. J. For. Res.* 8:220–227.

Van Wagner, C.E. 1983. Fire behavior in northern coniferous forests, pp. 65–80 in R.W. Wein and D.A. MacLean, eds. *The Role of Fire in Northern Circumpolar Ecosystems.* John Wiley & Sons, New York.

Van Wagner, C.E. 1987. *Development and Structure of the Canadian Forest Fire Weather Index System.* Technical Report 35. Canadian Forest Service, Ottawa, Ontario, Canada.

Van Wagner, C.E., and I.R. Methven. 1980. *Fire in the Management of Canada's National Parks: A Background Philosophy for Setting Strategy.* Occasional Paper 2. Department of the Environment, Parks Canada, National Parks Branch, Ottawa, Ontario.

Viereck, L.A. 1983. The effects of fire in black spruce ecosystems of Alaska and northern Canada, pp. 201–220 in R.W. Wein and D.A. MacLean, eds. *The Role of Fire in Northern Circumpolar Ecosystems.* John Wiley & Sons, New York.

Viereck, L.A., K. Van Cleve, and C.T. Dyrness. 1986. Forest ecosystem distribution in the taiga environment, pp. 22–43 in K. Van Cleve, F.S. Chapin III, P.W. Flanagan, L.A. Viereck, and C.T. Dyrness, eds. *Forest Ecosystems in the Alaskan Taiga.* Springer-Verlag, New York.

Ward, P.C., and A.G. Tithecott. 1993. *The Impact of Fire Management on the Boreal Landscape of Ontario.* Publication 305. Ontario Ministry of Natural Resources, Aviation, Flood, and Fire Management Branch, Sault Ste. Marie, Ontario, Canada.

Wein, R.W. 1983. Fire behaviour and ecological effects in organic terrain, pp. 81–95 in R.W. Wein and D.A. Maclean, eds. *The Role of Fire in Northern Circumpolar Ecosystems.* John Wiley & Sons, New York.

Wright, H.A., and A.W. Bailey 1982. *Fire Ecology: United States and Southern Canada.* Wiley-Interscience, New York.

Wright, H.E., Jr., and M.L. Heinselman. 1973. The ecological role of fire in natural conifer forests of western and northern North America—introduction. *Quat. Res.* 3:319–328.

Yarie, J. 1981. Forest fire cycles and life tables: a case study from interior Alaska. *Can. J. For. Res.* 11:554–562.

8. Extent, Distribution, and Ecological Role of Fire in Russian Forests

Anatoly Z. Shvidenko and Sten Nilsson

Introduction

Russia has a complex system for classification of land use and land cover of forested areas. All lands managed by the federal government are part of Russia's forest fund (FF), which in 1993 contained 1,180.9 million ha (64.4% of the total land area of Russia). The FF is divided into two main categories: (1) forestlands (886.5 million ha) that include areas designated for forests; and (2) nonforestlands (294.4 million ha) that include unproductive and unused lands (e.g., bogs, rocks, tundras) and land converted to human use (e.g., roads, pastures, farms). In turn, the forestland category consists of: (1) forested areas occupied by closed forests (763.5 million ha); and (2) unforested areas designated for forests but temporarily unforested, including burns and dead stands, natural sparse forests (open woodlands), and unregenerated clear-cut areas (123.0 million ha).

The vast territories of the Russian FF are characterized by a tremendous diversity of climate, soil, vegetation, and strength and peculiarities of anthropogenic impacts. The FF stretches through 11 time zones in the longitudinal direction, with the forests growing in ten vegetational zones and subzones—from forest-tundra zone in the north to the semideserts in the south (Kurnaev 1973). The distribution of the FF, forest land, and forest area categories within these zones is summarized in Table 8.1. At least four major gradients drive the distribution, structure, and productivity of Russian forests: temperature, precipitation, aridity,

Table 8.1. Data on Russian Forests by Vegetation Zones[a]

Vegetation zone	Area[b] ($\times 10^6$ ha)				Species composition[c]	Growing stock ($m^3 ha^{-1}$)
	Total	Forest Fund	Forestland	Forested area		
Subarctic and tundra	299.1	110.0	7.1	—[d]	—	—
Forest tundra, sparse taiga, and meadow forests	297.3	229.	185.	134.	7L1S1P1B	65
Northern taiga	128.2	123.1	78.7	65.6	4P3L3S	110
Middle taiga	394.0	380.5	334.2	316.9	4L2P2S1C1B	120
Southern taiga	236.9	218.7	176.8	155.7	3P3S2L1F1C	145
Mixed forests, deciduous forests, and forest steppe	247.8	109.0	93.7	84.8	3P2S2B1O1H1SD	155
Steppe, semidesert, and desert	106.5	9.5	7.9	6.1	5O1P1E2B1A	70
Total	1,709.8	1,180.9	884.1	763.5	8.2 Con0.3H1.5SD	112

[a]The primary source for the data in this table is the State Forest Account survey data as of January 1, 1993. Other definitions are from Goskomles SSSR (1990, 1991) and FSFMR (1995).
[b]Area does not include large inland water reservoirs.
[c]The letter denotes the dominant species, and the number represents the percent (×10) composition of the dominant species (e.g., 7L designates 70% of growing stock in stands dominated by larch). Species: P, pine; S, spruce; C, cedar (*Pinus sibirica* and *P. koraiensis*); F, fir; L, larch; B, birch; A, aspen; O, oak; E, elm; Con, coniferous; H, hard deciduous; SD, soft deciduous.
[d]There are no forests in the subarctic and tundra vegetation zone.

and land use. The major part (about 80%) of the forest area category is situated in the taiga zone, and about 95% of Russian forests are boreal forests.

Table 8.1 summarizes the tree species that exist within the different vegetation zones. Seven tree species make up about 87% of all Russian forests (pine 16.2%, spruce 10.7%, larch 37.3%, cedar 5.6%, fir 2.0%, birch 12.4%, and aspen 2.7%). Stands with these major forest species cover 90.3% of forest area in Russia (coniferous 71.9%, hard deciduous 2.4%, and soft deciduous 16.0%).

Basic Definitions and Classifications

Forest fire in the Russian boreal zone is a widespread and natural phenomenon, and its impact on the environment has local, regional, and global dimensions. The diversity of forest types, growing conditions, landscape peculiarities, structure and productivity of forests, and types of anthropogenic impacts define the nature or characteristics of fire. There are many ways to describe fires, including their distribution, intensity, and ecological impact on terrestrial ecosystems and landscapes.

The dual role of forest fires, both destructive and constructive, is well recognized. In the southern part of the boreal zone, forest fires are considered to be one of the most dangerous environmental phenomena, often resulting in significant losses in terms of human life and property and the economic value and productivity of mature forest stands. By contrast, in the unmanaged and unused forests of the northern and sparse taiga zones (particularly in permafrost regions), fires are viewed as both positive and negative. On the positive side, surface fires with a long cycle (80–100 years) prevent decreased productivity of forests and paludification of forestlands (Melekhov 1948; Sheshukov 1978; Matveev 1992; Sheshukov et al. 1992; Furayev 1996). On the negative side, frequent non–stand-replacing fires can significantly decrease the actual productivity and stability of forests (Utkin 1965). Under extreme climatic conditions, fire is often responsible for the impoverishment of forests and the shifting of tundra to the south. For example, there are estimates that the belt of relatively treeless areas in the taiga-tundra ecotone in northern Eurasia has been generated by fires (Krjuchkov 1987). This ecotone has a width of 100–250 km and an area that is increasing by 0.3 million ha annually (Krjuchkov 1987).

Forest fire in the boreal zone of northern Eurasia is a significant factor in altering physical, chemical, and biological characteristics of the landscape, including the microclimate conditions of the atmosphere's surface layer (light, humidity, wind), temperature and moisture of upper soil layers, and chemical properties of soils. The directions of the soil formation processes can be changed dramatically, especially in humid and wet sites. Generally, fire results in changes to the successional patterns found on specific landforms, including changes in species composition and other biological characteristics of forests (e.g., age structure, productivity) (Orphanitskaja and Orphanitskiy 1959; Komin 1967; Chertovsky et al. 1983; Popova 1983; Pozdnjakov 1983; Saposhnikov et al. 1993; Furayev 1996).

Some scientists argue for the crucial role of fire in the geographical distribution of forests over large areas (Chudnikov 1931) and even consider fire as a powerful geomorphic force that strongly shapes the development of taiga landscapes (Tumel 1939). Rojkov and co-workers (1996) found that different phases of postfire succession accounted for 40%–96% of the total forested area of the different vegetation zones described in Table 8.1.

An example of the overall influence of fire on boreal forest cover has been reported by Furyaev (1996) for a typical taiga landscape (of 165,000 ha) in the Kos-Yenisey plain of east Siberia. The total area of fire (including repeated fire at the same place) was estimated for the period 1700–1956 as being 5.38 times the total area of the landscape, or an average fire return interval of 48 years.

Several forest fire classification schemes have been suggested for Russia (Melekhov 1947; Kurbatsky 1970a; Sheshukov 1983; Telizin 1970). Vegetation fires can be divided into homogeneous fires occurring within a single vegetation type (i.e., tundra, forest, bog, and steppe fires) and mixed or landscape fires covering two or more vegetation types (Kurbatsky 1970b, 1972).

Forest fires can be simple (i.e., on-ground, crown, and soil) or composite. On-ground fires are divided into superficial, stable, and (fallen) wood fires. Crown fires are classified as running and stable fires. Finally, soil (underground) fires are usually divided into peat and litter fires (Far Eastern forest fire experts also identify turf fires, Sheshukov 1974). Although this classification is consistent with the needs for forest fire protection, it does not take into account ecological consequences of fires in any explicit way. Forest management practices of Russia have been using a simplified classification and official statistics identify only on-ground (superficial and stable or litter), crown, and underground fires (Shetinsky and Sergeenko 1996).

Fire is a landscape phenomenon, and for each forest landscape, a fire or pyrological regime can be defined in terms of regional climate and weather conditions, landscape structure, types of forest ecosystems (Kireev 1997), specific site characteristics (e.g., species composition, morphological structure of stands, amount of forest combustibles), human population density, and previous history of disturbance. From an ecological perspective, fire regimes can be divided into two groups: (1) digressive regimes, which through suppression of reproduction lead to the impoverishment of forests and eventual desertification of forested areas; and (2) tolerant regimes, which support the restoration of indigenous (or relatively indigenous) forest types and associations.

A fire regime (as a specific type of disturbance regime) can be quantified by a number of indicators explicitly defined in space and time (Shvidenko 1997). The extent of fire is described as an average area burned over several years within a definite landscape, management, or administrative unit. Indicators used within the Russian system of forest fire protection are actual burning or an index of actual burning. Actual burning is defined as an average over several years of the number of fires (n) for an area of 1 million ha. In practice, five classes are used: more than 56; 55–26, 25–13, 12–6, and less than 5. Index of actual burning is defined as $BI = nS^{-0.5}$, where S is the burned area for 1 million ha of the FF. Forest statistics present annual burnt areas in FA using the three types of fire (on-ground, crown, and underground) for each forest enterprise. For unforested areas and nonforest lands, only total area burned is reported.

Several indicators are used to describe the fire regime. The fire cycle or fire return interval (average number of years between two consequent fires at a specific location) is the most important indicator of the fire regime (McPherson et al. 1990). Fire frequency is defined as number of fires per unit time in a specified area. Intensity of burning is the amount of energy generated by a unit of burned area per time unit (Kurbatsky 1972). Furyaev (1996) classified on-ground fire by intensity into three classes: weak intensity, less than 21,000 kJ m^{-2}; medium intensity, from 21,000 to 63,000 kJ m^{-2}; and strong intensity, more than 63,000 kJ m^{-2}. He reported the maximum fire intensity in fire experiments in Siberia (the zone of southern taiga) to 82,700 kJ m^{-2}. Although intensity indicates the ecological effects of fire on a forest ecosystem, it is a difficult characteristic to quantify because it is not always possible to provide direct measurements of energy generated by fire, and there are also many other factors that significantly affect the postfire behavior of forest ecosystems.

Some indirect indicators of fire severity as a function of fire intensity and residence time have been suggested. For on-ground fires, the height of scorchings on tree trunks or the level of litter consumption has been proposed as an intensity measure (Voinov and Sofronov 1976; Sheshukov 1977). It has been shown for many regions and species that the height of scorchings is closely correlated with postfire mortality of trees. Specifics of soil (underground) fires can be characterized by the length of smoldering burning, speed of soil burning, and speed of the increase of smoldering area (Gundar 1978a).

The last official manual on forest fire protection in Russia (Shetinsky and Sergeenko 1996) introduced different indicators of three levels (low, medium, high) of fire intensity for different types of fires. For on-ground superficial fires, the height of burn marks on stems, fire spread rate, height of flame, and intensity of burning are used (respectively, <1 m, <1 m sec^{-1}, <0.5 m, <100 kW m^{-1} for low intensity; 1–2 m, 1–3 m sec^{-1}, 0.5–1.5 m, 101–750 kW m^{-1} for medium intensity, and > 2 m, >3 m sec^{-1}, >1.5 m, >750 kW m^{-1} for high intensity). Intensity of on-ground stable (litter) fires is classified by the level of combusted litter (only the A_{01} layer burns under low intensity; the A_{01}+A_{02} layers burn under medium intensity; and all litter up to mineral horizons burn under fire of high intensity). The intensity of peat (underground) fires is defined by the depth of the burnt layer (<7 cm over the entire area and <30 cm in the root zone; 7–25 cm and complete peat layer in the root zone; and all peat up to mineral horizons, respectively, for low, medium, and high intensity). Finally, low-intensity crown fires result in 30% mortality; a medium-intensity fire results in up to 60% mortality; and a high-intensity crown fire results in almost complete mortality (< 6% of the area has living trees).

The amount and status of forest combustibles are the most important factors contributing to fire type, intensity, and effects. Different classifications of forest combustible materials have been suggested (Kurbatsky 1962; Arzibashev 1974; Konev 1977; Melekhov 1979; Sheshukov 1988). The general classification of Sheshukov (1988) depends on the close connection between forest combustibles

(FC) and forest fire type. The classification includes initial (primary) and transformed (secondary) FCs. Primary FCs include crown and living aboveground plants (living forest floor vegetation). Secondary FCs include surface fuels (litterfall, coarse woody debris) and organic soils (litter, peat, turf).

These fuel categories can be further subdivided. For example, the classification suggested by Kurbatsky (1962) for Siberia considers seven categories: (1) mosses, lichens, and litterfall; (2) litter; (3) herbs and small shrubs; (4) coarse woody debris; (5) understory; (6) twigs, needles, and leaves of living trees; and (7) stems and branches.

The FCs that play a central role in the process of biomass burning are also called basic burning conductors (BBC) or actual fuel material (Vasilenko 1976). Based on the concept of BBC indicators, Sofronov (1978) divided on-ground forest combustibles for Siberian forests into ten fuel categories: (1) lichens; (2) dry mosses; (3) humid mosses; (4) long mosses; (5) sphagnum; (6) hypnum green mosses; (7) friable litterfall; (8) compact litterfall; (9) standing-dead grasses; and (10) standing-dead herbaceous plants.

Mortality (dieback), expressed as a percentage of dead growing stock (or number of trees), is a measure of the level of damage or ecological effects of a fire. To classify postfire damage of stands, Voinov and Sofronov (1976) suggested five classes of damage: (1) mortality 0–30% by number of stems and 0–25% by growing stock volume; (2) 31–70% and 26–60%, respectively; (3) 71–100% (all three classes are mainly a result of different intensities of on-ground fires); (4) completely dead stands (usually after strong crown fires); and (5) burns with fallen stems after underground (soil, peat) fires.

The classifications mentioned earlier are used as a basis to understand the nature of wild fires, their ecological consequences, and the organization of forest fire protection. Below, they are used to describe basic patterns of fire in the boreal forests of Russian northern Eurasia.

Basic Indicators of Fire Regimes in Northern Eurasia Forests

Two circumstances contribute to uncertainties in the estimates of the historical and current extent of forest fires in Russia. First, about 40% of the FF have never received nor are currently receiving any fire protection or monitoring. Second, for political reasons, official statistical data on forest fires before 1988 were purposely falsified.

For the most part, areas not protected against fires are located in the forest tundra and sparse and northern taiga of west Siberia (43 million ha in 1989), east Siberia (119 million ha), and the Far East (249 million ha). In 1992, only 64% of the total FF area was protected (Goskomstat 1993). Since 1992, the area of the protected zone has most likely decreased because of reductions of budgets within the federal government.

Annual forest fire statistics on fire-protected areas during 1986–1995 are presented in Table 8.2. During this period, the number of annually detected forest fires was between 13,000 and 26,000, with the majority of fires (86.3%) being

Table 8.2. Number and Area of Forest Fires on Fire Protected Territory of the Forest Fund During 1971–1995[a]

	Number of fires		Areas of fires (× 1,000 ha)						
			Total	Forest area					
Year	Total	Lightning caused	Forest Fund	Crown fire	On-ground fire	Underground fire	Total	Unforested area	Nonforestland
1986	16,353	2,860	1,159.7	207.4	487.2	1.1	695.7	10.0	453.9
1987	13,439	2,548	4,414.0	123.4	412.8	0.9	537.2	31.6	384.5
1988	18,573	3,661	1,011.3	143.8	613.2	0.5	757.5	29.4	224.4
1989	21,934	5,273	2,040.4	247.4	248.8	8.0	1,504.3	123.7	412.4
1990	17,672	38	167.0	273.9	1,043.2	0.9	1,318.0	48.2	303.6
1991	17,965	3,620	1,126.2	116.0	411.3	3.5	530.9	151.2	444.2
1992	25,777	2,523	1,142.8	56.0	544.2	3.3	603.5	87.9	451.3
1993	18,428	2,804	1,200.4	104.1	618.5	0.9	723.5	25.1	451.8
1994	20,287	1,957	723.1	61.4	464.4	2.0	527.8	9.0	186.3
1995	25,951	2,653	462.9	23.6	325.8	3.1	352.4	7.7	102.7
Average during 1986–1995									
Area	19,638	2,694	1,495.1	135.7	617.0	2.4	755.1	52.4	687.4
Percent	100	13.7	100	9.1	41.2	0.2	50.5	3.5	46.0
Average during 1971–1995									
Area	17,478	na	643.7	93.5	271.5	3.7	368.7	44.3	230.8
Percent	100	na	100	14.5	42.2	0.6	57.3	6.9	35.8

[a]Based on data from the Russian Federal Forest Service.

caused by human activities. Thus, natural factors (lightning) caused only 13.7% of the fires. The annual average total burned area in fire-protected territories was about 1.5 million ha, half of which occurred on forested areas. The data do not include prescribed burns (which were negligibly small) or fires on agricultural lands. For this latter category, no statistics exist.

For comparison purposes, the data from 1971 to 1985 estimate the total burned area to be 0.64 million ha annually, of which 0.27 million ha was in forested areas. These figures are less than half the annual area burned compared with 1986–1995, presenting evidence of the systematic bias of statistical data before 1988.

The temporal and geographic distributions of fires within Russia vary, with 60–90% of the annual burned areas usually concentrated in four to six regions. During average fire years in Siberia, about 1% of the fires are large fires (burned area >200 ha). In dry years, the percentage of large fires may rise to 10%. In any case, large fires typically result in 50–80% of the total area burned and cause up to 90% of the total damage (Valendik 1990). In 1989, 3,400 forest fires occupying 0.81 million ha were detected in the Tyumen region, 900 (0.47 million ha) in the Tomsk region, 100 (0.32 million ha) in the Sakhalin region, and 1,000 (0.17 million ha) in the Khabarovsk region. Thus, more than 80% of the burned areas in the country were concentrated in the territory of four administrative regions. For extremely dry years, a similar picture can be observed even for densely populated regions with intensive on-ground fire protection.

The annual burned area in unprotected territories can only be estimated indirectly. The long-term inventory data (which exist for all Russian FF areas for 1961–1993) and other information were used to develop this estimate, including (1) areas of burned forests and dead stands, and stands undergoing natural postfire regeneration; (2) data on the distribution of forests by age classes and age groups; (3) databases developed by the IIASA Forest Study (e.g., distribution of forests by the level of transformation in virgin, natural, and anthropogenic forests) or by categories of stand age structure; (4) data on the extent and distribution (other than fire) of stand-replacing and non–stand-replacing disturbances (e.g., harvest, insect outbreaks, and pollution); (5) data from different surveys, partly from the All-Russia Research and Information Center on Forest Resources (VNIIZlesresours); (6) numerous publications and archives; and (7) estimates derived from satellite imagery for some taiga regions.

The basic approaches used for the estimate were (1) the average stand age in unmanaged forests in the year of the last stand-replacing fire plus half of the period for postfire regeneration; (2) for specific sites, the ratio between different types of fires as well as other relative indicators of the fire regime; and (3) the type of stand age structure is mostly defined by the frequency of non–stand-replacing fires.

Based on these methods, the estimate of the average burned areas in the total Russian FF territory and northern areas of the land reserve varies between 2.6 and 4.3 million ha annually during the period 1988–1992. We chose this period as a representation for the past 30 years to make an historical comparison and use the latest state forest account, presenting the inventory data from January 1993.

The results of this analysis, aggregated by forest zones, are presented in Table 8.3. The area damaged by forest and tundra fires in Russia is estimated to be 3.5 million ha yr^{-1} for 1988–1992. The estimates for all the FF (3.0 million ha yr^{-1}) is about 2.7 times higher than the official estimates for the protected area for the same period. It is estimated that 0.5 million ha yr^{-1} occur on reindeer pastures and state land reserves in the extreme north (subarctic and tundra zones). The average area of burned forests is estimated to be 1.5 million ha yr^{-1}, fires on nonforest-lands (e.g., mires, tundra landscapes) burn 1.0 million ha yr^{-1}, and fires on unforested areas (postfire areas, cuttings, sparse forests, grassy glades) burn 0.5 million ha yr^{-1}.

There is evidence to support our conclusion that the official fire statistics underestimate the actual area burned in Russia. Stocks (1991) estimated the burned areas in the extremely dry and hot year of 1987 to be 10 million ha for the former Soviet Union (official statistics give 4.4 million ha; Table 8.3). Based on analysis of AVHRR data, Cahoon and associates (1991) determined the burned areas in 1987 of the territories north of the Amur River to be 3.6 million ha. Using a similar approach, Cahoon and co-workers (1994) estimated that the total burned area in the Russian Far East and east Siberia in 1987 was 12.0 million ha. Our estimate, by using Soviet satellite data for 1987 for central Siberia and the major part of the Far East, is only about 6 million ha. In the year 1915, with catastrophic weather conditions, forest fires were observed on a huge territory of 1.6 million km^2, and the total area of burnt forest area was estimated to about 12 million ha (Shostakovich 1924). In contrast with these large fire years, there have been years with a rather low level of fire activity in Russia. For example, using AVHRR imagery, Cahoon and colleagues (1996) estimated the burnt areas in the total Russian territory in 1992 to be about 1.5 million ha.

Based on average frequency of on-ground fires in pine and larch stands, Conard and Ivanova (1998) estimated the average fire area for Russia to be about 12 million ha. Our approach does not support this conclusion if we (1) distribute such areas proportionally including areas of burnt-out stands; (2) use regional ratios between stand-replacing and non–stand-replacing fires; and (3) account for postfire regeneration, areas of burned out forestland are estimated from 40 to 55 million ha for the 1990s. Our opinion is that overestimating of annual fire areas by Conard and Ivanova (1998) was caused by uncertainties of the terminology used, specifically using the occurrence of fire in different forest associations instead of the fire cycle.

The estimates presented in Table 8.3 should be considered as a first-order assessment. It is difficult to estimate the distribution of peat (soil) fires because there are no statistics reported on this topic. Treeless mires (mostly peatland) are estimated to be 137.7 million ha in FF territories under state forest management in Russia, and the area of peat soils covered by forests is at least this large (Vompersky et al. 1975; Glebov 1976; Goldin 1976). The estimates of the extent of wetlands in Russia show that 139 million ha is occupied by mires (defined as areas with a depth of peat >0.3 m), of which 95 million ha is located on tundra and frozen soils in the taiga. Additionally, 230 million ha is swamp lands (with a depth

Table 8.3. Estimates of Annual Forest Fire Area During 1988–1992 by Vegetational Zones and Types of Forests for the Total Forest Fund Area[a]

	Forest fire area by type of fire ($\times 10^6$ ha)				
		On-ground fire			
Vegetation zones	Crown fire	Forested area	Unforested area	Nonforest lands	Underground fire
Subarctic and tundra	—[b]	—	—	0.43	0.07
Forest tundra, sparse taiga, and meadow forests	0.055	0.26	0.10	0.18	0.08
Northern taiga	0.060	0.29	0.11	0.19	0.09
Middle taiga	0.074	0.36	0.14	0.27	0.08
Southern taiga	0.037	0.17	0.08	0.14	0.02
Mixed forests, deciduous forests, and forest steppe	0.014	0.07	0.03	0.07	—
Steppe, semidesert, and desert	—	0.01	0.01	0.01	—
Total	0.240	1.16	0.47	1.29	0.34

[a]Data from the Russian Federal Forest Service.
[b]—, No fires occurred in these categories.

of peat of up to 0.3 m), of which 179 million ha is located on tundra and frozen soils. Wetlands, located only in the taiga take up about 163 million ha (Vompersky 1994). Rojkov and associates (1997) estimate the total area of all Russian wetlands to be 418.4 million ha.

Ozhogin (1979) indicates that peat fires occur only when relative humidity is less than 50%. A higher risk for peat fire exists if the precipitation during a forest fire period is less than one-half of the long-term average. Kurbatsky (1962), Furyaev (1970), Arzibashev (1974), and Gundar (1978a,b) have identified a peat fire risk if the water table level in the peat is 0.3–0.5 m below the ground surface. Sofronov and Volokitina (1990) have identified a risk with a water table level lower than 0.6–0.9 m in southern and western Siberia. Depending on the level of mineralization, peat can burn with a water content of 200–230% (ash content from 22.4 to 33.2%) in the Far East (Gundar and Kostirina 1976) and with a water content of up to 400–500% in Siberia (Kurbatsky 1962). The critical water content under which on-ground fires can turn into deep peat fires is 200–250% (Gundar 1978b). Under some conditions, drained peatlands (5.3 million ha in Russia in 1989) are able to burn even with a snow cover (Rjabukha 1973). Isaev (1966) has illustrated that larch forests on peatlands are subject to higher fire risks than larch on dry sites.

The conditions for peat fire risks mentioned previously occur one to three times during a 10-year period in the majority of the major boreal regions of Russia (Chibisov 1974; Chervonny 1979). Generally, the frequency of destructive fires on wetlands is low (from about 70–80 to 150–300 years), but the strength, amount of consumed organic, and impact on postfire dynamics of forest ecosystems can be very high. Based on the considerations presented previously and regional analyses of peat fire-driving forces, we estimate the average annual burnt area of peatland and peat soil to be 0.34 million ha, or 9.7% of the estimated total annual burnt area in northern Eurasia.

Temporal Regularities of Fire

For a specific landscape, cyclic regularities of climate and weather, the seasonal and daily dynamic of atmospheric processes, forest cover, changes in vegetation phenology, fire ignition sources, and level of forest fire protection all contribute to the regularity of fire occurrence. There is evidence that anthropogenic factors affected fire frequency in boreal northern Eurasia during the past several hundred years (Furyaev 1996; Gromtsev 1996; Sannikov and Goldammer 1996). A relatively reliable quantification of fire frequency in Russian forests is possible for the past 50 years since fire statistics became available for a major part of country. Korovin (1996) presented a comprehensive aggregated analysis of fire frequency for the Russian protected zone for 1945–1993 and indicated that deterministic and cyclic components are intrinsic for these time series. If they are eliminated, the residuals could be estimated as a stationary stochastic process.

The most important general features on frequencies of forest fire in northern Eurasia can be briefly listed in the following way (Melekhov 1947; Moiseenko

1958; Furyaev and Kireev 1979; Valendik 1990; Sheshukov et al. 1992; Furyaev 1996; Korovin 1996): (1) there are evident forest fire frequency regularities at different spatial and temporal scales that are highly variable; (2) the basic driving forces of fire frequency are forest types, geographic locality, availability of fire sources, and level and reliability of forest fire protection; (3) light coniferous boreal forests (basically dominated by pine and larch) have an average fire frequency from 15–20 to 60–70 years, but it could vary from 4–15 on dry sites of densely populated areas up to 250–300 years in sphagnum pine stands dependent on forest type and infrastructure development; (4) fire in dark coniferous forests is more rare than in light coniferous ones, and as a very common average could be estimated from 70–90 to 150–180 years; (5) these trends could be significantly different for remote forests as well as during catastrophic fire seasons; and (6) the frequency of all types of fires for the taiga zone could be described by the Rayleigh's low or by the exponential distribution, dependent on the distance from populated areas or from roads.

Postfire Dieback

One of the more immediate effects of fire on forests can be expressed through analysis of tree mortality or dieback. Fires in the Russian boreal forests are different from those in North America in that a much higher percentage of fires in North America is crown fires that result in almost 100% mortality of trees. By contrast, most of the fires in the Russian boreal forest are on-ground fires that do not result in immediate tree mortality. However, for several (up to 10) years after a fire, increased tree mortality is observed.

The average period for tree dieback caused by fire is estimated to be 5 years, with a variation of 2–10 years. Evdokimenko (1985) reported that during the 5-year period, after prescribed burning, the postfire mortality exceeded the natural mortality by 6, 9, and 120 times, respectively, for weak, medium, and strong fires. After 10 years, these ratios were only 2, 3, and 12. Indirect consequences of the fires are observed over a longer period, specifically if frequent recurrent fires occur.

Postfire mortality for trees varies greatly and depends on the type and intensity of the fires, relief, presence of permafrost, weather conditions, species composition, age and diameter of trees, and many other factors. The following are average estimates of postfire mortality for forests in the taiga zone. A light superficial ground fire results in 6–12% mortality, whereas a more intense, steady ground fire results in 15–20% mortality. A fire that consumes the litter layer results in 30–50% mortality, whereas a fire that consumes much of the organic soil (turf) results in 60% mortality. A peat fire results in 70% mortality. Finally, crown fires result in 75% mortality. Mortality variations in these fires can be large, with ranges of 5–90% for litter fire and 35–85% for turf fire.

In many forest types in Siberia, peat and crown fires cause 100% mortality. This is very typical for forests found on permafrost sites (Matveev 1992). Many researchers report the full destruction of stands even after steady on-ground fires (Zvetkov 1988; AUIRCFR 1990). Sibirina (1989) observed a mortality rate of 60% during the first 4 years after a middle-intensity on-ground fire in cedar stands in the Far East. On-ground stable fires of low intensity resulted in the mortality of dominant species in spruce-fir and larch forests in the Sakhalin Island (southern and middle taiga) of 3–11% (by number of trees). In the same region medium-intensity fires resulted in mortality of 3–34%, and high-intensity fires resulted in mortality of 56–72% (Mikhel 1984). Sibirina (1989) showed that on-ground fires in cedar-broadleaf forests resulted in a 4-year mortality of 60% of the growing stock. The mortality in spruce-fir stands at the basin of the lower part of the Amur River was 35% of the growing stock after 3 years (Solovjov and Sheshukov 1976).

Site characteristics play a crucial role in postfire characteristics of forests. For instance, in forest tundra and northern and sparse taiga stands of all species (including larch) usually die almost completely after a on-ground fires because of superficial root systems (Zvetkov 1988; Matveev 1992).

The height of scorchings on stems and average diameter of stands can be used as general indicators of fire severity (strength of fire consequences) in some forest types (Solovjov 1972; Voinov and Sofronov 1976; Valendik et al. 1977). Figure 8.1 shows that in sites where the height of scorching reaches 6 m, all the pine trees died. A similar level of scorching in larch stands will only cause a mortality of 50–

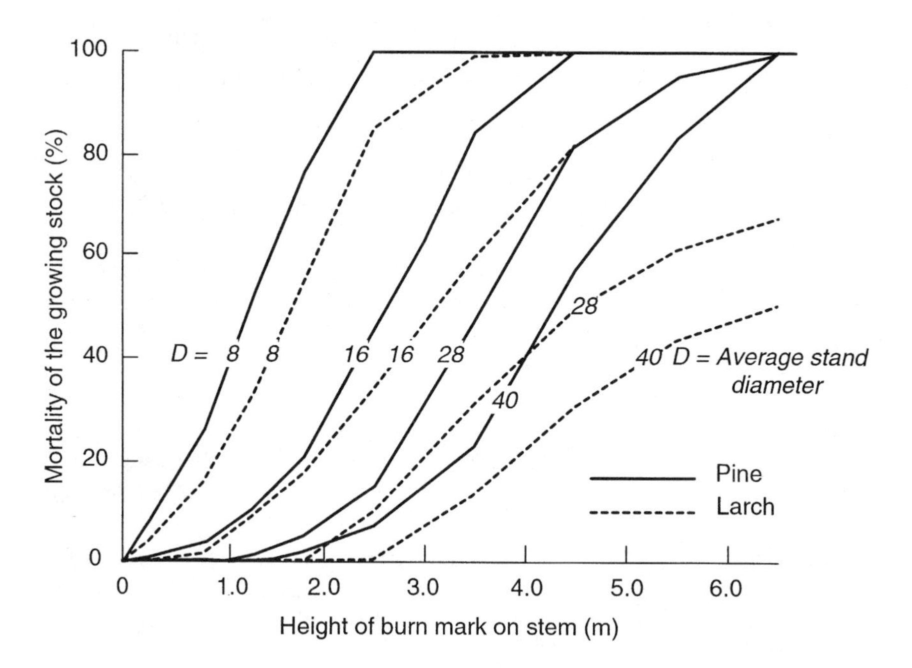

Figure 8.1. Postfire mortality of growing stock at different stem heights with fire burns and for different average stand diameters for pine and larch. (Based on Voinov 1986.)

60%. The conclusion is that the mortality is highly correlated with the diameter of trees. It has also been shown that in uneven-aged mature spruce stands, trees close to the average diameter (24–30 cm) have the least mortality, which increases with both thinner and thicker trees (Solovjov and Sheshukov 1976; Voinov 1986; Matveev 1992).

Generally, the rate of mortality is higher for larch than pine, but in some individual Siberian forest types, the opposite occurs. Mortality is higher in spruce than in cedar or birch (Solovjov 1973; Solovjov and Krokhalev 1973; Mishkov and Starodumov 1976; Solovjov and Sheshukov 1976; Mikhel 1984; Sibirina 1989; Valendik 1990; Matveev 1992). More complete relations have been presented for the Far East (Sheshukov et al. 1978).

Postfire Forest Regeneration

Natural reforestation after fires is a consequence of a great number of factors of different nature and different scales: geographic distribution and climate; structure of landscapes and locality of a burnt area in a landscape; type and peculiarities of relief; site characteristics (e.g., parent material, soil, drainage ability, type of moistening); biological and ecological properties of tree species; specifics of forest types and succession stages; type of fire regime and size of burns; and availability of seeds. Numerous regional publications are devoted to this problem (Starikov 1961; Utkin 1965; Manko and Voroshilov 1978; Pozdnjkov 1986; Chertovsky et al. 1987; Sokolov et al. 1994).

The process of postfire regeneration is strongly regional and site oriented. Features of natural postfire reforestation include the following: (1) the restoration ability of boreal forests is very high; areas of burnt and dead stands in Russia have decreased twofold during the past 30–35 years; (2) postfire reforestation in the extreme north (forest tundra, northern and sparse taiga) is slow and requires a rather long time (as long as 30–35 years) due to insufficient availability and quality of seeds; (3) productivity of forests of the first postfire generation on permafrost is two- to threefold higher than in undisturbed areas; (4) in all vegetation zones, excluding larch stands in the extreme north, stand-replacing fire results in successional patterns where a change in the dominant species occurs; and (5) recurrent fires often lead to impoverishment of forests and generation of grassy glades, paludification of sites, and finally, indefinite long periods of deforestation and green desertification. There are studies showing that a significant part of recurrent burns (as much as 30–50%) are not restored for many years; and (6) regeneration under canopy layer of existing stands after on-ground fire is dependent on the frequency of repeated fires on the same site.

References

Afanasjev, V.A. 1989. Geoclimatic peculiarities of forest fires in Kamchatka, pp. 119–120 in *Results of Study of the Far Eastern Forests and Their Multifunctional Use*. DalNIILKH [Far Eastern Forestry Research Institute], Khabarovsk, Russia. (in Russian)

Arzibashev, E.S. 1974. *Forest Fires and Fighting with Them.* Forest Industry, Moscow. (in Russian)

AUIRCFR. 1990. *Review of the Sanitary State of Forests in 1990.* All-Union Informational and Research Center for Forest Resources, Moscow. (in Russian)

Cahoon, D.R., Jr., J.S. Levine, and W.R. Cofer III. 1991. The great Chinese fire of 1987: a view from space, pp. 61–66 in J.S. Levine, ed. *Global Biomass Burning.* MIT Press, Cambridge, MA.

Cahoon, D.R., Jr., B.J. Stocks, J.S. Levine, W.R. Cofer III, and J.M. Pierson. 1994. Satellite analysis of the severe 1987 forest fires in northern China and southeastern Siberia. *J. Geophys. Res.* 99:18,627–18,638.

Cahoon, D.R., Jr., B.J. Stocks, J.S. Levine, W.R. Cofer III, and J.A. Barber. 1996. Monitoring the 1992 forest fires in the boreal ecosystem using NOAA AVHRR satellite imagery, pp. 795–802 in J.L. Levine, ed. *Biomass Burning and Climate Change. Vol. 2. Biomass Burning in South America, Southeast Asia, and Temperate and Boreal Ecosystems, and the Oil Fires of Kuwait.* MIT Press, Cambridge, MA.

Chertovsky, V.G., V.A. Anikeeva, N.I. Kubrak, and N.I. Kudrjavzeva. 1983. Change of forest litter properties after harvesting, pp. 211–212 in *Role of Litter in Forest Biogenesis.* Nauka, Moscow. (in Russian)

Chertovsky, V.G., B.A. Semenov, and V.F. Zvetkov. 1987. *Subtundra Forests.* Agropromizdat [Publishing House of the Agricultural Complex], Moscow. (in Russian)

Chervonny, M.G. 1979. *Aircraft Forest Protection.* Forest Industry, Moscow. (in Russian)

Chibisov, G.A., ed. 1974. *Increase of Forest Productivity in the European North.* AIFFCh [Arkhangelsk Institute of Forests and Forest Chemistry], Arkhangelsk, Russia. (in Russian)

Chudnikov, P.I. 1931. *Impact of Fire on Regeneration of Urals Forests.* Selkhozgiz [Agricultural State Publishing House], Moscow. (in Russian)

Conard, S.G., and G. A. Ivanova. 1998. Wildfire in Russian boreal forests—potential impacts of fire regime characteristics on emissions and global carbon balance estimates. *Environ. Pollut.* 98:305–313.

Evdokimenko, M.D. 1985. Specifics of growth of pyrogenic pine stands in the Lake Baikal region, pp. 35–47 in E.N. Elagin, ed. *Productivity and Structure of Forest Communities.* Institute of Forests and Timber, Krasnoyarsk, Russia. (in Russian)

Furyaev, V.V. 1970. Impact of fires and insect infestations on formation of forests between rivers Ket and Culim, pp. 408–421 in A.B. Shukov, ed. *Problems of Forestry. Vol. 1.* Institute of Forests and Timber, Krasnoyarsk, Russia. (in Russian)

Furyaev, V.V. 1996. *Role of Fires in the Forest Regeneration Process.* Nauka, Novosibirsk, Russia. (in Russian)

Furyaev, V.V., and D.M. Kireev. 1979. Study of Post–Forest Fire Dynamics on a Landscape Basis. Nauka, Novosibirsk, Russia. (in Russian)

Glebov, F.Z. 1976. Some tasks for forest bog investigations due to the peculiarities of forest and bog interaction in west Siberia, pp. 6—18 in *Theory and Practice of Forest Bog Science and Hydroforest Amelioration,* Institute of Forests and Timber, Krasnoyarsk, Russia. (in Russian)

Goldin, D.I. 1976. Development of forest drainage in the Tomsk region, pp. 113—119 in *Theory and Practice of Forest Bog Science and Hydroforest Amelioration.* Institute of Forests and Timber, Krasnoyarsk, Russia. (in Russian)

Goskomstat. 1993. *Forest Management in the Russian Federation.* Goskomstat [State Committee on Statistics of the Russian Federation], Moscow. (in Russian)

Gromtsev, A.N. 1996. Retrospective analysis of natural fire regimes in landscapes of eastern Fennoscandia and problems in their anthropogenic transformation, pp. 45–54 in J.G. Goldammer and V.V. Furyaev, eds. *Fire in Ecosystems of Boreal Eurasia.* Kluwer Academic Publishers, Dordrecht, The Netherlands.

Gundar, S.V. 1978a. On soil fire characteristics, pp. 142—144 in *Burning and Fires in Forests.* Institute of Forests and Timber, Krasnoyarsk, Russia. (in Russian)

Gundar, S.V. 1978b. *Soil Fire in the Nishne Amur River Basin: Prevention and Extinguishing.* Institute of Forests and Timber, Krasnoyarsk, Russia. (in Russian)

Gundar, S.V., and T.V. Kostirina. 1976. Prediction of peat fires, pp. 98–101 in *Increase of the Productivity of the Forests in the Far East.* Reports of the Far Eastern Forestry Research Institute 16. DalNIILKH [Far Eastern Forestry Research Institute], Khabarovsk, Russia. (in Russian)

Isaev, A.S. 1966. *Stem Insects on Dahurica Larch.* Nauka, Moscow. (in Russian)

Kireev, D.M. 1997. *Methods of Investigation of Forests by Air Photography.* Nauka, Novosibirsk, Russia. (in Russian).

Komin, G.E. 1967. Impact of forest fires on age structure and growth of northern taiga bog pine forests in the Ural Mountain region, pp. 67–76 in *Types and Dynamics of Forests in the Ural Mountain Region.* Sverdlovsk Technological Institute, Sverdlovsk, Russia. (in Russian)

Konev, E.V. 1977. *Physical Backgrounds of Vegetational Materials Burning.* Nauka, Novosibirsk, Russia. (in Russian)

Korovin, G.N. 1996. Analysis of the distribution of forest fires in Russia, pp. 112–128 in J.G. Goldammer and V.V. Furyaev, eds. *Fire in Ecosystems of Boreal Eurasia.* Kluwer Academic Publishers, Dordrecht, The Netherlands.

Krjuchkov, V.V. 1987. *North at the Boundary of the Third Millennium.* Mysl, Moscow. (in Russian)

Kurbatsky, N.P. 1962. *Techniques and Tactics of Forest Fire Extinguishing.* Goslesbumizdat [State Publishing House for the Forest, Pulp, and Paper Industry], Moscow. (in Russian)

Kurbatsky, N.P. 1970a. Investigations of quantity and properties of forest combustible materials, pp. 3—58 in *Problems of Forest Pyrology.* Institute of Forests and Timber, Krasnoyarsk, Russia. (in Russian)

Kurbatsky, N.P. 1970b. Classification of forest fires, pp. 384–407 in A.B. Shukov, ed. *Problems of Forestry. Vol. 1.* Institute of Forests and Timber, Krasnoyarsk, Russia. (in Russian)

Kurbatsky N.P. 1972. Terminology of forest pyrology, pp. 171–231 in N.P. Kurbatsky and E.V. Konev, eds. *Questions of Forest Pyrology.* Institute of Forests and Timber, Krasnoyarsk, Russia. (in Russian)

Kurnaev, S.F. 1973. *Forest Growth Division of the USSR.* Nauka, Moscow. (in Russian)

McPherson, G.R., D.D. Wade, and C.B. Phillips. 1990. *Glossary of Wildland Fire Management Terms Used in the United States.* University of Arizona, Tucson.

Manko, J.I., and V.P. Voroshilov. 1978. *Spruce Forests of Kamchatka.* Nauka, Moscow. (in Russian)

Matveev, P.M. 1992. *Post-Fire Impacts on Larch Biogenesis on Permafrost.* Marijsky Polytechnical Institute, Joshkar-Ola, Russia. (in Russian)

Melekhov, I.S. 1947. Nature of Forests and Forest Fires. Arkhangelsk Institute of Forests and Forest Chemistry, Arkhangelsk, Russia. (in Russian)

Melekhov, I.S. 1948. *Influence of Fires on Forests.* Goslestekhizdat [State Forest Publishing House], Moscow, Russia. (in Russian)

Melekhov, I.S. 1979. *Forest Pyrology 2.* Moscow Forest Technical Institute, Moscow. (in Russian)

Mikhel, V.A. 1984. Impact of major factors on trees' mortality under on-ground fires, pp. 124–129 in V.T. Chumin, ed. *Utilization and Regeneration of Forest Resources in the Far East,* Issue 26. Far Eastern Forestry Research Institute, Khabarovsk, Russia. (in Russian)

Mishkov, F.F., and A.M. Starodumov. 1976. Influence of fires on the shrub cedar stand with ash, pp. 106–107 in *Increase of the Productivity of the Forests in the Far East.* Reports of the Far Eastern Forestry Research Institute 18. DalNIILKH [Far Eastern Forestry Research Institute], Khabarovsk, Russia. (in Russian)

Moiseenko, S.N. 1958. Natural regeneration of pine and larch stands in the Amur oblast, pp. 158—176 in *Natural Regeneration of Forest in the Far East.* DalNIILKH [Far Eastern Forestry Research Institute], Dolinsk, Russia. (in Russian)

Orphanitskaja, V.G., and J.A. Orphanitskiy. 1959. Influence of the burning of clearcut areas on some soil properties, pp. 85–89 in *Report of the Arkhangelsk Forest Technical Institute.* Arkhangelsk Institute of Forests and Forest Chemistry, Archangelsk, Russia. (in Russian)

Ozhogin, I.M. 1979. Dependence between air humidity and forest fires. *For. Manage.* 8:27–32. (in Russian)

Popova, E.N. 1983. Role of the forest litter in the nutrition of phytocenoses in the Middle Anagara River Region, pp. 160—161 in *Role of Litter in Forest Biogenesis.* Nauka, Moscow. (in Russian)

Pozdnjakov, L.K. 1983. *Forest on Permafrost.* Nauka, Novosibirsk, Russia. (in Russian)

Pozdnjakov, L.K. 1986. *Frozen Forestry.* Nauka, Novosibirsk, Russia. (in Russian)

Rjabukha, A.A. 1973. Contra-fire measures on drained areas. *For. Manage.* 10:55—66. (in Russian)

Rojkov, V.A., D. Efremov, S. Nilsson, V.N. Sedykh, A.Z. Shvidenko, V. Sokolov, and V.B. Wagner. 1996. *Siberian Landscape Classification and a Digitized Map of Siberian Landscapes.* WP-96–111. International Institute for Applied Systems Analysis, Laxenburg, Austria.

Rojkov, V.A., V. Vagner, S. Nilsson, and A. Shvidenko. 1997. Carbon of Russian wetlands, pp. 112–113 in *Fifth International Carbon Dioxide Conference.* Cairns, Queensland, Australia, 8–12 September 1997. Extended abstracts. CSIRO Division of Atmospheric Research, Aspendale, Australia.

Sannikov, S.N., and J.G. Goldammer. 1996. Fire ecology of pine forests of northern Eurasia, pp. 151–167 in J.G. Goldammer and V.V. Furyaev, eds. *Fire in Ecosystems of Boreal Eurasia.* Kluwer Academic Publishers, Dordrecht, The Netherlands.

Saposhnikov, A.P., G.A. Selivanova, and T.M. Iljna. 1993. *Soil Generation and Peculiarities of Turnover in Mountain Forests of Southern Sikhote-Alin.* DalNIILKH [Far Eastern Forestry Research Institute], Khabarovsk, Russia. (in Russian)

Sheshukov, M.A. 1974. Turf fires and some of their peculiarities, pp. 158–161 in V.T. Chumin, ed. *Increased Productivity of Forests of the Far East. Vol. 16.* Far Eastern Forestry Research Institute, Khabarovsk, Russia. (in Russian)

Sheshukov, M.A. 1977. Types, intensity and driving forces of forest fire. *For. Manage.* 5:68–72. (in Russian)

Sheshukov, M.A. 1978. Impact of fires on development of taiga biogenesis, pp. 166–167 in N.P. Kurbatsky, ed. *Burning and Fire in Forests.* Institute of Forests and Timber, Krasnoyarsk, Russia. (in Russian)

Sheshukov, M.A., ed. 1983. *Recommendation of Forest Fire Preventive Measures and Extinguishing in the Zone of On-Ground Fire Protection of Far Eastern Forests.* Far Eastern Forestry Research Institute, Khabarovsk, Russia. (in Russian)

Sheshukov, M.A. 1988. Classification of forest combustibles, pp. 55–65 in G.D. Glavazkiy, ed. *Forest Fires and Fighting with Them.* All-Union Research Institute on Forestry and Forest Mechanization, Pushkino, Russia. (in Russian)

Sheshukov, M.A., V.I. Solovjov, and I.B. Naikrug. 1978. Influence of different factors on the damage of stands and tree species by fires, pp. 145–150 in *Use and Regeneration of Forest Resources in the Far East.* Reports of the Far Eastern Forestry Research Institute. DalNIILKH [Far Eastern Forestry Research Institute], Khabarovsk, Russia. (in Russian)

Sheshukov, M.A., A.P. Savtchenko, and V.V. Peshkov. 1992. *Forest Fires and Their Fighting in the North of the Far East.* DalNIILKH [Far Eastern Forestry Research Institute], Khabarovsk, Russia. (in Russian)

Shetinsky, E.A. and V.N. Sergeenko, eds. 1996. *Forest Fire Protection.* Federal Forest Service of the Russian Federation, Moscow, Russia. (in Russian)

Shostakovich, V.B. 1924. Forest fires in Siberia in 1915. *Trans. East-Siberian Div. Russian Geog. Soc.* 47:1–9. (in Russian)
Shvidenko, A. 1997. Global Significance of Disturbance in Boreal Forests. Paper presented at the 8th Annual IBFRA Conference, 4–8 August 1997, Duluth, Minnesota.
Sibirina, L.A. 1989. Post-fire break-up of the cedar-broadleaved forests, pp. 118–119 in *Results of Research on the Far Eastern Forests and Problems of Improvement of Multipurpose Use of Forests.* DalNIILKH [Far Eastern Forestry Research Institute], Khabarovsk, Russia. (in Russian)
Sofronov, M.A. 1978. Forest pyrological regionalization of the USSR State Forest Fund, pp. 108–109 in N.P. Kurbatsky, ed. *Burning and Fire in Forests.* Institute of Forests and Timber, Krasnoyarsk, Russia. (in Russian)
Sofronov, M.A., and A.B. Volokitina. 1990. *Recommendations on Forest Fire Protection for the Southern Taiga Bog Forests in Siberia.* Institute of Forests and Timber, Krasnoyarsk, Russia. (in Russian)
Sokolov, V.A., A.S. Atkin, and S.K. Farber. 1994. *Structure and Dynamics of Taiga Forests.* Nauka, Novosibirsk, Russia. (in Russian)
Solovjov, V.I. 1972. Some data on the impact of forest fires on stands, pp. 229–231 in *Utilization and Regeneration of Forest Resources of the Far East.* Far Eastern Forestry Research Institute, Khabarovsk, Russia. (in Russian)
Solovjov, V.I. 1973. Consequences of fires in the pine and larch forests of the Amur district, pp. 173–181 in *Utilization and Regeneration of Forest Resources in the Far East.* Reports of the Far Eastern Forestry Research Institute. DalNIILKH [Far Eastern Forestry Research Institute], Khabarovsk, Russia. (in Russian)
Solovjov, V.I., and A.K. Krokhalev. 1973. Some changes in pine stands due to post fire, pp. 182–189 in *Utilization and Regeneration of Forest Resources in the Far East.* Reports of the Far Eastern Forestry Research Institute. DalNIILKH [Far Eastern Forestry Research Institute], Khabarovsk, Russia. (in Russian)
Solovjov, V.I., and M.A. Sheshukov. 1976. Dynamics of the post-fire dieback in a spruce stand with green mosses, pp. 85–90 in *Increase of the Productivity of the Forests in the Far East.* Reports of the Far Eastern Forestry Research Institute. DalNIILKH [Far Eastern Forestry Research Institute], Khabarovsk, Russia. (in Russian)
Starikov, G.F. 1961. *Forests on the Northern Part of Khabarovsk Kray.* Far Eastern Forestry Research Institute, Khabarovsk, Russia. (in Russian)
Stocks, B.J. 1991. The extent and impact of forest fires in northern circumpolar countries, pp. 197–202 in J.L. Levine, ed. *Global Biomass Burning: Atmospheric, Climatic, and Biospheric Implications.* MIT Press, Cambridge, MA.
Telizin, G.P. 1970. Characteristics of forest combustible materials and their connection with peculiarities of burning. *Sci. Rep. DalNIILKH* 10:248—252. (in Russian)
Tumel, V.F. 1939. On some changes of frozen regimes of soils in connection with burning out of vegetation cover, pp. 64–79 in *Reports of the Permafrost Commission, Issue VIII.* Academy of Sciences of the USSR, Moscow. (in Russian)
Utkin, A.I. 1965. *Forests of Central Jakutia.* Nauka, Moscow. (in Russian)
Valendik, E.N. 1990. *Fighting Large Forest Fires.* Nauka, Novosibirsk, Russia. (in Russian)
Valendik, E.N., O.B. Vorobiov, and A.M. Matveev. 1977. Prediction of forest fire contours, pp. 52–66 in *Characteristics of the Processes of Forest Fires.* Institute of Forests and Timber, Krasnoyarsk, Russia. (in Russian)
Vasilenko, A.V. 1976. Role of fire in forestry, pp. 98–102 in V.G. Chertovsky, ed. *Current Research in Forest Typology and Pyrology.* Arkhangelsk Institute of Forests and Forest Chemistry, Arkhangelsk, Russia. (in Russian)
Voinov, G.S., ed. 1986. *Forest Inventory Reference Book for the North-East of the European Part of the USSR.* Arkhangelsk Institute of Forests and Forest Chemistry, Arkhangelsk, Russia. (in Russian)

Voinov, G.S., and M.A. Sofronov. 1976. Predicting tree mortality in a stand after on-ground fire, pp. 115–121 in V.G. Chertovsky, ed. *Current Research in Forest Typology and Pyrology.* Arkhangelsk Institute of Forests and Forest Chemistry, Arkhangelsk, Russia. (in Russian)

Vompersky, S.E. 1994. The role of mires in carbon circulation, pp. in 5–37 in *Biogenic Peculiarities of Mires and Their Rational Use.* Report from the XI Sukachev Memory Meetings. Nauka, Moscow. (in Russian)

Vompersky, S.E., E.D. Sabo, and A.S. Formin. 1975. *Forest Drainage Amelioration.* Forest Industry, Moscow. (in Russian)

Zvetkov, P.A. 1988. Post-fire regeneration of *Laris dahurica* in Evenkia, pp. 117–126 in *Forest Fires and Their Fighting.* Institute of Forest and Timber, Siberian Division, Academy of Sciences of the USSR, Krasnoyarsk, Russia. (in Russian)

9. Long-Term Perspectives on Fire-Climate-Vegetation Relationships in the North American Boreal Forest

Ian D. Campbell and Michael D. Flannigan

Introduction

The ecological role of fire in the modern boreal forest has been reviewed by several authors (Wright and Heinselman 1973; Wein and MacLean 1983; Johnson 1992, Payette 1992). Wildfire and climate are intimately linked (Swetnam 1993). Shorter-term weather patterns are clearly an important determinant of fire activity at daily-to-monthly time scales. At longer time scales of decades and centuries, Swetnam (1993) has shown that fire occurrence in Giant Sequoia groves was related to climate. However, other factors such as topography and fuel play a role in the fire regime. Fuel type, structure, moisture, accumulation, density, flammability, and continuity can affect the fire regime.[1] An extreme example highlighting the role of fuel can be found in many deserts where fires should be frequent due to the hot and dry environment but are absent because there is no fuel. Therefore, climate controls fire directly through opportunities for ignition and spread but also through regulation of the accumulation and structure of fuel at decadal or greater time scales.

Fire, in turn, plays a large role in determining forest stand patterning both in space and in time, and hence in fuel characteristics. There is also a further obvious follow-through effect on carbon storage, both in the tree biomass and other boreal

[1]Fire regime has four components: fire intensity, fire frequency, timing (season) of the fire, and fire size (Malanson 1987).

carbon pools such as peats and soils. Further, fire injects aerosols into the atmosphere and thus affects climate directly (Clark et al. 1996; Levine and Cofer, this volume). At long time scales, there are therefore feedbacks linking fire, forest structure, carbon storage, and climate (Fig. 9.1).

This chapter reviews the record of past and present North American boreal fire regimes at various time scales and how they relate to both climate and vegetation. From this review, we draw some possible conclusions regarding possible future fire regimes and their likely linkages with forest structure and carbon storage over longer time scales.

Fire and Fuel

Quinby (1987) found that in Algonquin Park, in the Great Lakes–St. Lawrence forest (transitional between boreal and eastern deciduous forest), the fire incidence probability (FIP) differed markedly between different species, with the highest FIP occurring in pine stands, second highest in shade-intolerant hardwoods (mainly birch and aspen), and lowest in the shade-tolerant hardwoods (mainly sugar maple). Quinby (1987) concludes that this reflects both a greater flammability of pines and intolerant hardwoods and a greater likelihood of postfire colonization by pines and intolerant hardwoods. Also, Flannigan and Wotton (1995) found that jack pine volume was strongly correlated with fire intensity.

This may seem to raise the specter of a chicken-or-the-egg argument—when climate changes, does an increase in fire promote fire-prone vegetation, or vice versa? In some cases, they do promote each other and so may create a positive feedback loop of increasing fire and increasing fire-prone vegetation until a new equilibrium is reached.

Most intolerant hardwoods are readily able to invade burned areas where competition is absent. In addition, aspen and birch are able to reproduce vegetatively

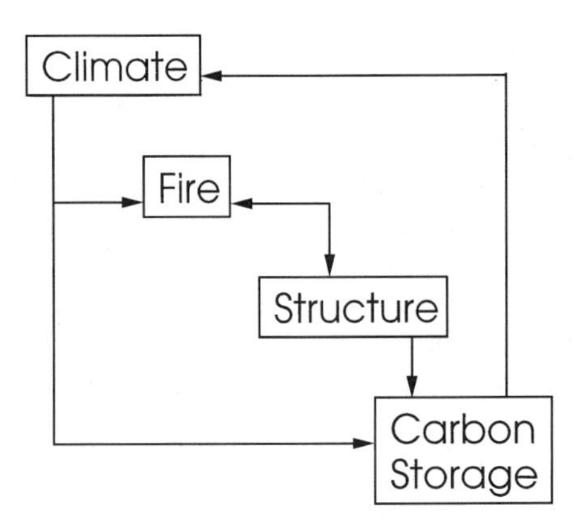

Figure 9.1. Conceptual interactions between climate, fire, forest structure, and carbon storage at long time scales. At shorter time scales, fire does exert a direct influence on carbon storage, but this becomes negligible when compared with the impact of changing forest structure at long time scales (Kurz and Apps 1994). Forest structure may also have a direct effect on climate through albedo and surface roughness (Bonan et al. 1992).

after fire. Several conifers, however, are adapted specifically to fire; in the case of jack pine, the serotinous cones open readily when heated by a fire, so that abundant jack pine seed is often available following a fire. Many boreal species also prefer a mineral seedbed of the type found after a deep-burning fire for germination and establishment.

In Canada, fire weather follows a broadly zonal distribution, with fire weather index values increasing southward (see Fig. 20.1). Actual area burned by fire, however, is only poorly geographically correlated with average fire weather (see Fig. 15.4), and average fire weather values are poor predictors of area burned, as most of the area burned occurs on a few days with extreme fire weather conditions. Fire is more abundant in the drier western boreal forest than in the moister eastern boreal. Fire may regulate the southern border of the boreal forest in the west, where grass fires have been proposed as a limiting factor in seedling establishment (Bird 1961; Strong 1977; Looman 1979; Archibold and Wilson 1980), although it has also been suggested that fire alone may not have been effective in this regard (Campbell et al. 1994). However, in some dry regions that cannot support continuous vegetation cover, fire declines in importance, as in the open tundra north of the boreal forest or in the bunchgrass steppe to the southwest. Thus there is clearly a fuel limitation on fire in Canada.

In most of the boreal forest, fuel abundance is not a limiting factor except at short time scales (less than a few decades) and small spatial scales (the area of a stand). In the more open woodland to the north of the boreal forest, into the open tundra, and also in the southwestern grassland and aspen parkland, fuel abundance may be a limiting factor at very short (annual to decadal) time scales.

For instance, at a site near Red Deer, Alberta, in the modern aspen parkland, the paleocharcoal record suggests an increase in fire with the recent development of the aspen forest. It appears that the aspen forest developed in the late 1800s, not as a result of climate, fire suppression, or agriculture activities but rather as a result of the extirpation of bison. Prehistorically, bison would have controlled aspen by destroying new shoots that developed after a grass fire had killed mature trees; with the elimination of the bison, an aspen forest was able to develop along the southern margin of the western boreal forest (Campbell et al. 1994).

If in the extreme cases of tundra and desert the fire regime is partly controlled by fuel parameters, it is reasonable to suppose that this must also be the case in the main boreal forest, albeit to a lesser extent. Using fire models and fuel data from subalpine forests, Bessie and Johnson (1995) have shown that fuel considerations are relatively unimportant to fire behavior when compared with weather. However, these subalpine forests are dominated by conifers, and the results from Bessie and Johnson (1995) would not apply to the western boreal where significant portions of the landscape vegetation is deciduous. Clark (1989) has suggested a logarithmic increase, through time, in the probability of a surface fire following a fire in Minnesota. This probability curve is explained as a fuel accumulation function. Others have noted that there is a strong probability of surface fire in the few years following a crown fire in spruce and pine stands (Johnson 1992)—presumably as the trees killed by the first fire may provide ample fine fuel along with grasses and forbs. Nevertheless, after this initial period, fire probability is reduced until the stand reaches maturity, after which it follows an exponential rise.

In large parts of the boreal forest, stand age distributions suggest an equal probability of fire for stands of all ages (Johnson 1992). This may reflect the tendency of these regions to be dominated by very large fires. The probability of a fire starting may not be equal in all stands, but once started, the probability of propagation into a stand may be nearly stand age independent where large fires dominate. However, the probability of a fire occurring in the landscape as a whole rather than in a given stand may still be partly a function of the landscape fraction covered with a vegetation type allowing for a high ignition probability.

If Clark's (1989) suggestion of a logarithmic increase in ignition probability with fuel accumulation is correct, we would expect to find that different fuel accumulation rates result in different curves of postfire probability of ignition. Different fuel accumulation rates would depend, at the scales of landscapes rather than of an individual stand, as well as on the climate as much as on anything else, warmer and moister climates could be expected to both generate and decompose fuel more rapidly. However, the need for the fuel to be dry would also affect the probability of fire, so that the greater fuel accumulation rate in a warmer moister forest would be at least partially offset by the lower likelihood of ignition and spread due to fuel moisture. Just how these two factors interact and under what climate the resulting peak in fire frequency (number of fires per unit time at a site) occurs are unclear.

Although the grassland climate is typically drier than the boreal forest, with much lower fuel accumulation rates, the fire return interval is much shorter due to the abundance of fine fuels and low moisture storage in the vegetation. More to the dry extreme, however, is the desert, where the lack of fuel excludes fire. At the other extreme, the luxuriant rainforest also is relatively free of wildfire. We should perhaps envisage an inverted parabolic relationship between climatic moisture and fire frequency—with the western boreal near the apex, where moisture is sufficient for rapid fuel accumulation but dry periods frequent and severe enough to allow abundant opportunity for fire ignition and spread (Fig. 9.2). It is worth noting in this regard that the distribution of precipitation in time is critical, as a wet climate may nevertheless experience a pronounced dry season during which the vegetation would be fire prone. Flannigan and Harrington (1988) found in a study relating area burned in Canada to meteorological variables that distribution of precipitation (i.e., the frequency of dry spells), and not precipitation amount, was the critical factor.

The eastern boreal region is more humid than the western boreal region and also has longer fire return intervals. In eastern Canada, there is an increased abundance of relatively fire-intolerant species such as balsam fir. In the eastern boreal region, we could therefore expect a warmer/drier climate to cause an increase in fire activity. Bergeron and Flannigan (1995) used General Circulation Model (GCM) outputs to predict future Fire Weather Index (FWI) values. Despite increased precipitation in the GCM output, the FWI in western Canada was estimated to be

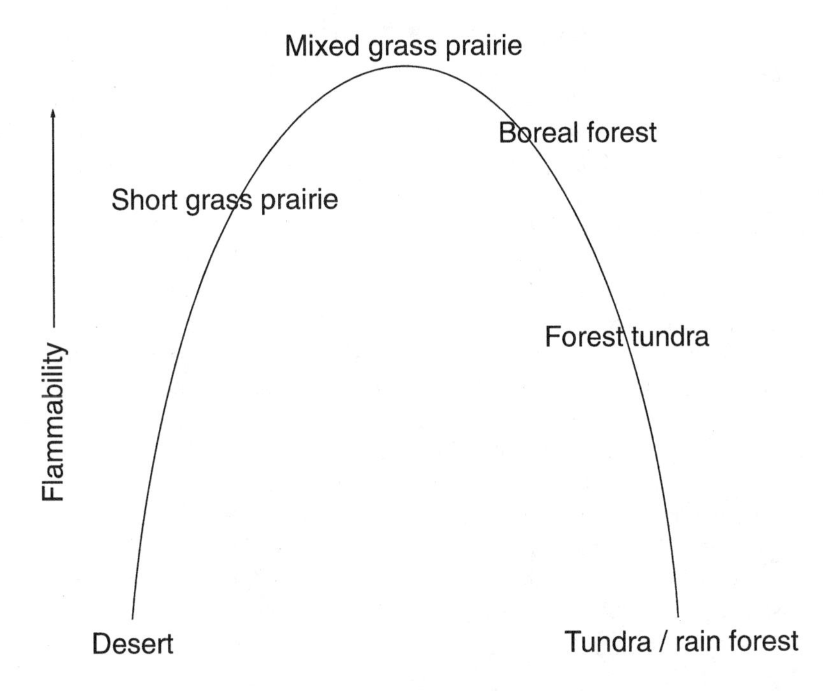

Figure 9.2. Conceptual relationship between major bioclimates and fire regime.

generally higher than today because the increased temperatures more than made up for the increased precipitation. In eastern Canada, the situation was reversed: lower FWI values are predicted, because the increase in temperature is insufficient to outweigh the increase in precipitation. Climate, of course, is not the only determinant of fire, and the GCM outputs are still relatively crude, particularly with regard to precipitation.

Such geographic comparisons are fraught with difficulties when we attempt to use them for predicting the impacts of climate change. Not only do species assemblages, topography, soils, geology, and history of land use differ from region to region, but so does the response of the regional climate to global forcing factors. Records of past fire regimes at the same location can be used to examine the effects of climate change.

Recent human activities in North American boreal forests have likely modified fire regimes, although in what way is difficult to determine and likely site-specific. For instance, anthropogenic fires clearly cluster around population centers and transportation routes, whereas fire suppression activities focus on the commercial forest and inhabited areas. Determining how these activities have affected fire regimes is complicated because of the lack of detailed reliable data on prehistoric fire regimes (Barney and Stocks 1983). Also, although there is little evidence for it, many authors have assumed that prehistoric people may have had a significant role in the fire regime (Day 1953).

One aspect of the fire–fuel interaction that has received considerable attention is flammability. Mutch (1970) put forward a hypothesis that fire-dependent plant communities are more flammable than non–fire-dependent communities due to natural selection. Some believe this hypothesis is flawed from an evolutionary perspective (Snyder 1984), whereas others have put forward an individualistic argument for the evolution of flammability (Bond and Midgley 1995). For our present purposes, we are concerned not with the evolution of flammability but with the following questions: does fire promote flammable species, and do flammable species promote fire? The answers are not clear. Some flammable species such as jack pine and black spruce thrive in an environment disturbed by fire. However, there are other boreal species that are not as flammable but also thrive in a fire-disturbed environment, such as aspen, birch, and pin cherry. Conversely, some species are flammable but are eliminated from sites after a fire. Balsam fir is one such example. Intuitively, the fire regime should depend on the percentage cover of flammable species at the landscape level. Regions with higher flammability would promote fire through easier ignition and through a longer smouldering period that would increase the likelihood of experiencing conditions conducive to fire growth. In this chapter, we review the Canadian paleofire record with specific reference to elucidating the roles of vegetation and climate in determining fire activity levels.

Methods for Studying Historical Fire Regimes

With fire return intervals of a century or more in much of the boreal forest, historic data are not sufficient to assess the effects of changing climates on fire regimes and forest cover. At best, we have noted the increases in fire incidence during warm/dry periods, such as the 1920s and 1930s and the 1980s and 1990s (Van Wagner 1988). Because these climate events were of short duration, forest composition and structure would not have had time to equilibrate with the climate. It is possible that the disequilibrium effects of such abrupt climate changes masked what would eventually have been the long-term effects had the changed climate persisted.

Tolonen (1983) discussed three sources of long-term fire regime data: fire scars, other vegetation indicators, and charcoal in peat or lake sediments. Fire scar studies allow us to extend the record prehistorically, but this extension is limited by the longevity of the trees (from one to three centuries in most of the boreal forest). The death of scarred trees, from other causes or from subsequent fires, degrades the quality of the record with time, so that high-quality fire scar records can only reasonably be constructed for a period of less than 200 years in most of the boreal forest.

The only way to reconstruct fire records spanning more than a few hundred years is through analysis of lake sediments or other charcoal-accumulating stratigraphic deposits. Filion and co-workers (1991) have reconstructed changes in fire regimes in Quebec from charcoal beds buried in sand dunes. Charcoal beds in peat have also been used (Khury 1994), as has charcoal abundance in lake sediments (Winkler 1985; Patterson et al. 1987; Clark 1988).

Charcoal influx studies in lake sediments are most often carried out by using slides prepared for pollen analysis (most pollen preparation procedures do not eliminate charcoal from the sediment). In this approach, all charcoal or charcoal fragments above a certain size are counted by microscopic examination or through a chemical assay method (Winkler 1985). Although the results are not directly comparable, they all give very rough approximations of fire activity (MacDonald et al. 1991). The fine fraction of charcoal appears to reflect regional fire activity, whereas macroscopic charcoal reflects fire within the drainage basin of the site. The nitric acid assay technique is found to be very promising and to reflect mainly regional fire activity because the bulk of the charcoal is in the fine fraction. Progressively coarser fractions can be expected to represent progressively more local fires (Clark and Royall 1996). A difficulty with the nitric acid assay method lies in the possibility of historic contamination by fly ash and other industrial emissions; however, this does not appear to be a concern except downwind of industrial centers. A greater problem in some regions may be contamination of the sediment with Cretaceous or Tertiary coal. This would also affect optical charcoal counts to the extent that some coal fragments may not be distinguishable from charcoal.

A more rigorous technique involves the use of varved (annually laminated) lake sediments, which are plasticized, thin-sectioned, and used to identify years of high coarse charcoal influx into the lake (Clark 1988). This method gives a reasonable estimate of the years of fire within the lake basin, although varved lakes are not available in all areas. A major advantage of this method is its annual resolution. Samples prepared for pollen analysis are rarely contiguous annual increments, so that fire years could fall between samples, or be mixed with low charcoal influx years, smoothing the signal. Even with nitric acid assay, sampling is not likely to yield annual resolution due to the amount of sediment required for accurate assay.

Paleofire Histories

Holocene Climate History of Canada

Figure 9.3 represents in broad terms the climate changes that have occurred in Canada over the past 10,000 years. The reality is inevitably much more complex, and interested readers are referred to Matthews and associates (1989) and Ritchie (1987) for an introduction to the complexities of Canadian paleoclimate and paleovegetation reconstructions.

By the start of the Holocene at 10 KYBP (thousands of years before present), glacial ice had receded from large parts of Canada. The climate at the edge of the retreating ice sheet was undoubtedly cold. Although in many places (particularly in the eastern Great Lakes region) there is evidence for forest-tundra in fairly close proximity to the ice, there is also abundant evidence of periglacial features such as

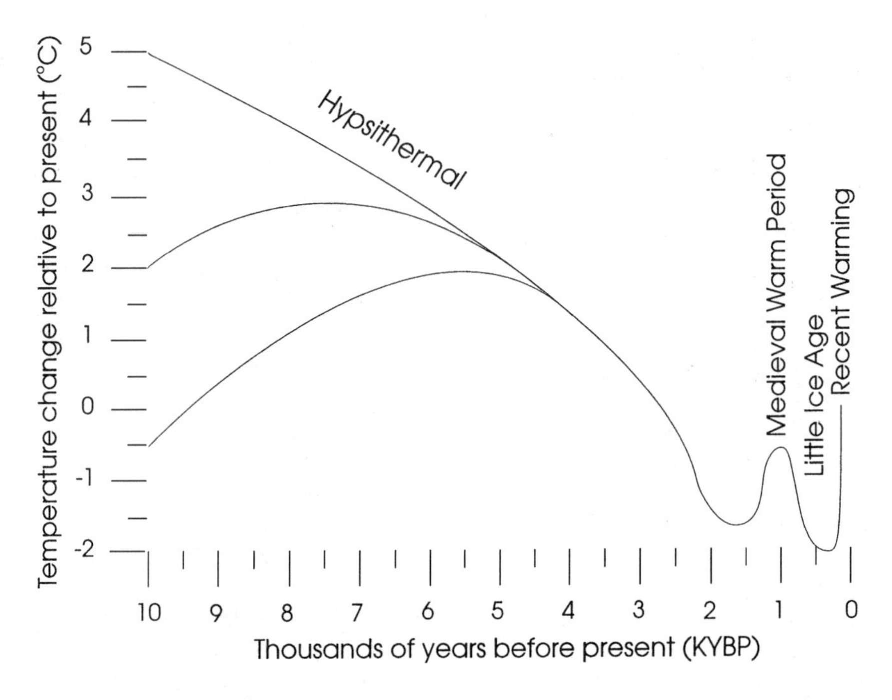

Figure 9.3. Generalized climate history of Canada. The early to mid-Holocene was generally warmer and drier than the present; one recent estimate suggests that the mean annual temperature was as much as 5°C higher at 6 KYBP (Zoltai, 1995). Although the broad climate change from the early Holocene to the late Holocene is adequately explained by variations in the earth's orbital parameters (COHMAP 1988), the minor climate fluctuations such as the Medieval Warm Period and the Little Ice Age are as yet poorly understood, although several forcing mechanisms have been proposed for the Little Ice Age (Grove 1988).

fossil ice wedges and patterned ground, indicating that the mean annual temperature was below 0°C. It is likely that a several-hundred-kilometer-wide swath of nearly treeless tundra dominated newly deglaciated terrain.

The early Holocene in most of Canada is marked by a rapid postglacial warming, ending between 12 and 6 KYBP in different regions. Temperatures reached up to 5°C warmer than present (Zoltai 1995). The summer insolation maximum occurred between 12 and 9 KYBP. As much of Canada was still ice covered or influenced by regional ice masses or cold ocean surfaces, this insolation maximum is not reflected as a thermal maximum in much of Canada. Instead, the thermal maximum lags the insolation maximum by varying intervals in different places, due to regional climate cooling by the waning ice mass and cold ocean surfaces. Therefore, in a region that was deglaciated early, the thermal maximum occurs early. In a late-deglaciated region, the thermal maximum occurs late. In a region such as parts of southeastern Canada, which were deglaciated fairly early but remained under the influence of upwind ice masses, warming may have started early but not reached a peak until the regional ice mass was gone.

Following the early Holocene thermal maximum (hypsithermal), each region experienced a gradual cooling until about 2 KYBP, at which time there was a minor cool/wet period. This period is poorly represented in most sites partly because it was of short duration (perhaps 500 years) and partly because it was of low magnitude. A warm period about A.D. 900–1100 corresponding to the European Medieval Warm Period shows in most high-resolution and high-sensitivity proxy records, as does a cool/moist period about A.D. 1450–1850 corresponding to the European Little Ice Age. The past 100 years (the period of instrumental record in most of Canada) is marked by a significant 1.1°C warming trend (Gullett and Skinner 1992), although there are some regions (particularly east and north of Hudson Bay) that show a weak cooling or little change.

Eastern Canada

Foster and King (1986) described birch forests in southeastern Labrador. Birch occurs only on sites that have burned within the past 110 years, and the border between the birch and the adjacent conifers is distinct. In this area, the interval between fires is believed to be longer than in most of the rest of the boreal forest, perhaps by several hundred years, allowing a thicker organic soil horizon to develop and affecting the vegetation succession after a fire does occur (Foster 1985). One effect is the establishment of a multiaged forest on some sites where the presence of charred organic material prolongs the period of seedling establishment by several years or even decades; another effect is the development of mixed stands as initially pure paper birch stands are progressively invaded by conifers.

In the 110-year period A.D. 1870–1980, major fires in southeastern Labrador occurred in 1890, 1898, 1951, 1959, and 1975–1976 (Foster 1983), the fire rotation period (time between stand-replacing fires at a given site) is estimated at 500 years. There was a significant absence of fire from 1899 to 1950 and again from 1960 to 1971. The years of high fire activity were also low summer precipitation in the study area but were not necessarily low-precipitation years or high-fire-incidence years elsewhere in the eastern boreal forest. Thus the conditions suitable for fire may be local. Interestingly, young stands were most frequently burned, with mature stands often acting almost as fire breaks. This is due to the moist climate, allowing progressive paludification of older stands.

Green (1982) investigated charcoal and pollen in a Nova Scotia lake. He found greatest fire frequencies when spruce, pine, and birch were most abundant on the landscape and lowest fire frequencies during periods of dominance of beech, hemlock, and fir. Fire was most abundant in the past 2000 years, despite this being a relatively cool period. This period also showed a moistening trend throughout eastern North America (Ritchie 1987). This may be the effect of cooler climate promoting pine and spruce at the expense of beech, which would tend to suggest a fuel-determined, rather than directly climate-determined, fire regime at millennial time scales.

Quebec

Fire at the treeline in northern Quebec shows patterns relating to vegetation and possibly climate. In the shrub tundra, fire is infrequent, and average fire sizes are small. In the open forest, fire is more frequent, and average fire sizes are larger; in the closed forest, fire activity is still greater (Payette et al. 1989). The gradient is steep; in the northern boreal forest, the fire rotation period is estimated at about 100 years, whereas in the shrub tundra, it is nearly 10,000 years (Payette et al. 1989). Payette and colleagues (1989) conclude that this gradient in fire regimes is partly a result of climate but also largely a result of vegetation type.

Filion and associates (1991) investigated fire activity in northern Quebec, using paleosols and buried charcoal beds in sand dunes. They found strong differences between the tundra, forest tundra, and boreal forest over the past 4650 (^{14}C) years. In the tundra, fire was very rare. In the forest tundra, fire was rare until 1650 B.P., at which time fire frequencies in the forest tundra became greater than the mean for the boreal forest. Forest-tundra fire frequencies have remained high and are presumed to be the cause of continuing large-scale deforestation. In the boreal forest, fire frequencies fluctuated through time but were generally greater during periods of cool climate than during periods of warm climate.

One possible mechanism that could explain the findings of Filion and coworkers (1991) is that a rapidly cooling climate would cause an increase in individual tree mortality in this cold-stressed region. This, in turn, would increase the available fuel load. Stocks (1987) found that forest fire potential in spruce budworm–killed balsam fir was significantly higher for more than 10 years after stand mortality. As the last few thousand years have been dominantly a time of cooling, the strong increase in fire in the forest tundra would be easily explained by this mechanism. If this was the case, then we could expect future climate warming/drying to cause an increase in individual tree mortality near the southern margin of the boreal forest. This could increase fire in this region because of increased fuel loading, independently of increased ignition opportunities or fire weather.

Sirois and Payette (1991) have found that recent fires in the forest tundra resulted in deforestation, as the burned areas are reverting to tundra. In the northern boreal forest, some burns are regenerating as forest tundra. This is consistent with the earlier findings of Payette and Gagnon (1985) that indicated that increased fire activity was the immediate cause of a period of southward treeline recession in northern Quebec, starting about 3 KYBP. This implies a southward retreat of the ecotones, which is consistent with the generally cooling climate of the past few thousand years. Here, fire is the mechanism by which the forest is moving more rapidly toward equilibrium with the new climate than it otherwise might.

Farther south, in the southern boreal Abitibi region of western Quebec, Dansereau and Bergeron (1993) found that the distribution of small fires is strongly influenced by factors such as topography and water bodies. This involves not only limitations on fire spread due to the presence of the water bodies but also a concentration of lightning activity on the emergent topography. This was also seen in the research of Bergeron and Brisson (1990) and Bergeron (1991), in which the fire regime on islands is very different from that on the mainland. In any modeling or extrapolation to the future of fire regimes, these topographical influences must be considered.

Bergeron (1991) also found that there was reduction in fire activity in western Quebec at around A.D. 1850. This corresponds to the end of the Little Ice Age and the subsequent climatic warming. Bergeron and Archambault (1993) showed that a decrease in the number and extent of fires correlated with a reduced drought frequency.

Northern Quebec also provides evidence that the spruce forest can regenerate in the absence of fire. Payette and Morneau (1993) demonstrated that lichen-spruce woodland relicts, initiated some 3000–4000 years ago, continue to self-perpetuate through layering in the absence of fire.

Ontario

Clark and Royall's (1995) charcoal record from Crawford Lake (southern Ontario) does not show a strong variation in charcoal influx. Clark and Royall (1995) show variations in the data record to anthropogenic influences, although this has been debated (Campbell and McAndrews 1995; Clark 1995). Indian-set fire has also been invoked to explain the presence of oak savannas in southern Ontario (embedded in the deciduous forest) at the time of European settlement. Szeicz and MacDonald (1991) have demonstrated that the savannas were edaphically controlled and were not caused by anthropogenic fire, although natural fire in these excessively well-drained sites may have played a role. Campbell and Campbell (1994) have suggested that the prehistoric anthropogenic impact on the forest landscape of southern Ontario was minimal, due in large part to the small prehistoric population.

Farther north, in the Great Lakes–St. Lawrence forest in Algonquin Park, Cwynar (1978) found that during the period A.D. 770–1270 fires occurred roughly every 80 years. Modern fire activity as determined from fire scars shows that a major fire now occurs every 45 years on average. Because Cwynar's (1978) study spanned only the 500-year period from A.D. 770 to 1270, it is difficult to interpret this change in fire regime. Nevertheless, the pollen diagram indicates that at some time since A.D. 1270, the percentage of pine in the forest increased. This is presumed to be in response to the climatic cooling of the Little Ice Age (Campbell and McAndrews 1991). If so, then the scenario would have to be that the climate change brought about a change in the vegetation, and it was the change in vegetation that brought about the change in fire regime. This would be very similar to the scenario discussed below for Minnesota.

Minnesota and Wisconsin

At several sites in Minnesota, in the oak savanna (transitional between the boreal forest and the prairie), Clark (1990) determined the annual rate of charcoal influx to varved lakes. Assuming that the charcoal flux is a proxy for fire activity, Clark (1990) was able to reconstruct several interesting features of past fire regimes in the area. Most notably, in the past 750 years, the background charcoal influx peaked in the period A.D. 1400–1600 (Clark 1990).

The pollen records from Minnesota suggest that as climate cooled from the maximum warmth of several thousand years ago, the study area was invaded first by oak savanna, then white pine. About A.D. 1400, red/jack pine became more common (Grimm 1983; Clark 1990). Fire increased along with red/jack pine, and aspen also peaked in the period A.D. 1400–1600. After A.D. 1600, red/jack pine was temporarily reduced but regained its importance after about A.D. 1800. The question this poses is whether the fire regime was responding to the climate and the vegetation was responding to the fire or whether the vegetation was responding to the climate and the fire regime was responding to the vegetation.

Clark (1990) compared this charcoal record with a varve thickness record from Hector Lake (Leonard 1986) in the Canadian Rocky Mountains, which shows glacial advance after A.D. 1600. However, Gajewski (1988), Bernabo (1981), and Grimm (1983) show climate cooling in the Great Lakes region starting as early as A.D. 1400, which agrees with the general Little Ice Age chronology.

Because the fire regime peaked during a period of cooling, it is most likely that the fire regime was responding to the change in vegetation, which was, in turn, responding to the change in climate. However, there was likely a complex interplay between fire regime and vegetation; the aspen increase was probably a response to the fire regime rather than directly to climate; and the competition between white pine and red/jack pine was also likely influenced by the fire regime. After A.D. 1600, increases in precipitation may have outweighed vegetation effects with respect to the fire regime. Another possibility is that the vegetation and therefore the fire regime slightly lagged behind climate, and the peak in red/jack pine, and consequently in fire activity, was a delayed response to the preceding Medieval Warm Period.

Gajewski and co-workers (1985) found generally higher levels of charcoal in three Wisconsin sites during the Little Ice Age. At another site farther northwest in Minnesota, Swain (1973) found a peak in charcoal and hence fire activity in the late 17th century during the coldest, wettest period of the Little Ice Age. Altogether, however, the Little Ice Age at this site had a lower charcoal influx than did the preceding Medieval Warm Period. Nearby, in Wisconsin, Swain (1978) found a decrease in fire activity as measured by charcoal on pollen slides during the Little Ice Age. Interestingly, Clark's (1990) site falls very near the range limit of jack pine, Swain's (1973) Minnesota site falls well within the range of jack pine, and the Wisconsin sites fall outside the range of jack pine. It may be that the dynamics of jack pine in response to the climate change drove the fire activity.

Western Canada and Alaska

Studies of charcoal layers in peat deposits in Saskatchewan, Manitoba, and Keewatin suggest a peak in fire activity about 3.7 KYBP (Bryson et al. 1965; Nichols 1967a,b). Nichols (1967a) has suggested that this peak, which is part of a broader period of increased fire activity from 6–1.5 KYBP, may be related to the global posthypsithermal cooling, which would have brought increased incursions of cold and dry arctic air into the boreal forest. However, continued cooling since 3.7 KYBP has not further increased fire activity.

Johnson and associates (1990) studied the fire regimes of Glacier National Park in British Columbia through fire scars and stand origin maps. They found that there had been a decrease in fire frequency after A.D. 1760, which they attribute to the Little Ice Age. They also found that despite an active fire suppression policy in the park, there has been no observable change in fire activity in recent decades.

MacDonald (1987) found an increase in charcoal in lake sediment through the Holocene near the Alberta/British Columbia border northwest of Edmonton. This increase corresponds with the migration of pine into the area about 6500 B.P. At that time, although climate was starting to become cooler and moister, pine had only recently reached this part of western Canada after several millennia of postglacial migration from the southeast. It seems that this is one clear case in which the increase in pine brought the increase in fire rather than the fire bringing the pine.

Farther east, Khury (1994), working with charcoal beds in peatlands of Alberta, Saskatchewan, and Manitoba, found that there was much more frequent fire during the hypsithermal period prior to 5 KYBP than subsequently, by a factor of about 2. The sites at which this is found are today all in the middle or high boreal. During the hypsithermal, they would likely have still been in the boreal forest. This agrees with the results of charcoal counting on pollen slides from sites near Edmonton, where the charcoal influx was greater during the hypsithermal than after about 4 KYBP (Vance et al. 1983).

MacDonald and co-workers (1991) also found a peak in fire activity in Wood Buffalo Park in northern Alberta during the mid- to late 1800s and another peak in the past 40 years. Their published record only extends 200 years into the past but was studied by several methods, including microscopic counts on pollen slides, macroscopic counts, and nitric acid assay.

Timoney and Wein (1991) examined the modern fire pattern in northwestern Canada through aerial photography. Larger burns occurred in the closed forest than in the forest tundra, demonstrating the impact of fuel continuity on fire extent.

Landhaeusser and Wein (1993) studied the area of a recent burn near Inuvik. They found that the burn had allowed partial invasion of former shrub tundra by paper birch and balsam poplar, which had also partially replaced spruce in previously treed areas. They conclude that recent climate warming has made treeline advance into the tundra possible but that this advance can be hastened by clearing the shrub tundra by fire.

Three sites in Alberta and Saskatchewan have recently been analyzed by using the Winkler method of charcoal assay. These are Pine Lake, in the aspen parkland of southern Alberta; Amisk Lake, in the southern boreal of central Alberta; and Opal Lake, in the midboreal of central Saskatchewan. At Pine Lake, the recent charcoal record suggests a change in fire activity with the decline of bison and the increase in aspen in the mid-1800s (Fig. 9.4; C. Campbell, personal communication). This lake is large enough that it can be expected, by analogy with pollen dispersal and representation (Prentice 1985), to carry a reasonably regional fire signal.

Farther north, the Amisk Lake charcoal record shows a strong increase in fire activity corresponding with the recent warming starting in the mid-1800s (Fig. 9.5). This lake is larger than Pine Lake, and the signal is likely representative of much of the central Alberta southern boreal forest. It is unlikely that the increase in charcoal in the recent period is due to industrial emissions, as there is no heavy industry upwind of this area and no opaque spherules diagnostic of industrial emissions (Goldberg, 1986) were found on pollen slides. The charcoal minimum at this site about A.D. 1870 is difficult to interpret in the absence of a longer prehistoric record to place it in climatic context, but certainly there was little or no anthropogenic cause for this minimum in this region at that time.

To the east, in the midboreal of central Saskatchewan, the record at Opal Lake shows a decline in fire in the mid-1800s and again in the late 20th century (Fig. 9.6). This lake is small, only about 2 ha in area, and the charcoal signal is likely dominated by production from the local area of less than 100 ha. The area around the lake is presently dominated by fairly open pine-lichen and aspen stands on well-drained coarse soils. It may be that during the Little Ice Age there was more pine and less aspen and so a more fire-prone landscape. The warming at the end of the Little Ice Age in the mid-1800s would have caused an opening of the stands and an increase in aspen. The 20th-century reduction may be related to

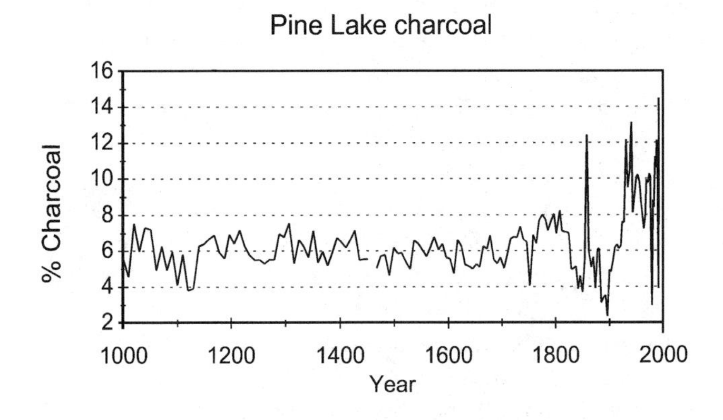

Figure 9.4. Charcoal abundance record from Pine Lake, Alberta (C. Campbell, personal communication). Dating control from ^{210}Pb, volcanic ash, and geochemical stratigraphy.

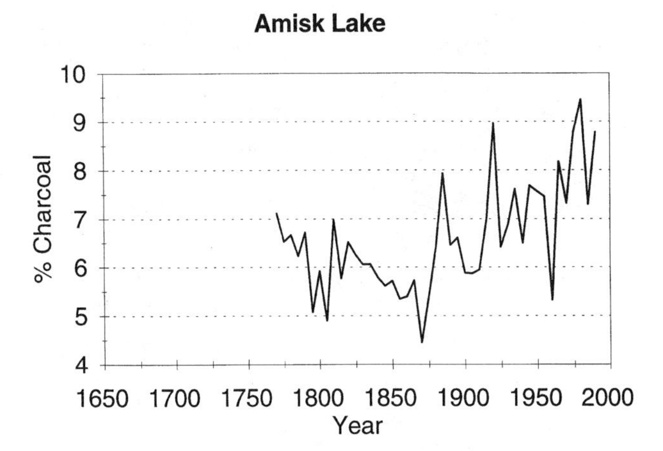

Figure 9.5. Charcoal abundance record from Amisk Lake, Alberta. Dating control from varve counting.

logging and fire suppression. Thus even within this relatively short time period and in this small geographic region, the fire response to climate change has not been consistent from one site to another.

To the south, in Yellowstone National Park, the Little Ice Age was a time of relatively few but large fires. Before and after the Little Ice Age, the park's fire regime was dominated by smaller but more frequent fires (Millspaugh and Whitlock 1995). This may suggest that the cooler moister climate effectively

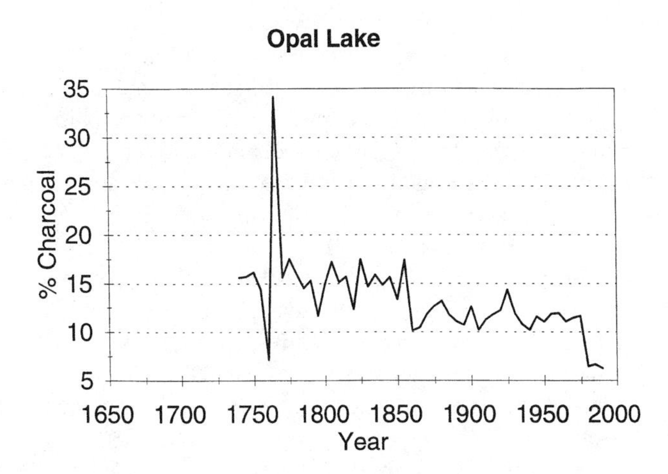

Figure 9.6. Charcoal abundance record from Opal Lake, Saskatchewan. Dating control from varve counting.

suppressed fire long enough for large accumulations of fuel to produce large fires, analogous to the recent anthropogenic fire suppression resulting in the large fire of 1988.

At Wien Lake (4° west of Fairbanks), the early Holocene thermal maximum corresponds with peak *Populus* pollen abundance and low incidence of macroscopic charcoal. Charcoal incidence rises later, during the subsequent cooling, when spruce becomes dominant (Hu et al. 1993). This suggests that during the thermal maximum, the climate was too dry to support a coniferous forest and instead *Populus* dominated the landscape. Despite the dryness during this period, fire was not a major feature, presumably due to the vegetation type. Later, as climate cooled and became moister, spruce became dominant, and fire became a common occurrence.

Discussion and Conclusions

Two themes recur throughout this review: (1) an increase in warm dry weather may lead to an increase in fire; and (2) an increase in conifers, whether prompted by warmer and drier or cooler and moister climate, may also lead to an increase in fire. Thus the picture that emerges is of a more complex, dynamic system than is often assumed. Future warming may cause a decrease in conifers and hence fire in some regions, particularly in the southern boreal forest. In areas where the abundance of conifers is expected to remain relatively unchanged, the warmer climate, if also drier, will probably lead to an increase in fire activity. Similarly, at the northern margin of the boreal forest, where warming is expected to lead to an increase in conifer densities, fire may also increase.

Global circulation models used to project future fire danger levels have suggested that future warming will result in a longer and more severe fire season throughout Canada (Flannigan and Van Wagner 1991; Wotton and Flannigan 1993). However, these models do not take into account change in vegetation. The interactions between climate, fire, and vegetation are complex. A simplified diagram of these interactions is shown in Figure 9.7. The influence of fire on climate is small compared with the influence of climate on fire. Likewise, the influence of vegetation on climate is less than the influence of climate on vegetation. The relative influence of fire on vegetation as compared with the influence of vegetation on fire needs to be addressed.

Further complexities appear when we consider that these interactions are operating at different time scales. For example, the fire regime responds almost immediately to weather,[2] whereas the vegetation response may be a reflection of the climate of the past century. Finally, other factors may be clouding the determination of the relative importance of the fire–vegetation interactions when using actual field data including topography, soils, ignition agents, competition, and anthropogenic effects (e.g., agricultural clearing, fire management, logging).

---

[2]Fire behavior can change dramatically over a few days.

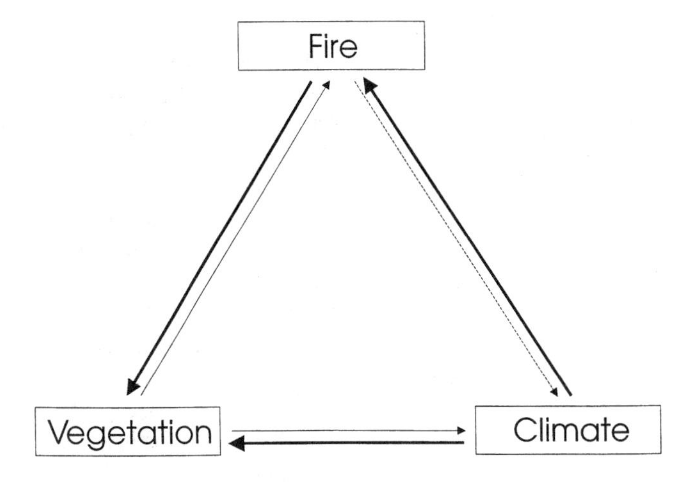

Figure 9.7. Conceptual interactions between climate, fire, and vegetation. Thick lines, thin lines, and dashed line indicate progressively weaker interactions.

In the short term, the disequilibrium response of the system to a rapid climate change may be to increase deadwood through increased individual tree mortality. This may lead to a temporary increase in fire activity in affected areas (Wein 1989; Campbell and McAndrews 1993). Any increase in fire activity will result in a net shift of age class distributions toward younger age classes, which will, in turn, have repercussions on both the short- and long-term carbon storage capacity of the forest (Kurz et al. 1995; Kurz and Apps 1996). Although young trees grow more vigorously, young trees and soils contain less total carbon than do older trees and soils.

Knowledge Gaps

The geographic and temporal density of paleofire studies in boreal North America is very low. The greatest density lies on a north-south transect through western Quebec and southern Ontario, then westward through Wisconsin to Minnesota. Given the complexity of the relationships suggested by this limited available network of studies, the collection of long (i.e., 1,000+ years), high-resolution (i.e., ≤ 20-year sampling interval) fire regime records throughout boreal North America seems necessary.

We also lack a basic understanding of the degree to which fire regimes change in response to vegetation change and the magnitude of the feedbacks that operate in this system. Although it is clear that fire allows vegetation to change and adapt to changing climate more rapidly, at least in some circumstances such as the treeline, it is not known to what degree the fire regime can change without there first being a change in the vegetation. In other words, we do not yet know whether it is vegetation or climate (directly, as opposed to indirectly via its effect on the vegetation) that exerts a dominant influence on the fire regime.

For useful predictions of future fire regimes, it would also be important to have a better grasp of the degree to which standing deadwood contributes to the fire regime. Rapid climate change may temporarily result in large amounts of standing deadwood, which may create an unprecedented situation for fire.

There are also methodological knowledge gaps, which must be resolved prior to tackling some of the theoretical knowledge gaps discussed above. These include issues such as the relative dispersal of charcoal by air, streams, or surface runoff, charcoal redeposition, and the differences between large and small lakes in sensing regional and local fires. Although these issues are related to similar difficulties encountered and largely successfully overcome in pollen analysis (Faegri et al. 1989, MacDonald and Edwards 1991), they must still be addressed for charcoal analysis. It is as yet unclear to what degree charcoal dispersal and deposition are analogous to pollen dispersal and deposition. Certainly, the density of charcoal is more variable, and the lofting of charcoal by the heat of the fire must also affect its dispersal. Similarly, charcoal taphonomy must be investigated to determine the degree to which different species produce different amounts and size classes of charcoal and the stability of the larger charcoal fragments though the depositional and analytical processes.

Acknowledgments

We thank Marty Alexander, Celina Campbell, Jim Clark, Bill DeGroot, Dave Martell, Brian Stocks, Mike Weber, and Stephen Zoltai for discussions of the topics covered in this chapter. The charcoal data from Amisk, Opal, and Pine lakes were developed in collaboration with Celina Campbell.

References

Archibold, O.W., and M.R. Wilson. 1980. The natural vegetation of Saskatchewan prior to agricultural settlement. *Can. J. Bot.* 58:2031–2042.

Barney, R.J., and B.J. Stocks. 1983. Fire frequencies during the suppression period, pp. 45–62 in R.W. Wein and D.A. MacLean, eds. *The Role of Fire in Northern Circumpolar Ecosystems.* J. Wiley & Sons, New York.

Bergeron, Y. 1991. The influence of island and mainland lakeshore landscapes on boreal forest fire regimes. *Ecology* 72:1980–1992.

Bergeron, Y., and S. Archambault. 1993. Decreasing frequency of forest fires in the southern boreal zone of Quebec and its relation to global warming since the end of the "Little Ice Age." *The Holocene* 3:255–259.

Bergeron, Y., and J. Brisson. 1990. Fire regime in red pine stands at the northern limit of the species' range. *Ecology* 71:1352–1364.

Bergeron, Y., and M.D. Flannigan. 1995. Predicting the effects of climate change on fire frequency in the southeastern Canadian boreal forest. *Air Water Soil Pollut.* 82:437–444.

Bernabo, J.C. 1981. Quantitative estimates of temperature changes over the last 2700 years in Michigan based on pollen data. *Quat. Res.* 15:143–159.

Bessie, W.C., and E.A. Johnson. 1995. The relative importance of fuels and weather on fire behaviour in subalpine forests. *Ecology* 76:747–762.

Bird, R.D. 1961. *Ecology of the Aspen Parkland of Western Canada in Relation to Land Use.* Publication 1006, Research Branch, Canadian Department of Agriculture, Ottawa, Ontario, Canada.

Bonan, G.B., D. Pollard, and S.L. Thompson. 1992. Effects of boreal forest vegetation on global climate. *Nature* 359:716–718.
Bond, W.J., and J.J. Midgley. 1995. Kill thy neighbour: an individualistic argument for the evolution of flammability. *Oikos* 73:79–85.
Bryson, R.A., W.N. Irving, and J.A. Larsen. 1965. Radiocarbon and soil evidence of former forest in the southern Canadian tundra. *Science* 147:46–48.
Campbell, C., I.D. Campbell, C.B. Blyth, and J.H. McAndrews. 1994. Bison extirpation may have caused aspen expansion in western Canada. *Ecography* 17:360–362.
Campbell, I.D., and C. Campbell. 1994. The impact of late woodland land use on the forest landscape of southern Ontario. *Great Lakes Geog.* 1:21–29.
Campbell, I.D., and J.H. McAndrews. 1991. Cluster analysis of late Holocene pollen trends in Ontario. *Can. J. Bot.* 69:1719–1730.
Campbell, I.D., and J.H. McAndrews. 1993. Forest disequilibrium caused by rapid Little Ice Age cooling. *Nature* 366:336–338.
Campbell, I.D., and J.H. McAndrews. 1995. Charcoal evidence for Indian-set fires: a comment on Clark and Royall. *The Holocene* 5:369–370.
Clark, J.S. 1988. Particle motion and the theory of charcoal analysis: source area, transport, deposition, and sampling. *Quat. Res.* 30:67–80.
Clark, J.S. 1989. Ecological disturbance as a renewal process: theory and application to fire history. *Oikos* 56:17–30.
Clark, J.S. 1990. Fire and climate change during the last 750 yr in northwestern Minnesota. *Ecol. Mon.* 60:135–159.
Clark, J.S. 1995. Climate and Indian effects on southern Ontario forests: a reply to Campbell and McAndrews. *The Holocene* 5:371–379.
Clark, J.S., and P.D. Royall. 1995. Transformation of a northern hardwood forest by aboriginal (Iroquois) fire: charcoal evidence from Crawford Lake, Ontario, Canada. *The Holocene* 5:1–9.
Clark, J.S., and P.D. Royall. 1996. Particle-size evidence for source areas of charcoal accumulation in late Holocene sediments of eastern North America. *Quat. Res.* 43:80–89.
Clark, J.S., B.J. Stocks, and P.J.H. Richard. 1996. Climate implications of biomass burning since the 19th century in eastern North America. *Global Change Biol.* 2:433–442.
COHMAP members. 1988. Climatic changes of the last 18,000 years: observations and model simulations. *Science* 241:1043–1052.
Cwynar, L.C. 1978. Recent history of fire and vegetation from laminated sediment of Greenleaf Lake, Algonquin Park, Ontario. *Can. J. Bot.* 56:10–21.
Dansereau, P-R., and Y. Bergeron. 1993. Fire history in the southern boreal forest of northwestern Quebec. *Can. J. For. Res.* 23:25–32.
Day, G.M. 1953. The Indian as an ecological factor in the northeastern forest. *Ecology* 34:329–346.
Faegri, K., Kaland, P.E., and Krzywinski, K. 1989. *Textbook of Pollen Analysis,* 4th ed. John Wiley & Sons, Chichester, UK.
Filion, L., D. Saint-Laurent, M. Desponts, and S. Payette. 1991. The late Holocene record of aeolian and fire activity in northern Quebec, Canada. *The Holocene* 1:201–208.
Flannigan, M.D., and J.B. Harrington. 1988. A study of the relation of meteorological variables to monthly provincial area burned by wildfire in Canada 1953–80. *J. Appl. Meteorol.* 27:441–452.
Flannigan, M.D., and C.E. Van Wagner. 1991. Climate change and wildfire in Canada. *Can. J. For. Res.* 21:66–72.
Flannigan, M.D., and B.M. Wotton. 1995. Fire regime and the abundance of jack pine, pp. 625–636 in *Proceedings of the 2nd International Conference on Forest Fire Research.* Domingo Xavier Vegas, Coimbra, Portugal.
Foster, D.R. 1983. The history and pattern of fire in the boreal forest of southeastern Labrador. *Can. J. Bot.* 61:2459–2471.

Foster, D.R. 1985. Vegetation development following fire in *Picea mariana* (black spruce)–*Pleurozium* forests of south-eastern Labrador, Canada. *J. Ecol.* 73:517–534.

Foster, D.R., and G.A. King. 1986. Vegetation pattern and diversity in S.E. Labrador, Canada: *Betula papyrifera* (birch) forest development in relation to fire history and physiography. *J. Ecol.* 74:465–483.

Gajewski, K. 1988. Late Holocene climate changes in eastern North America estimated from pollen data. *Quat. Res.* 29:255–262.

Gajewski, K., M.G. Winkler, and A.M. Swain. 1985. Vegetation and fire history from three lakes with varved sediments in northwestern Wisconsin (U.S.A.). *Rev. Palaeobot. Palynol.* 44:277–292.

Goldberg, E.D. 1986. *Black Carbon in the Environment: Properties and Distribution.* Wiley, New York.

Green, D.G. 1982. Fire and stability in the postglacial forests of southwest Nova Scotia. *J. Biogeog.* 9:29–40.

Grimm, E.C. 1983. Chronology and dynamics of vegetation change in the prairie-woodland region of southern Minnesota, U.S.A. *New Phytol.* 93:311–350.

Grove, J.M. 1988. *The Little Ice Age.* Methuen, New York.

Gullett, D.W., and W.R. Skinner. 1992. *The State of Canada's Climate: Temperature Change in Canada 1895–1991.* SOE Report 92–2. Environment Canada, Ottawa, Canada.

Hu, F.S., L.B. Brubaker, and P.M. Anderson. 1993. A 12,000 year record of vegetation change and soil development from Wien Lake, central Alaska. *Can. J. Bot.* 71:1133–1142.

Johnson, E.A. 1992. *Fire and Vegetation Dynamics: Studies from the North American Boreal Forest.* Cambridge University Press, Cambridge, UK.

Johnson, E.A., G.I. Fryer, and M.J. Heathcott. 1990. The influence of man and climate on frequency of fire in the interior wet belt forest, British Columbia. *J. Ecol.* 78:403–412.

Khury, P. 1994. The role of fire in the development of Sphagnum-dominated peatlands in western boreal Canada. *J. Ecol.* 82:899–910.

Kurz, W.A., and M.J. Apps. 1994. The carbon budget of Canadian forests: a sensitivity analysis of changes in disturbance regimes, growth rates, and decomposition rates. *Environ. Pollut.* 83:55–61.

Kurz, W.A., and M.J. Apps. 1996. Retrospective assessment of carbon flows in Canadian boreal forests, pp. 173–182 in M.J. Apps and D.T. Price, eds. *Forest Ecosystems and Forest Management and the Global Carbon Cycle.* NATO ASI Series, Series I: Global Environmental Change. Springer-Verlag, Berlin.

Kurz, W.A., M.J. Apps, S.J. Beukema, and T. Lokstrum. 1995. Twentieth century carbon budgets of Canadian forests. *Tellus* 47B:170–177.

Landhaeusser, S.M., and R.W. Wein. 1993. Postfire vegetation recovery and tree establishment at the arctic treeline: climate-change—vegetation-response hypotheses. *J. Ecol.* 81:665–672.

Leonard, E.M. 1986. Varve studies at Hector Lake, Alberta, Canada, and the relationship between glacial activity and sedimentation. *Quat. Res.* 25:199–214.

Looman, J. 1979. The vegetation of the Canadian prairie provinces. 1. An overview. *Phytocoenologia* 5:347–366.

MacDonald, G.M. 1987. Postglacial development of the subalpine-boreal transition forest of western Canada. *J. Ecol.* 75:303–320.

MacDonald, G.M., and K.J. Edwards. 1991. Holocene palynology: I. Principles, population and community ecology, palaeoclimatology. *Prog. Phys. Geog.* 15:261–289.

MacDonald, G.M., C.P.S. Larsen, J.M. Szeicz, and K.A. Moser. 1991. The reconstruction of boreal forest fire history from lake sediments: a comparison of charcoal, pollen, sedimentological, and geochemical indices. *Quat. Sci. Rev.* 10:53–71.

Malanson, G.P. 1987. Diversity, stability and resilience: effects of fire regime, pp. 49–63 in L. Trabaud, ed. *The Role of Fire in Ecological Systems.* SPB Academic Publishing, The Hague.

Matthews, J.V., Jr., T.W. Anderson, M. Boyko-Diakonow, R.W. Mathewes, J.H. McAndrews, R.J. Mott, P.J.H. Richard, J.C. Ritchie, and C.E. Schweger. 1989. Quaternary environments in Canada as documented by paleobotanical case histories, pp. 483–539 in R.J. Fulton, ed. *Quaternary Geology of Canada and Greenland.* The Geology of North America, vol. K-1. Geological Society of America, Boulder, CO.

Millspaugh, S.H., and C. Whitlock. 1995. A 750-year fire history based on lake sediment records in central Yellowstone National Park, U.S.A. *The Holocene* 5:283–292.

Mutch, R.W. 1970. Wildland fires and ecosystems—a hypothesis. *Ecology* 51:1046–1051.

Nichols, H. 1967a. The post-glacial history of vegetation and climate at Ennadai Lake, Keewatin, and Lynn Lake, Manitoba (Canada). *Eiszeitalter Ggw.* 18:176–197.

Nichols, H. 1967b. Pollen diagrams form sub-arctic central Canada. *Science* 155:1665–1668.

Patterson, W.A., K.J. Edwards, and D.J. Maguire. 1987. Microscopic charcoal as a fossil indicator of fire. *Quat. Sci. Rev.* 6:3–23.

Payette, S. 1992. Fire as a controlling process in the North American boreal forest, pp. 144–169 in H.H. Shugart, R. Leemans, and G.B. Bonan, eds. *A Systems Analysis of the Global Boreal Forest.* Cambridge University Press, Cambridge, UK.

Payette, S., and R. Gagnon. 1985. Late Holocene deforestation and tree regeneration in the forest tundra of Quebec. *Nature* 313:570–572.

Payette, S., and C. Morneau. 1993. Holocene relict woodlands at the eastern Canadian treeline. *Quat. Res.* 39:84–89.

Payette, S., C. Morneau, L. Sirois, and M. Desponts. 1989. Recent fire history of the northern Quebec biomes. *Ecology* 70:656–673.

Prentice, I.C. 1985. Pollen representation, source area, and basin size: toward a unified theory of pollen analysis. *Quat. Res.* 23:76–86.

Quinby, P.A. 1987. An index to fire incidence. *Can. J. For. Res.* 17:731–734.

Ritchie, J.C. 1987. *Postglacial Vegetation of Canada.* Cambridge University Press, Cambridge, UK.

Sirois, L., and S. Payette. 1991. Reduced postfire tree regeneration along a boreal forest-forest-tundra transect in northern Quebec. *Ecology* 72:619–627.

Snyder, J.R. 1984. The role of fire: much ado about nothing? *Oikos* 43:404–405.

Stocks, B.J. 1987. Fire potential in the spruce budworm–damaged forests of Ontario. *For. Chron.* 63:8–14.

Strong, W.L 1977. Pre- and post-settlement palynology of southern Alberta. Rev. *Palaeobot. Palynol.* 23:373–387.

Swain, A.M. 1973. A history of fire and vegetation in northeastern Minnesota as recorded in lake sediments. *Quat. Res.* 3:423–443.

Swain, A.M. 1978. Environmental changes during the past 2000 years in north-central Wisconsin: analysis of pollen, charcoal, and seeds from varved lake sediments. *Quat. Res.* 10:55–68.

Swetnam, T.W. 1993. Fire history and climate change in giant Sequoia groves. *Science* 262:885–889.

Szeicz, J., and G.M. MacDonald. 1991. Postglacial vegetation history of oak savanna in southern Ontario. *Can. J. Bot.* 69:1507–1519.

Timoney, K.P., and R.W. Wein. 1991. The areal pattern of burned tree vegetation in the subarctic region of northwestern Canada. *Arctic* 44:223–230.

Tolonen, K. 1983. The post-glacial fire record, pp. 21–44 in R.W. Wein and D.A. MacLean, eds. *The Role of Fire in Northern Circumpolar Ecosystems.* J. Wiley & Sons, New York.

Vance, R.E., D. Emerson, and T. Habgood. 1983. A mid-Holocene record of vegetative change in central Alberta. *Can. J. Earth Sci.* 20:364–376.

Van Wagner, C.E. 1988. The historical pattern of annual area burned in Canada. *For. Chron.* 64:182–185.

Wein, R.W. 1989. Climate change and wildfire in northern coniferous forests: climate change scenarios, p. 169 in D.C. MacIver, H. Auld, and R. Whitewood, eds. *Proceedings of the Tenth Conference on Fire and Forest Meteorology.* Toronto.

Wein, R.W., and D.A. MacLean, eds. 1983. *The Role of Fire in Northern Circumpolar Ecosystems.* J. Wiley & Sons, New York.

Winkler, M.G. 1985. Charcoal analysis for paleoenvironmental interpretation: a chemical assay. *Quat. Res.* 23:313–326.

Wotton, B.M., and M.D. Flannigan. 1983. Length of the fire season in a changing climate. *For. Chron.* 69:187–192.

Wright, H.E., Jr., and M.L. Heinselman. 1973. The ecological role of fire in natural conifer forests of western and northern North America. *Quat. Res.* 3:317–513.

Zoltai, S.C. 1995. Permafrost distribution in peatlands of west-central Canada during the Holocene Warm Period 6000 years BP. *Geog. Phys. Quat.* 49:45–54.

10. Controls on Patterns of Biomass Burning in Alaskan Boreal Forests

Eric S. Kasischke, Katherine P. O'Neill, Nancy H.F. French, and Laura L. Bourgeau-Chavez

Introduction

As discussed in the introduction to this section, fire serves an important ecological role in the boreal forest, especially in those processes controlling the exchange of carbon dioxide and other greenhouse gases with the atmosphere. One of the key requirements for quantifying the effects of fire on the carbon cycle in boreal forests is estimating the amount of biomass consumed during fire.

During the late 1970s and 1980s, a series of controlled burns were carried out in Canadian forests (Stocks 1980, 1987, 1989; Alexander et al. 1991; Quintillo et al. 1991). These studies were used to develop geographically distributed fire prediction and behavior models (Stocks et al. 1989; Alexander and Quintillo 1990). In addition, Auclair (1985) studied the patterns of burning in a spruce-lichen forest, and Dyrness and Norum (1983) studied patterns of ground-layer biomass burning in controlled fires in black spruce forests in interior Alaska. Although these studies represent a significant body of research, they still do not present a basis to estimate the patterns of biomass burning in North American boreal forests. With the exception of the Dyrness and Norum (1983) study, previous research was conducted in forests that had relatively low levels of ground-layer[1] carbon levels (between 10 and 30 t ha^{-1}). In areas underlain by permafrost within the North American boreal forests, the amount of ground-layer carbon in mature forests can

[1]The term *ground layer* includes litter, mosses, lichen, and organic soil.

be between 50 and 150 t ha^{-1}. In addition, the controlled burns did not necessarily capture the conditions that naturally occur during wildfires because they were conducted at times when fire conditions were low so that fire fighters were available to control the purposely set fires.

Estimates of the amount of carbon released during boreal forest fires have been made by using two approaches. First, Stocks (1991) and Cahoon and co-workers (1994) use an estimate of 12 t ha^{-1} based on data collected during controlled fires in Canada (Stocks and Kauffman 1997). Others have used a higher value of 25–30 t ha^{-1} based on field observations and model results (Kasischke et al. 1995).

Observations on the patterns of biomass burning in the black spruce forests of interior Alaska have led to the initial conclusion that in boreal forests with deep organic soils, large amounts of carbon can be released during fires. The focus of this study is to begin to address two basic, but important questions: (1) How much biomass is burned (and carbon released) in the ground layers of forests with deep organic soils in interior Alaska? (2) What are the factors influencing patterns of biomass burning in these forests?

Methods

Test Sites

Sites located in eight different areas were used in this study. These sites included forests burned in natural wildfires between 1954 and 1996, as well as forest stands adjacent to unburned stands. The ages of the forest stands in these sites were determined either by the date of the latest fire or through counting of tree rings from cross-sections obtained from the base of the boles from five to ten overstory trees. The fires were all located in or near the Tanana River Valley of Alaska (Fig. 10.1, Table 10.1).

The studies discussed in this chapter began in the summer of 1993 at the Tok test area. The Tok test sites all burned within a 5-day period in early July 1990 during an extremely intense ground/crown fire. Initial studies at this site focused on two issues: (1) quantifying the amounts of biomass consumed during fires in the black spruce forests of this region; and (2) comparing the patterns of biomass burning in black spruce forests to those found in other forest types in this area (white spruce, aspen, and willow-black spruce stands). The overall sampling protocol was to locate a burned forest stand immediately adjacent to an unburned forest stand. However, because of the large scale of the 1990 Tok fire, there were no unburned black spruce sites near the burned sites. The unburned sites for the area were located within several kilometers of the burned sites. Finally, no unburned willow-black spruce stands were located in the Tok region. Analysis of the burning patterns in the ground layer of this forest type was based on analysis of data collected at the 1994 Hajdukovich Creek fire.

With the exception the Hajdukovich Creek area, at each of the other study areas one burned and one unburned stand were sampled. Three black spruce sites were

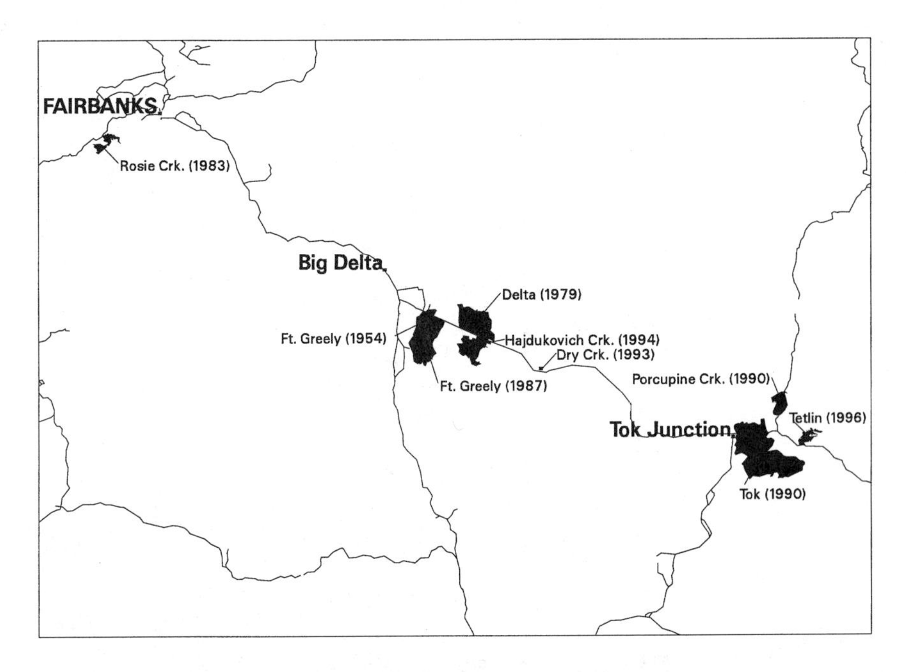

Figure 10.1. Location of test sites in the Tanana River Valley of Alaska used in study of the patterns of biomass burning.

located at Hajdukovich Creek, along with one willow-black spruce stand. Black spruce sites labeled "a" and "b" were burned in late August–early September, whereas site "c" was burned in June 1994. Moreover, site b was located in a dense, closed-canopy black spruce stand with a feathermoss ground cover, whereas Sites a and c were located in an open-canopy forest with a sphagnum moss ground cover.

Measurement of Burning of Ground-Layer Biomass/Carbon

Several different sampling procedures were used to measure patterns of ground-layer biomass/carbon remaining in the burned test sites. In addition to the study discussed in this chapter, the Tok site was also used to investigate methods for using satellite remote sensors to monitor biomass burning and the effects of fires on boreal forests (French et al. 1996; Bourgeau-Chavez et al. 1997; Harrell et al. 1995; Kasischke and French 1997). To support both the biomass burning and the remote sensing research at the Tok site, 200 × 200-m grids were established in each of the 14 burned study sites and divided into four quadrants of equal dimensions. The grids were located in areas that were homogeneous in terms of forest cover and burn severity (based on review of preburn aerial photography and ground surveys). Two 10 × 10-m plots were randomly located within each quadrant (eight total at each site). Ground-layer/organic soil depths were measured at

Table 10.1. Fire Locations and Test Sites Used in Study

			Number of test sites					
			Black spruce		White spruce		Aspen	
Fire location	Year of fire	Identifier	Burned	Unburned	Burned	Unburned	Burned	Unburned
Fort Greely	1956	FG56		1				
Rosie Creek	1982	RC82	1	1				
Fort Greely	1987	FG87	1	1				
Tok	1990	TK90	12[a]	4	2	2	1	1
Porcupine Creek	1990	PC90	1	1				
Dry Creek	1993	DC93	1	1				
Hajdukovich Creek	1994	HC94	4[b]	4[b]				
Tetlin	1996	TL96	1	1				

[a]Includes two willow–black spruce stands.
[b]Includes one willow–black spruce stand.

the corner and center of each plot (five samples per plot), resulting in 40 soil depth samples per study site.

A different sampling design was used in the remaining sites. Two 150- to 200-m-long transects were located in each plot, with the first transect beginning 50 m from one edge of the plot and the second beginning at a distance of 150 m. Twenty sample points were located at random locations along each transect.

The organic soil horizons at all sample points were described following the nomenclature of the Canadian Soil Classification System (Agriculture Canada Expert Committee on Soil Survey 1987). The uppermost portion of the soil profile (O_f or fibric horizon) consisted primarily of materials that were recently derived from mosses and herbaceous plants (with small amounts of leaves, needles, and twigs) that had undergone relatively little decomposition and whose botanical origin was still recognizable. These horizons contain more than 40% rubbed fiber content; that is, more than 40% of the organic fibers remained intact after rubbing a sample between the thumb and forefinger about ten times. In the black spruce stands in interior Alaska, we recognized three different classes of fibric soil, with each class having a distinctly different bulk density. Humic horizons (O_h) are those organic materials found at the base of the organic profile in a very advanced stage of decomposition. The botanical origin of these materials cannot be visually determined, and the rubbed fiber content was less than 10% by volume. Humic material also has a bulk density significantly higher than that of the fibric horizon. Materials of intermediate decomposition, which had been altered both physically and chemically from plant material yet did not meet the requirements for either a fibric or humic horizon, were classified as mesic (O_m).

Although these profile designations correspond approximately to the fibric (O_i), hemic (O_e), and sapric (O_a) horizons described under the U.S. Soil Classification System, the Canadian system requires that O horizons be derived from mosses, rushes, or woody materials. They do not include significant amounts of leaf or woody litter (Buol et al.1989).

Finally, in ground layers overlain by mosses, the upper photosynthetic moss tissue was classified as "green moss," and the underlying nonphotosynthetic plant tissue was described as "brown moss."

At a selected number of sites, samples of the forest floor were collected for determination of bulk density and percentage carbon content following the methods of Harden and associates (1997). Samples were collected from the unburned aspen stand and one of the unburned white spruce stands at the 1990 Tok fire site, from two of the unburned black spruce stands (sites a and b) at the 1994 Hajdukovich Creek fire site, and from the unburned black spruce stand at the 1996 Teltin fire site. The three black spruce sites were selected because they were representative of the range of organic soil densities encountered at all sites.

At each sampling point, a square (approximately 20 cm) profile was excavated from the ground/organic soil layer by using a shovel and/or saw. The depth of the different layers or horizons was then measured in the resultant hole to the nearest 0.5 cm. The willow-black spruce stands also contained isolated grass tussocks that rose 30–60 cm above the organic soil and were 20–50 cm in diameter. These tussocks were partially consumed during the fires. To estimate their level of consumption, we measured the diameter and height of 25 randomly selected tussocks in both the burned and unburned sites.

The resultant soil cores were retained and taken back to the laboratory. The profiles were then subdivided into genetic horizons (fibric, mesic, humic, mineral) and two cubes (about 4 × 4 cm in size) cut from the center of each horizon by using an electric knife with two reciprocating blades to minimize the potential of compaction. The dimensions of the cubes were measured to the nearest 0.1 cm and the samples weighed, dried in a 65°C oven for 48 hours, and then reweighed to determine volumetric moisture content and bulk density.

The oven-dry samples were then ground in a Wiley Mill until all material was able to pass through a 60-mesh sieve. Roots were not removed prior to grinding. Total percentage carbon and nitrogen were determined by using a Perkins Elmer 2400 Series II CHNS/O Analyzer. This instrument uses a combustion method to convert sample elements to a gaseous form (CO_2, H_2O, N_2) that can be separated and detected as a function of their thermal conductivity. The instrument is calibrated against the concentrations of carbon and nitrogen detected in standards of known chemical composition. For this study, calibrations were made by using acetanalide ($C_6H_5NHCOCH_3$), peach leaves, and municipal digested sludge (US EPA #1333). Calibration standards were run every 15 samples and blank tin capsules run every ten samples. Instrument drift was corrected as needed.

Bulk density (g soil m^{-3}) and fraction of carbon (g C g^{-1} soil) were used to calculate the weight of carbon per unit area depth of each soil horizon (g C m^{-3}). These carbon densities were then multiplied by the average thickness of each soil horizon measured in the field to determine the average carbon density (t ha^{-1}) for the test sites. The difference between carbon stored in the burned and unburned sites provides an estimate of the forest floor carbon consumed by fire.

Measurement of Burning of Aboveground Biomass/Carbon

To estimate the amount of aboveground biomass consumed during the 1990 Tok fire, we compared the biomass present in unburned test sites to that present in burned test sites. This approach was possible because ground surveys conducted during the summer of 1992 showed that all understory biomass had been consumed during the fire and because the remnants of the burned trees were present in all the plots.

Different sampling approaches were used in the burned and unburned sites. In the burned sites, within each 10 × 10-m plot,[2] the species of all trees present were identified and counted. The basal diameter was measured for all black spruce and willow trees, and the basal diameter and diameter at breast height were measured for white spruce and aspen trees. Height measurements were collected for a subsample of trees to determine the relationship between diameter and height.

[2]For several plots with high tree densities, only one-half of the plot was sampled.

The amount of biomass contained in the different tree species and tree components (e.g., leaves, needles, branches, and boles) was estimated by using allometric equations of Barney and associates (1978) and Stocks (1980) for black spruce, Yarie and Van Cleve (1983) for white spruce, Peterson and colleagues (1970) for aspen, and Telfer (1969) for willow.

The degree of burning of biomass was estimated by using the following relative scale that allowed for efficient sampling of all trees present at a site:

1. Tree killed but no needles and leaves consumed during the fire.
2. All needles and leaves and up to 50% of the secondary and smaller branches consumed during the fire.
3. All secondary branches and up to 50% of the primary branches consumed during the fire.
4. Between 50% and 100% of the primary branches consumed during the fire.
5. All branches and foliage consumed during fire.

For some trees, not only was all the foliage consumed, but part of the tree boles also burned. In these cases, the percentage of bole consumption was estimated.

To estimate the biomass of the subcanopy vegetation, all vegetation in several 1-m^2 areas in unburned stands were harvested, dried, and weighed.

Results

Patterns of Carbon Storage and Biomass Burning in the Ground Layer

Table 10.2 summarizes the bulk density and percentage carbon content of organic horizons in the different forest types. These data show that within a specific horizon, the mean bulk density is similar between forest types and also that bulk density increases dramatically as a function of depth. However, the percentage carbon shows greater variability between forest types. Overall, there is a decrease in percentage carbon with depth.

Patterns of carbon density per unit depth are similar for the different forest types, and there is an increase in carbon storage with depth (Fig. 10.2). Note that although there is a difference in carbon density within the different fibric soil categories in Figure 10.2, the difference in Fibric I and II is relatively small. For future studies, these two categories can be combined. The data in Figure 10.2 were used to estimate the amount of carbon present in the burned and unburned stands based on the average thicknesses of the forest floor horizons (Fig. 10.3). Figure 10.4 shows the average carbon storage as a function of stand age and forest type for burned and unburned stands.

Combining the bulk density with the percentage carbon data generated the plot presented in Figure 10.2. These data show that as depth into the ground layer increases, there is an increase in carbon storage. These data were used to estimate the amount of carbon present in the ground layers in both the burned and unburned stands based on the measurement of the depths of the different layers in the test

Table 10.2. Ground-Layer Bulk Densities and Carbon Percentages for Alaskan Forest

Site	Forest type		Moss/litter	Fibric I	Fibric II	Fibric III	Humic
Bulk density							
HC94b	Black spruce	n	20	20	2		14
		Average	0.036	0.049	0.061		0.220
		Standard deviation	0.017	0.016	0.001		0.106
TL96	Black spruce	n	10	6	10		14
		Average	0.028	0.038	0.071		0.193
		Standard deviation	0.012	0.011	0.018		0.046
HC94a	Black spruce	n	12		12	12	
		Average	0.027		0.055	0.149	
		Standard deviation	0.005		0.026	0.059	
Black spruce average			0.030				
		n	42	26	4	12	28
		Average	0.032	0.046	0.062	0.149	0.207
		Standard deviation	0.016	0.016	0.023	0.059	0.083
TK94	Aspen	n	20		20		
		Average	0.051		0.086		
		Standard deviation	0.015		0.017		
TK94	White spruce	n	10		13		
		Average	0.025		0.108		
		Standard deviation	0.011		0.080		

Percentage carbon content							
HC94b	Black spruce	*n*	18	18	1		11
		Average	35.61	40.35	18.52		28.11
		Standard deviation	10.65	13.70	0.00		7.22
TL96	Black spruce	*n*	5	3	6		7
		Average	40.69	41.72	47.96		38.64
		Standard deviation	5.30	2.82	16.03		4.16
HC94a	Black spruce	*n*	6		6	6	
		Average	26.61		34.44	25.74	
		Standard deviation	3.66		3.59	4.31	
Black spruce average		*n*	29	21	13	6	18
		Average	34.62	40.55	39.45	25.74	32.21
		Standard deviation	9.91	12.74	14.26	4.31	8.06
TK94	Aspen	*n*	10		10		
		Average	43.92		41.37		
		Standard deviation	0.73		1.52		
TK94	White spruce	*n*	2		4		
		Average	31.65		33.12		
		Standard deviation	1.06		1.83		
Tons of carbon per hectare per centimeter layer depth							
	Black spruce		1.11	1.87	2.46	3.83	6.66
	White Spruce		2.24		3.56		
	Aspen		0.79		3.58		

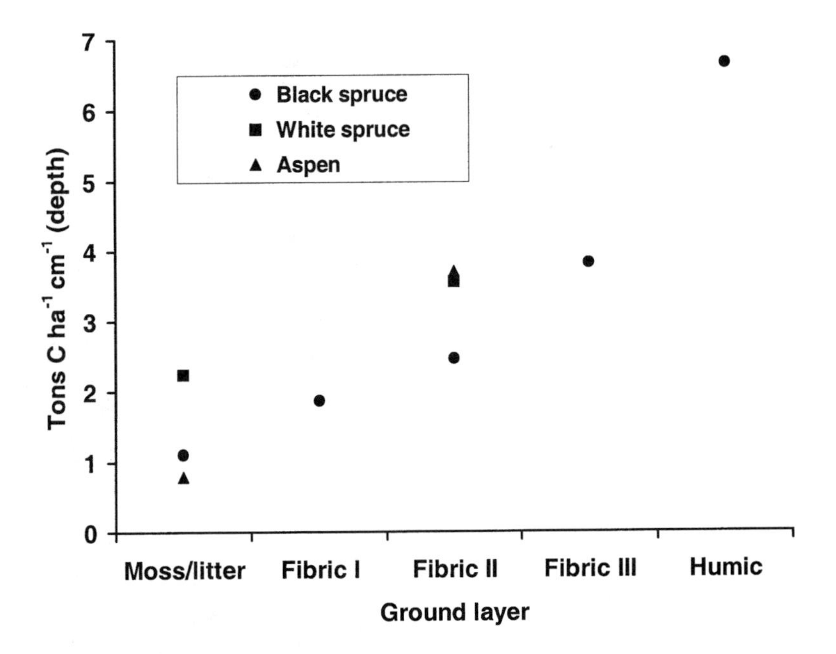

Figure 10.2. Average level of carbon per unit depth in different layers of Alaskan boreal forests.

stands. Data from the willow/black spruce stands in the 1994 Hajdukovich Creek fire site are also presented in Figure 10.4. Soil depth data for these stands were not presented in Figure 10.3 because of the presence of hummocks.

The data in Figure 10.4 show that more carbon is present in the ground layer of black spruce forests than other two forest types. There was an average of 11.1 t C ha^{-1} in the aspen stand, 30.1 t C ha^{-1} in the white spruce stands, 81.6 t C ha^{-1} in the unburned black spruce stands (average age of 116 years), and 166.2 t C ha^{-1} in the unburned willow-black spruce stand. These ground-layer carbon inventories are consistent with observations from other studies conducted in interior Alaska (Van Cleve et al. 1983).

Although fire removed most carbon from forest floors of white spruce and aspen (Fig. 10.4), in most cases there was a significant level of carbon left in the ground layer of black spruce forests. On average, fires consumed 43.0% of the ground-layer carbon in the black spruce stands (ranging from 19.6 to 88.6%), 66.5% of the ground-layer carbon in white spruce stands, and 82.1% of the ground-layer carbon in the aspen stands. Because of the wet ground conditions, a lower percentage (7.2%) of the carbon was consumed in the willow-black spruce forest type.

The amount of carbon released during these fires averaged 33.5 t ha^{-1} in the black spruce stands (range: 13.6–75.7 t C ha^{-1}), 20.0 t ha^{-1} in the white spruce stands, 9.1 t ha^{-1} in the aspen stands, and 12.2 t ha^{-1} in the willow-black spruce

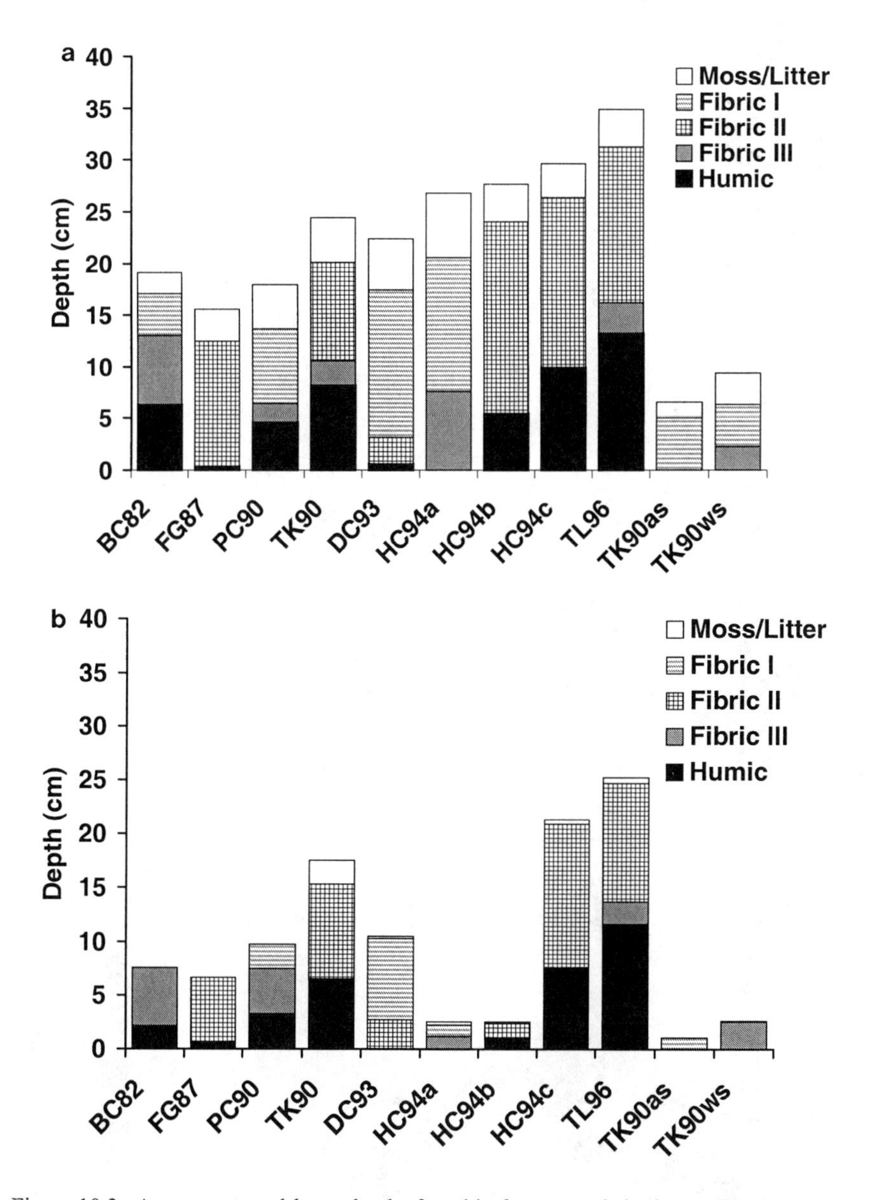

Figure 10.3. Average ground-layer depths found in forest stands in the study test sites for (a) unburned sites and (b) burned sites. Note that all sites are black spruce sites with the exception of TK90as (aspen) and TK90ws (white spruce).

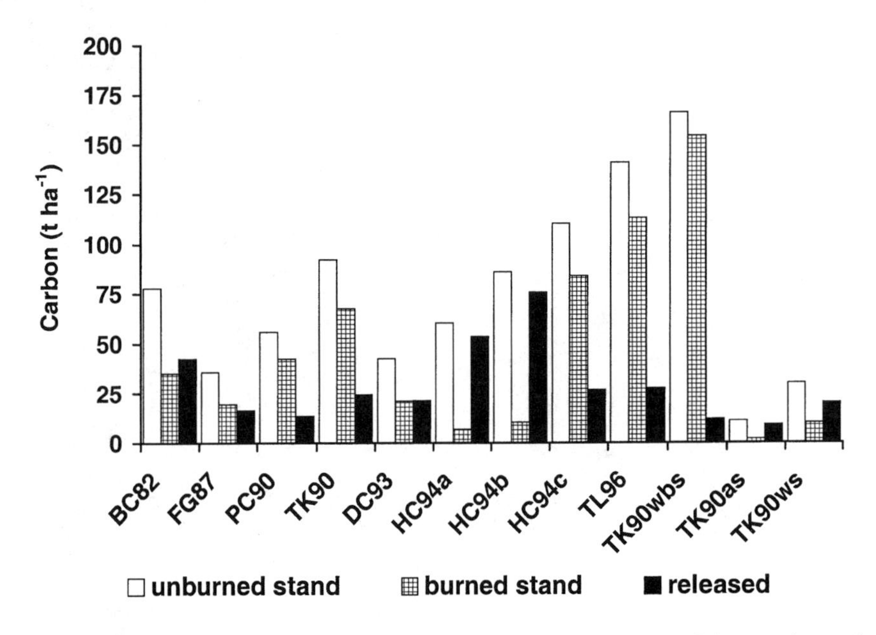

Figure 10.4. Average amounts of carbon present in the ground layers of the test sites and amounts of carbon released during fires. Site TK90wbs is a willow-black spruce stand.

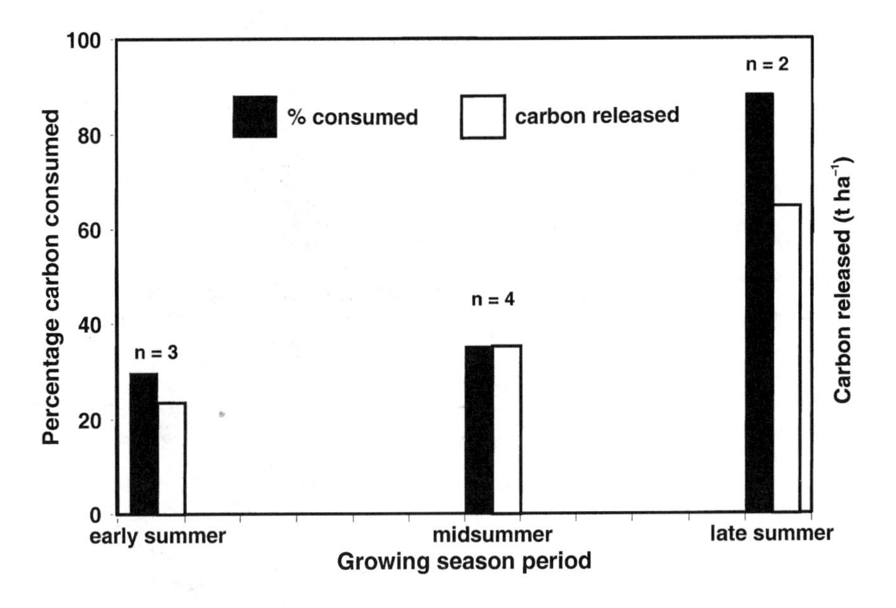

Figure 10.5. Seasonal patterns of ground-layer biomass burning and carbon release in Alaskan black spruce forests.

stands. These levels are higher than those reported for Canadian boreal forests (6 t ha^{-1}) by Stocks (1991).

Figure 10.5 presents a plot of fire-released carbon as a function of time during the growing season for the black spruce stands. Although the number of samples is limited, the data show that as the growing season progresses, the percentage of ground-layer biomass consumed and carbon released increases. Because black spruce forests are underlain by permafrost that often extends well into the organic soil profile, the amount of organic soil available for burning during fires early in the growing season is low. As the growing season advances and the active layer increases, the deeper organic soil layers thaw and can be consumed during fires if soil moisture is low.

Figure 10.6 presents a plot of total ground-layer carbon as a function of stand age for the different black spruce stands used in this study. The age of the mature black spruce forest stands ranged between 70 and 205 years and had ground-layer carbon levels ranging between 35.6 and 140.7 t C ha^{-1}. Although there is a correlation between ground-layer carbon and stand age ($r^2 = 0.43$, $P < .003$), many of the recently burned stands contain high levels of carbon. In stands less than 10 years old, the average carbon level is 44.4 t ha^{-1}, ranging between 6.9 and 113.1 t C ha^{-1}. Although carbon levels in the ground layers of black spruce forests increase as the stands age, age is not the sole control on the amounts of carbon present in these forests. It is clear from these data that the severity of the most

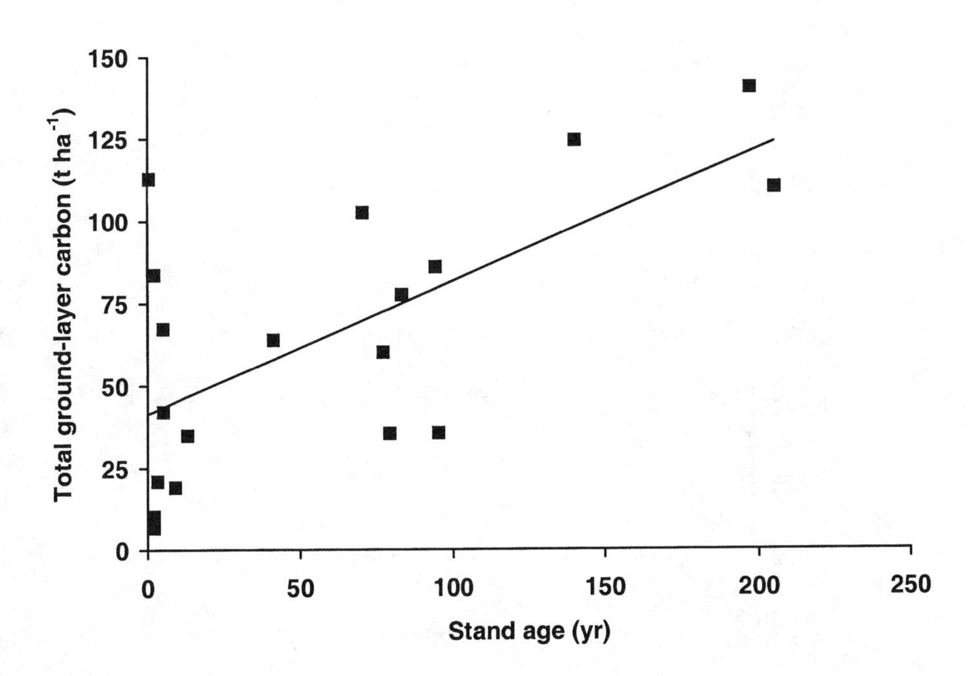

Figure 10.6. Average ground-layer carbon in black spruce forests plotted as a function of stand age.

recent fire is one of the principal factors controlling the longer-term patterns of carbon storage in the ground layers of boreal forests.

A more clear indicator of the amount of carbon present in the ground layer of black spruce forests is the overall depth of the ground layer, which integrates a number of factors, including stand age, organic soil remaining after the previous fire, and the overall drainage of the site. Figure 10.7 presents a plot of total ground-layer carbon as a function of total ground thickness. A greater degree of the variation in the ground-layer carbon storage is explained by variations in humic soil thickness ($r^2 = 0.85, P < .0001$) than is explained by variations in the total ground-layer thickness ($r^2 = 0.64, P < .001$). This result should not be surprising because of the high carbon density of humic soils (Fig. 10.2). In addition, the presence of deep humic layers is an indication of poorly drained soils overlying permafrost. Carbon-dating studies show that the humic soils in Alaskan black spruce forests are often in excess of 1,000 years in age (Mann et al. 1995). The accumulation of carbon in the organic soils of these forests, therefore, is an extremely long-term process.

Patterns of Carbon Storage and Biomass Burning in Aboveground Biomass

Table 10.3 summarizes the estimates of aboveground biomass and biomass consumed during fires based on measurements made at the 1990 Tok fire site. Atkins (1995) measured understory vegetation in the different unburned test sites in the

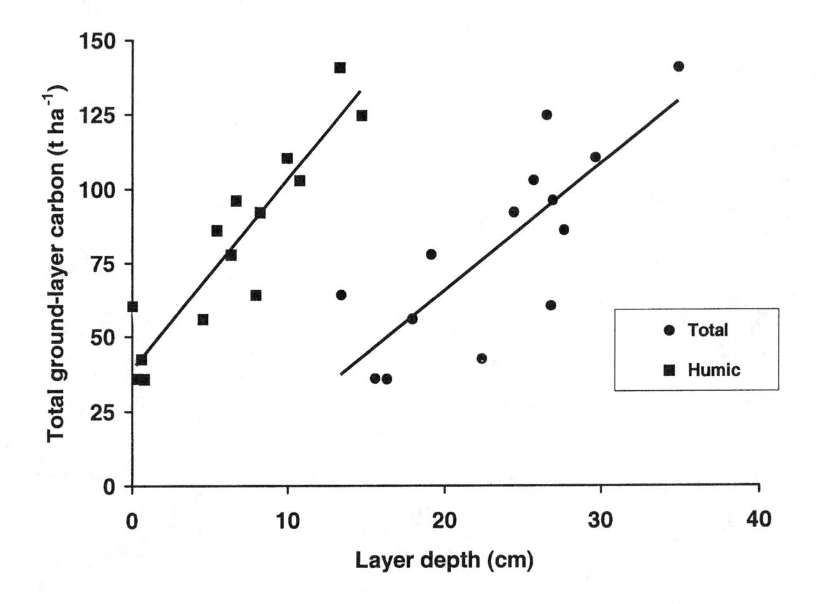

Figure 10.7. Average ground-layer carbon in mature black spruce forests plotted as a function of total ground-layer depth and humic soil-layer depth.

Table 10.3. Aboveground Biomass Data from Tok Fire Site (TK90)

Site	Tree species	Trees sampled	Basal diameter		Density	Stand biomass	Biomass burned	
			Average	Standard deviation	(stems ha^{-1})	(t ha^{-1})	(%)	(t ha^{-1})
Black spruce stands								
1	Black spruce	554	6.1	3.6	6,925	48.4	30.4	14.7
2	Black spruce	601	5	3.5	8,586	52.3	28.4	14.9
3	Black spruce	834	4.6	2.4	10,425	29.6	35.4	10.5
4	Black spruce	625	4.6	2.5	11,364	33.4	36.3	12.1
5	Black spruce	612	5.3	3	7,650	32.8	41.1	13.5
6	Black spruce	676	4.8	3	9,657	36.9	22.6	8.3
7	Black spruce	863	4.3	2.7	10,788	30.4	37.1	11.3
8	Black spruce	1,042	3.3	2.4	13,025	24.1	48.2	11.6
9	Black spruce	824	5.9	3.5	10,300	65.9	46.6	30.7
10	Black spruce	362	5.6	5.2	4,525	47.7	29.2	13.9
					Average	40.2	35.5	14.2
White spruce stands								
1	Spruce	827	4.7	3.3	10,388	46.5	15.6	7.3
	Aspen	20	14.9	4.2	250	8.1	11.5	0.9
2	Spruce	382	5.8	4.6	4,775	45.3	17.4	7.8
	Aspen	20	11.7	4.7	213	4.4	17.3	0.8
					Average	52.2	16.2	8.4
Willow/black spruce stands								
1	Black spruce	189	3.4	2.4	2,363	4.5	49.4	2.2
	Willow	2,216	0.7	—[a]	—	0.67	95	0.63
2	Black spruce	150	2.9	2.3	3,000	4	37.6	1.52
	Willow			Not sampled				
					Average	4.9	51	2.5
Aspen stands								
1	Aspen	52	11.9	3.7	1,785	53.2	13.6	7.2

[a]—, No data.

Tok region and found there was 4 t ha^{-1} in black spruce stands, 2 t ha^{-1} in the aspen and white spruce stands, and 1 t ha^{-1} in willow-black spruce stands. All understory vegetation was consumed by fire at this site.

Figure 10.8 presents the patterns of carbon levels in the aboveground vegetation of the different forest types for the Tok test site. The data in this figure were derived by assuming that carbon percentage of aboveground biomass is 53%, based on measurements made by Susott and co-workers (1991). Biomass consumption was greatest in the willow-black spruce stands (59.3%), next highest in the black spruce stand (42.3%), and lowest in the white spruce and aspen stands (19.2 and 16.7%, respectively). The overall levels of carbon released from burning of aboveground biomass were much lower than those observed from burning of ground-layer biomass: 1.9 t ha^{-1} in the willow-black spruce stands, 9.9 t ha^{-1} in the black spruce stands, 5.5 t ha^{-1} in the white spruce stands, and 4.4 t ha^{-1} in the aspen stands. These values are consistent with those reported by Stocks and associates (1991) for Canadian boreal forests (6.0 t C ha^{-1}).

A Model of Biomass Burning for Alaska Boreal Forests

Even though the results of the studies presented in this chapter present an improved understanding of the patterns of biomass burning that occur in the boreal forest region of Alaska, they are still insufficient to estimate the total amount of carbon released during fires at landscape scales. In most cases, the areas burned by

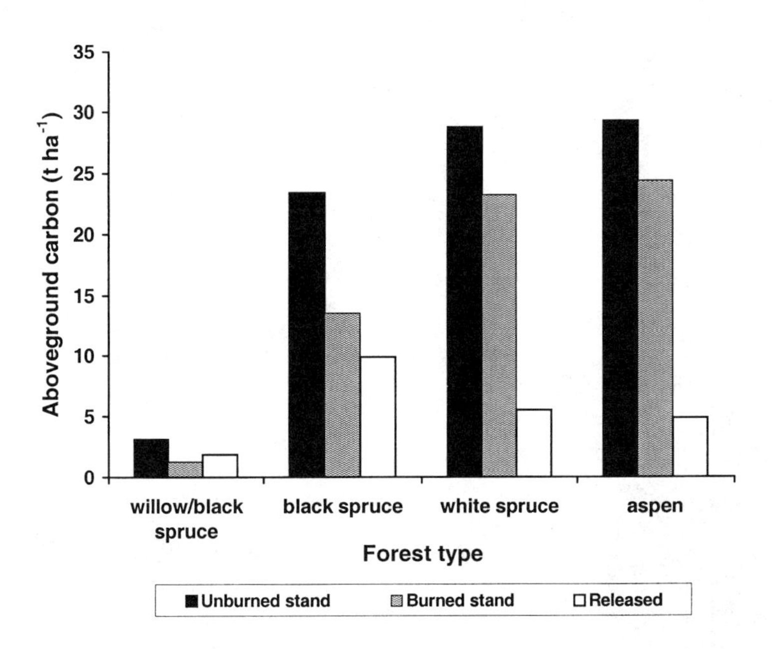

Figure 10.8. Average carbon levels in different forest types in the Tok study region, including unburned stands and burned stands.

larger fires in Alaska rarely contain a single forest type, but a continuum of different forest types. For example, interpretation of aerial photography collected over the Tok study region prior to the fire that occurred in 1990 has shown that there were seven discrete forest types occupying this landscape (Table 10.4).

To overcome this limitation, an approach was developed to extrapolate the data collected at the Tok test sites. The primary factor controlling forest ecosystem distribution in the Tok test site region is the permeability or drainage of the mineral soil substrate. The mineral soil profiles varied throughout this site. The upper soil horizons contained fine-textured, poorly drained silts and clays, whereas the lower horizons consisted of a coarse, more well-drained matrix of silt, sand, and gravels. The depth of the silt/clay layers ranged from 20 to 30 cm in the more well-drained sites that contained aspen and white spruce and was more than 200 cm in the more poorly drained sites containing black spruce willow, and a black spruce-willow mixture. These latter areas are typically underlain by permafrost that not only further impedes drainage but also is indicative of much cooler soil temperatures than are found in the areas with aspen and white spruce forests.

Figure 10.9 presents of plot of the average carbon levels for the unburned forest stands (y-axis) as a function of the overall drainage conditions of the forest floor (x-axis) (the spruce-willow data are from the Hajdukovich Creek site). In this representation, the warmest/driest soils are at the lower end of the scale while the wettest/coldest soils are at the higher end of the scale. The ecosystems in the Tok study region sampled in this study (Table 10.4) can be arrayed along this scale as follows: aspen (warmest/driest)—aspen/spruce—white spruce—white spruce-black spruce—black spruce—black spruce-willow—willow (coolest/wettest). From Figure 10.9, it can be seen that there is strong correlation between soil moisture/temperature and the patterns of carbon present in the different forest types. Vegetation biomass decreases linearly as soil wetness increases, and ground-layer biomass increases in a quadratic fashion.

A similar plot can be created for the fraction of carbon consumed during fire (Fig. 10.10). In this case, the fraction or percentage of carbon/biomass consumed during fire increases linearly as a function of soil wetness for aboveground vegetation, and it decreases linearly for ground-layer biomass.

Table 10.4. Forest Cover and Patterns of Carbon Release from the 1990 Tok Burn

Forest cover type	Percentage of coverage	Carbon release during fire ($t\ ha^{-1}$)
Aspen	13.1	14.2
Aspen-white spruce	17.2	16.1
White spruce	8.3	20.2
White spruce-black spruce	0.2	29.7
Black spruce	34.1	32.5
Black spruce-willow	20.3	18.5
Willow	6.7	2.1
	Weighted average	21.4

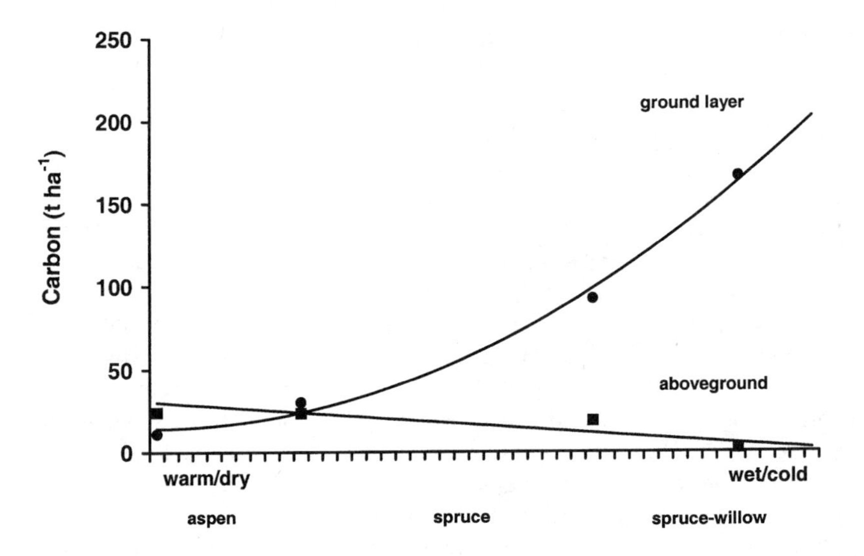

Figure 10.9. Model of carbon present in the aboveground vegetation and ground-layer biomass as a function of the soil temperature/moisture gradient found in the Tok study region test sites.

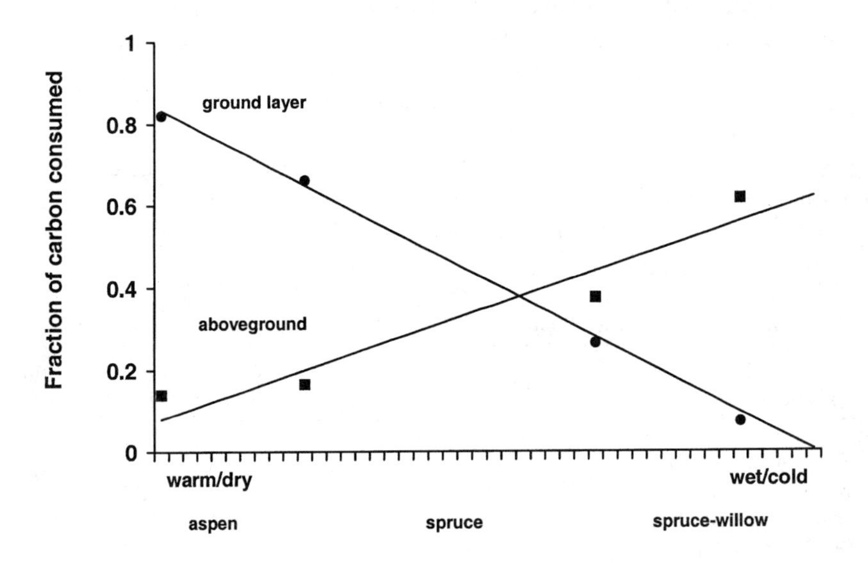

Figure 10.10. Model of the fraction of biomass/carbon consumed during forest fires in the aboveground vegetation and ground-layer biomass as a function of the soil temperature/moisture gradient found in the Tok study region test sites.

To create a simple model of how site drainage conditions influence biomass accumulation and biomass burning, we assigned a value of 0.1 to the dry/warm category and 1.0 to the cold/wet category in Figures 10.9 and 10.10. Using this gradation, we assigned values of 0.1, 0.3, 0.7, and 0.9 to the aspen, white spruce, black spruce, and willow-black spruce forest types, respectively. A regression then run for the biomass levels and fractions of biomass consumed as a function of the ground drainage rating. Despite the limited number of points in this model, the correlation coefficients for the aboveground carbon levels was $r^2 = 0.75$ ($P < .14$) and for the ground-layer carbon $r^2 = 0.99$ ($P < .1$) assuming a quadratic model. The correlation coefficients for the fraction of carbon consumed was $r^2 = 0.91$ ($P < .05$) for the aboveground carbon and $r^2 = 0.99$ ($P < .001$) for the ground-layer carbon.

By combining the information in Figures 10.9 and 10.10, it is possible to estimate the amount of carbon released during fire as a function of soil-drainage class (Fig. 10.11) to estimate the amount of carbon released during fires in the different forest ecosystems in this region (Table 10.4). Using this approach, we estimate that 21.4 t C ha^{-1} was released on average during the 1990 Tok fire.

The data in Figure 10.5 show that there is a considerable degree of seasonal variability in the amounts of biomass consumed during fires. This leads to the question: how would the average level of carbon released change if the fire occurred at a different time during the growing season or during a season that had different temperature and precipitation patterns than were present in the Tok region in the summer of 1990? To address this question, an additional model was

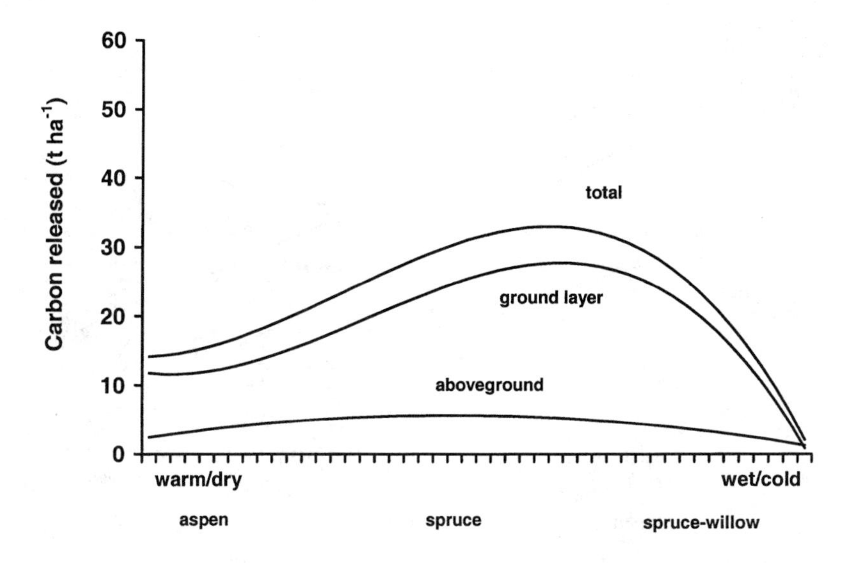

Figure 10.11. Model of the levels of carbon released during biomass burning as a function of forest types organized along the soil temperature/moisture gradient found in the Tok study region.

created assuming that the Tok fire consumed a moderate amount of biomass. Two additional burning scenarios were considered: a lighter burn and a more severe burn (Fig. 10.12). The consumption fractions for the vegetation layer (tree and understory biomass) were based on the observation that regardless of fire severity, the fraction of vegetation consumed in aspen and white spruce forests is always considerably less than that consumed in black spruce and willow forests. It was

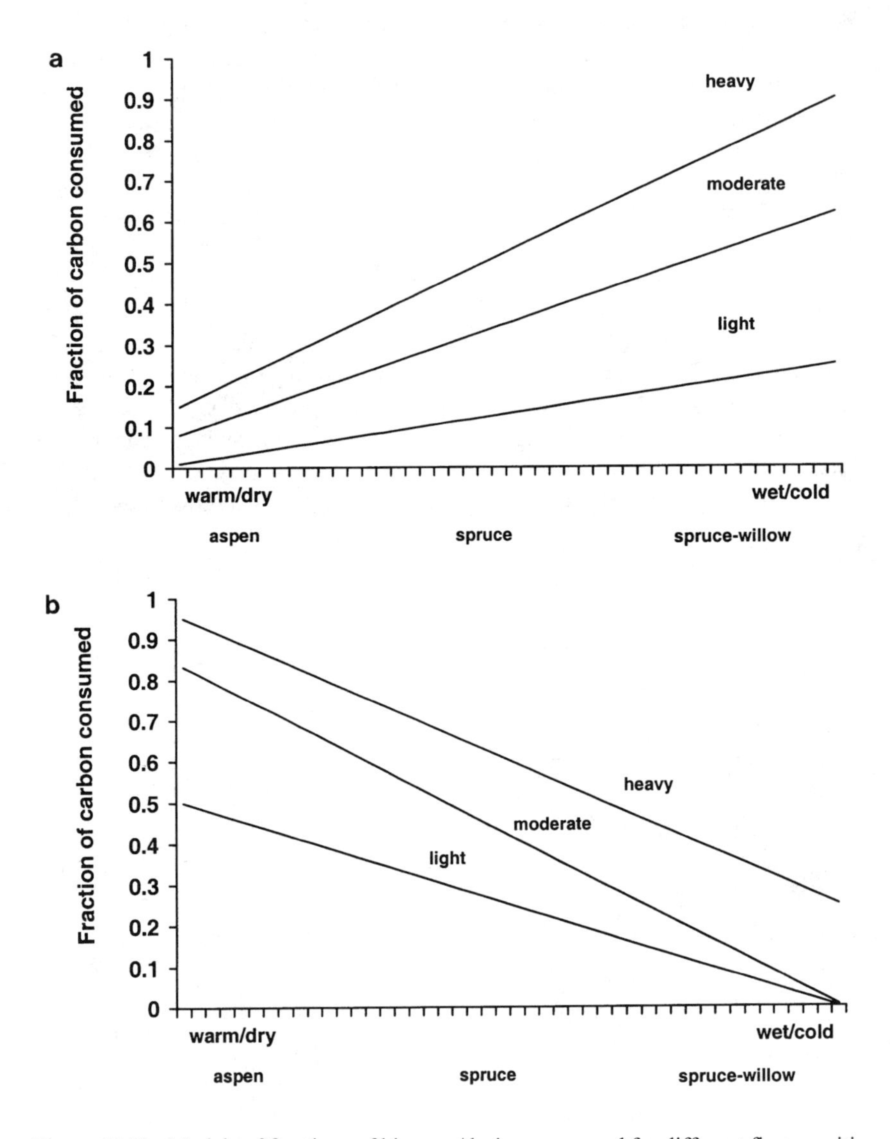

Figure 10.12. Models of fractions of biomass/during consumed for different fire severities as a function of the temperature/moisture gradient found in the Tok study region test sites: (a) aboveground vegetation and (b) ground-layer biomass.

assumed that the maximum amount of vegetation consumed in an aspen forest would be 15% of the total during a severe burn and 2% of the total in a light burn. For the willow type, we assumed 90% of the total vegetation would be consumed during a severe burn and 25% consumed during a light burn. The ground-layer biomass consumption models in Figure 10.12 were based largely on the observations made in this study and assuming that (1) at most, 50% of the biomass of the aspen forest was consumed during a light fire; and (2) 25% of the willow biomass was consumed during a heavy fire because of the high soil moisture content of the organic soils of these sites throughout the growing season.

Using the carbon curves for the Tok site (Fig. 10.9), Figure 10.13 presents models of the patterns of carbon release during fires based on the light and severe fire scenarios (the moderate scenario is presented in Fig. 10.11). Using the areal coverage of forest types presented in Table 10.4, the model predicts an average of 11.7 t C ha^{-1} during a light fire and 42.2 t C ha^{-1} would be released during a severe fire. Regardless of fire severity, the models predict that most of the carbon released from fire (> 80%) originates from the ground layer.

Summary

The boreal region of interior Alaska contains a wide range of forest types, ranging from aspen-white spruce forests on warmer drier sites to black spruce forests on cooler/wetter sites. Much of the carbon stored in the forests of this region is found in the thick organic soils of the black spruce forests. In the ground layers, on average 11 t C ha^{-1} was present in the aspen test site, 30 t C ha^{-1} in the white spruce test sites, and 82 t C ha^{-1} in the black spruce test sites. In terms of aboveground carbon storage, at the Tok site there was an average of 28.2 t C ha^{-1} in the aspen stand, 27.7 t C ha^{-1} in the white spruce stands, and 21.3 t C ha^{-1} in the black spruce stand.

The studies reported in this chapter illustrate that the amount of carbon released during fires in these forests varies a great deal, depending on forest type, the moisture/temperature conditions of the ground layer, and the timing of a fire during the growing season. It was shown that the patterns of biomass/carbon accumulation and biomass burning could be modeled along the soil temperature/moisture gradient controlling the distribution of forest types in this region. The lowest level of vegetation consumption during fires occurred on the forests occupying the warmest/driest sites (aspen and white spruce forests), whereas the highest levels of vegetation consumption occurred in the forests occupying the coldest/wettest sites (black spruce and willow). For ground-layer biomass, the highest fractions consumed during fires occurred in the warmest/driest sites, and the lowest fractions occurred in the coldest/wettest sites.

Based on variable burning scenarios, the amount of carbon released from a landscape similar to one found in the Tok region would range from 11.7 t C ha^{-1} during poor fire conditions to 42.2 t C ha^{-1} during severe fire conditions. Although the lowest value corresponds well with estimates produced by other re-

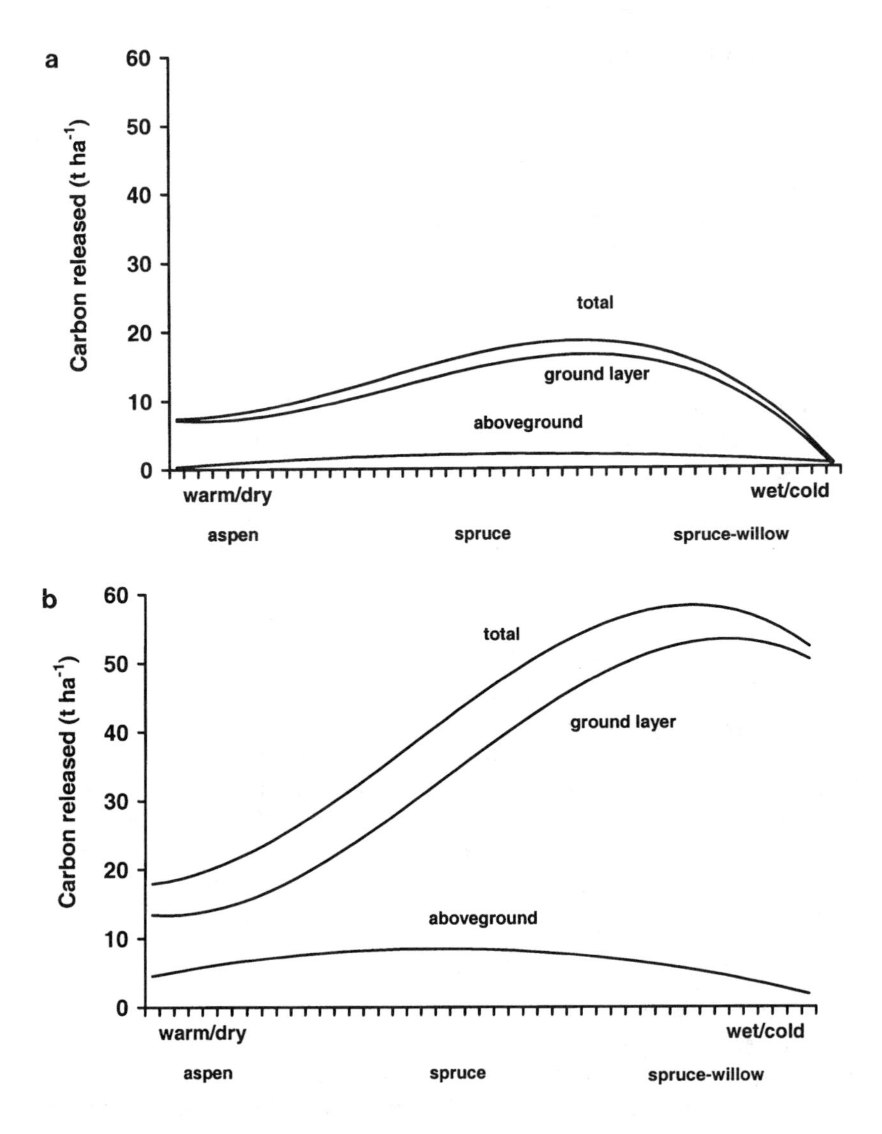

Figure 10.13. Models of levels of carbon released during biomass burning for different fire severities as a function of the temperature/moisture gradient found in the Tok study region test sites: (a) light-fire scenario and (b) severe-fire scenario.

searchers (Stocks and Kauffman 1997), the higher value is almost four times greater than previous estimates. This higher value is due to two factors. First, the forests in interior Alaska have much thicker organic soil profiles than the Canadian forests described by Stocks and Kauffman (1997) to derive their estimates. The average amount of ground-layer carbon released from black spruce fires was 33.5 t ha^{-1}. Second, the highest values of carbon consumption (as high as 75.7 t C

ha^{-1}) originate from a severe fire event under conditions that may not have occurred during the controlled burns used to generate the data in previous studies.

Fire represents an extremely important mechanism for the direct transfer of carbon from the terrestrial surface to the atmosphere in the boreal forest. The data from this study show that almost 90% of the organic soil profile in black spruce forests can be consumed during late-season fires. This level of consumption has been noted in other fires as well. Whether this level of biomass consumption is typical remains to be determined by future research.

Because thicker organic soils, especially the humic soils, take hundreds to thousands of years to form, fires burning deep into the organic soil profiles can represent a long-term net loss of carbon from the terrestrial surface. Because forests with thick organic soils are found throughout the boreal forest, this long-term carbon loss would not be an event localized to the forests of Alaska but would occur on a global basis. If climate warming results in increased fire frequency and/or severity, because of these facts the release of carbon during fires from deeper organic soils will represent a net loss of carbon from the boreal forest and could significantly increase the levels of greenhouse gases present in the atmosphere.

References

Agriculture Canada Expert Committee on Soil Survey. 1987. *The Canadian System of Soil Classification.* Canadian Government Publishing Centre, Ottawa, Canada.

Alexander, M.E., and D. Quintillo. 1990. Perspectives on experimental fires in Canadian forestry research. *Math. Comput. Model.* 13:17–26.

Alexander, M.E., B.J. Stocks, and B.D. Lawson. 1991. *Fire Behavior in Black Spruce–Lichen Woodland: The Porter Lake Project.* Northwest Region Information Report NOR-X-310. Canadian Forest Service, Yellowknife, NWT, Canada.

Atkins, T.L. 1995. Carbon release from a wildfire in the Alaskan taiga. M.S. thesis, University of Virginia, Charlottesville, VA.

Auclair, A.N.D. 1985. Postfire regeneration of plant and soil organic pools in a *Picea mariana–Cladonia stellaris* ecosystem. *Can. J. For. Res.* 15:279–291.

Barney, R.J., K. Van Cleve, and R. Schlentner. 1978. Biomass distribution and crown characteristics in two Alaskan Picea mariana ecosystems. *Can J. For. Res.* 8:36–41.

Bourgeau-Chavez, L.L., P.A. Harrell, E.S. Kasischke, and N.H.F. French. 1997. The detection and mapping of Alaskan wildfires using a spaceborne imaging radar system. *Int. J. Remote Sens.* 18:355–373.

Buol, S.W., F.D. Hole, and R.J. McCracken. 1989. *Soil Genesis and Classification,* 3rd ed. Iowa State University Press, Ames, IA.

Cahoon, D.R., Jr., B.J. Stocks, J.S. Levine, W.R. Cofer III, and J.M. Pierson. 1994. Satellite analysis of the severe 1987 forest fires in northern China and southeastern Siberia. *J. Geophys. Res.* 99:18,627–18,638.

Dyrness, C.T., and R.A. Norum. 1983. The effects of experimental fires on black spruce forest floors in interior Alaska. *Can J. For. Res.* 13:879–893.

French, N.H.F., E.S. Kasischke, L.L. Bourgeau-Chavez, P.A. Harrell, and N.L. Christensen, Jr. 1996. Monitoring variations in soil moisture on fire disturbed sites in Alaska using ERS-1 SAR imagery. *Int. J. Remote Sens.* 17:3037–3053.

Harden, J.W., K.P. O'Neill, S.E. Trumbore, H. Veldhuis, and B.J. Stocks. 1997. Moss and soil contributions to the annual net carbon flux of a maturing boreal forest. *J. Geophys. Res.* 102:28,805–28,816.

Harrell, P.A., L.L. Bourgeau-Chavez, E.S. Kasischke, N.H.F. French, and N.L. Christensen. 1995. Sensitivity of ERS-1 and JERS-1 radar data to biomass and stand structure in Alaskan boreal forest. *Remote Sens. Environ.* 54:247–260.

Kasischke, E.S., and N.H.F. French. 1997. Natural limits on using AVHRR imagery to map patterns of vegetation cover in boreal forest regions. *Int. J. Remote Sens.* 18:2403–2426.

Kasischke, E.S., N.H.F. French, L.L. Bourgeau-Chavez, and N.L. Christensen, Jr. 1995. Estimating release of carbon from 1990 and 1991 forest fires in Alaska. *J. Geophys. Res.* 100:2941–2951.

Mann, D.H., C.L. Fastie, E.L. Rowland, and N.H. Bigelow. 1995. Spruce succession, disturbance, and geomorphology on the Tanana River floodplain, Alaska. *Ecoscience* 2:184–195.

Peterson, E.B., Y.H. Chan, and J.B. Cragg. 1970. Aboveground standing crop and caloric value in an aspen clone near Calgary, Alberta. *Can. J. Bot.* 48:1459–1469.

Quintillo, D., M.E. Alexander, and R.L. Ponto. 1991. Spring fires in a semimature trembling aspen stand in central Alberta. Northwest Region Information Report NOR-X-323. Canadian Forest Service, Yellowknife, NWT, Canada.

Stocks, B.J. 1980. Black spruce crown fuel weights in northern Ontario. *Can J. For. Res.* 10:498–501.

Stocks, B.J. 1987. Fire behavior in immature jack pine. *Can J. For. Res.* 17:80–86.

Stocks, B.J. 1989. Fire behavior in mature jack pine. *Can J. For. Res.* 19:783–790.

Stocks, B.J. 1991. The extent and impact of forest fires in northern circumpolar countries, pp. 197–202 in J.S. Levine, ed. *Global Biomass Burning: Atmospheric, Climatic and Biospheric Implications.* MIT Press, Cambridge, MA.

Stocks, B.J., and J.B. Kauffman. 1997. Biomass consumption and behavior of wildland fires in boreal, temperate, and tropical ecosystems: parameters necessary to interpret historic fire regimes and future fire scenarios, pp. 169–188 in J.S. Clark, H. Cachier, J.G. Goldammer, and B.J. Stocks, eds. *Sediment Records of Biomass Burning and Global Change.* NATO ASI Series, Subseries 1, Global Environmental Change, Vol. 51, Springer-Verlag, Berlin.

Stocks, B.J., B.D. Lawson, M.E. Alexander, C.E. Van Wagner, R.S. McAlpine, T.J. Lynham, and D.E. Dube. 1989. The Canadian forest fire danger rating system: an overview. *For. Chron.* 65:258–265.

Susott, R.A., D.E. Ward, R.E. Babbitt, and D.J. Latham. 1991. The measurement of trace gas emissions and combustion characteristics for a mass fire, pp. 245–257 in J.S. Levine, ed. *Global Biomass Burning: Atmospheric, Climatic and Biospheric Implications.* MIT Press, Cambridge, MA.

Telfer, E.S. 1969. Weight-diameter relationships for 22 woody plant species. *Can J. Bot.* 47:1851–1855.

Van Cleve, K., L. Oliver, R. Schlentner, L.A. Viereck, and C.T. Dyrness. 1983. Productivity and nutrient cycling in taiga forest ecosystems. *Can J. For. Res.* 13:747–766.

Yarie, J., and K. Van Cleve. 1983. Biomass and productivity of white spruce stands in interior Alaska. *Can J. For. Res.* 13:767–772.

11. Postfire Stimulation of Microbial Decomposition in Black Spruce (*Picea mariana* L.) Forest Soils: A Hypothesis

Daniel D. Richter, Katherine P. O'Neill, and Eric S. Kasischke

Introduction

Wildfires and Boreal Forests

Across northern latitudes, the modern boreal forest extends over about 1.2×10^7 km^2, an extraordinarily vast area that spans northern Europe, Asia, and North America (Van Cleve et al. 1983a; Nikolov and Helmisaari 1992). Periodic wildfires are common to this forest. Areas burned have large year-to-year variation, depending on climatic conditions; on average, on the order of 10^5 km^2 of boreal forest burns each year (Stocks et al. 1996). Individual fires occasionally burn extensive areas, sometimes covering greater than 1,000,000 ha in a single burn (Cahoon et al. 1994; Murphy et al., this volume).

Wildfires in the black spruce forest are often intense crown and ground fires, killing trees and consuming large fractions of the tree and moss biomass and surficial soil organic matter. As a result, enormous quantities of chemical compounds are released to the atmosphere as gases and particulates. These pyrogenic releases during boreal fires contribute substantially to the chemistry of the global atmosphere and the global cycling of chemical elements (Cahoon et al. 1994; Kasischke et al. 1995; Levine and Cofer, this volume; French et al., this volume).

Combustion losses of C range from 10 to more than 50 t ha^{-1} (Stocks and Kauffman 1997; Kasischke et al., this volume, Chapter 10). Occasionally, intense fires may combust nearly all aboveground organic matter and leave behind a nearly bare, ash-covered mineral soil. Even in the most intense conflagrations, however, in which fire consumes nearly all biomass and O horizons, organic matter remains in the mineral soil, material that is potentially susceptible to postfire decomposition.

Boreal forests are well adapted to regenerate in burned-over areas, and they typically grow for 50–200 years after being burned (Viereck 1983). Boreal forests are characterized by low net primary productivity (NPP) and low decomposition rates of the organic matter that is produced. Low decomposition (i.e., low soil microbial respiration) leads to large accumulations of the organic matter. Low decomposition is controlled by low soil temperature, low soil fertility, permafrost, and if the soil is not frozen, water saturation that limits root and microbial activity. Large amounts of organic C accumulate in organic matter of both the forest floor (O horizons) and mineral soils. About 30% of the world's soil C is stored in these northern soils Tables 2.1 and 2.2), and changes in the soil disturbance regimes of these soils could have major impacts on global C cycling (Kasischke et al. 1995).

Black spruce forests are the most extensive forest ecosystem of the Alaskan taiga, tend to develop on the coolest wettest soils, and have large amounts of C stored in the upper organic profile (Fig. 10.4). When dry, these thick organic layers plus moss and tree biomass can provide an enormous fuel source for wildfire. The prominent moss and forest floor layers accumulate at rates up to more than 100 t C ha^{-1} in a century of forest development following wildfire (Van Cleve and Viereck 1981; O'Neill et al. 1997; Goulden et al. 1998). The thick organic layers (sometimes >30 cm) retain large amounts of moisture and promote anaerobic conditions. The wet organic layers have extremely low thermal conductivities and effectively insulate mineral soil, thereby ensuring not only low soil temperatures but also permafrost. All these conditions increase the vulnerability of the forest to burning during dry weather.

The Hypothesis

Immediately following fires in black spruce forests of interior Alaska, insulating moss and the thick blanket of organic matter are at least partially oxidized, and an ash-covered soil is much more exposed to solar radiation. As a consequence, heat diffusion into the soil is enhanced, and permafrost may melt 1- or 2-m depths below the mineral-soil surface. This soil warming may last for several decades. Viereck (1983) reports studies that observed increases in soil temperature 25 years following fire. Not only do postburn soils become warmer, but the quality of the organic substrate is enhanced as the uppermost layers are initially enriched in nutrients from ash particulates, and they also become increasingly well aerated. This warmer, more aerobic soil surface is hypothesized to stimulate microbial decomposition of unburned surficial organic matter and newly unfrozen mineral-soil organic matter as well.

This chapter presents data to support a hypothesis that biogenic CO_2 emissions are increased in boreal forests in the initial decades following wildfires. Postfire stimulation of microbial decomposition and CO_2 emissions is hypothesized to be similar in magnitude to pyrogenic CO_2 emissions themselves.

Methods

Research Area, Ecosystems, and Soils

The black spruce stands used in this study were located on broad alluvial flats in the vicinity of Tok and Delta Junction, Alaska. All are supported by poorly drained soils, most of which are newly classified as Gelisols. Soils are derived from sediments of the Tanana River, a major tributary of the Yukon River, and have received varying amounts of loessal material from nearby sources. Soils are Pleistocene in age and consist of silt loams underlain by deposits of sand and gravel. Mottling is common and prominent, generally bright orange to red on a gray to gray-blue matrix. Both soil profiles and the ground surface have evidence of cryoturbation, with occasional earth hummocks that are 1–2 m in diameter (see Fig. 12.4). Some profiles exhibit vertic properties with shrink-swell cracks that range up to 2 cm wide at the soil surface and extend into the soil 10–30 cm in depth. Shrinkage cracks may be a mechanism for introducing surface C into the deeper soil profile where it can be immobilized by permafrost. Topography in the region is relatively flat (0–5% slope), suggesting little horizontal percolation of water and minimal loss of C through erosion and runoff. More than 80% of Alaska is underlain by permafrost, and soil of mature black spruce stands typically remains frozen in the upper meter of profile until the end of August.

The climate in the Alaskan interior is generally cold and dry. Bounded to the north by the Brooks Range and to the south by the Alaska Range, the Tanana River Valley experiences a strong continental climate. Average annual maximum and minimum air temperatures at Tok (1954–1998) are 2.0°C and −11.4°C, respectively, with lowest monthly temperatures in January (mean −30.5°C) and highest monthly temperatures in July (mean 21.5°C). The region averages 22.2 cm (± 7.5 cm) of precipitation annually, with one-third falling as snow usually during October–March and one-third falling as rain during July and August (National Climate Data Center 1998). Potential evapotranspiration is 47.5 cm, resulting in large moisture deficits in most years. The summer season (late May–August) is characterized by long periods of sunlight (about 21 h day^{-1} in June).

Soils in mature black and white spruce forests have highly developed litter-fibric-humic horizons that overlie the mineral soil (Chapter 10). The upper portion of this profile is composed principally of living and recently dead mosses that are relatively undecomposed. The lower part of the profile consists of decomposed and charred material that has accumulated over decades to millennia (Harden et al. 1997). During many ground fires, upper moss and fibric horizons burn completely whereas humic layers are combusted only in the most severe burns.

Recently burned spruce stands have a cover of early successional moss species such as *Ceratodon prupureus, Polytrichum juniperinum,* and *Pohlia nutans.*

These mosses are relatively thin and may provide little control over soil moisture and thermal conditions, although they do contribute to soil respiration. In mature black spruce stands, moss and lichen species are commonly *Hylocomium splendens* and *Pleurozium schreberi* and, to a lesser extent, *Sphagnum* spp. The latter mosses play a critical role in mediating organic matter, thermal, and water budgets of the soil systems.

Field and Laboratory Methods

To evaluate the hypothesis that microbial decomposition is stimulated in the postfire environment of black spruce (*Picea mariana* L.) ecosystems, we measured soil respiration of an age sequence of burned and unburned forests and evaluated ecological processes that control soil emissions of CO_2. We also measured soil temperature and soil chemistry in paired burned and unburned black spruce forests. A model was used to simulate heat diffusion into burned black spruce forests.

To investigate the below-ground soil environment, soil pits were excavated in ten black spruce stands. Pits were excavated to a depth of 1 m unless restricted by permafrost or gravel; average depth was 0.74 m. Soil temperature was measured with a digital thermometer inserted horizontally into the organic and mineral soil at the same time that pits were excavated. Soil respiration was estimated by using a dynamic closed gas exchange system (EGM-1 environmental gas monitor and SRC-1 soil respiration chamber, PP Systems). A detailed description of this system and its operation may be found in Parkinson (1981). Living moss and vegetation were not removed prior to measurement to limit micro-site disturbance and the subsequent potential for enhanced decomposition and disturbance of root biomass. This means that respiration from mosses is included as part of soil efflux.

Soil chemistry was examined from samples of organic and mineral soils collected from both burned and unburned sites. Soil pH, exchangeable cations, and total C and N were estimated in O horizons and mineral soils by conventional methods (Page et al. 1982). Heat diffusion and its alteration by wildfire were simulated with a heat diffusion model.

Results

Postfire Soil Carbon Levels

Although wildfires are typically intense conflagrations in boreal spruce forests, oxidizing 10 to more than 70 t C ha^{-1} (Chapter 10), they combust only a fraction of the organic C in the ecosystem. The boreal forest may have low NPP, but during the decades following fires, organic C accumulates in great amounts due to low decomposition. Thus, accumulation of potentially flammable organic matter (i.e., net ecosystem productivity) is relatively high in these ecosystems.

A chronosequence of black spruce stands in the Tok-Delta Junction region indicated that moss plus O horizons may accumulate nearly 100 t ha^{-1} of organic

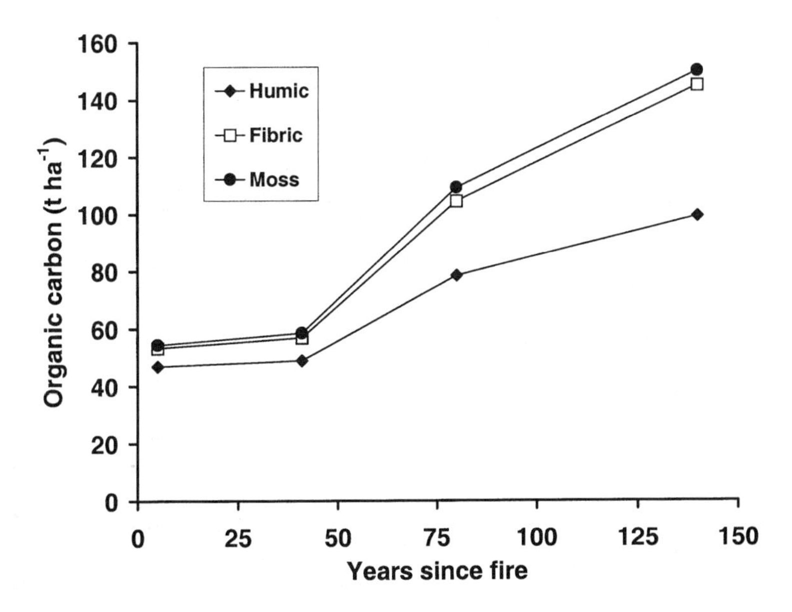

Figure 11.1. Accumulations of organic C in moss biomass and O horizons of black spruce forests that range in age from 5 to 140 years since the last wildfire in interior Alaska.

C over 140 years of postfire forest growth (Fig. 11.1). This chronosequence study of time since fire may well be confounded by burn severity that has affected remnant soil C on each site. In particular, this may result in the rate of increase in humic materials to appear higher than it actually is. In the chronosequence illustrated in Figure 11.1, about 50 t ha^{-1} of organic C resides on the surface of the mineral soil following fires and is potentially available for postburn decomposition.

Large amounts of organic C are stored below ground in mineral soil, some of which may become decomposable following fire. Organic C averaged 152.1 t ha^{-1} (SE $\pm$ 26.7) in the upper 1 m of mineral soil of ten soil pits excavated in black spruce stands in the vicinity of Tok and Delta Junction. Although many mineral soil profiles lacked a distinct A horizon, most organic C was stored in the upper 50 cm of the profile (Table 11.1). To the extent that melting of the previously frozen mineral soil follows fires, even soil C contained in well-humified organic matter may be subject to enhanced decomposition. Freezing and melting may also have physical effects on soil organic matter that enhance its decomposability.

These accumulation rates are supported by C inventories in other boreal black spruce forests. In one study of a black spruce forest in northern Manitoba, Canada, ecosystem C totaled 175 t ha^{-1} after 120 years of forest growth since the last fire. Of this amount, tree biomass contained 40 t ha^{-1}, live and dead moss 45 t ha^{-1}, and the soil about 90 t ha^{-1} (Harden et al. 1997; Goulden et al. 1998). Studies of eight black spruce stands in interior Alaska contained similarly large amounts of

Table 11.1. Chemical Characteristics of Two Soil Profiles in Poorly Drained Black Spruce Forests near Tok, Alaska[a]

Soil horizon	Soil depth (cm)	Carbon %	Carbon t ha^{-1}	C/N ratio	pH $CaCl_2$	ECEC (cmolc kg^{-1})	CECt (cmolc kg^{-1})	BSt (%)	Mehlich ext P (μg g^{-1})
Tok Stand 11									
O	11	35.1	41.9	31.8	4.91	52.3	131.1	39.6	106.8
B1	0–15	1.26	24.6	14.0	5.26	7.1	13.1	54.0	7.0
B2	15–35	1.11	28.9	12.3	6.66	9.2	15.4	59.4	3.2
BC	35–65	0.78	30.4	11.1	6.91	8.1	8.6	93.8	7.6
BC	65–95	0.91	35.5	13.0	6.84	8.2	8.6	94.8	3.7
Tok Stand 13									
O	10	31.0	41.2	25.2	4.49	50.5	139.5	35.7	12.8
A	0–16	2.05	42.6	15.8	6.64	12.6	17.8	70.8	1.9
B1	16–35	2.94	72.6	16.3	6.80	17.7	23.2	76.3	3.5
B2	35–45	1.97	25.6	16.4	6.79	13.3	16.3	81.5	3.0
B2	45–55	0.57	7.4	11.4	6.83	6.8	10.1	67.1	1.5
II C	55–90	0.50	22.8	10.0	6.63	6.1	8.1	74.8	0.9

[a]Their permafrost has melted since 1990, the date of the Tok fire. ECEC, effective cation exchange capacity; CECt, total cation exchange capacity; BSt, base saturation on a total cation exchange capacity basis; Mehlich ext P, Mehlich extractable phosphorus.

soil C (Van Cleve et al. 1983b). Mineral soil C is a component not well estimated in many studies, mainly due to limitations on soil sampling imposed by permafrost.

Postfire Observations of Soil Temperature and Permafrost Melting

Mature black spruce forests are typically underlain by permafrost that thaws annually to form a shallow biologically active layer by the end of many growing seasons. Both the depth and rate of thaw can be influenced by recent fire activity because the radiation balance of burned black spruce stands changes rapidly following fire and leads to marked increases in soil temperature that may last many years following fire. For example, in a mature black spruce forest near Tetlin Junction, Alaska, the organic profile has just begun to thaw by mid-July 1997, and by mid-August soils were still frozen in the upper 15 cm of the mineral soil (Fig. 11.2a). By contrast, in an adjacent stand that burned the previous year, the organic horizons were completely thawed by mid-June, and by mid-August the depth to permafrost exceeded 35 cm.

In 1996 at Tok, soils in burned spruce stands had permafrost that had melted to 50-cm depth by early June 1996, whereas even by mid-August the permafrost in unburned stands had not melted to 30 cm (Fig. 11.2b). In 1997 at Tok, permafrost was actively melting at a 1-m depth by mid-June, and by mid-August soil temperatures exceeded 15°C at a 10-cm depth and 6°C in soil at 1 m (Fig. 11.3). At these sites, live fine roots were found throughout the 1-m profile of newly melted soil.

This postfire soil warming has been observed at other sites and may last for decades after burning (Viereck and Dyrness 1979; Van Cleve et al. 1983a; Dyrness et al. 1986; Auclair and Carter 1993; Trumbore and Harden 1997). Viereck and Dyrness (1979) observed that in the first year following the 1971 Wickersham Dome fire, soil temperature at a 50-cm soil depth increased by up to 6°C over unburned soils. Long-term studies of permafrost following the Wickersham Dome Fire further demonstrate decades-long permafrost melting and biologically significant increases in soil temperature.

Such increases in temperature and melting of permafrost may be hypothesized to greatly stimulate microbial activity. Presuming a Q_{10} of 2.0 and a temperature increase of 5–10°C would suggest a 50–100% increase in rate of decomposition in the absence of postfire shifts in microbial populations. Considering that permafrost may melt and increase the depth of the soil's biologically active zone by 1 m or more and that these effects may well last more than a decade, postfire microbial emissions of CO_2 may well be substantial.

Simulated Heat Diffusion into the Soil Profile

Transfer of heat into the soil profile is dependent on both the temperature gradient at the soil surface and the ability of a given soil material to transfer heat energy downward. The change in soil temperatures following fire can be described mathematically as a function of (1) the temperature gradient at the soil surface; (2) the

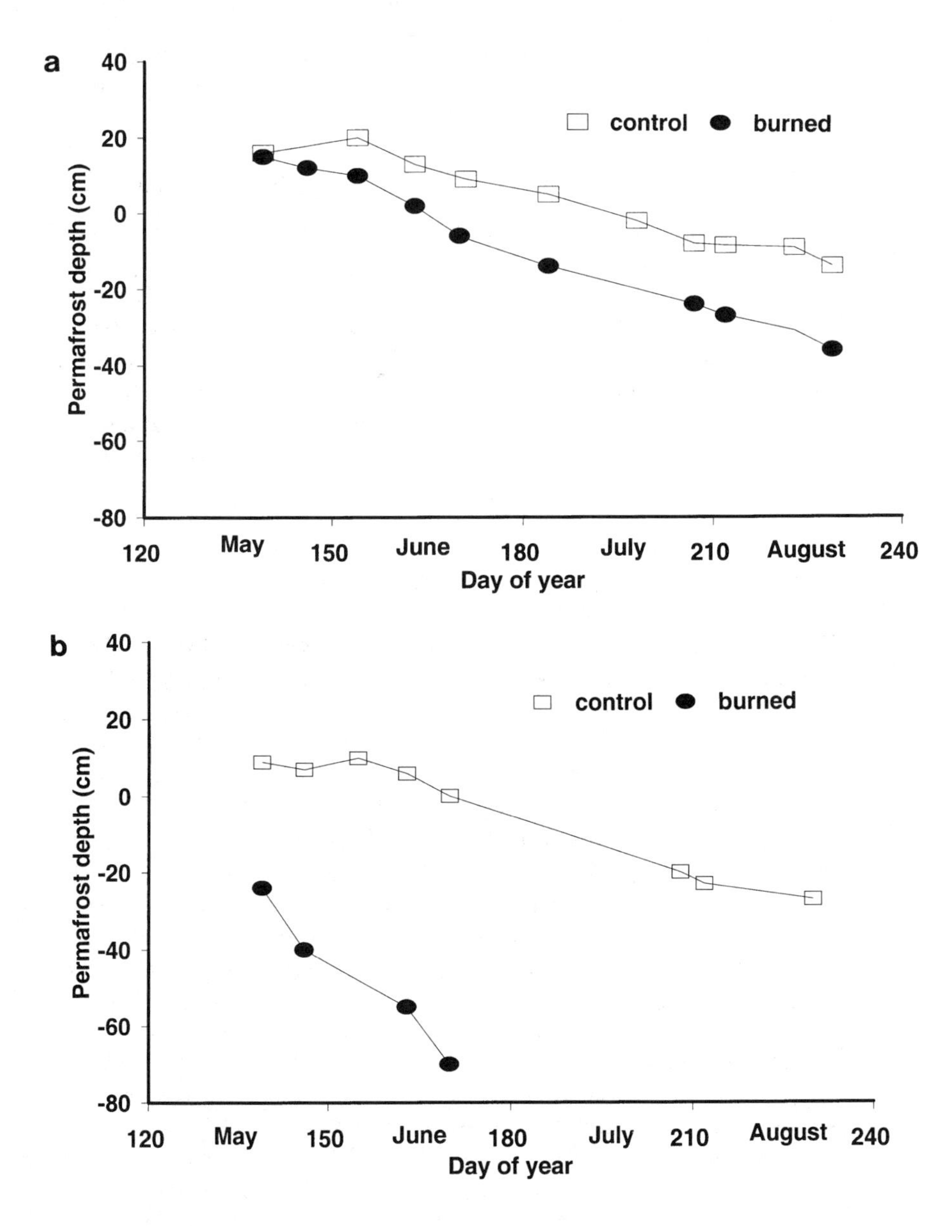

Figure 11.2. Seasonal increases in depth to permafrost in paired burned and unburned (control) black spruce forests in the summer of 1997: (a) 1996 Tetlin, Alaska, fire and (b) 1990 Tok, Alaska, fire. The depth of 0 represents the interface between the mineral soil and organic soil. Positive values are organic soil depths, and negative values are mineral soil depths.

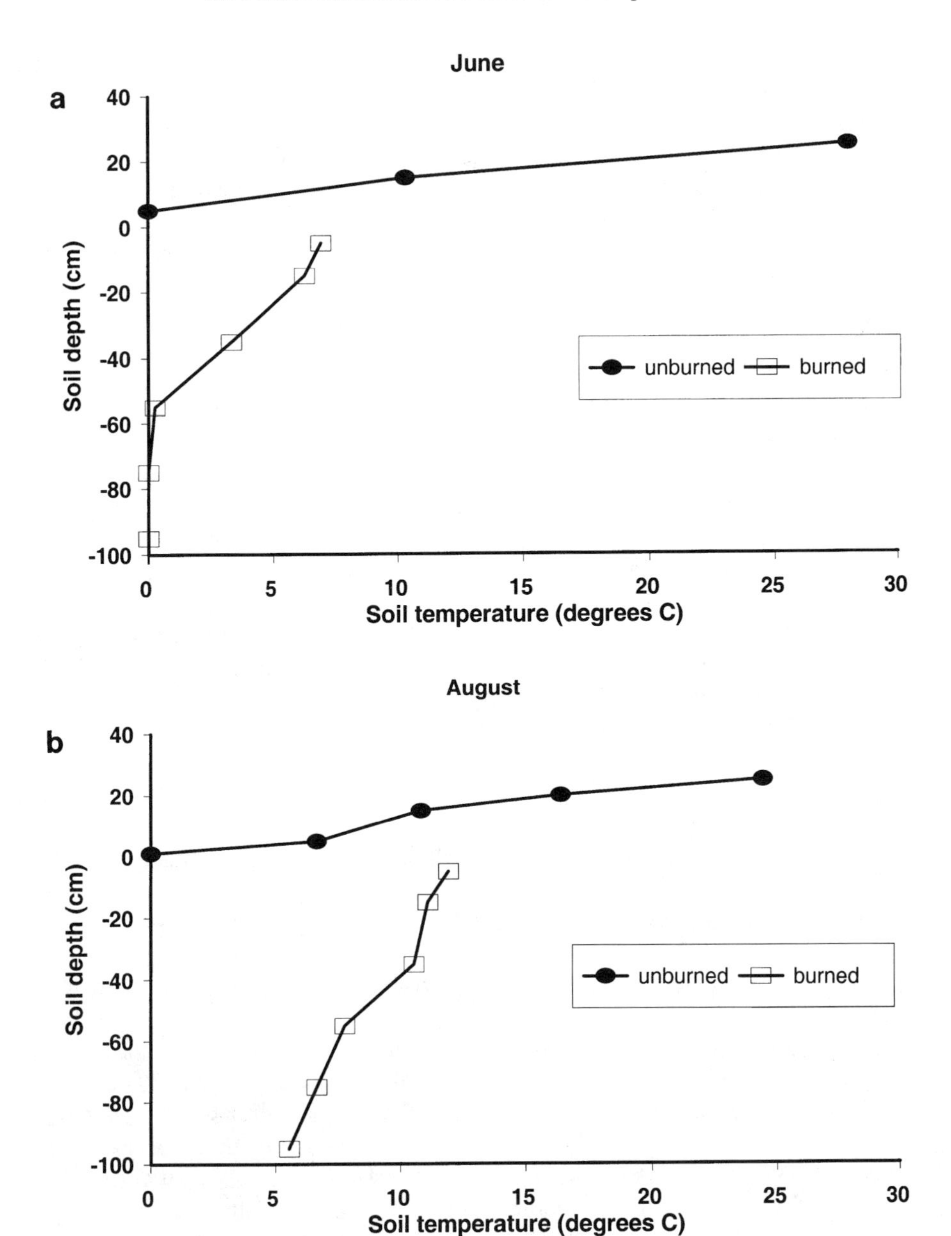

Figure 11.3. Measured temperature profiles in a burned and unburned black spruce soil in June and August 1997 at Tok, Alaska.

thermal diffusivity; and (3) the amount of organic matter remaining on the soil surface following a fire. Organic matter effectively insulates underlying mineral soil from heating and limits downward conduction of heat due to relatively high porosity and to frequently high moisture content. The high heat capacity of soil water requires that a greater amount of heat energy be added to a material with a higher water content to effect the same change in temperature.

Under steady-state conditions, changes in soil temperature (δT) over time (δt) may be described by the heat flow equation:

$$\delta T/\delta t = K_T \, \delta^2 T/\delta z^2 \qquad (11.1)$$

where K_T is the soil thermal diffusivity (the ratio of thermal conductivity to heat capacity for a given moisture content and z is depth. The thermal diffusivity of a moist loam soil is approximately 6.0×10^{-3} cm^2 sec^{-1} (Nakshabandi and Kohnke 1965) while that for a moist peat soil is 1.2×10^{-3} cm^2 sec^{-1} (Geiger 1965). Given these values, Equation (11.1) indicates that, all else being equal, heat will be conducted five times more rapidly in a purely mineral soil than in an organic soil.

However, the surface temperature of soil changes over the course of a day. To determine diurnal heat transfer in a nonsteady-state system, Equation (11.1) must be solved for temperature with the additional assumption that as the depth approaches infinity, the soil temperature will equilibrate at the annual mean temperature. Carslaw and Jager (1959) have described the solution to this equation as

$$T\,(z,\, t) = T_a + A \exp\,(z/d) \sin\,(\phi t + z/d) \qquad (11.2)$$

where T_a is average air temperature, A is the amplitude of surface fluctuations, z is depth, ϕ is the angular frequency of the surface change, and d is $(2\,K_T/\phi)^{1/2}$. This equation describes a sine wave whose amplitude decreases with depth.

To simulate the effect of wildfire on summer soil temperatures, we first assumed that the only effect of fire on heat transfer was to reduce the thickness of the surface organic horizon. We then solved Equation (11.2) at six temperatures throughout a 24-hour day to a depth of 1 m for a "burned" and "unburned" soil profile. The unburned soil was modeled as 30 cm of peat overlying 70 cm of silt loam, and the burned soil was modeled as 5 cm of peat overlying 95 cm of silt loam. Based on temperature measurements recorded during the summer of 1997, we assumed a daily temperature fluctuation at the soil surface of 20°C (A = 10), and an average surface temperature of 15°C (T_a = 15). Results of this simulation are shown in Fig. 11.4.

In the unburned soil, daily fluctuations in temperature of more than 10^{-1}°C do not penetrate beyond the upper 25 cm; daily change is less than 10^{-3}°C at 90 cm. Daily temperature changes do not fully penetrate the organic profile, and the temperature of the mineral soil underneath remains constant. Following fire, daily fluctuations in temperature may reach a depth of 60 cm into the mineral soil. These simulations are conservative because (1) burning often leaves less than 5 cm of organic material at the surface; (2) the upper organic materials (especially in

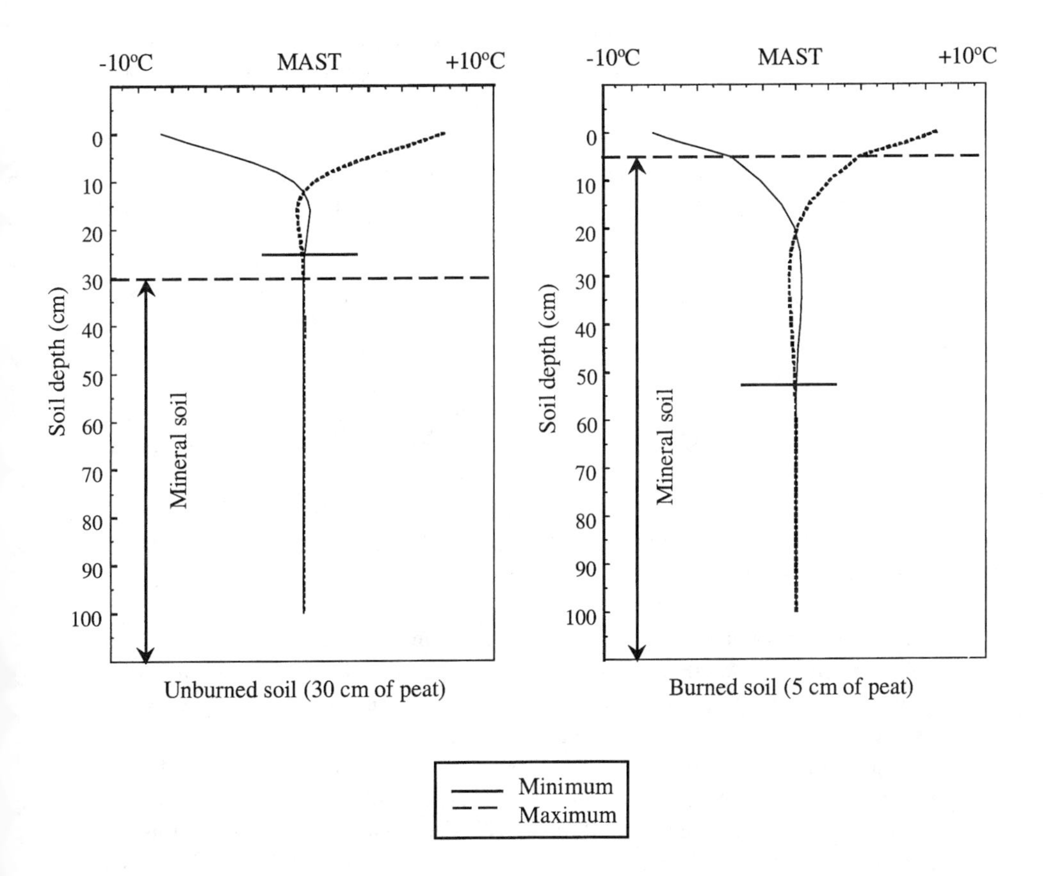

Figure 11.4. Diurnal temperature flux modeled for a burned and unburned soil. The daily temperature varies by 20°C, and the center of the temperate scale has been represented as mean annual soil temperature (MAST). The dashed horizontal line represents the organic–mineral soil interface; the horizontal solid line is the lowest depth at which temperature fluctuates by more than 0.1°C over 1 day. In the unburned soil, the organic matter damps temperature changes within 25 cm of the surface; daily temperature variations do not penetrate below the depth of the organic matter. In the burned soil, daily temperature changes penetrate to below 50 cm. Nearly 0.5 m of mineral soil experiences daily temperature variation. Reducing the thickness of the organic layer by fire increases seasonal warming of mineral soil and may promote the decomposition of C previously immobilized by permafrost.

burned sites) are subject to drying during most summers, resulting in a much lower K_T for the organic soil; and (3) we did not include postfire increases in surface temperature due to decreased soil albedo and loss of canopy shading.

This model simplifies the soil system in that it does not account for permafrost development during winter months and the effects that this permanently frozen soil will have on summer soil temperatures. In addition, Equation (11.2) was developed to describe penetration of annual temperature waves; caution should be used before using this approach quantitatively. Nonetheless, the Jager and Carslaw model does provide a useful illustration of the insulating capacity of organic material and the way in which fire can change the temperature profile at depth. Observations and theoretical simulation indicate that following fire, greater penetration of heat into the mineral soil profile increases the rates of organic matter decomposition at depth.

Postfire Changes in Soil Chemistry

Black spruce forests have notably low soil fertility, and fire can have a pronounced effect not only on increasing soil-rooting volume but also on O horizon chemistry. High-pH ash and charcoal are created by fire, and ash elements (P, Ca, Mg, and K) are mineralized to bioavailable forms. Soluble electrolytes are leached from the ash into unburned or partly burned H horizons and upper mineral soil. Such postfire effects are well documented in forest fires throughout the world.

In June 1996, a forest burned near Tetlin, Alaska, and in August the upper soil layers of burned and unburned profiles were sampled to estimate soil chemical properties. Soil pH, exchangeable cations, effective cation exchange capacity (ECEC), and base saturation were relatively low in the superficial horizons of the unburned moss-covered forest soils. The pH of these highly organic unburned layers was less than 4, base saturation was 15% or less, and ECEC was very low due to pH-dependent charge of acid surfaces that increased in the burned O horizons (Table 11.2). In 2 months following fire, rain had leached at least some soluble nutrients into the unburned F and H horizons.

The soil pH, exchangeable cations, ECEC, and base saturation were all increased following fire. Soil ECEC nearly rose to approximate the total CEC due to pH-dependent charge of organic matter (Table 11.2). Base saturation was 15% or

Table 11.2. Soil Chemistry in Paired Burned and Unburned Plots following 1996 Fire in Black Spruce O Horizons in the Tetlin Vicinity in Interior Alaska[a]

Horizon	pH $CaCl_2$	ECEC (cmolc kg^{-1})	CECt (cmolc kg^{-1})	BSt (%)
Unburned				
Moss[b]	3.83	28.0	160.7	13.9
Fibric 1 (Oe)	3.98	26.6	142.9	15.2
Fibric 2 (Oe)	4.71	39.8	118.2	38.2
Humic (Oa)	6.21	46.8	88.1	53.9
Burned				
Fibric (Oe)	6.84	68.5	85.9	64.0
Humic (Oa)	6.52	75.5	95.0	67.2

[a]Collections made August 1996. ECEC, effective cation exchange capacity; CECt, total cation exchange capacity; BSt, base saturation on a total cation exchange capacity basis.
[b]Live plus dead materials. Probably *Hylocomium splendens*.

less in unburned surficial moss and F layers and more than 60% in burned layers. The chemical changes represent responses to fire, mainly associated with the neutralization of acidic unburned O horizons due to inputs of ash and charcoal. The changes appear to be substantial, and such enrichments may well benefit microbial decomposition in the postfire environment.

In forests that have not been recently burned, H horizons (O_a) of the forest floor are mixtures of humified organics as well as an unknown amount of charcoal and ash from previous fires. The H layer often contains a large accumulation of C and appears to be relatively higher in pH; bioavailable ash nutrients such as Ca, Mg, K, and P; and cation exchange capacity than layers of unburned materials that have accumulated since the last fire (Table 11.2). Some of the C in the basal H layers can be very old, as ^{14}C analyses substantiate (Harden et al. 1992). Although burned F and H horizons may be largely composed of relatively recalcitrant charcoal and well-humified organic matter, elevated temperatures and bioavailability of nutrients may facilitate organic matter's decomposition in the postfire environment.

Postfire Soil Respiration

Soil respiration was measured during the growing season in paired plots in adjacent burned and unburned black spruce forests in the Tok-Delta Junction vicinity. Forests selected had been burned at different times in the past, ranging from 1 to 10 years. Respiration was measured weekly from May through August 1997. Ten individual measurements were usually made at each site.

Soil CO_2 flux from forest soils burned within the past 10 years averaged 0.42 g CO_2 m^{-2} h^{-1} in the growing season of 1997 (SE = 0.008). By contrast, soil CO_2 flux from paired controls averaged 0.82 g CO_2 m^{-2} h^{-1} (SE < 0.016). Across this sample of different sites and times of measurement, soil respiration of these burned sites averaged about 50% of that from unburned sites (Table 11.3).

The contributions of respired CO_2 from plant roots, moss, and microbes differ among these burned and unburned ecosystems. In the initial years after fire, microbe respiration probably represents the majority of the soil CO_2 flux. By contrast, in unburned systems the majority of the CO_2 is derived from plant root and moss respiration. In the forest stands at the Tetlin site (burned the previous year), there were no significant plants or moss, and thus the soil respiration could be assumed to be 100% microbial in origin.

Schlentner and Van Cleve (1985) estimated that in mature black spruce forests about 20% of soil respiration was derived from microbial decomposition and 80% was from plant root respiration. A 20/80% proportionality suggests that in the unburned systems at Tok and Delta Junction, microbial respiration may account for 0.16 g m^{-2} h^{-1} of the total soil efflux of 0.82 g m^{-2} h^{-1}. Using this proportionality in the Tetlin ecosystems (the only pair of systems with minimal to no NPP) suggests that microbial respiration accounted for a flux of 0.114 g CO_2 m^{-2} h^{-1} from the unburned forest. This rate of microbial respiration is about 36% of the soil respiration in the burned Tetlin site (Table 11.3). Thus, in the burned eco-

Table 11.3. Soil Respiration from Paired Burned and Unburned Black Spruce Forests in the Summer of 1997

Forest stand	Years since fire	Soil respiration ($g\ CO_2\ m^{-2}\ h^{-1}$)	CV[a] (%)	Sample size
Tetlin burn (1996)	1	0.32	46.9	149
Tetlin control		0.57	40.4	150
Hijukavic burn (1994) (moderate burn)	3	0.38	197.5	128
Hijukavic control		0.72	54.0	132
Buffalo burn (1994) (severe burn)	3	0.47	46.8	130
Buffalo control		0.86	43.0	130
Tok burn (1990)	7	0.50	44.7	165
Tok control		1.19	50.0	180
1987 burn	10	0.42	40.5	70
Control		0.74	42.9	65

[a]Coefficient of variation percentage of all samples collected May–August 1997. CV percentage includes temporal and spatial variations plus sampling error.

system with no moss or plant growth, gross soil respiration was measured at 0.32 $g\ m^{-2}\ h^{-1}$, almost threefold higher than an estimate of microbial respiration using the factor of Schlentner and Van Cleve (1985). These data suggest that the postfire environment may greatly stimulate microbial respiration over rates in unburned ecosystems.

Although separating plant from microbial respiration in these flux measurements is problematic at this stage of our research (except at the Tetlin burn, where there was no moss or plants following the previous year's fire), it appears that we can readily hypothesize that because plant production and respiration are greatly diminished in burned systems, microbial decomposition of partly burned organic matter on the soil surface and within the mineral soil profile is stimulated as the biologically active zone is warmed and increased greatly in volume.

Initial Estimate of Soil Carbon Loss Due to Stimulation of Microbial Decomposition

To evaluate the consequences of the hypothesis, we made an initial and conservative estimate of postfire stimulation of microbial respiration. In this estimation, we base our calculations on the 20/80% partition of Schlentner and Van Cleve (1985) between microbial and plant contributions to soil respiration in mature black spruce ecosystems. We then estimate that of the gross soil respiration in the 1-year-old burned ecosystem at Tetlin (0.32 $g\ CO_2\ m^{-2}\ h^{-1}$), 0.206 $g\ CO_2\ m^{-2}\ h^{-1}$ is attributed to postfire stimulation of microbial decomposition. Over a 120-day growing season, this hypothetical stimulation amounts to a soil C efflux of 1.62 t ha^{-1}. Accounting for an additional efflux in the transition months between the summer and winter and assuming half the rate of the growing season flux (0.10 g $CO_2\ m^{-2}\ h^{-1}$ stimulation for 90 days per year), microbial respiration might be stimulated by as much as 2.22 t $ha^{-1}\ yr^{-1}$ in the postburn ecosystem. Because permafrost melts and mineral soil profiles increase in temperature for over a decade following fire, these hypothetical rates may suggest a stimulation of microbial respiration on the order of 20 t C ha^{-1} from terrestrial ecosystem to the atmosphere. For comparison purposes, in Chapter 10, Kasischke and associates estimated that 33.5 t C ha^{-1} was released from the burning of ground-layer biomass during fires in black spruce forests.

Future Research to Test the Hypothesis

Although no one has directly measured postfire stimulation of microbial mineralization of organic matter, these hypothetical estimates are supported at least in part by observations and calculations of many others. Some of these include Dyrness and Norum (1983), Van Cleve and associates (1983a), Viereck (1983), Bonan and Van Cleve (1992), Auclair and Carter (1993), Burke and colleagues (1997), and Zepp and colleagues (1997). Field and laboratory-based research in combination with computer simulation should be able to test the hypothesis presented here in the short-term future.

Radiocarbon estimates of soil CO_2 can help discriminate root from microbial respiration (Goulden et al. 1998; Trumbore and Harden 1997). In unburned ecosystems, measurements of the ^{14}C and ^{12}C in soil respiration can isolate microbial respiration from plant respiration. Hypothetically, in mature black spruce ecosystems with a preponderance of soil CO_2 attributable to plant root respiration (80% according to Schlentner and Van Cleve 1985), flux of CO_2 from the soil will bear a signature of near-current ^{14}C of the aboveground atmosphere ($\delta\ ^{14}C\text{-}CO_2$ will be strongly positive). In recently burned ecosystems, $^{14}C\text{-}CO_2$ in soil respiration will be a function of the ^{14}C content of relatively old C in soil organic matter (i.e., to C that was fixed probably hundreds and thousands of years in the past). Burned systems will hypothetically have a signature of ^{14}C in their soil CO_2 flux that is depleted in ^{14}C due to radioactive decay ($\delta\ ^{14}C\text{-}CO_2$ will be strongly negative).

Moreover, the isotopic signature of soil CO_2 flux hypothetically varies between the two systems during the course of the growing season. On the one hand, soil respiration in both burned and unburned systems hypothetically have $\delta\ ^{14}C\text{-}CO_2$ values that turn more negative during the growing season as microbes in deeper layers O horizons and mineral soils become increasingly active as summer progress. This is well documented in mature spruce systems that have not been recently burned (Goulden et al. 1998). On the other hand, compared with unburned systems, recently burned forests hypothetically have soil respiration fluxes of CO_2 with $\delta\ ^{14}C$ values that turn more rapidly negative, due to respiration of old soil organic matter within the soil biologically active zones that may be 1 to 2 m in depth.

Chronosequence studies of black spruce stands of different ages will also lend a considerable amount of evidence to the question of microbial and pyrogenic transfers of terrestrial C to the atmosphere. Refinements in temperature-moisture controls on microbial decomposition can be made in laboratory incubation studies. A variety of approaches to estimate postfire stimulation of decomposition is well within our grasp, and several are being used to more directly quantify the hypothesis of postfire stimulation of decomposition.

Conclusions

The evaluation of the fire stimulation hypothesis leads to several conclusions. First, the dynamics of C in soils and ecosystems of the boreal zone are clearly important to the global C cycle. Second, although organic matter's decomposition is relatively slow in boreal soils and storage of soil C relatively large, the ecological processes that control C turnover are notably dynamic, especially due to the periodic role of wildfires. Third, fires drastically alter many of the factors controlling soil C storage and microbial decomposition of soil C. Soil temperature, thickness of the biologically active layer, and nutrient availability are all changed by fire. Soil temperature and the depth of the biologically active soil zone may be greatly increased for more than a decade following fire. And fourth, heterotrophic microbial activity in burned sites can be hypothesized to be greatly enhanced following fire. Fire-stimulated microbial respiration in the postburn environment may transfer as much terrestrial C to the atmosphere as that due to wildfire's combustion itself.

References

Auclair, A.N.D., and T.B. Carter. 1993. Forest wildfires as a recent source of CO_2 at northern latitudes. *Can. J. For. Res.* 23:1530–1536.

Bonan, G.B., and K. Van Cleve. 1992. Soil temperature, nitrogen mineralization, and carbon source-sink relationships in boreal forests. *Can. J. For. Res.* 22:629–639.

Burke, R.A., R.G. Zepp, M.A. Tarr, W.L. Miller, and B.J. Stocks. 1997. Effect of fire on soil-atmosphere exchange of methane and carbon dioxide in Canadian boreal forest sites. *J. Geophys. Res.* 102:29,289–29,300.

Cahoon, D.R., B.J. Stocks, J.S. Levine, W.R. Cofer III, and J.M. Pierson. 1994. Satellite analysis of the severe 1987 forest fire in northern China and southeastern Siberia. *J. Geophys. Res.* 99:18,627–18,638.

Carslaw, H.S., and J.C. Jager. 1959. *Conduction of Heat in Solids.* Oxford University Press, London.

Dyrness, C.T., and R.A. Norum. 1983. The effects of experimental fires on black spruce forest floors in interior Alaska. *Can. J. For. Res.* 13:879–893.

Dyrness, C.T., L.A. Viereck, and K. Van Cleve. 1986. Fire in taiga communities of interior Alaska, pp. 74–86 in K. Van Cleve, F.S. Chapin III, P.W. Flanagan, L.A. Viereck, and C.T. Dyrness, ed. *Forest Ecosystems in the Alaskan Taiga.* Springer-Verlag, New York.

Geiger, R. 1965. *The Climate Near the Ground.* Harvard University Press, Cambridge, MA.

Goulden, M.L., S.C. Wofsy, J.W. Harden, S.E. Trumbore, P.M. Crill, S.T. Gower, T. Fries, B.C. Daube, S-M. Fan, D.J. Sutton, A. Bazzaz, and J.W. Munger. 1998. Sensitivity of boreal forest carbon balance to soil thaw. *Science* 279:214–217.

Harden, J.W., E.T. Sundquist, R.F. Stallard, and R.K. Mark. 1992. Dynamics of soil carbon during deglaciation of the Laurentide Ice Sheet. *Science* 258:1921–1924.

Harden, J.W., K.P. O'Neill, S.E. Trumbore, H. Veldhuis, and B.J. Stocks. 1997. Moss and soil contributions to the annual net carbon flux in a maturing boreal forest. *J. Geophys. Res.* 102:28,805–28,816.

Kasischke, E.S., N.L. Christensen, and B.J. Stocks. 1995. Fire, global warming, and the carbon balance of boreal forests. *Ecol. Appl.* 5:437–451.

Nakshabandi, G.A., and H. Kohnke. 1965. Thermal conductivity and diffusivity of soils as related to moisture tension and other physical properties. *Agr. Meteorol.* 2:271–279.

National Climate Data Center. 1998. Alaska climate summaries. Available on the World Wide Web at http://www.wrcc.sage.dri.edu/summary/climsmak.html.

Nikolov, N., and H. Helmisaari. 1992. Silvics of the circumpolar boreal forest tree species, pp. 13–84 in H.H. Shugart, R. Leemans, and G.B. Bonan, eds. *A Systems Analysis of the Global Boreal Forest.* University Press, Cambridge, UK.

O'Neill, K.P., E.S. Kasischke, D.D. Richter, and V. Krasovic. 1997. Effects of fire on temperature, moisture, and CO_2 emissions from Tok, Alaska—an initial assessment, pp. 295–303 in I.K. Iskandar, E.A. Wright, J.K. Radke, B.S. Sharratt, P.H. Groenevelt, and L.D. Hinzman, eds. *International Symposium on Physics, Chemistry, and Ecology of Seasonally Frozen Soils.* June 10–12, 1997, University of Alaska, Fairbanks, Alaska. U.S. Army Cold Regions Research and Engineering Laboratory, Hanover, NH.

Page, A.L., R.H. Miller, and D.R. Keeney, eds. 1982. *Methods of Soil Analysis,* Part 2: *Chemical and Microbiological Properites,* 2nd ed. *Agronomy* 9(2). Soil Science Society of America, Madison, WI.

Parkinson, K.J. 1981. An improved method for measuring soil respiration in the field. *J. Appl. Ecol.* 18:221–228.

Schlentner, R.E., and K. Van Cleve. 1985. Relationships between CO_2 evolution from soil, substrate temperature, and substrate moisture in four mature forest types in interior Alaska. *Can. J. For. Res.* 15:97–106.

Stocks, B.J., and J.B. Kauffman. 1997. Biomass consumption and behavior of wildland fires in boreal, temperate, and tropical ecosystems: parameters necessary to interpret historic fire regimes and future fire scenarios, pp.169–188 in J.S. Clark, H. Cachier, J.G. Goldammer, and B.J. Stocks, eds. *Sediment Records of Biomass Burning and Global Change.* NATO ASI Series, Subseries 1, Global Environmental Change, Vol. 51. Springer-Verlag, Berlin.

Stocks, B.J., B.S. Lee and D.L. Martell. 1996. Some potential carbon budget implications of fire management in the boreal forest, pp. 89–96 in M.J. Apps and D.T. Price, eds. *Forest Management and the Global Carbon Cycle.* NATO ASI Series, Subseries 1, Global Environmental Change, Vol. 40. Springer-Verlag, Berlin.

Trumbore, S.E., and J. Harden. 1997. Accumulation and turnover of carbon in soils of the BOREAS NSA: 1. Methods for determining soil C balance in surface and deep soil. *J. Geophys. Res.* 102:28,805–28,816.

Van Cleve, K., and L. Viereck. 1981. Forest succession in relation to nutrient cycling in the boreal forest of Alaska, pp. 184–211 in D.C. West, H.H. Shugart, and D.B. Botkin, eds. *Forest Succession, Concepts and Application.* Springer-Verlag, New York.

Van Cleve, K., C.T. Dyrness, L.A. Viereck, J. Fox, F.S. Chapin, and W. Oechel. 1983a. Taiga ecosystems in interior Alaska. *Bioscience* 33:39–44.

Van Cleve, K., L. Oliver, R. Schlentner, L.A. Viereck, and C.T. Dyrness. 1983b. Productivity and nutrient cycling in taiga forest ecosystems. *Can. J. For. Res.* 13:747–766.

Viereck, L. 1983. The effects of fire in black spruce ecosystems of Alaska and northern Canada, pp. 201–220 in R.W. Wein and D.A. MacLean, eds. *The Role of Fire in Northern Circumpolar Ecosystems.* John Wiley & Sons, New York.

Viereck, L., and C.T. Dyrness. 1979. Ecological effects of the Wickersham Dome fire near Fairbanks, Alaska. General Technical Report PNW-90. USDA Forest Service, Portland, OR.

Zepp, R.G., W.L. Miller, M.A. Tarr, and R.A. Burke. 1997. Soil-atmosphere fluxes of carbon monoxide during early stages of postfire succession in upland Canadian boreal forests. *J. Geophys. Res.* 102:29,301–29,311.

12. Influence of Fire on Long-Term Patterns of Forest Succession in Alaskan Boreal Forests

Eric S. Kasischke, Nancy H.F. French, Katherine P. O'Neill, Daniel D. Richter, Laura L. Bourgeau-Chavez, and Peter A. Harrell

Introduction

The cold climate and resulting low decomposition rates in the ground layers of boreal forests (such as those found in interior Alaska) result in the development of deep organic soils. In turn, these soils have an important role in many physical, chemical, and biological processes (Van Cleve et al. 1986). In combination with the slope, aspect, elevation, and composition of the underlying mineral soil profile of a specific site (Swanson 1996), organic soils are particularly influential in regulating ground temperature and moisture. As a general rule, the presence of a deep organic soil layer serves to insulate the forest floor during the growing season, causing colder temperatures than would otherwise occur. In many forested sites, autogenic cooling resulting from deepening organic soil layers eventually leads to the formation of permafrost, which, in turn, impedes drainage and substantially increases soil moisture (Van Cleve and Viereck 1981; Van Cleve et al. 1983a, b).

It has been shown that the distribution of different boreal forest ecosystems in Alaska is strongly correlated with a gradient of soil moisture and temperature (Fig. 12.1, modified from Van Cleve and Viereck 1981). In general, black spruce (*Picea mariana* [Mill.], BSP) forests inhabit sites with cooler, wetter conditions, such as north-facing hill slopes, all slopes at higher elevations, and low-relief areas with poorly drained soils. White spruce (*Picea glauca* [Moench] Voss) forests (which include successional stages dominated by deciduous trees [*Populus*

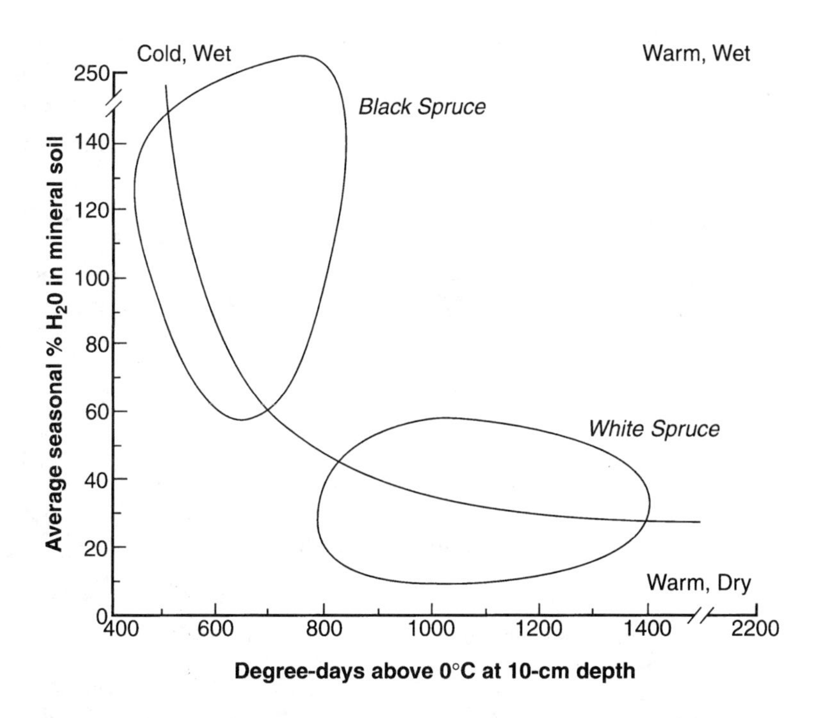

Figure 12.1. Relationship between black spruce and white spruce ecosystem distribution in interior Alaska and soil temperature and moisture. (Modified from Van Cleve and Viereck 1981.)

balsamifera L., *Populus tremuloides* Michx., and *Betula papyrifera* Marsh.]) inhabit warmer, drier sites including south-facing slopes, river floodplains, and low-relief areas with well-drained soils. Furthermore, it has been hypothesized that the autogenic cooling of the forest floor during primary succession is partially responsible for the mosaic of white and black spruce forests found on river floodplains in interior Alaska (Drury 1956; Van Cleve and Viereck 1981; Viereck et al. 1983, 1986, 1993).

Fire also plays an important role in the regulation in ground temperature and moisture. The paradigm presented in Figure 12.2 was developed to explain variations in ground temperatures and moisture following fires in black spruce forests of interior Alaska (Van Cleve et al. 1983a; Viereck 1983). Studies have shown that for the first several years immediately after a fire there is a significant increase in soil temperature from increased solar insolation, decreased shading, and decreased surface albedo. In addition, there is a decrease in soil moisture during the growing season (due to a recession of the permafrost layer resulting in improved soil drainage) (Dyrness and Norum 1983; Dyrness et al. 1986; Viereck et al. 1983). Although this paradigm can be used to explain patterns of soil temperature and moisture over the short term, the longer-term effects of fire on these

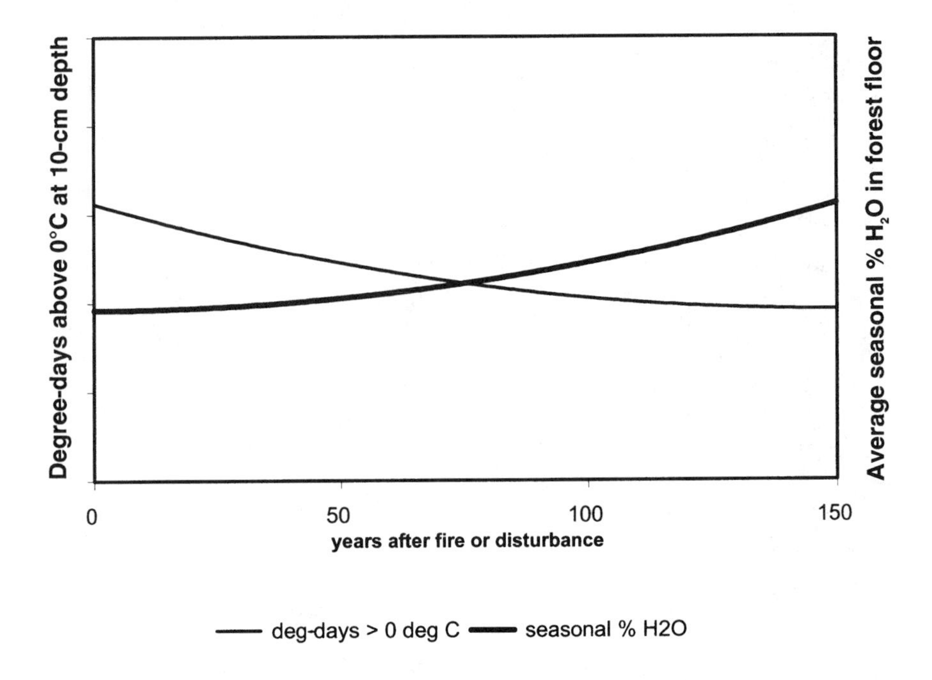

Figure 12.2. General model of the patterns of relative soil temperature and moisture after fires in Alaskan black spruce forests. (Modified from Van Cleve et al. 1983a.)

factors may, in fact, be more complex. In turn, long-term variations in soil moisture and temperature would be expected to influence patterns of forest succession.

In this chapter, we present evidence collected in burned and unburned forests in the upper Tanana River Valley of Alaska that suggests the longer-term patterns of forest succession at some sites in this region include transitions from white spruce to black spruce forests, as well as the reverse (i.e., black spruce to white spruce forests). A theory explaining a pathway for these transitions is presented based on recent observations. In this theory, an important factor in the transitions between the two major forest types is the severity of burning of the organic soil profiles of these forests.

Study Setting

The forest succession theory presented in this chapter evolved from a number of observations made during investigations begun in 1991. The initial studies were focused on developing approaches to use satellite imagery to study the effects of fire on Alaskan boreal forests (see Chapters 17, 18, and 23). This research included a series of field studies to understand and properly interpret the remotely sensed signatures (Kasischke et al. 1995a,b; French et al. 1996a,b; O'Neill et al. 1997; Michalek 1999). The eventual goal is to develop approaches to estimate

(1) the direct release of greenhouse gas emissions during fires through consumption of biomass, particularly in the organic soil layer; and (2) the indirect fluxes of greenhouse gases from organic soils resulting from variations in the patterns of soil respiration after fire.

More specifically, the evidence supporting a new pattern of forest succession arose out of data collected at three different study sites in the upper Tanana River Valley of Alaska (Fig. 10.1): (1) test sites near Tok (63° 18′ north latitude, 142° 45′ west longitude) that burned during a 40,000-ha fire in the summer of 1990; (2) test sites near Hajdukovich Creek, Alaska (63°50′ north latitude, 145°15′ west longitude) that burned in a 10,000-ha fire the summer of 1994; and (3) forested sites located in a fire near Fort Greely (63°55′ north latitude, 145°31′ west longitude) that have grown since a 6,600-ha fire that occurred in 1956.

The longer-term climatic conditions of the upper Tanana River Valley are typical of those found throughout interior Alaska, with the area being dominated by low winter temperatures, warmer temperatures and long periods of daylight during the growing season and overall low levels of precipitation. Within the Tanana River Valley, there is an east-to-west gradient in temperature and precipitation conditions, with the western region being slightly warmer and wetter than the eastern region. The climate at Fairbanks is typical of that found in the western reaches of the valley (Slaughter and Viereck 1986). The average annual temperature at Fairbanks is –3.4°C, with the coldest month being January (–24.4°C) and the warmest month being July (17.7°C). Annual precipitation is 287 mm. Weather in the eastern region of the valley is typified by longer-term climatic data from Tok, where the average annual temperature is –5.0°C. The coldest month is January (–27.1°C), the warmest month is July (14.3° C), and the annual precipitation is 222 mm. Throughout the entire valley, less than one-third of the precipitation occurs as snowfall. Throughout the Tanana River Valley, the wettest months of the year are July and August, when one-third of the precipitation occurs as rainfall.

Observations Supporting a Successional Chronosequence

Over the past several years, several observations have been made that supported the development of a new successional paradigm for some physiographical settings common to interior Alaska.

Observation 1: Evidence of White Spruce Preceding Black Spruce at a Test Site

During initial site surveys within the area burned during the 1994 Hajdukovich Creek fire, several burned black spruce forest stands were visited. In one area where the fire had completely consumed the organic soil, several large stumps were discovered (Fig. 12.3a). These stumps were 15–20 m apart and had diameters between 30 and 45 cm at a distance of 20 cm above the ground surface. Several partial (e.g., 3–7 m long) tree boles were lying on the ground adjacent to

Figure 12.3. Ground photographs of (a) a white spruce trunk located within a recently burned black spruce forest in the Hajdukovich Creek fire site; (b) an unburned black spruce stand adjacent to the burned forest where the white spruce trunks were discovered; and (c) a mature white spruce stand with trees whose diameters are the same as those located within the burned black spruce stand.

Figure 12.3. (*continued*)

the stumps. Only six tree species are found in the interior region of Alaska: two coniferous (white spruce and black spruce) and four deciduous (balsam poplar, paper birch, trembling aspen, and larch). Examination of these stumps showed that they were most likely from coniferous trees because of the presence of clearly defined growth rings (which are absent in the deciduous tree species found in this area: aspen, balsam poplar, and birch) and because the trunks of most deciduous tree species decompose much more rapidly than the coniferous species, especially when lying on the ground. Because of their large size, it was concluded these stumps were from white spruce trees rather than black spruce trees.

Figure 12.3b presents a photograph of an unburned, medium-density black spruce stand within 300 m of the area where the large stumps were discovered. For comparison purposes, Figure 12.3c presents a photograph of a mature white spruce forest whose trees have a basal diameter of 30–35 cm (trees of this stature are common at sites 5 km to the northeast of the burned area). Ground surveys within the unburned black spruce stand revealed the existence of numerous standing-dead tree boles with the same diameter of the larger living trees in this stand. These dead snags are similar to those found in recent burned black spruce forests and most likely originated from the previous fire at this site. This indicates that black spruce forests existed on this site for at least the past 150–200 years. Ground surveys also showed the unburned stand had a well-developed, 28-cm-deep organic soil profile with 8 cm of humic soil (site HC94b in Fig. 10.3a). Surveys during the summers of 1996 and 1997 showed that permafrost was well established in the unburned site, with the active layer depths in late August being 25–30 cm.

Observation 2: Presence of Soil Features Typically Associated with Permafrost in Sites Without Permafrost

Earth hummocks are cryogenic features resulting from heaving of the surface mineral soil layer through trapping of a lens of water between permafrost on the bottom and downward freezing soil on the top during winter freeze-up (Zoltai and Pettapiece 1974; Tarnocai and Zoltai 1978). The path of least resistance for expansion of the trapped water is upward, resulting in a bulging of the mineral soil layer. This type of freezing typically results in earth hummocks 50–300 cm in diameter, which project upward 20–80 cm (Tarnocai and Zoltai 1978). These earth hummocks are associated with terrain underlain by permafrost.

Field surveys showed that there were a number of black spruce stands within the area burned during the Tok fire containing earth hummocks (Fig. 12.4a). The presence of earth hummocks in sites inhabited by black spruce is consistent with observations made in Canadian forests, as confirmed during a visit to these sites by Steven Zoltai in the fall of 1995.

However, observations of earth hummocks were also made within burned and unburned aspen forests located in and adjacent to the 1994 Tok fire (Fig. 12.4b and c). Data collected during the summers of 1995–1997 showed that no permafrost existed in the mature, unburned aspen stand. Our conclusions from these observations are twofold: (1) the aspen sites at one time were underlain by permafrost; and (2) at the time when permafrost was present, black spruce forests, not aspen forests, were most likely present on these sites.

Observation 3: There Is a Strong Connection Between Climate and Fire in the North American Boreal Forest

Several lines of evidence support the theory that patterns of biomass burning in the North American boreal forest are controlled by climatic factors. First, most of the total area burned in the boreal forest of North America occurs during a relatively few number of years when drought conditions are prevalent (Fig. 1.2 and 15.4). Since 1971, more than 50% of the total area burned in this region (62 million ha) has occurred in the five most severe fire years. Modeling studies by Wotton and Flannigan (1993) showed that the expected rise in temperature projected by most general circulation models would increase the length of the growing season in the boreal region by 25%, which should result in a corresponding increase in the annual area burned. Theoretical studies by Flannigan and Van Wagner (1991) showed that the same climate changes would result in increases in the fire weather index for boreal forests. From these observations, they concluded that the probabilities for wildfires would increase in boreal regions based on the climate change predictions.

An argument can be made that the predictions of increased fire activity from a warming climate by Wotton and Flannigan (1993) and Flannigan and Van Wagner (1991) have been borne out by recent climate and fire patterns in North American boreal forests. An analysis of climate records by Hansen and co-workers (1996) shows that the region covered by the North American boreal forest has experienced a 1–1.6°C warming over the past 30 years. Based on analysis of changes in the seasonal patterns of the atmospheric concentration of carbon dioxide at different latitudes, Keeling and associates (1996) estimate that the length of the growing season in the Northern Hemisphere has increased by about 1 week over the same time period. This observation is supported by analyses of seasonal vegetation patterns on satellite imagery (Myneni et al. 1997). The observations of increased temperature and a lengthening growing season are the most likely explanations for the recent increases in the annual area burned in the boreal forests of North America (see Chapter 20). During the 1970s, the average area burned was 1.64 million ha yr^{-1}, whereas during the 1990s, the average area burned has almost doubled, to 3.16 million ha yr^{-1}.

Observation 4: There Is a High Degree of Variability in the Patterns of Burning of Organic Soil in Black Spruce Forests During Fires

Although it has long been recognized that fires have the potential to consume large amounts of organic matter from the floors of boreal forests with deep organic soils (Dyrness and Norum 1983; Viereck 1983; Dyrness et al. 1986), it has only been recently that the patterns of this biomass burning have been quantified. As an example, the studies presented in Chapter 10 illustrated that there is a high degree of variability in the levels of burning of organic soil in the black spruce forests of interior Alaska, with the levels of consumption ranging between 20 and 90% of the total biomass present. In addition, the data presented in Figure 10.5 suggest that fires in black spruce forests occurring later in the growing season result in a greater burning of organic soil. Because a warming climate is expected to result in a longer growing season, it should result in a higher level of burning of organic soil as well.

Observation 5: Effects of Fire Severity on Ground Temperature

Numerous researchers have documented the overall effects of fire on the ground temperature regime in boreal forests (e.g., that fires result in an increase in ground temperature and active layer depth for several decades after a fire) (Brown 1983; Dyrness and Norum 1983; Viereck 1983; Dyrness et al. 1986; Mackay 1995; Swanson 1996). It has also been reported that there is an initial decrease in soil

Figure 12.4. Ground photographs of earth hummocks located in the 1990 Tok, Alaska, fire site: (a) site containing black spruce forests prior to the fire; (b) site containing an aspen forest prior to the fire; and (c) cross-sectional view of aspen site earth hummock.

Figure 12.4. (*continued*)

moisture after a fire, which eventually increases as temperatures decrease and permafrost becomes reestablished (Van Cleve et al. 1983a; Viereck 1983).

In June 1994, a forest fire started near Hajdukovich Creek, Alaska, and swept through forests located on a flat (slope <<< 0.5%) alluvial outwash bounded on the southeast by the Alaska Range and the northwest by the Tanana River and bisected by the Alaskan Highway. The prefire mosaic of forests on this outwash plain consisted primarily of 70- to older than 150-year-old black spruce stands along with a number of stands of aspen, mixed aspen-white spruce, and peatlands dominated by willow and low-density black spruce. The lightning-started 1994 Hajdukovich Creek fire burned primarily through the black spruce stands covering 84% of the burned area.

This fire started near an inactive U.S. Army training site. Because of the remote possibility of the existence of undetonated munitions in several locations within this area, this fire was not suppressed. Weather conditions (primarily the lack of rain) allowed the fire to continue to burn and smolder throughout the summer of 1994, eventually consuming 10,000 ha of forest. During the latter stages of the Hajdukovich Creek fire in mid- to late August, very dry conditions led to deep smoldering fires in the organic soil layers of the black spruce forests. These late-season smoldering fires were unique in that they consumed the entire ground layer, including the mosses, lichens, litter, and organic soil. The black spruce areas that burned early in this fire still had a significant amount of organic soil remaining (e.g., Site HC93c in Fig. 10.3).

Between August 15 and 25, 1996, soil pits to a depth of 1 m were dug in three burned black spruce stands within the 1994 Hajdukovic, as well as in adjacent unburned stands. Soil temperatures were obtained at 5-cm increments in all soil pits. In two of the sites (sites HC94a and HC94b in Fig. 10.3), nearly 90% of the organic soil was consumed during the fire, but in the third site (HC94c), only 20% of the organic soil profile was consumed. Although permafrost was present in all the unburned black spruce stands, the active layer was greater than 1 m deep in all the burned stands as a result of the ground warming that occurred because of the effects of the fire. The sites where most of the organic soil was removed were on average 6° warmer than the unburned stand, whereas the area with a deeper remnant organic soil profile only warmed by an average of 2°C. The warmer temperatures are attributed to the removal of the organic soil that serves to insulate the ground layer (Fig. 11.3). These observations show the importance of the severity of burning of the organic soil on ground temperature.

Observation 6: Invasion of Aspen into Black Spruce Forest Sites After Fire

Figure 12.5 presents two ground photographs collected at black spruce sites in the 1994 Hajdukovich Creek fire. Aspen seedlings were invading sites where most of the organic soil profile was consumed during the fire by the summer of 1997 (Fig. 12.5a), while they were entirely absent in areas where a deeper organic soil profile remained (Fig. 12.5b). Observations made in areas of the 1990 Tok fire show that aspen seedlings do become established in sites where deeper organic soils are present, but in very low densities. These observations show that given the right conditions (presence of a warm mineral soil surface), aspen seedlings can invade sites that were previously occupied by black spruce stands underlain by permafrost.

Observation 7: The Presence of Transitional Forest Types

If the theory that changes in forest cover are driven by factors other that site physiography (e.g., climate and fire severity) is true, then transitional forest types should exist. Recent field observations do show that transitional forest types are present in the upper Tanana River Valley and other regions of Alaska. First, observations at the 1990 Tok fire site show that there are spruce stands that contain a mixture of white spruce and black spruce trees. Second, numerous stands of mixed aspen-black spruce forests are throughout the upper Tanana River Valley. Finally, Yarie (1983) reports the presence of mixed white spruce-black spruce and aspen-black spruce forests in the Porcupine River, which is located in northeastern Alaska, south of the Brooks Range.

Figure 12.6 presents ground photographs from within the boundaries of the 1954 Fort Greely fire. The two stands were adjacent to one another, with one stand having an upper canopy of aspen and a lower canopy of black spruce and the second having a canopy of pure black spruce. Field measurements collected in the summer of 1997 showed that the total organic soil depth in the mixed aspen-black spruce stand was only 6 cm, whereas in the pure black spruce stands it was more

Figure 12.5. Effects of fire severity on patterns of forest succession after the 1994 Hajdukovich Creek, Alaska, fire: (a) sites where fire removed most of the organic soil are being invaded by wind-dispersed aspen seedling; and (b) areas where deep organic soils remain do not have aspen seedlings. Black spruce seedlings are present in both sites.

Figure 12.6. Ground photographs collected in the summer of 1997 in a region that burned in 1954. (a) An area with a deep organic profile contains only black spruce; (b) an area with a shallower profile contains a two-tiered canopy, with aspen in the upper canopy layer and black spruce in the lower layer.

than 13 cm deep. Ground temperatures collected in August 1998 showed that the average temperature in the top 1 m of mineral soil was 1.1°C warmer in the aspen-black spruce stand than in the pure black spruce stand.

Observation 8: Climate in the Boreal Forest Is Not Constant

Over the longer term (e.g., decade to century time scales and longer), climate has never been constant in the North American boreal forest, based on a variety of data sources. Hansen and colleagues (1996) analyzed surface climate records to show that there has been a 1–1.6°C rise in temperature throughout the North American boreal region over the past 30 years. Jacoby and D'Arrigo (1995) showed there has been a 2–3°C rise in temperature in the Alaskan boreal forest region over the past century based on analysis of tree-ring data. Over the past 6,000 years, paleoecological studies have shown that the average temperature has varied by at least ± 1°C when compared with current temperatures (Ritchie and Hare 1971).

Influence of Fire on Longer-Term Patterns of Forest Succession

The above observations suggest that successional changes in forest cover in the study region may be occurring over time scales longer than described in the successional paradigms developed to date for the upland white spruce and upland and lowland black spruce forests in interior Alaska (Van Cleve and Viereck 1981; Van Cleve et al. 1983a; Viereck et al. 1983, 1986). The evidence suggests that in addition to the physiographical factors used to account for succession in existing theories, others must also be considered.

The study presented in Chapter 10 shows that there is a wide degree of variation in patterns of burning of ground-layer biomass in black spruce forests, both within season and between season. Because the patterns of fire and biomass burning in Alaskan spruce forests are climate controlled, the average fire severity should be linked with the longer-term patterns of cooling and warming that have occurred in this region. As discussed above, the amount of ground-layer biomass consumed during fires has a strong influence on the patterns of postfire ground temperature and, by association, most likely controls soil moisture as well. Therefore, long-term patterns of postfire ground temperature and moisture in boreal forest are not in a steady state but vary in response to changes in climate and fire severity.

If these factors are considered, then a longer-term successional pattern might be present in the white and black spruce forest ecosystems in interior Alaska, as illustrated in Figure 12.7. In this paradigm, a cooling climate is associated with decreasing fire severity, and a warming climate is associated with increasing fire severity. The former scenario leads to cooler and wetter soil conditions, which favor the eventual establishment of black spruce forests over the continuation of white spruce forests. The latter scenario leads to warmer and drier soil conditions, which ultimately favors white spruce forests over black spruce forests.

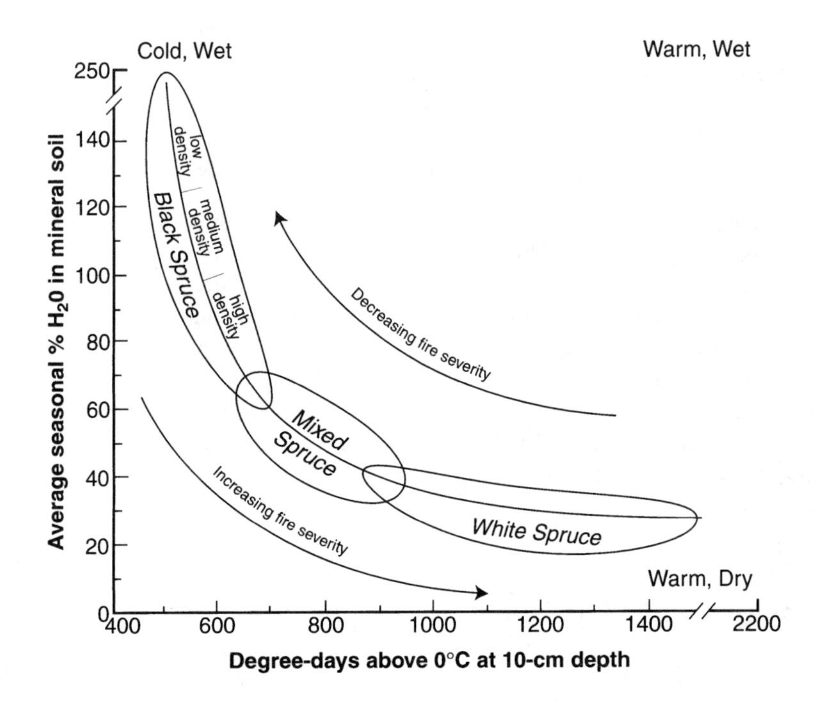

Figure 12.7. Model accounting for the effects of fire severity and climate change on the patterns of soil temperature and moisture and patterns of forest succession in interior Alaska.

The paradigm in Figure 12.7 suggests that if an upland site inhabited by white spruce is left undisturbed, it will eventually develop a deep humic layer and become cold and wet enough that the site eventually becomes inhabited by black spruce. This is similar to the successional paradigm described for floodplain white spruce forests in interior Alaska (Drury 1956; Van Cleve and Viereck 1981; Viereck et al. 1983, 1986, 1993). Although this process can be used to explain the situation observed at the Hajdukovich Creek site (Fig. 12.3), it certainly cannot be used to explain the situation observed at the Tok site (Fig. 12.4). In addition, the theory of autogenic cooling by the accumulation and building up of a deep organic soil layer does not account for the likely effects of fire on site conditions.

The data show that variations in burn severity have a strong effect on postfire patterns of ground temperature. In terms of the overall pattern of ground temperature exhibited in Figure 12.1, by reducing the amount of organic soil, a more severe fire should increase the number of degree-days above 0° at a depth of 10 cm, whereas a less severe fire should decrease this value (Fig. 12.8a). At the same time, the warming of the ground layer as a result of the more severe fire would be expected to decrease the average seasonal soil moisture to a greater degree than a less severe fire, and a less severe fire would be expected to increase this value (Fig. 12.8b).

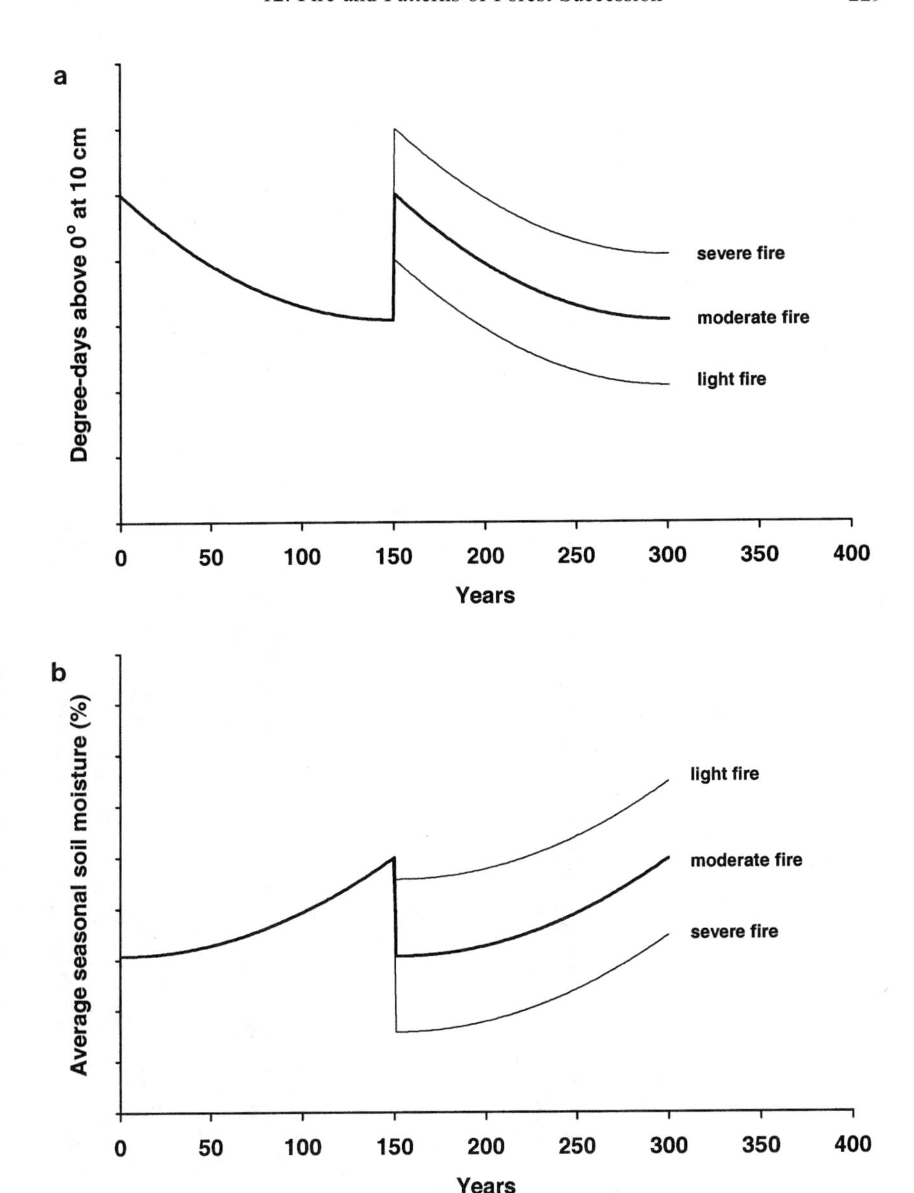

Figure 12.8. Theoretical effects of fire severity on patterns of (a) relative soil temperature and (b) relative soil moisture.

Based on the models presented in Figure 12.8, we can speculate as to what happens to ground temperatures and moisture levels if there is a series of fires with increased fire severity or decreased fire severity over a number of successional cycles (Fig. 12.9). If the fire severity or frequency were to increase (relative to a nominal level that resulted in a stable temperature/moisture regime over the

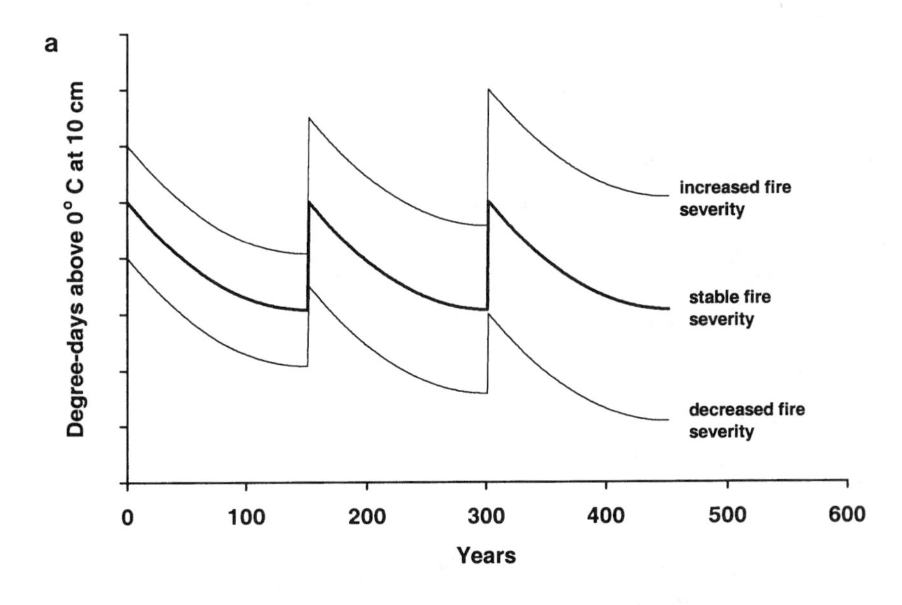

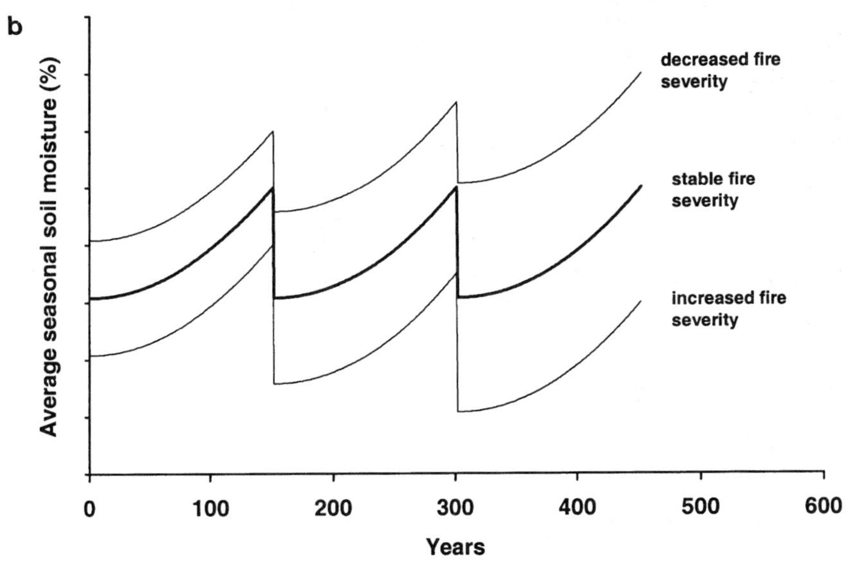

Figure 12.9. Effects of long-term changes in fire severity on patterns of (a) relative soil temperature and (b) relative soil moisture.

long term), then the amount of organic soil present would decrease and the site would become warmer and drier, conditions that favor the eventual establishment of white spruce forests (Fig. 12.2 and 12.7). If the fire severity or frequency were to decrease over time, then the amount of organic soil would increase and the site would become colder and wetter, which favors the continuing presence of black spruce forests.

The continuing buildup of organic soils associated with continuous low levels of fire severity have additional implications on forest succession. The continuing aggradation of organic soils and cooling soil conditions will have a strong effect on the lower organic soil profile. The continuing decomposition and compression of this soil layer will eventually lead to denser humic soils typical of older black spruce ecosystems. In addition, the cooler, wetter soil conditions will lead to less vigorous growth of black spruce, with the canopy trees of these ecosystems having a lower density, height, and diameter.

Longer-term variations in burn intensity may account for the different forest stand characteristics at sites HC94a and HC94b. These sites are located within 1 km of one another and have very similar mineral soil profiles: a 20-cm-deep A horizon consisting of silt; a 30-cm-deep B horizon consisting of a silty loam; and a more than 50-cm-deep IIC horizon consisting of gravel outwash. The black spruce forests occupying these two sites are dramatically different (Table 12.1). Site HC94b has a well-developed humic soil profile (Fig. 10.3), with this layer and the underlying mineral soil above permafrost being saturated with water at the time of sampling in mid-August 1996. By contrast, there is no humic soil present in the organic soils of site HC94a, and the organic soils and entire mineral soil profile were decidedly drier at the time of sampling. Even though the forests occupying these two sites were nearly identical in age (90 versus 77 years), the cooler, wetter conditions at site HC94b led to the development of a much less dense forest canopy with smaller and shorter trees. The hypothesis outlined in Figure 12.7 holds that at some point in the past, the fires at site HC94b were less intense than those at site HC94a, which led to the cooler, wetter ground conditions at the former site.

Discussion

It is well established that in the boreal forests of Alaska, differences in seasonal soil moisture and temperature patterns are strongly correlated with forest type (Van Cleve et al. 1981, 1983b). Many theories of succession developed for this region use this correlation as a basis for explaining succession in different forest

Table 12.1. Summary of Black Spruce Forest Stand Characteristics

Stand characteristic	HC94a Closed-canopy forest	HC94b Open-canopy forest
Stand age	90	77
Tree density (stems ha^{-1})	22,000	9,400
Basal diameter (cm)	4.5	3.2
Tree height (m)	5.3	2.5
Ground layer depth (cm)	26.8	27.8
Moss layer	6.2	3.7
Fibric I	13.0	18.6
Fibric II	7.6	
Humic		5.5

types (Drury 1956; Van Cleve and Viereck 1981; Viereck et al. 1983, 1986). A significant portion of the research behind these theories has been devoted to understanding how allogenic site conditions (e.g., soil type, slope, aspect, and elevation) influence patterns of ground temperature and moisture, resulting in a set of theories that describes forest succession as a function of physiographical characteristics. Although the role of fire and the development of the organic soil profiles have both been recognized as important factors in these theories, the interaction between them over the longterm and the role of this interaction in succession have not been fully explored.

Ecologists have recognized the importance of variations in climate (temperature and moisture) on longer-term patterns of succession in the North American boreal forest. Paleobotanical studies have clearly shown that there have been significant changes in the composition of forest ecosystems occupying different sites, with these changes being associated with changing climatic conditions (Nichols 1969; Ritchie and Hare 1971). Since the Holocene maximum 6,000 years ago, there have been periodic changes in climate with temperatures changing by at least ± 1°C compared with current temperatures. The paleobotanical record shows that these variations in climate also are associated with variations in the fire regime (Clark 1988).

In addition, studies of the effects of future climate change have also recognized that increased air temperatures will result in warming and drying air conditions, causing changes in ecosystem composition (Bonan et al. 1990). These modeling studies did not explicitly link the effects of climate to fire severity; therefore, the changes in ecosystem composition may occur at a different rate than currently projected.

The studies at the Hajdukovich Creek and Tok fire test sites indicate that black spruce and white spruce forests existed at the same location at different times. Measurements collected within these test sites and others in the Tanana River Valley show the degree of burning of ground-layer biomass influences postfire patterns of ground temperature and active layer depth. It is inferred from these measurements that the temporal patterns of soil moisture will also vary as a function of severity of ground-layer burning. The ground measurements needed to more fully test this hypothesis will be collected during future field studies. Because seasonal averages of ground temperature and moisture are highly correlated with patterns of forest distribution in boreal forests, a new theory for succession between white spruce and black spruce forests can be developed based primarily on the degree of fire severity over the longer term. Periods of low-severity fires will result in a transition from white spruce to black spruce forests or the perpetuation of existing black spruce forests. Periods of higher-frequency or higher severity fires would result in the perpetuation of existing white spruce forests or the succession of a black spruce to a white spruce forest.

Although a general theory of how fire severity affects forest succession through its influence on the depth of the organic soil layer has been developed, the specifics of the overall process have not been quantified. In particular, although there is a general understanding of factors that influence the patterns of fire severity, outside of a few experimental burns (Dyrness and Norum 1983), the spatial and temporal variations of ground-layer biomass burning in black spruce forests are not well known. In addition, relatively few measurements of the effects of fire on patterns of soil temperature and moisture in these ecosystems have been collected (Viereck 1982). Finally, there are very few measurements of the dependence of soil respiration on soil moisture and temperature in Alaskan black spruce forests (Schlentner and Van Cleve 1985; O'Neill et al. 1997; Richter et al., this volume). Increased soil respiration not only reduces the amount of organic matter present in the ground layer but is a very important consideration in nutrient cycling, which, in turn, is important for plant and tree growth.

References

Bonan, G.B., H.H. Shugart, and D.L. Urban. 1990. The sensitivity of some high-latitude boreal forests to climate parameters. *Clim. Change* 16:9–29.

Brown, R.J.E. 1983. The effects of fire on the permafrost ground thermal regime, pp. 97–110 in R.W. Wein and D.A. MacLean, eds. *The Role of Fire in Northern Circumpolar Ecosystems.* J. Wiley & Sons, New York.

Clarke, J.S. 1988. Effect of climate change on fire regimes in northwestern Minnesota. *Nature* 334:233–235.

Drury, W.H. 1956. Bog flats and physiographic processes on the upper Kuskokwim River Region, Alaska. *Contrib. Gray Herbarium Harvard Univ.* 138. Cambridge, MA.

Dyrness, C.T., and R.A. Norum. 1983. The effects of experimental fires on black spruce forest floors in interior Alaska. *Can. J. For. Res.* 13:879–893.

Dyrness, C.T., L.A. Viereck, and K. Van Cleve. 1986. Fire in taiga communities of interior Alaska, pp. 74–86 in K. Van Cleve, F.S. Chapin III, P.W. Flanagan, L.A. Viereck, and C.T. Dyrness, eds. *Forest Ecosystems in the Alaskan Taiga.* Ecological Studies 57. Springer-Verlag, New York.

Flannigan, M.D., and C.E. Van Wagner. 1991. Climate change and wildfire in Canada, *Can. J. For. Res.* 21:61–72.

French, N.H.F., E.S. Kasischke, R.D. Johnson, L.L. Bourgeau-Chavez, A.L. Frick, and S.L. Ustin. 1996a. Using multi-sensor satellite data to monitor carbon flux in Alaskan boreal forests, pp. 808–826 in J.S. Levine, ed. *Biomass Burning and Climate Change,* Vol. 2: *Biomass Burning in South America, Southeast Asia, and Temperate and Boreal Ecosystems, and the Oil Fires of Kuwait.* MIT Press, Cambridge, MA.

French, N.H.F., E.S. Kasischke, L.L. Bourgeau-Chavez, P. Harrell, and N.L. Christensen, Jr. 1996b. Monitoring variations in soil moisture on fire disturbed sites in Alaska using ERS-1 SAR imagery. *Int. J. Remote Sens.* 17:3037–3053.

Hansen, J., R. Ruedy, M. Sato, and R. Reynolds. 1996. Global surface air temperature in 1995: return to pre-Pinatubo level. *Geophys. Res. Lett.* 23:1665–1668.

Jacoby, G.C., and R.D. D'Arrigo. 1995. Tree ring width and density evidence of climatic and potential forest change in Alaska. *Global Biogeochem. Cycles* 9:227–234.

Kasischke, E.S., N.H.F. French, L.L. Bourgeau-Chavez, and N.L. Christensen, Jr. 1995a. Estimating release of carbon from 1990 and 1991 forest fires in Alaska. *J. Geophys. Res.* 100:2941–2951.

Kasischke, E.S., N.L. Christensen, Jr., and B.J. Stocks. 1995b. Fire, global warming and the mass balance of carbon in boreal forests. *Ecol. Appl.* 5:437–451.

Keeling, C.D., J.F.S. Chin, and T.P. Whorf. 1996. Increased activity of northern vegetation inferred from atmospheric CO_2 records. *Nature* 382:146–149.

Mackay, J.R. 1995. Active layer changes (1968 to 1993) following the forest-tundra fire near Inuvik, N.W.T., Canada. *Arct. Alp. Res.* 27:323–335.

Michalek, J.L., N.H.F. French, E.S. Kasischke, R.D. Johnson, and J.E. Colwell. 1999. Using Landsat TM data to estimate carbon release from burned biomass in an Alaskan spruce forest complex. *Int. J. Remote Sens.* (in press).
Myneni, R.B., C.D. Keeling, C.J. Tucker, G. Asrar, and R.R. Nemani. 1997. Increased plant growth in the northern high latitudes from 1981 to 1991. *Nature* 386:698–702.
Nichols, H. 1969. The late quarternary history of vegetation and climate at Porcupine Mountain and Clearwater Bog, Manitoba. *Arct. Alp. Res.* 1:155–167.
O'Neill, K.P., E.S. Kasischke, D.D. Richter, and V. Krasovic. 1997. Effects of fire on temperature, moisture, and CO_2 emissions from Tok, Alaska—an initial assessment, pp. 295–303 in I.K. Iskandar, E.A. Wright, J.K. Radke, B.S. Sharratt, P.H. Groenevelt, and L.D. Hinzman, eds. *International Symposium on Physics, Chemistry, and Ecology of Seasonally Frozen Soils.* June 10–12, 1997, University of Alaska Fairbanks, Alaska. U.S. Army Cold Regions Research and Engineering Laboratory, Hanover, NH.
Ritchie, J.C., and F.K. Hare. 1971. Late-quarternary vegetation and climate near the arctic treeline of northwestern North America. *Quat. Res.* 1:331–342.
Schlentner, R.E., and K. Van Cleve. 1985. Relationships between CO_2 evolution from soil, substrate temperature, and substrate moisture in four mature forest types in interior Alaska. *Can. J. For. Res.* 15:97–106.
Slaughter, C.W., and L.A. Viereck. 1986. Climatic characteristics of the taiga in Interior Alaska, pp. 9–21 in K. Van Cleve, F.S. Chapin III, P.W. Flanagan, L.A. Viereck, and C.T. Dyrness, eds. *Forest Ecosystems in the Alaskan Taiga.* Ecological Studies 57. Springer-Verlag, New York.
Swanson, D.K. 1996. Susceptibility of permafrost soils to deep thaw after forest fires in interior Alaska, U.S.A., and some ecological implications. *Arct. Alp. Res.* 28:217–227.
Tarnocai, C., and S.C. Zoltai. 1978. Earth hummocks of the Canadian Arctic and subarctic. *Arct. Alp. Res.* 10:581–594.
Van Cleve, K., and L. Viereck. 1981. Forest succession in relation to nutrient cycling in the boreal forest of Alaska, pp. 184–211 in D.C. West, H.H. Shugart, and D.B. Botkin, eds. *Forest Succession: Concepts and Application.* Springer-Verlag, New York.
Van Cleve, K., C.T. Dyrness, L.A. Viereck, J. Fox, F.S. Chapin III and W. Oechel. 1983a. Taiga ecosystems in interior Alaska. *Bioscience* 33:39–44.
Van Cleve, K., L. Oliver, R. Schlentner, L.A. Viereck, and C.T. Dyrness. 1983b. Productivity and nutrient cycling in taiga forest ecosystems. *Can. J. For. Res.* 13:747–766.
Van Cleve, K., F.S. Chapin, III, P.W. Flanagan, L.A. Viereck, and C.T. Dyrness, eds. 1986. *Forest Ecosystems in the Alaskan Taiga.* Ecological Studies 57. Springer-Verlag, New York.
Viereck, L.A. 1982. Effects of fire and fire lines on active layer thickness and soil temperatures in interior Alaska, pp. 123–135 in H.M. French, ed. *Proceedings of the Fourth Canadian Permafrost Conference.* National Research Council of Canada, Ottawa, Ontario, Canada.
Viereck, L.A. 1983. The effects of fire in black spruce ecosystems in Alaska and northern Canada, pp. 201–220 in R.W. Wein and D.A. MacLean, eds. *The Role of Fire in Northern Circumpolar Ecosystems.* John Wiley & Sons, New York.
Viereck, L.A., C.T. Dyrness, K. Van Cleve, and M.J. Foote. 1983. Vegetation, soils, and forest productivity in selected forest types in interior Alaska. *Can. J. For. Res.* 13:703–720.
Viereck, L.A., K. Van Cleve, and C.T. Dyrness. 1986. Forest ecosystem distribution in the taiga environment, pp. 23–43 in K. Van Cleve, F.S. Chapin III, P.W. Flanagan, L.A. Viereck, and C.T. Dyrness, eds. *Forest Ecosystems in the Alaskan Taiga.* Ecological Studies 57. Springer-Verlag, New York.
Viereck, L.A., C.T. Dyrness, and M.J. Foote. 1993. An overview of the vegetation and soils of the floodplain ecosystems of the Tanana River, interior Alaska. *Can. J. For. Res.* 23:889–898.

Wotton, B.M., and M.D. Flannigan. 1993. Length of the fire season in a changing climate. *For. Chron.* 69:187–192.

Yarie, J. 1983. *Forest Community Classification of the Porcupine River Drainage, Interior Alaska, and Its Application to Forest Management.* U.S. Forest Service Technical Report PNW-154. U.S. Department of Agriculture, Portland, Oregon.

Zoltai, S.C., and W.W. Pettapiece. 1974. Tree distribution on perenially frozen earth hummocks. *Arct. Alp. Res.* 6:403–411.

III. SPATIAL DATA SETS FOR THE ANALYSIS OF CARBON DYNAMICS IN BOREAL FORESTS

Eric S. Kasischke and Brian J. Stocks

In the chapters of Section II, many of the processes influencing carbon cycling in the boreal forest were reviewed, with a particular focus on the effects of fire. The information presented in these chapters was based primarily on studies that were carried out over very small spatial scales at individual plots. The material contained in the chapters of this section offers a different perspective on patterns of fire and carbon storage in the boreal forest, specifically, presenting approaches and data sets for deriving estimates of fire and carbon budgets at continental scale and for monitoring the surface of the boreal forest at landscape and regional scales.

The first two chapters describe large-scale (continental) data sets that describe the distribution of carbon in the Eurasian and North American boreal forest. In Chapter 13, Alexeyev and co-workers describe the different approaches that have been developed to estimate the amounts of carbon stored in the Russian boreal forest. Over the past decade, several researchers have estimated the carbon stores of the different ecosystems of this region. All the data sets used in these studies reside in the Russian-language literature and therefore are inaccessible to Western scientists. In addition, many of the Russian studies carried out by using these data have been published in Russian-language reports and journals. In their chapter, Alexeyev and colleagues present an overview of the different approaches used in these studies and compare and contrast the results.

In Chapter 14, Bourgeau-Chavez and associates describe the different ecozones found throughout the North American boreal forest and summarize estimates of carbon present in these regions based on surveys conducted by the Canadian Forest Service and Canadian Department of Agriculture and field studies conducted by a variety of researchers in Alaska. These regional carbon estimates provide the basis for estimating carbon release from fires in Chapter 21 and changes to the North American boreal forest carbon budget as a result of changing climates in Chapter 25.

In Chapter 15, Murphy and colleagues discuss the fire data sets developed for the North American boreal forest region. Not only does this chapter present background on how these data sets were generated, but it presents two types of data: (1) annual area burned for the time period of 1920–present (1940–present for Alaska); and (2) boundaries of large fires (>250 ha) for the time period of 1980–present. The generation of this latter data set allows for analysis of regional patterns of fire, as well as estimation of carbon release from fires at this scale as well.

In Chapter 16, Shvidenko and Nilsson present information of fire statistics for the Russian boreal forest. This data record is neither as complete nor historically extensive as the North American data record, and the reasons for this are presented in this chapter. Shvidenko and Nilsson use this data set to estimate carbon release from fires in this region.

The last two chapters of this section focus on information that can be obtained through the analysis of satellite imagery. Although the launch of the Landsat satellite started the modern era of remote sensing, the past decade has seen an explosion in the number and diversity of satellite remote sensing systems available to the earth science community. During the next decade, even more systems will be launched, presenting the fire, forestry, and climate change community even more opportunities to monitor the boreal region. In Chapter 17, Ahern and co-workers review the different sensors operating in the visible and near-infrared region of the electromagnetic spectrum and the different types of information that can be derived from these satellites. In Chapter 18, Kasischke and associates review the various information products that can be derived from imaging radar systems.

The data sets and collection techniques discussed in this section play a central role in efforts to quantify the effects of fire and climate change on the boreal forest carbon budget. These data sets/techniques provide the baseline information required to apply the understanding of specific processes developed through the site-specific studies discussed in Section II. These site-specific studies provide the basis for developing landscape and regional-scale models of carbon flux processes, several of which are described in Section IV. The data sets/data collection techniques described in this section are needed for providing inputs for many of these models.

13. Carbon Storage in the Asian Boreal Forests of Russia

Vladislav A. Alexeyev, Richard A. Birdsey, Victor D. Stakanov, and Ivan A. Korotkov

Introduction

According to the most recent national inventory, Russia has 1180.9 million ha of land classified as "Forest Fund" of the country (Forest Fund of Russia 1995). These lands include 763.5 million ha of forested areas, 123.0 million ha of nonstocked lands (e.g., woodlands, clear-cuttings, burnt stands of trees), and 294.3 million ha of nonforestlands (e.g., peatlands, water, rocks, wastelands). According to the most recent global forest inventory, Russia's forests comprise 22% of the earth's forest area or 43% of the earth's temperate and boreal forest area (Anonymous 1992).

Understanding the earth's carbon cycle and climate system requires better knowledge of Russia's current forest conditions and dynamics. These forests have the potential to sequester or emit large quantities of CO_2 over short periods of time. Because Russian forests are so vast, their health and management have global significance. Only recently have statistics on characteristics of the Russian forest become widely available. As a result, there have been numerous studies to estimate the quantity of carbon stored in Russian forests and to estimate how that quantity is changing.

Forests of Siberia and Far East burn more frequently than forests in the European region of Russia. For example, in 1988, 26.4 million ha of forestland is classified as having not yet recovered from fire, whereas only 0.2 million ha was still in a burned-over condition in European Russia (Forest Fund of the USSR

1990). As a rule, in the area where fires are monitored (about 50% of forested Asian Russia), 0.2–2.7 million ha of forests is thought to burn annually (Korovin 1996). Occasionally, the area of forest fires can be several times larger (Goldammer and Furyaev 1996). To estimate actual and potential CO_2 emissions from forest fires, it is necessary to know the current carbon storage of forest vegetation, mortmass (coarse woody debris and litter), soil, and peatlands.

The goals of this chapter are to (1) discuss the approaches to estimate carbon storage in Russian forests; and (2) describe the carbon content of forest ecosystems in the Asian part of Russia. There are two main approaches to estimate phytomass and carbon storage of the Russian vegetation at regional and national scales: the geobotanical method, and the forest inventory method. These approaches are described and their results compared in the following sections. An overview of the patterns of carbon storage in Russian boreal forests is also included.

Approaches to Estimate Phytomass and Carbon Storage

The Geobotanical Method

The method of using information from a network of research plots to estimate the biological productivity of ecosystems for a whole region is common in geobotany and soil science. To estimate the quantity of phytomass (or any other variable such as soil organic matter) in a region, the average of observations from research plots within the region is multiplied by the area of the region. Usually, these research plots are established without regard to any particular sampling design in preselected vegetation types and conditions. If the network of research plots is located in vegetation conditions that are average for the region, estimates from data collected at the plots may be representative of the current vegetation in the region. But if statistical sampling procedures were not used to select the research plots, the estimates can represent something different from the average vegetation conditions. Research plots often represent potential rather than current vegetation because they are usually established in relatively mature, closed-canopy forest conditions rather than more typical disturbed plot locations. The geobotanical method was frequently used in the former USSR because most statistical data sets (including National Forest Inventory data) were not available to scientists.

The most successful application of the geobotanical method was by N.I. Bazilevich. Her first book about the cycling of nutrients in different types of vegetation (Rodin and Bazilevich 1965) was based on 150 sample plots (the description of the USSR's vegetation was based on 50 plots). In a recent book about the productivity of the USSR's vegetation, Bazilevich (1993) used bibliographic references and original data from more than 2,500 research plots, including 1,210 forest sample plots, aggregated into 303 points. The description of each point included data about the phytomass of vegetation, mortmass, and productivity. In accordance with the goals of her research studies, Bazilevich used data about phytomass and annual production of the vegetation for whole geographic regions without attention to specific details about the current status of forests, meadows, and other areas within the regions. Thus, Bazilevich (1993) estimated phytomass and mortmass for restored or potential vegetation but not for the current vegetation conditions of the USSR.

There have been several attempts to estimate carbon content for potential forest vegetation of the USSR and Russia by using Bazilevich's database. Kolchugina and Vinson (1993a) used Bazilevich's database (Bazilevich et al. 1986) and different geographic maps to calculate carbon storage of the forest vegetation in ecoregions of the former USSR. According to their estimates, forest phytomass in the former USSR consists of 87.7 Gt C, and mortmass (litter and coarse woody debris) totals 40.1 Gt C. Shvidenko and associates (1996) digitized maps of Bazilevich (1993) and estimated carbon content in different ecoregions of terrestrial vegetation of Russia. They concluded that carbon in phytomass of Russian vegetation equals 86.5 Gt and that carbon in mortmass equals 59.9 Gt.

Forest Inventory Method

A second approach to determine the storage of phytomass in a country or region is to combine two kinds of data: (1) statistical forest inventory databases; and (2) data from research plots in different ecoregions that contain estimates of forest phytomass for various ecosystem components such as tree boles, branches, and understory vegetation. Although the Russian National Forest Inventories do not include data collected specifically to estimate the phytomass of forests, they are designed to estimate (using statistical sampling techniques) the area, condition, volume, and other information important for managing forestlands. Data from research plots are used to determine coefficients for converting volume of timber estimated by forest inventories to the stock of phytomass and carbon of forest ecosystems.

One of the earliest applications of the forest inventory method was by Johnson and Sharpe (1983) for a region of the United States. Brown and co-workers (1989) presented a methodology based on large-scale inventories in four tropical countries. The forest inventory method has recently been used in temperate and boreal forests: in the United States (Birdsey 1990, 1992), Europe (Kauppi et al. 1992), Russia (Makarevskiy 1991), and Canada (Kurz et al. 1992). This approach was also used by Isaev and colleagues (1993, 1995), Kolchugina and Vinson (1993b,c, 1995), Turner and colleagues (1995), Alexeyev and Birdsey (1994), and Lakida and associates (1995, 1996) to estimate the carbon stored in the forests of Russia. These investigators developed their own models of carbon estimation, but in general their calculations were based on the general approach of combining forest inventory data with data from intensive research plots.

The paper by Shvidenko and colleagues (1996) contains information about the exact forest areas within the Russian ecoregions. In our studies, we multiplied their area estimates times our own estimates of average carbon density (Table C-2 in Alexeyev and Birdsey 1994) to estimate the carbon content for the current forest vegetation in different ecoregions of Russia (Table 13.1). Using this com-

Table 13.1. Carbon Content in Forest Vegetation of Russia Using the Inventory Method[a]

Zone	Area (million ha)	Phytomass		Mortmass	
		(million t)	(t ha^{-1})	(million t)	(t ha^{-1})
Zone of forest-tundra	108.4	3,424	31.6	3,910	36.2
Zone of boreal forests	508.8	35,108	69.0	23,596	46.4
Zone of mixed forests	13.1	1,626	124.1	2,061	42.9
Zone of sub-boreal forest	134.5	13,867	103.1	1,005	33.5
Zone of steppes	3.6	41	11.4	28	7.9
Subarid and arid zone	2.7	21	7.7	8	3.0
All zones	771.1	54,087	70.1	30,608	43.0

[a]Adapted from Alexeyev and Birdsey (1994) and Shvidenko et al. (1996).

bined methodology, the current total carbon storage of the Russian forest phytomass equals 54.1 Gt and the carbon density is 70.1 t ha^{-1}. In this case, the inventory method gives an estimate of current carbon storage that is 63% of the potential carbon storage in forests estimated by Shvidenko and associates (1996).

The most detailed description of methods and data for carbon storage in the Russian forests (stands, understory, mortmass, and soils) and peatlands (vegetation and peat) was published in the monograph of Alexeyev and Birdsey (1994). The data are given for each of the 71 administrative territories of Russia and for the 63 natural ecoregions of the country. Part of these data is compared with published materials of other authors (Table 13.2). For the most part, we are in agreement with the estimates derived by other researchers.

Carbon in Forest Soils

The organic matter of forest soils for all of Russia has not been inventoried by using statistical sampling techniques. Estimates of soil carbon are available from research plots and from soil surveys, which are nonstatistical characterizations of the soils of a region based on soil type mapping and purposely selected (nonrandom) sample characterizations of the properties of representative soil types. Thus a methodology very similar to the geobotanical method described earlier is commonly used for estimating forest soil carbon.

Shugalei and associates (1998) used data published by soil scientists to estimate the average carbon storage in forest soils for each ecoregion of Russia, then multiplied this average times the forest area within each ecoregion. They estimated that Russian forest soils contain 74 Gt C. Kolchugina and Vinson (1993c) used a similar method to estimate that Russian forest soils contain 100 Gt C. Both estimates exclude peatlands.

Carbon in Peatlands

Estimates of carbon in peatlands are based on the same methodology as estimates of carbon in forest soils. However, the published values are very inconsistent

Table 13.2. Carbon Storage and Density of the Forest Phytomass in Ecoregions, Economic Regions, and Geographic Regions of Russia, Estimated with the Forest Inventory Method[a]

Territory	Area (million ha)	Phytomass (million t)	Density (t ha^{-1})	Reference
Ecoregions				
Zone of forest-tundra	108.4	1,640	15.1	Alexeyev and Birdsey 1994
Zone of boreal forests	508.8	18,940	37.2	Alexeyev and Birdsey 1994
Zone of mixed forests	13.1	640	48.9	Alexeyev and Birdsey 1994
Zone of sub-boreal forest	134.5	6,510	48.4	Alexeyev and Birdsey 1994
Zone of steppes	3.6	130	36.1	Alexeyev and Birdsey 1994
Subarid and arid zones	2.7	110	40.7	Alexeyev and Birdsey 1994
All ecoregions	771.1	27,970	36.3	
Economic regions				
Pre-Baltic	0.3	15.6	58.5	Alexeyev and Birdsey 1994
	0.3	14.7	55.2	Lakida et al. 1996
Northern	76.0	27,33.1	35.9	Alexeyev and Birdsey 1994
	76.0	3,166.6	41.6	Lakida et al. 1996
Northwestern	10.4	551.8	53.1	Alexeyev and Birdsey 1994
	10.4	557.5	53.7	Lakida et al. 1996
Central	20.3	1,044.4	51.4	Alexeyev and Birdsey 1994
	20.3	1,039.9	51.2	Lakida et al. 1996
Volgo-Vyatsky	13.3	614.2	46.2	Alexeyev and Birdsey 1994
	13.3	619.8	46.6	Lakida et al. 1996
Central Chernozemny	1.5	78.6	53.5	Alexeyev and Birdsey 1994
	1.5	69.8	47.6	Lakida et al. 1996
Povolshsky	4.8	221.9	46.5	Alexeyev and Birdsey 1994
	4.8	205.0	43.0	Lakida et al. 1996
North Caucasus	3.7	281.2	76.8	Alexeyev and Birdsey 1994
	3.7	254.9	69.6	Lakida et al. 1996
Ural	35.8	1,647.2	46.1	Alexeyev and Birdsey 1994
	35.8	1,711.1	47.9	Lakida et al. 1996
Geographic regions				
European Russia	166.0	7,188	43.3	Alexeyev and Birdsey 1994
	166.0	7,639	46.0	Lakida et al. 1996
West Siberian Russia	90.1	3,401	37.7	Alexeyev and Birdsey 1994
East Siberian Russia	234.5	9,587	40.9	Alexeyev and Birdsey 1994
Far Eastern Russia	280.5	7,805	27.8	Alexeyev and Birdsey 1994
Russia total	771.1	27,980	36.3	Alexeyev and Birdsey 1994
	771.1	39,757	51.6	Isaev et al. 1993
	771.1	35,070	45.5	Isaev et al. 1995
	771.1	25,000	32.4	Kobak, personal communication
Former Soviet Union	799.9	50,200	62.7	Kolchugina and Vinson 1993c

[a]From Alexeyev and Birdsey (1994).

because estimates of the area of peatlands in Russia vary widely, and the different types of peatlands have not been studied comprehensively. There is confusion about what is included in different statistical compilations. For example, data on peatlands without commercial value may be omitted from official statistics but may be included within other land categories such as forest. In some studies, estimates of the carbon in peatland vegetation are not separated from estimates of carbon in the peat deposits.

Efremov and co-workers (1998) divided peatlands into nonforest, sparsely wooded, and excessively moist formations. He then estimated the average density of carbon within these categories and multiplied the averages times the estimated area within ecoregions. They estimated that peatlands of Russia contain 275 Gt C, a significant portion of the earth's carbon pool. Botch and colleagues (1995) interrelated a number of regional databases and maps to estimate a total quantity of 215 Gt C in peatlands of the former Soviet Union, a larger area.

Patterns of Carbon Storage in Russian Boreal Forests

The statistics in this section have been compiled from data presented by Alexeyev and Birdsey (1994, 1998). The methods to estimate carbon storage are described in detail by Alexeyev and co-workers (1995). A brief summary is presented here.

To generate the data presented in this section, we used the forest inventory method described above. Our estimates are based on the more detailed and diverse data available from the 1988 national inventory (for the USSR) rather than the older data published in 1995. According to the 1988 national inventory, the total area of Forest Fund amounts to 1,182 million ha, with a forested area of 771.1 million ha (Forest Fund of the USSR 1990). The difference from the 1995 inventory (about 1%) is much less than typical sampling errors of forest inventories. In addition, we used unpublished data from 98 forestry enterprises in Krasnoyarsk and the Republic of Yakutia because these areas were not sampled intensively enough in the national inventory. We developed maps of natural ecoregions. To estimate the amount of carbon in various ecosystem components for each of the ecoregions, we used data from many published research studies.

We developed equations to convert the volume of growing stock to carbon in phytomass and estimated the appropriate coefficients by region, species, and age class. The equations and coefficients are described in detail in Alexeyev and co-workers (1995). We included understory vegetation in the phytomass, which consisted of tree seedlings, shrubs, dwarf shrubs, herbs, forbs, and grasses. We also included mosses and lichens in phytomass.

Estimates of coarse woody debris were derived from published growth tables and published and unpublished forest inventory data. Estimates of carbon in litter were derived from the studies of soil scientists, as were estimates of carbon in soils. We attempted to estimate the biomass of forest consumers (animals, microorganisms, and fungi) from available data. Finally, we conducted a special study of peatlands following the general methodology described above. These methods and resulting estimates are described in detail in Alexeyev and Birdsey (1998).

Carbon in the Forest Vegetation of the Economic Regions of Asian Russia

The main forest areas of Russia are located in the Asian region—Siberia and the Far East (Fig. 13.1). According to the National Inventory data of 1988 (Forest Fund of the USSR 1990, 1991), forestlands of Asian Russia consist of 605.1 million ha of tree stands, 113.0 million ha of nonstocked land, and 262.6 million ha of nonforest area. Coniferous forests (larch, spruce, Scotch and Siberian pines, and fir) occupy 448.7 million ha (74.2% of forested area). Deciduous hardwoods comprise 2.2% of the area of tree stands in the Far Eastern economic region (excluding Yakutia, which is geographically part of eastern Siberia). The remainder of the area is covered by deciduous softwoods (13.5% of forested area) and dwarf birches and krummholz of *Pinus pumila* (10.2% of area) (Table 13.3). The administrative territory with the most wetlands is Tyumen oblast, where the area of peatlands exceeds the forested area (Table 13.3).

The total volume of growing stock in forests of Asian Russia equals 61.37 billion m^3. Larch (*Larix* sp.) stands contain 41.9% of this volume; Scots pine

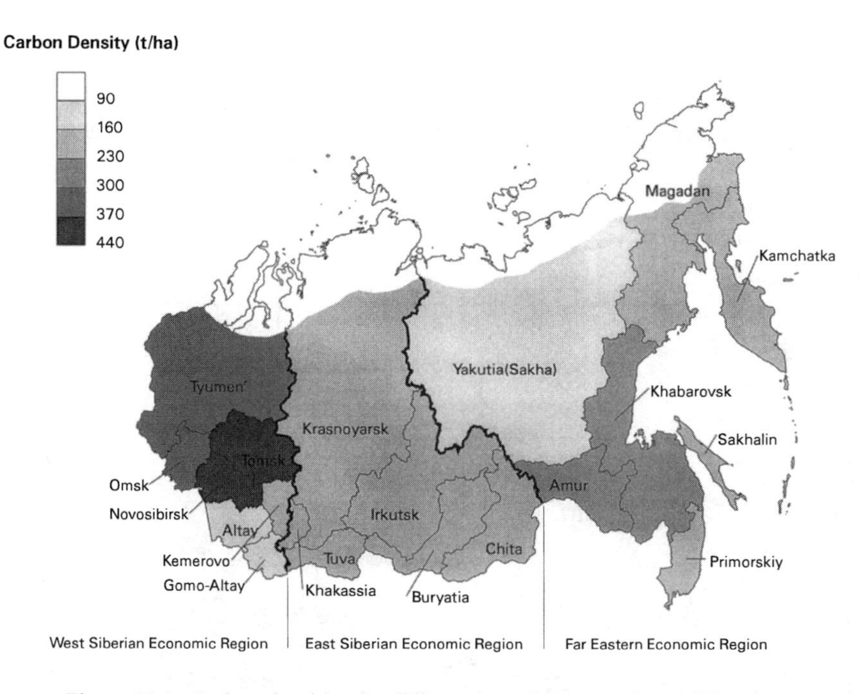

Figure 13.1. Carbon densities for different boreal forest regions of Russia.

Table 13.3. Distribution of Forest and Peatland Areas (in thousands of hectares) in Administrative Territories of Asian Russia[a]

Administrative conifer territory	Hardwood forest	Softwood forest	Deciduous shrub forest[b]	Deciduous total forest	Forest	Peatlands
West Siberian economic region						
Altai kray	4,629	2	2,440	292	7,363	609
Kemerovo oblast	3,051	—	2,556	8	5,615	274
Novosibirsk oblast	965	1	3,264	19	4,249	4,149
Omsk oblast	1,147	—	3,225	4	4,376	3,088
Tomsk oblast	10,492	—	8,386	5	18,883	15,491
Tyumen oblast	39,152	—	10,111	347	49,610	55,501
Subtotal	59,436	3	29,982	675	90,096	79,112
East Siberian economic region						
Krasnoyarsk kray	95,837	—	19,831	1,094	116,762	25,114
Irkutsk oblast	46,083	—	8,575	3,874	58,532	12,470
Chita oblast	20,413	1	5,122	3,352	28,888	3,623
R. Buryatia	17,131	4	1,640	3,389	22,164	1,060
R. Tuva	7,609	—	276	233	8,118	1,399
Subtotal	187,073	5	35,444	11,942	234,464	43,666
Far Eastern economic region						
Primor'ye kray	6,971	3,661	2,010	47	12,689	476
Khabarovsk kray	36,778	1,670	5,017	5,952	49,417	12,112
Amur oblast	14,822	760	4,992	1,968	22,542	6,588
Kamchatka oblast	1,171	5,995	1,381	11,258	19,805	1,781
Magadan oblast	10,033	8	355	12,582	22,978	17,244
Sakhalin oblast	3,967	944	383	336	5,630	929
R. Sakha (Yakutia)	128,409	—[c]	2,029	17,053	147,491	42,230
Subtotal	202,151	13,038	16,167	49,196	280,552	81,360
Total	448,660	13,046	81,593	61,813	605,112	204,138

[a]Estimates for forests from Alexeyev and Birdsey (1994). Peatland data from Efremov et al. (1998).
[b]Shrubs include mainly *Pinus pumila* and dwarf birches.
[c]—, No data.

(*Pinus sylvestris* L.) 16.9%; Siberian pine (*P. siberica*) 12.3%; birch (*Betula* sp.) 10.4%; spruce (*Picea* sp.) 8.0%; fir (*Abies* sp.) 4.1%; aspen (*Populus* sp.) 3.4%; oak (*Quercus* sp.) 0.7%; subarctic shrubs (*Pinus pumila* and dwarf birches) 2.2%; and other species, 0.1% of the total volume (Forest Fund of the USSR 1990).

According to our calculations, the carbon pool of the forest vegetation of Asian Russia is 20.8 Gt C, including 18.9 Gt C in tree stands, 0.5 Gt C in krummholz and shrub communities, and 1.4 Gt C in the understory of forests (Table 4 in Alexeyev and Birdsey 1994). The weighted average carbon density of the forest vegetation is 34.9 t ha^{-1}.

More than 80% of the carbon that has accumulated in the forest communities is contained in coniferous forests (Table 13.4), with larch stands accounting for 55% of the carbon of all other conifers together. The carbon storage of the hardwood forests is very small, consisting of only 3% of the total. The Asian part of Russia is characterized not only by the largest amount of carbon storage but also the largest portion (63.4%) of mature forests.

Carbon in the Forest Vegetation of Ecoregions of Asian Russia

Estimates for forest ecosystems are presented in this chapter are for the administrative territories that account for the economic activity in Russia, including forestry management. However, these economic and administrative territorial borders do not match the boundaries of the natural ecoregions and do not reflect the peculiarities of nature. Manipulation of data within an economic or political framework hinders understanding of processes that govern formation of stock and transport of carbon. Therefore, in addition to estimating the carbon stock for administrative territories, we estimated carbon for the main ecoregion types found in Asian Russia (Table 13.5).

Forest-Tundra and Mountain Subarctic Forests

The stocked area of the forest-tundra and mountain subarctic forests is more than 104.7 Million ha (17% of total forested area), with the mountain forests totaling 65.9 million ha. The simple composition of the stands (larches prevail) combined with relatively low productivity and low stand densities (0.3–0.4 of the standard density) leads to a low carbon accumulation rate and a low average density. The average density of carbon in vegetation is 15 t ha^{-1} (Table 13.5). Carbon of lower layers of plants in subarctic forests amounts to 20% of the vegetation carbon.

Boreal Forests

The stocked area of the boreal (taiga) zone within Asian Russia is 391.9 million ha, of which 149.0 million ha is covered by mountain forests. This part of the forest biome makes the largest contribution to the carbon pool of forest vegetation, containing 14.3 Gt C (66.4%) with an average density of 36.5 t ha^{-1}. It is well known that the northern, middle, and southern subzones of boreal forests are essentially different in structure, productivity, and composition of stands. These

Table 13.4. Carbon (in millions of tons) in Stands of Different Age Classes for the Dominant Species in Asian Russia[a,b]

Dominant tree species	Young stands Class 1[c]	Young stands Class 2[d]	Middle-aged	Maturing	Mature	Total
West Siberian economic region						
Larix spp.	1.3	4.1	41.2	46.5	141.0	234.0
Pinus sylvestris	11.2	17.1	161.4	180.5	528.2	898.5
Pinus sibirica	1.0	7.3	92.3	110.0	341.0	551.6
Picea obovata	1.6	2.4	32.1	32.8	125.5	194.4
Abies sibirica	2.0	4.8	32.1	28.9	65.9	133.7
Betula spp.	8.7	20.8	153.3	152.0	601.6	936.4
Populus tremula	2.7	6.1	46.6	43.9	163.2	262.6
East Siberian economic region						
Larix spp.	20.0	90.7	339.7	787.9	2,724.0	3,962.4
Pinus sylvestris	20.5	56.6	383.1	231.4	1,095.1	1,786.6
Pinus sibirica	6.8	37.7	182.3	313.8	784.0	1,324.6
Picea obovata	2.0	5.3	44.6	33.2	401.7	486.8
Abies sibirica	2.2	7.3	62.5	52.6	303.0	427.6
Betula spp.	11.9	27.1	213.8	165.7	454.4	872.9
Populus tremula	2.8	5.9	33.5	24.2	122.7	189.1
Far Eastern economic region						
Larix spp.	51.2	158.1	876.9	510.5	2,818.0	4,414.7
Pinus sylvestris	2.7	8.1	68.3	38.9	223.1	341.0
Pinus koraiensis	0.7	3.6	55.8	38.7	111.6	210.4
Picea spp.	6.3	20.8	124.6	103.8	366.2	621.9
Abies spp.	0.9	2.5	16.9	10.9	49.1	80.3
Betula spp.	6.4	14.7	114.9	51.0	96.7	283.7
Quercus mongolica	4.6	13.0	65.8	38.1	129.8	251.2
Betula ermanii	0.4	3.2	16.7	23.2	283.1	326.6
Populus tremula	1.6	3.2	38.8	14.6	44.2	101.4
Chosenia arbutifolia	0.0	0.1	4.6	1.0	7.9	13.6
Tilia amurensis	0.0	0.1	1.1	0.5	0.2	1.9
Asian part of Russia						
Larix spp.	72.5	252.9	1,257.8	1,344.9	5,683.0	8,611.1
Pinus sylvestris	34.4	81.8	612.8	450.8	1,846.4	3,026.2
Pinus sibirica	7.8	45.0	274.6	423.8	1,125.0	1,876.2
Pinus koraiensis	0.7	3.6	55.8	38.7	111.6	210.4
Picea spp.	9.9	28.5	201.3	169.8	893.4	1,302.9
Abies spp.	5.1	14.6	111.5	92.4	418.0	641.6
Betula ermanii	0.4	3.2	16.7	23.2	283.1	326.6
Quercus mongolica	4.6	13.0	65.8	38.1	129.8	251.3
Betula spp.	27.0	62.6	482.0	368.7	1,152.7	2,093.0
Populus tremula	7.1	15.2	118.9	82.7	330.1	554.0
Chosenia arbutifolia	0.0	0.1	4.6	1.0	7.9	13.6
Tilia amurensis	0.0	0.1	1.1	0.5	0.2	1.9
Total	169.5	520.6	3,202.9	3,034.6	11,981.2	18,908.8

[a]Estimates are from Alexeyev and Birdsey (1994).
[b]Carbon in shrubs for the four regions totals 2.4, 104.2, 421.5, and 528.1 millions of tons, respectively.
[c]Early regeneration.
[d]Advanced regeneration.

Table 13.5. Carbon Storage and Density of Forest Vegetation in Ecoregions of Asian Russia[a]

	Western Siberia		Eastern middle Siberia		Siberia and Yakutia		Far East		Total	
Ecoregions	(Gt)	($t\ ha^{-1}$)	(Gt)	($t\ ha^{-1}$)	(Gt)	($t\ ha^{-1}$)	(Gt)	($t\ ha^{-1}$)	(Gt)	($t\ ha^{-1}$)
Plains										
Forest-tundra zone	0.14	12	0.41	15	—[b]	—	—	—	0.55	14
Boreal forest zone										
Northern taiga	0.66	32	0.79	24	—	—	—	—	1.45	27
Middle taiga	1.75	42	1.09	45	1.95	29	—	—	4.79	36
Southern taiga	1.50	50	1.55	62	—	—	—	—	3.05	55
Forest-steppe zone	0.34	50	0.17	43	—	—	0.20	34	0.71	43
Steppe zone	0.07	40	—	—	—	—	—	—	0.07	40
Subtotal	4.46	39	4.01	35	1.95	29	0.20	34	10.6	35
Mountains										
Subarctic zone	—	—	0.13	15	0.67	17	0.25	14	1.05	16
Boreal zone	—	—	0.89	39	1.78	28	2.33	37	5.00	34
Sub-boreal zone	—	—	2.29	50	1.14	43	1.32	49	4.75	48
Subarid zone	—	—	0.10	39	—	—	—	—	0.10	39
Subtotal	—	—	3.41	43	3.59	28	3.90	36	10.9	34
Total	4.46	39	7.42	39	5.54	29	4.10	36	21.5	34

[a]From Alexeyev and Birdsey (1994).
[b]—, No data.

parts of the boreal zone also have varying carbon density (Table 13.5). The lower layers of the taiga forests are less developed under the closed-stand canopy and add up to no more than 1–3% of the carbon storage of the stands.

The relative productivity of the different geographic ecoregions is uneven, but productivity and carbon of forests are determined at present not only by natural factors but also by anthropogenic activity. Areas little affected by clear-cutting such as the southern taiga forests of middle Siberia contain the maximum amount of carbon in the boreal zone—62 t ha^{-1} (Table 13.5). In the European part of Russia, the carbon density in the same subzone amounts to 14 t ha^{-1} less (Alexeyev and Birdsey 1994).

One of the factors determining the amount of carbon storage in the forests of Asian Russia is the presence of permafrost. About 60% of Siberian and Far Eastern forests of Russia grow on soils where permafrost is present. The forest communities growing on permafrost of middle Siberia, eastern Siberia and Yakutia contain 9.3 Gt C, with an average density of 26 t ha^{-1}.

In most climatic sectors, an increasing amount of accumulated carbon is observed from forest-tundra (and mountain subarctic forests) to the forest-steppe and mountain sub-boreal forests, with productivity decreasing in the zone of steppes and deserts (Alexeyev and Birdsey 1994). Exceptions are the forests of Asian Russia. The increasing continentality of the climate in Siberia and Yakutia precludes growth of high-yield hardwood deciduous species, and therefore the carbon density is highest in the southern taiga (Table 13.5).

Forest-Steppes and Mountain Sub-Boreal Forests

Forests of these ecoregions cover an area of 115.8 million ha (18.8% of the stocked territory). The major vegetation community is mountain forest (99.3 million ha; Table 13.5). The total storage of carbon accumulated in vegetation is 5.5 Gt, with an average density of 47.2 t ha^{-1}. In the plains, the carbon stock is substantially less than it might have been without the impact of human disturbance of the forests.

Forest-Steppes and Mountain Forests of Subarid Zones

The forests of steppes and mountain stands of subarid regions cover a total of 0.7% of the forested area and do not make any considerable contribution to the total storage of carbon in the forests of Asian Russia (Table 13.5). The forests of the steppe biome generally grow on lands not fit for agricultural crops.

Carbon Storage in the Forest and Peatland Ecosystems of Asian Russia

Carbon in vegetation represents only part of the total carbon present in forest ecosystems. Other important components are mortmass (coarse woody debris and litter) and soil organic matter. The forest ecosystems of Asian Russia have 94.1 Gt C (Table 13.6). The amount of carbon in the forest vegetation equals 20.8 Gt (22.1%); carbon in the mortmass is 12.9 Gt (13.7%); and carbon in the soils is estimated to be 60.4 Gt (64.2%).

Table 13.6. Carbon Storage (in millions of tons) of Forests and Peatlands in Administrative Territories of Asian Russia[a]

Economic region and administrative territory	Forested areas	Peatlands				
	Phytomass	Mortmass	Soil	Vegetation	Peat	Total
Western Siberia						
Altai kray	323.0	111.8	533.5	2.8	218.3	1,189.4
Kemerovo oblast	175.2	88.8	623.8	0.4	106.8	995.0
Novosibirsk oblast	151.9	78.9	789.9	20.1	2,613.4	3,654.2
Omsk oblast	188.4	85.0	741.2	12.9	1,536.4	2,563.9
Tomsk oblast	825.4	357.6	2,554.7	73.2	10,850.9	14,661.8
Tyumen oblast	1,736.6	1,174.1	7,725.0	183.8	27,003.2	37,822.7
Total	3,400.5	1,896.2	12,968.1	293.2	42,329.0	60,887.0
Eastern Siberia						
Krasnoyarsk kray	4,501.0	2,922.7	11,874.7	141.0	9,602.4	29,041.8
Irkutsk oblast	3,047.3	1,221.3	5,024.3	12.7	4,430.7	13,736.3
Chita oblast	944.9	575.7	2,872.6	9.0	1,739.4	6,141.6
R. Buryatia	761.7	436.0	2,364.9	3.0	437.8	4,003.4
R. Tuva	332.1	137.5	696.5	6.7	907.6	2,080.4
Total	9,587.0	5,293.2	22,833.0	172.4	17,117.9	55,003.5
Far East						
Primor'ye kray	667.4	298.5	1,605.1	2.3	249.4	2,822.7
Khabarovsk kray	1,848.0	1,211.3	5,055.4	71.3	6,916.5	15,102.5
Amur oblast	746.0	477.8	2,426.5	58.3	4,038.8	7,747.4
Kamtchatka oblast	570.3	635.1	1,729.4	14.4	897.5	3,846.7
Magadan oblast	292.7	416.2	1,314.3	90.0	6,092.6	8,205.8
Sakhalin oblast	229.3	139.3	494.0	5.1	550.3	1418.0
R. Sakha (Yakutia)	3,451.3	2,497.5	12,005.7	67.5	14,913.8	32,935.8
Total	7,805.0	5,675.7	24,630.4	308.9	33,658.9	72,078.9
Total, Asian Russia	20,792.5	12,865.1	60,431.5	774.5	93,105.8	18,7969.4

[a]Estimates for forests from Alexeyev and Birdsey (1994). Peatland data from Efremov et al. (1998).

The greatest stock of carbon is on 204.1 million ha of Siberian and Far East mires: they contain 93.9 Gt C, most of which (99.2%) is in the peatlands (Table 13.6). More than half of the carbon in the peatlands is distributed in the shallow peatlands without commercial resource importance (Efremov et al. 1998).

The highest carbon density is found in the southern taiga subzone of the boreal forest ecoregion, where forests are dense and relatively mature and where peat deposits are most common (Table 13.7). The lowest carbon density is in the forest-tundra zone. We estimate that Asian Russia contains a total of 189.1 Gt C in forests and peatlands combined.

Comparison of Results from Various Studies

The method of geobotany produces a significantly higher estimate of carbon storage in phytomass (Tables 13.1 and 13.2). Makarevskiy (1991) analyzed published data about productivity of forest ecosystems and showed that most of the research plots in the Karelian Republic of Russia were located in more productive (up to two to three times) forest stands than the average, which is consistent with our own observations and conclusions. There are three reasons why research plots are usually established in forest stands that are different from the average. First, according to regulations, the density of trees on sample plots must be as uniform as possible. So, scientists generally avoid gaps and openings within the area of the sample plot. Second, many scientists wanted to highlight the most productive forest ecosystems of the country. Third, data collection is easier and more accurate in compact, small-size stands.

Data on phytomass in Russia started to appear frequently after the beginning of the International Biological Program (in Russia in 1968–80). However, up to this time the total number of Russian forest research plots with phytomass data scarcely exceeded 3,000. The total area of these experimental plots is not more than 800 ha, which is 0.000001% of the Russian forested area. Although this number of sample plots may be sufficient for some national-level statistics, it is insufficient for most regional or subregional data needs given the variability of ecological conditions and stand dynamics due to harvesting and natural disturbance.

Because the research plots were not established in all forest conditions, critical information for estimating current carbon storage was missing, such as area of different forest types within regions, stand age distribution, areas of burns, and harvested areas. Nevertheless, critical information on percentage distribution of phytomass fractions (leaves, branches, stem, roots, and understory) for different forest classifications is available only from these research plots.

It is useful to review the similarities and differences between the various estimates of phytomass of the Russian forests developed with the method of forest inventory. The results from Alexeyev and Birdsey (1994) and Lakida and associates (1996) for European Russia and its nine economic regions are in general agreement (Table 13.2). The differences are within about 0.4–13.7%. Both groups of authors worked independently and used their own models.

Table 13.7. Carbon Storage and Density of Forest and Peatland Ecosystems of Asian Russia[a,b]

Ecoregions	Western Siberia		Middle Siberia		Eastern Siberia and Yakutia		Far East		Total	
	(Gt)	(t ha^{-1})	(Gt)	(t ha^{-1})	(Gt)	(t ha^{-1})	(Gt)	(t ha^{-1})	(Gt)	(t ha^{-1})
Plains										
Forest-tundra zone	3.8	165	3.7	131	—[c]	—	—	—	7.5	146
Boreal forest zone										
Northern taiga subzone	10.3	237	5.6	162	—	—	—	—	15.9	205
Middle taiga subzone	21.0	315	5.8	216	15.8	195	—	—	42.6	244
Southern taiga subzone	34.9	564	8.2	277	—	—	—	—	43.1	472
Forest-steppe zone	2.8	280	1.8	355	—	—	2.0	126	6.6	213
Steppe zone	0.7	182	—	—	—	—	—	—	0.7	182
Subtotal	73.5	352	25.1	202	15.8	195	2.0	126	116.4	270
Mountains										
Subarctic zone	—	—	1.0	113	8.0	155	11.1	308	20.1	208
Boreal zone	—	—	7.9	252	8.7	134	14.1	180	30.7	175
Sub-boreal zone	—	—	9.0	189	5.6	198	7.6	235	22.2	205
Subtotal	—	—	17.9	204	22.3	154	32.8	223	74.1	195
Total	73.5	352	43.0	203	38.1	168	34.8	214	189.1	233

[a]Estimates from Alexeyev and Birdsey (1998) and Efremov et al. (1998).
[b]Includes carbon in forest vegetation, coarse woody debris, litter, soil, and peat.
[c]—, No data.

The estimates of Alexeyev and Birdsey (1994) and the estimates of Isaev and co-workers (1993, 1995) are rather different. Isaev and colleagues used conversion ratios for transformation of wood volume to phytomass for the main forest species and different age groups for Russia as a whole country. Our conversion ratios for the main forest species and different age groups were calculated separately for European and Asian Russia and, within these broad regions, separately for northern, middle, and southern taiga ecoregions. The difference between ecoregional conversion ratios and ratios for the whole country may have real significance.

Some part of the estimated carbon stock in Isaev and associates (1993, 1995) must have a lower value than we estimated. We suspect that the carbon stock and density of the subarctic shrubs of *Pinus pumila* and *Betula nana* are much less than 8.9 Gt C (136 t ha^{-1}) (Isaev et al. 1993) or 2.74 Gt C (48 t ha^{-1}), respectively (Isaev et al. 1995). According to these estimates, carbon density of the subarctic shrubs is higher than the carbon density of tree stands in the northern and some other economic regions of Russia (Table 13.2). Our estimates (Alexeyev and Birdsey 1994) show that these subarctic shrubs have 0.54 Gt c of the carbon stock and 8.6 t ha^{-1} of the carbon density.

K. Kobak (1996, private communication) estimated the carbon pool of the Russian forest vegetation as 25 Gt C and carbon density as 32.4 t ha^{-1}. This value is 10.3% lower than our estimate and within the limits of our errors.

The largest difference is between our data for Russia and Kolchugina and Vinson's (1993c) data for the former Soviet Union. The main reason is that the authors used a conversion ratio of 0.53 Mg C m^{-3} calculated by Sampson (1992) for the marketable wood of the U. S. forests. Later, Kolchugina and Vinson (1995) used the ratio from the paper of Isaev and associates (1995), which reduced their earlier estimate by about 25%, making it closer to our results. The shortcomings of the geobotany method and available databases were the reasons that Kolchugina and Vinson (1995) and A. Shvidenko (Lakida et al. 1995, 1996) revised their carbon storage estimates for the former Soviet Union.

In summary, the direct extrapolation of estimates from research plots to large regions of Russia has been abandoned as a methodology and replaced by a combination of methods that use the statistical estimates from forest inventories as the basis for classifying forest areas according to current condition. Data from research plots are used within this framework to estimate mortmass, soil carbon, and carbon for a large part of forested peatland in Russia. Comparing the different estimates that are based on the method of forest inventory confirmed that our estimates of carbon storage in the forest ecosystems have the predicted precision of ±10–15% (Alexeyev and Birdsey 1994).

Conclusions

Two approaches to estimate the carbon stock of phytomass in Russia give estimates that differ by a factor of 2. The geobotanical method, which estimates carbon in potential vegetation by using nonstatistical sampling techniques, yields an average estimate of about 70 t ha^{-1} for all of Russia. The forest inventory method, which estimates carbon in current vegetation by using a statistical forest inventory as a base, yields an average estimate of 32–52 t ha^{-1}. Disturbances such as fire and timber harvest account for much of the difference between the two methods. Differences associated with statistical versus nonstatistical sampling procedures are also important.

The biggest need to improve estimates, regardless of the methodology used, is additional inventory data to reduce uncertainties, particularly in remote areas of Asian Russia where sampling has been less intense than other areas. Monitoring disturbances such as timber harvest and wildfire are also critical, not only for estimating current carbon pools but also for estimating the exchange of CO_2 between forests and the atmosphere.

There are many conversion factors used to estimate carbon, and each of these includes some error. For example, estimates of timber density depend on species, growth conditions, age, and decay. There are insufficient studies available to accurately estimate timber density for a country as vast and diverse as Russia.

The error of estimating the large quantity of carbon in soils (including peatlands) is unknown but likely large. Because inventorying soil carbon by using a statistical sample is not practical, it may be more useful to investigate ways to estimate carbon fluxes directly from soils on a regional basis and ignore the inevitable errors in estimating the total amount of carbon stored at any given time.

There is a great deal of variability in carbon storage of Asia Russian, depending on the geographic region, forest type, and history of disturbance. These factors affect the forest composition and stand age structure, both strong determinants of carbon storage. The forests of Asian Russia, with their large proportion of the world's stock of carbon, in the future can be either a source or sink of atmospheric CO_2, depending on forest harvesting and forest management activity and natural factors such as fire frequency and insect outbreaks.

Acknowledgments

This study was supported by grant 96-04-48344 of the Russian Fund for Fundamental Investigations and by the Northern Global Change Research Program of the U.S. Department of Agriculture, Forest Service.

References

Alexeyev, V.A., and R.A. Birdsey, eds. 1994. *Carbon in Ecosystems of Forests and Peatlands of Russia.* Institute for Forest Research, Krasnoyarsk, Russia. (in Russian)

Alexeyev, V.A., and R.A. Birdsey, eds. 1998. *Carbon Storage of Forests and Peatlands of Russia.* General Technical Report NE-244. U.S. Department of Agriculture, Forest Service, Northeastern Forest Experiment Station, Radnor, PA.

Alexeyev, V.A., R.A. Birdsey, V.D. Stakanov, and I.A. Korotkov. 1995. Carbon in vegetation of Russian forests: methods to estimate storage and geographical distribution. *Water Air Soil Pollut.* 82:271–282.

Anonymous. 1992. *The Forest Resources of the Temperate Zones, 1992. Main Findings of the UN-ECE/FAO 1990 Forest Resource Assessment.* United Nations, New York.

Bazilevich, N.I. 1993. *Biological Productivity of Ecosystems in Northern Eurasia.* Nauka, Moscow. (in Russian)

Bazilevich, N.I., O.S. Grebenshikov, and A.A. Tishkov. 1986. *Geographical Regularities of Structure and Function of Ecosystems.* Nauka, Moscow. (in Russian)

Birdsey, R.A. 1990. Inventory of carbon storage and accumulation in U.S. forest ecosystems, pp. 24–31 in H.E., Burkhart, G.M. Bonnor, and J.J. Lowe, eds. *Research in Forest Inventory, Monitoring, Growth and Yield.* Publication FWS-3–90. Virginia Polytechnic Institute, Blacksburg, VA.

Birdsey, R.A. 1992. *Carbon Storage and Accumulation in United States Forest Ecosystems.* General Technical Report WO-59. U.S. Department of Agriculture, Forest Service, Washington, DC.

Botch, M.S., K.I. Kobak, T.S. Vinson, and T.P. Kolchugina. 1995. Carbon pools and accumulation in peatlands of the former Soviet Union. *Global Biogeochem. Cycles* 9:37–46.

Brown, S., A.J.R. Gillespie, and A.E. Lugo. 1989. Biomass estimation methods for tropical forests with applications to forest inventory data. *For. Sci.* 35:881–902.

Efremov, S.P., T.T. Efremova, and N.V. Melentyeva. 1998. Carbon storage in peatland ecosystems, pp. 69–76 in V.A. Alexeyev and R.A. Birdsey, eds. *Carbon Storage of Forests and Peatlands of Russia.* General Technical Report NE-244. U.S. Department of Agriculture, Forest Service, Northeastern Forest Experiment Station, Radnor, PA.

Forest Fund of the USSR. 1990. In M.M. Drozhalov, ed. *Statistical Collection,* Vol. 1. State Forestry Committee, Moscow. (in Russian)

Forest Fund of the USSR. 1991. In M.M. Drozhalov, ed. *Statistical Collection,* Vol. 2. State Forestry Committee, Moscow. (in Russian)

Forest Fund of Russia. 1995. *Statistical Collection.* State Forestry Committee, Moscow. (in Russian)

Goldammer, J.G., and V.V. Furyaev. 1996. Fire in ecosystems of boreal Eurasia: ecological impacts and links to the global system, pp. 1–20 in J.G. Goldammer, and V.V. Furyaev, eds. *Fire in Ecosystems of Boreal Eurasia.* Kluwer Academic Publishers, Dordrecht, The Netherlands.

Isaev, A.S., G.N. Korovin, A.L. Utkin, A.A. Pryashnikov, and D.G. Zamolodchikov. 1993. The estimation of storage and annual carbon deposition in phytomass of forest ecosystems of Russia. *Lesovedenie* 5:3–10. (in Russian)

Isaev, A., G. Korovin, D. Zamolodchikov, A. Utkin, and A. Pryashnikov. 1995. Carbon stock and deposition in Russian forests. *Water Air Soil Pollut.* 82:247–256.

Johnson, W.C., and D.M. Sharpe. 1983. The ratio of total to merchantable biomass and its application to the global carbon budget. *Can. J. For. Res.* 13:372–383.

Kauppi, P.E., K. Mielikainen, and K. Kuusela. 1992. Biomass and carbon budget of European forests, 1971 to 1990. *Science* 256:70–74.

Kolchugina, T.P., and T.S. Vinson. 1993a. Equilibrium analysis of carbon pools and fluxes of forest biomes in the Soviet Union. *Can. J. For. Res.* 23:81–88.

Kolchugina, T.P., and T.S. Vinson. 1993b. Carbon sources and sinks in forest biomes of the former Soviet Union. *Global Biogeochem. Cycles* 7:291–304.

Kolchugina, T.P., and T.S. Vinson. 1993c. Comparison of two methods to assess the carbon budget of forest biomes in the former Soviet Union. *Water Air Soil Pollut.* 70:207–221.

Kolchugina, T.P., and T.S. Vinson. 1995. Role of Russian forests in the global carbon balance. *Ambio* 24:258–264.

Korovin, G.N. 1996. Analysis of the distribution of forest fires in Russia, pp. 112–128 in J.G. Goldammer, and V.V. Furyaev, eds. *Fire in Ecosystems of Boreal Eurasia.* Kluwer Academic Publishers, Dordrecht, The Netherlands.

Kurz, W.A., M.J. Apps, T.M. Webb, and P.J. McNamee. 1992. *The Carbon Budget of the Canadian Forest Sector: Phase I.* Information Report NOR-X-326. Forestry Canada, Northwest Region, Northern Forestry Centre, Edmonton, Alberta, Canada.

Lakida, P., S. Nilsson, and A. Shvidenko. 1995. *Estimation of Forest Phytomass for Selected Countries of the Former European USSR.* WP-95–79. International Institute for Applied Systems Analysis, Laxenburg, Austria.

Lakida, P., S. Nilsson, and A. Shvidenko. 1996. *Forest Phytomass and Carbon in European Russia.* WP-96–28. International Institute for Applied Systems Analysis, Laxenburg, Austria.

Makarevskiy, M.F. 1991. Carbon storage and balance in forest and peatland ecosystems of Karelia. *Ecologia* 3:3–10. (in Russian)

Rodin, L.E., and N.I. Bazilevich. 1956. *Dynamics of Organic Matters and Rotation of Ash Elements and Nitrogen in Main Types of Vegetation of Earth.* Nauka, Moscow. (in Russian)

Sampson, R.N. 1992. Forestry opportunities in the United States to mitigate the effects of global warming. *Water Air Soil Pollut.* 64:1–2.

Shugalei, L.S., E.P. Popova, and V.A. Alexeyev. 1998. Organic carbon storage in soils of Russian forests, pp. 54–64 in V.A. Alexeyev and R.A. Birdsey, eds. *Carbon Storage in Forests and Peatlands of Russia.* General Technical Report NE-244. U.S. Department of Agriculture, Forest Service, Northeastern Forest Experiment Station, Radnor, PA.

Shvidenko, A.S., S. Nilsson, V.A. Rojkov, and V.V. Strakhov. 1996. Carbon budget of the Russian boreal forests: a system analysis approach to uncertainty, pp. 145–162 in M.J. Apps and D.T. Price, eds. *Forest Ecosystems, Forest Management, and the Global Carbon Cycle.* NATO ASI Series, vol. I40. Springer-Verlag, Berlin.

Turner, E.P., G.J. Koerper, M.E. Harmon, and J.J. Lee. 1995. A carbon budget for forests of the conterminous United States. *Ecol. Appl.* 5:421–436.

14. Characteristics of Forest Ecozones in the North American Boreal Region

Laura L. Bourgeau-Chavez, Eric S. Kasischke, James P. Mudd, and Nancy H.F. French

Introduction

In the chapters of Section II of this book, the various processes controlling the storage and release of carbon to the atmosphere from the North American boreal forest were discussed in some detail. The primary focus of the discussions in those chapters was on processes that occur over very small spatial scales. To estimate carbon storage and release over regional and continental scales it is necessary to scale, extrapolate, and aggregate observations made at plot and landscape scales. These regional carbon flux studies also require a baseline description of the distribution of forest types and the carbon they contain.

One means to fulfill these requirements is to define specific geographic units that have similar characteristics. In this chapter, we describe the development of a discrete number of forest ecozones for the North American boreal region. We discuss the characteristics of each ecozone (physiography, soils, climate, and dominant forest cover) and present two continental-scale data sets that describe the spatial distribution of carbon present in living vegetation and organic soils.

Ecozones of the North American Boreal Forest

One of the problems often faced in creating continental-scale maps is that scientists working in different countries use slightly different criteria in defining similar or identical geographic units. Several maps have been produced that define the different terrestrial ecoregions or ecozones of the United States and Canada. In this chapter, we use maps developed from these previous efforts to create a forest ecozone map for the entire North American boreal region. In generating this map, we reviewed efforts carried out separately for both Canada and Alaska. In Alaska, a single map was produced by a joint working group (Gallant et al. 1995), whereas in Canada two separate projects resulted in the creation of two forest ecozone maps: the Ecoclimatic Regions of Canada map generated by the Ecoregions Working Group (1989) and the Terrestrial Ecozones of Canada map created by Wiken (1986).

Each of these maps was generated through integration of a variety of factors. The Ecoclimatic Regions of Canada map was based on broad categories of vegetation physiognomy and climatic gradients. The Terrestrial Ecozones map of Wiken (1986) was based on these two factors but also took into account physiography, soils, major wildlife populations, and the distribution of water bodies.

For the purposes of this study, we selected the Terrestrial Ecozones map of Wiken (1986) to merge with the Ecoregions of Alaska map (Gallant et al. 1995). The Terrestrial Ecozones of Canada map was chosen because it was based on similar classification criteria as the map of Alaska by Gallant and associates (1995).

For our map the North American boreal forest was divided into nine forest ecozones (Fig. 14.1[1]). The Alaska Boreal Interior ecozone was created by combining the Interior Forested Lowlands and Uplands, Interior Bottomlands, and Yukon Flats regions from the Alaska ecoregions map with the Old Crow Flats and Old Crow Basin regions of the Taiga Cordillera ecozone of Canada. The Interior Highlands and Copper Plateau regions of the Alaska ecoregions map were merged into Canada's Boreal Cordillera ecozone.

The remaining ecozones are entirely within Canada and include the Taiga Plains, Taiga Shield, Hudson Plains, Boreal Plains, and Boreal Shield. The Boreal Shield and Taiga Shield are very large ecozones, covering a wide climatic range. The region east of Hudson Bay has higher precipitation, a longer growing season, and less extreme cold temperatures than the region to the west. There are also significant differences in the fire history of the western and eastern portions. Because of this diversity, we divided the Boreal and Taiga Shield ecozones into two halves, one to the east of Hudson Bay and the other to the west.

Descriptions of the characteristics of the Canadian ecozones are taken from Wiken (1986) and Rowe (1972), which contains a more detailed description of the different ecosystems found in the Canadian boreal forest. For the Alaska ecozones, the descriptions of Viereck and associates (1986) and Gallant and coworkers (1995) were used.

Although the terms *taiga* and *boreal forest* are often used interchangeably, they are not synonyms. *Taiga* is a Russian word referring to the subarctic boreal-tundra

[1]Figure 14.1 will be found in the color insert.

ecotone. It is defined by *Webster's Ninth New Collegiate Dictionary* more vaguely as "a moist subarctic coniferous forest that begins where the tundra ends and is dominated by spruce and fir." The term *boreal forest* includes the taiga and is defined by Webster as "the northern biotic area characterized by a dominance of coniferous forests." The ecozones described below have the term *boreal* in their title when describing closed-crown forests and the term *taiga* when referring to open-canopy forests and boreal-tundra ecotones.

Boreal Cordillera

The Canadian Boreal Cordillera ecozone is located in the middle of the Canadian Rocky Mountains. It covers the southern portion of the Yukon Territory and northern British Columbia. The bulk of Alaska's Boreal Cordillera is an extension of the Canadian Boreal Cordillera into the low mountains north of the Alaska Range and east of Fairbanks. Patches of Boreal Cordillera in Alaska also exist south of the Brooks Range, in the intermontane region north of Glenallen (the Copper Plateau), as well as in the Kuskokwim Mountains and Nulato Hills. The physiography of the ecozone consists of mountain ranges with high peaks, extensive plateaus, and intermontane plains. The Canada Boreal Cordillera was heavily glaciated during the Pleistocene epoch and has also been shaped by erosion, volcanic ash depositions, and mud slides. The surface material consists of glacial drift, colluvium, and exposed bedrock. In Alaska, many mountain peaks and the intermontane region were also glaciated in the Pleistocene epoch, but other parts of this region were not. Geological formations include proglacial lake deposits (Copper Plateau), Paleozoic and Precambrian metamorphic rocks, felsic volcanic rocks, and intrusive rocks as well as Cretaceous and Lower Paleozoic sedimentary rocks.

The climate of the Boreal Cordillera is characterized by long, cold winters and short, warm summers. Weather data for the mountainous parts of Alaska do not exist, but weather data for the Copper Plateau and Canada were available. In Canada, the mean temperatures in winter are typically −13 to −23°C and 9.5–11.5°C in summer. The Copper Plateau weather is similar, but summer temperatures range between 4 and 21°C. In the intermontane plateaus, the annual precipitation averages 30–60 cm (25–46 cm for the Copper Plateau). The Canadian mountains in the east of this ecozone average 100–150 cm of precipitation per year, and the western mountains average much higher. In general, precipitation increases and summer temperatures decrease with elevation. Correspondingly, there is an increase in litter accumulation from low to high latitudes attributable to the decreasing biological activity (Bray and Gorham 1964).

Tree species found in this ecozone include black spruce and lodgepole pine on less productive sites. Sites with dry sandy soils are dominated by lodgepole pine. Pure stands of white spruce or balsam poplar are common on floodplains. Warmer, dryer sites are dominated by white spruce, trembling aspen, and white birch. Boggy areas are dominated by tamarack and black spruce. Alpine fir occurs near the treeline.

In Canada, the closed and open forests are largely confined to the plateau regions and valleys. In the northwestern portion of the ecozone, forests are typically open and lodgepole pine and alpine fir are generally absent. Alpine tundra exists at high elevations. Here, sedge-dominated meadows and lichen-dominated rock fields prevail. Permafrost is widespread in the northern part of the ecozone and at higher elevations. Discontinuous permafrost occurs in the southern portions of the ecozone.

Taiga Plains

The Taiga Plains ecozone is located in the southwest corner of Northwest Territories, the northeast section of British Columbia, and northern Alberta. There are a large number of wetlands with deep organic soils in this ecozone, second only to the Hudson Plains ecozone. The physiography consists of level to gently rolling plains. Permafrost is extensive and perches water on the low-angled slopes, resulting in extensive seasonally waterlogged soils. The climate is cold and semiarid. The average annual precipitation ranges from 20 cm in the north to 50 cm in the south. The mean daily temperatures for winter range from –26 to –15°C and from 6.5 to 14°C in summer. Typically, there are 60–100 frost-free days. The cold temperatures, permafrost conditions, and poor drainage lead to the formation of cryosolic, gleysolic, and organic soils. Associate species typical in this ecozone include dwarf birch (*Betula nana*), labrador tea (*Ledum palustre*), willows, bearberry (*Arctostaphylos* spp.), mosses, and sedges. The uplands, foothills, and southern exposures in this ecozone are better drained and warmer and are dominated by mixed forests of white spruce, black spruce, tamarack, white birch, trembling aspen, balsam poplar, and lodgepole pine. Pine and balsam fir are absent in the far north. Open-canopied black spruce forests (taiga) dominate in the central portion of the Taiga Plains. Large expanses of swamp and peat exist in the southern Taiga Plains. The western Taiga Plains are dominated by permafrost and are mostly nonforested.

West Taiga Shield

The West Taiga Shield ecozone includes northern Manitoba, northern Saskatchewan, and the south central Northwest Territories. It is on the Canadian Shield where the geology consists of Precambrian bedrock outcrops and glacial moraine mantles, with some lacustrine deposits. Its physiography is dominated by rolling plains. Permafrost is discontinuous but widespread, which promotes lateral drainage; thus, lowlands are often waterlogged and contain a mosaic of lakes. Soils consist of Brunisols in the south and Cryosols in the north, with Gleysols in the poorly drained areas. The climate of the West Taiga Shield is subarctic continental with 20–50 cm yr^{-1} of precipitation (Wiken 1986). For the entire Taiga Shield, the mean daily winter temperatures range from –11 to –24.5°C and 6 to 11° C in summer, with approximately 70–100 frost-free days. The summers are short with extensive daylight and the winters long with little daylight. The northern boundary of this region contains open forests of spruce-lichen woodlands that grade into tundra. The central region is dominated by stunted coniferous and deciduous stands. Open-canopy, stunted black spruce with alders, willows, and tamarack dominate in fens and bogs. Balsam fir is absent from the western Taiga Shield. The uplands and river floodplains are dominated by mixed forests of white spruce, trembling aspen, balsam poplar, and white birch.

East Taiga Shield

The East Taiga Shield ecozone covers an area in central Quebec and Labrador. The surface material is similar to that of the West Taiga Shield but also includes marine deposits. The only difference in the composition of mixed-forest stands on uplands and along floodplains is the occurrence of balsam fir. Precipitation is much higher in this ecozone relative to the West Taiga Shield, with less extreme temperatures. Average annual precipitation ranges from 50 to 80 cm, with local areas exceeding 100 cm on the Labrador coast. The dominant tree species along the Atlantic Coast and Hudson Bay is white spruce, and balsam fir is rare but not absent. The southern boundary of this ecozone is marked by a decrease in open forests and an increase in closed-canopy forests. A patchwork of lakes, rivers, bogs, swamps, and muskeg mixed with upland forest and barrens occurs along the southern edge of the East Taiga Shield. Tamarack tends to frame the lakes. Trembling aspen, balsam poplar, and white birch are infrequent.

West Boreal Shield

The West Boreal Shield ecozone consists of northern Saskatchewan, central Manitoba, and western and southern Ontario. According to climatic data from Rowe (1972), the mean minimum temperatures of the West Boreal Shield range from –23 to –29°C in January, and the maximum July temperatures range from 23 to 26°C. Precipitation ranges from 43 to 76 cm yr^{-1} (Rowe 1972).

This ecozone is dominated by rolling terrain with Precambrian outcrops and
This ecozone is dominated by rolling terrain with Precambrian outcrops and
mantles of glacial moraine. The morainal areas are covered by closed stands of black and white spruce, balsam fir, and tamarack. The dominance of deciduous species (white birch, trembling aspen, and balsam poplar) increases southward. Eastern white pine, red pine, and jack pine are also found in the southern range. Podzol soils dominate in the south and Brunisols in the north. Throughout the region, exposed bedrock gives way to lichen, shrub, and forb communities. Small to medium-sized lakes are common across the region. The northern portion of this ecozone has very frequent fires and is thus dominated by jack pine and tamarack. The southern portion of the West Boreal Shield has lower fire frequency with a heterogeneous forest cover. The forest types range from mixed forests with lush shrub understories to floristically poor coniferous forests. Species include trembling aspen, white spruce, balsam fir, black spruce, balsam poplar, white birch, jack pine, and tamarack.

East Boreal Shield

The East Boreal Shield ecozone covers southern Ontario, southern Quebec, and Newfoundland. Its climate is characterized by higher precipitation, longer growing seasons, and less severe temperature extremes than the West Boreal Shield. Therefore, the climate is more favorable for forest growth. Extreme climates do occur in certain localities. The mean minimum temperature in January ranges from −7 to −27°C, and the mean maximum July temperature ranges from 21 to 24°C (Rowe 1972). Precipitation ranges from 81 to 135 cm annually in the East Boreal Shield, with areas of Newfoundland receiving 160 cm annually. The geology is similar to that of the West Boreal Shield.

In the southern East Boreal Shield, the fire cycle is longer and forests are typically closed crown. The southern region is within the clay belt that is characterized by endless stretches of black spruce over lowland flats and gently rising uplands. Tamarack is scarce in the southern portion, where a spruce and eastern white cedar mix is more common. Species from the Great Lakes–St. Lawrence region include eastern white pine, red pine, yellow birch, eastern hemlock, sugar maple, red spruce, black ash, and white elm. In the northern region, the fire cycle is shorter and upland sites are dominated by balsam fir. Species of the northern portion of the East Boreal Shield include white spruce, black spruce, balsam fir, jack pine, trembling aspen, tamarack, eastern white pine, balsam poplar, and eastern white cedar.

Hudson Plains

The Hudson Plains ecozone contains an extensive wetland complex extending west from Hudson Bay and south and west from James Bay. The area covers northeastern Manitoba, northern Ontario, and western Quebec. Large expanses of muskeg, swamp, and bog dominate the landscape, with a few areas of upland forest cover. Rivers run across the coastal plain, and their low alluvial banks support forest stands consisting of white spruce, balsam fir, trembling aspen, balsam poplar, and paper birch. In these well-drained alluvium soils and ridges, weak eutric Brunisol soil profiles are developed. Away from the rivers lie immense swampy, boggy areas that are poorly drained with gentle slopes.

The climate is similar to the Boreal Plains and Boreal Shield ecozones. However, the high water table, underlying permafrost, and low relief result in a colder and wetter environment. The bedrock is Paleozoic sediment covered mainly by marine clay. Beach sand deposits occur near the coasts of Hudson and James bays. The soils are mainly peat (organic Cryosol soils) with some raised beaches and river deposits. The ecozone is strongly affected by the cold wet Hudson Bay low air masses and the Polar high air masses. Climate is cold continental with 40–80 cm yr^{-1} precipitation. Winter mean daily temperatures are between −19 and −16°C and in summer range between 10.5 and 11.5°C. The average number of frost-free days is 70–100. Most of the area is treeless due to the harsh climate and highly acidic, nutrient poor soils. The vegetation appears subarctic due to the dominance of open woodlands of black spruce and tamarack in the muskegs. Often the organic surface is covered with a dense lichen cover. Dense sedge-moss-lichen covers dominate on the poorly drained areas. White elm, black ash, and eastern white cedar are sometimes scattered along the riverbanks. Jack pine is rarely present. From southeast to northwest, the amount of peatland and number of lakes increase, and the dominant forest becomes stunted open black spruce and tamarack. Black spruce does not appear to be well adapted to the coastal environment found along the bays. This area is dominated by white spruce.

Boreal Plains

The Boreal Plains ecozone includes the northeastern portion of British Columbia, northern half of Alberta, south central Saskatchewan, and southwestern corner of Manitoba. The Boreal Plains lie just to the north of the Prairie ecozone and have similar physical features but a cooler, moister climate. It differs greatly from the Boreal Shield because it is not located on the Canadian Shield and does not have many rock outcrops or lakes. The topography of this region is level to gently rolling, consisting primarily of glacial moraine and lacustrine deposits. Soils are primarily Luvisols with Black Chernozems in the south and Brunisols and Organics toward the north. The climate is moist with cold winters and fairly warm summers. The mean daily winter temperature ranges from −17.5 to −11°C, and the summer mean is from 13 to 15.5°C. The precipitation ranges from 30 to 62 cm yr^{-1}. The average number of frost-free days is 80–130 days yr^{-1}. This ecozone lies in the rainshadow of the moist Pacific orographical rainfall but is also influenced by continental climatic conditions.

The dominant conifers in this ecozone include white spruce, black spruce, jack pine, and tamarack. The deciduous species include white birch, trembling aspen, and balsam poplar. The deciduous species are most prevalent in the transitional regions to prairie grassland. Black spruce and tamarack become more important northward.

Alaska Boreal Interior

The boreal forests of Alaska are confined primarily to the interior region of the state, which is bounded to the north by the Brooks Range and to the south by the Alaska Range. The Alaska Boreal Interior ecozone has a continental climate with short, warm summers and long, very cold winters. There is much variation in temperature and precipitation from the western to eastern ranges of this ecozone. The maximum summer temperature increases from west to east (17–22°C), and the minimum winter temperature decreases (−18 to −35°C). Precipitation typically increases with elevation and ranges from 17 to 55 cm annually.

The topography includes flat bottomlands, rolling lowlands, and dissected plateaus with rounded low to high hills. The variable physiography plays an important role in determining forest cover given the influence of slope and aspect on ground temperature.

The ecozone was mostly unglaciated during the Pleistocene. The predominant geological features were from Mesozoic and Paleozoic sedimentary rocks and volcanic deposits, Quarternary alluvial and loess deposits, and alluvial and loess deposits of mixed origin. There are outwash gravel and morainal deposits in some parts of the Interior Bottomlands.

Soils are generally shallow and often underlain by permafrost. Permafrost is thin to moderately thick in the west and discontinuous in eastern Alaska but continuous in the Canada portion of this ecozone. There are few bedrock outcrops in this ecozone. Lakes are scattered throughout, and wetlands are abundant in the Yukon Flats, Interior Bottomlands, Old Crow Flats, and Old Crow Basin.

The Alaskan boreal forest has a high diversity of ecosystem types because of the topographical variability within the area (Viereck et al. 1986; Elliot-Fisk 1988). Forest ecosystems are characterized by differences in temperature, moisture, insolation, and disturbance (e.g., flooding or fire). Sites with underlying permafrost are covered with species adapted to cold soil temperatures, high soil moisture, and low nutrient availability (e.g., black spruce). By contrast, forests on well-drained and warmer sites, including south slopes and river levees where permafrost is absent, are generally dominated by white spruce. The early successional hardwood species, paper birch, trembling aspen, balsam poplar, willows, and alders (*Alnus* spp.) precede white spruce on these sites (Viereck 1983). Despite these generalizations, black spruce will occupy floodplain terraces as well as flat to rolling uplands of well- to poorly drained soils. However, these sites are generally less favorable and are often exposed or have very thin soils. The Yukon Flats, Interior Bottomlands, Old Crow Flats, and Old Crow Basin are dominated by low scrub bogs, fens, marshes, and meadows interspersed with a mix of forest types. Stunted black spruce and tamarack occur on better-drained portions of the Old Crow Flats and Old Crow Basin. White and black spruce forests, as well as a mixed type, dominate the bottomlands. Associates of tamarack, paper birch, willow, alder, and balsam poplar occur. Generally, mosses blanket the ground of the interior forests, but they are patchy in some areas. Most of the forests are closed-canopy types, but some open black spruce stands occur where drainage is very poor.

Disturbance, caused by fire and flooding, is important in shaping the forest ecosystems in Alaska's interior (Van Cleve et al. 1983). Flooding drives succession in the river floodplain ecosystems. As river alluvium is deposited, willow and alder colonize mineral deposits. Willow and alder succeed to balsam poplar and eventually to highly productive white spruce. With periodic flooding, permafrost is kept at bay, and white spruce can persist for 100 years or more (Viereck 1970). In the upland and lowland black spruce types, fire is instrumental in determining the distribution of forest ecosystems. The low annual precipitation, high summer temperatures, low relative humidity, frequent lightning storms, and trees adapted to fire make these forests prone to wildfire. The fire season is typically June–August; however, fires do occur earlier and later than this.

Distribution of Forest Cover and Forest Types

Detailed forest inventory data from across Canada were compiled into one large database by the Canadian Forest Service (CFS) in 1991 (Lowe et al. 1994; Gray and Power 1997). These inventory data have been used to describe the distribution of forests in the different boreal forest ecozones of Canada (Lowe et al. 1996) and the carbon distribution throughout the region (Penner et al. 1997). Although an equivalent effort has not been carried out for Alaska, a synthesis of forest types and cover as a function of different physiographic regions in the state was performed by Kasischke and co-workers (1995) based on field observations of Viereck and colleagues (1996) and Yarie (1983).

The inventory database of the CFS (Lowe et al. 1996) and that of Kasischke et al. (1995) were grouped into three broad categories (productive forest, unproductive forest, and nonforested lands) to obtain a general distribution of forest productivity by ecozone for boreal North America (Table 14.1). Unfortunately much of the unproductive land in Canada was not included in the CFS inventory. Thus, to generate the information in Table 14.1, a land cover category needed to be assigned to the noninventoried areas of Canada. For the Taiga Plains, Taiga Shield, and Boreal Shield ecozones, we assumed that all noninventoried land was proportionally split between the unproductive forest and nonforest categories. This split was based on the relative fraction of these categories within the areas that were inventoried. For the Hudson Plains ecozone, we assumed that 80% of the noninventoried land was in the nonforest category and the remaining 20% was in the unproductive forest category. Note that Table 14.1 presents the total land area (excluding water bodies greater than 100 ha in size) as well as average temperature ranges and precipitation for each ecozone.

In Table 14.2, we list the fraction of forest ecosystem types in productive versus unproductive forest categories by ecozone. The data for the productive forests are based on estimates of forest distribution summarized by Lowe et al. (1996). For the unproductive forests, we used the descriptions of forest ecosystems presented in this chapter.

Distribution of Aboveground Biomass

Two different sources were used to estimate the distribution of aboveground biomass throughout the North American boreal forest region. Penner et al. (1997) calculated aboveground biomass estimates for Canada based on the 1991 forest inventory datasets developed by the CFS (Lowe et al. 1996). Estimates for Alaska were based on data from Kasischke et al. (1995).

Lowe et al. (1991) developed a geographically explicit data set for aboveground biomass distribution based upon detailed forest inventory and the use of models. These models included allometric equations that converted stand level information on average tree height, diameter, and density (based on forest type and stand age distributions) to biomass estimates. The product available from

Table 14.1. Total Land Area, Fraction of Forest Cover, and Carbon Densities for the Forest Ecozones of the North American Boreal Forest Region[a]

Ecozone	Area (× 1,000 ha)	Fraction coverage			Mean annual temperature[b] (°C)	Mean annual precipitation[b] (cm)
		Productive forest	Unproductive forest	Nonforest		
Boreal Cordillera	60,621	0.30	0.33	0.37	1 to 5.5	30 to 150
Taiga Plains	58,415	0.30	0.58	0.12	−10 to −1	20 to 50
West Taiga Shield	52,612	0.06	0.53	0.41	−8	20 to 50
East Taiga Shield	67,021	0.15	0.44	0.41	−5 to −1	50 to 80
West Boreal Shield	71,720	0.60	0.33	0.07	−2 to 1.5[c]	43 to 76[c]
East Boreal Shield	98,847	0.70	0.23	0.07	−1 to 2[c]	81 to 135[c]
Hudson Plains	36,655	0.04	0.29	0.67	−4 to −2	40 to 80
Boreal Plains	67,653	0.52	0.25	0.23	−2 to 2	30 to 62
Alaska Boreal Interior	41,630	0.10	0.60	0.30	−3.5[d]	17 to 55[e]

[a]Note that total land area does not include water bodies greater than 100 ha in size.
[b]From Wiken (1986) except where noted.
[c]From Rowe (1972).
[d]From Slaughter and Viereck (1986).
[e]From Gallant et al. (1995).

Table 14.2. Summary of Forest-Type Distribution (Fractions) in the North American Boreal Forest Ecozones

Ecozone/forest type	Productive forest	Unproductive forest
Boreal Cordillera		
White spruce	0.55	
Lodgepole pine	0.35	0.30
Alpine fir	0.10	
Black spruce		0.70
Taiga Plains		
White spruce	0.90	
Lodgepole pine	0.10	0.20
Black spruce		0.80
West Taiga Shield		
White spruce	0.65	
Lodgepole pine	0.35	0.30
Black spruce		0.70
East Taiga Shield		
White spruce	0.70	
Lodgepole pine	0.25	0.30
Balsam Fir	0.05	
Black spruce		0.70
West Boreal Shield		
White spruce	0.75	
Jack pine	0.20	0.50
Balsam fir	0.05	
Black spruce		0.50
East Boreal Shield		
White spruce	0.50	
Jack pine	0.10	0.30
Balsam fir	0.10	
Mixed woods	0.30	
Black spruce		0.70
Hudson Plains		
White spruce	0.85	
Jack pine	0.15	
Black spruce		1.00
Boreal Plains		
White spruce	0.55	
Mixed deciduous	0.20	
Jack pine	0.25	0.20
Black spruce		0.80
Alaska Boreal Interior		
White spruce	1.00	0.20
Black spruce		0.80

this effort is a set of 47,000 polygons that are variable in size with the biomass data binned into five discrete classes.

Kasischke et al. (1995) estimated aboveground biomass for the Alaskan boreal forest region by extrapolating estimates from field observations over 64 distinct physiographic regions defined by Wahrhaftig (1965). The size of these regions ranged in size from 7100 to 6.9 million ha.

Distribution of Below-Ground Carbon

A soil carbon database has been developed for the Canadian boreal forest region (Lacelle 1997; Tarnocai 1997), which was recently expanded to include Alaska (Lacelle et al. 1997). This database is available in a format that can be directly imported into a geographic information system and consists of some 15,000 polygons. Each polygon contains a number of attributes; those of interest for carbon studies are total ground carbon content (in kg m^{-2}) and surface carbon content (carbon content of the top 30 cm of soil in kg m^{-2}).

Results

In Chapter 15, Murphy and colleagues present databases documenting the spatial and temporal distribution of fires throughout the North American boreal forest. These databases were combined with the ecozone boundaries to estimate average fire cycle for each region based on data collected between 1980 and 1994. These estimates are summarized in Table 14.3.

The relatively short 15-year record used to estimate the fire cycles in Table 14.3 results in an overestimation of fire cycle. For example, using a 49-year record

Table 14.3. Summary of Climate and Carbon Characteristics for the Different Ecozones of the North American Boreal Forest

Ecozone	Carbon density (t ha^{-1}) Aboveground	Carbon density (t ha^{-1}) Ground	Average area burned (ha yr^{-1})	Fire cycle (years)
Boreal Cordillera	32.9	90.1	153,064	396
Taiga Plains	27.3	125.9	326,190	179
West Taiga Shield	15.3	47.8	329,630	160
East Taiga Shield	11.5	82.3	206,599	324
West Boreal Shield	30.1	73.6	789,191	91
East Boreal Shield	35.9	109.5	109,265	905
Hudson Plains	29.1	118.8	72,406	506
Boreal Plains	34.8	92.9	373,480	181
Alaska Boreal Interior	22.7	90.4	225,741	184
Average	28.4	90.1	2,585,567	215

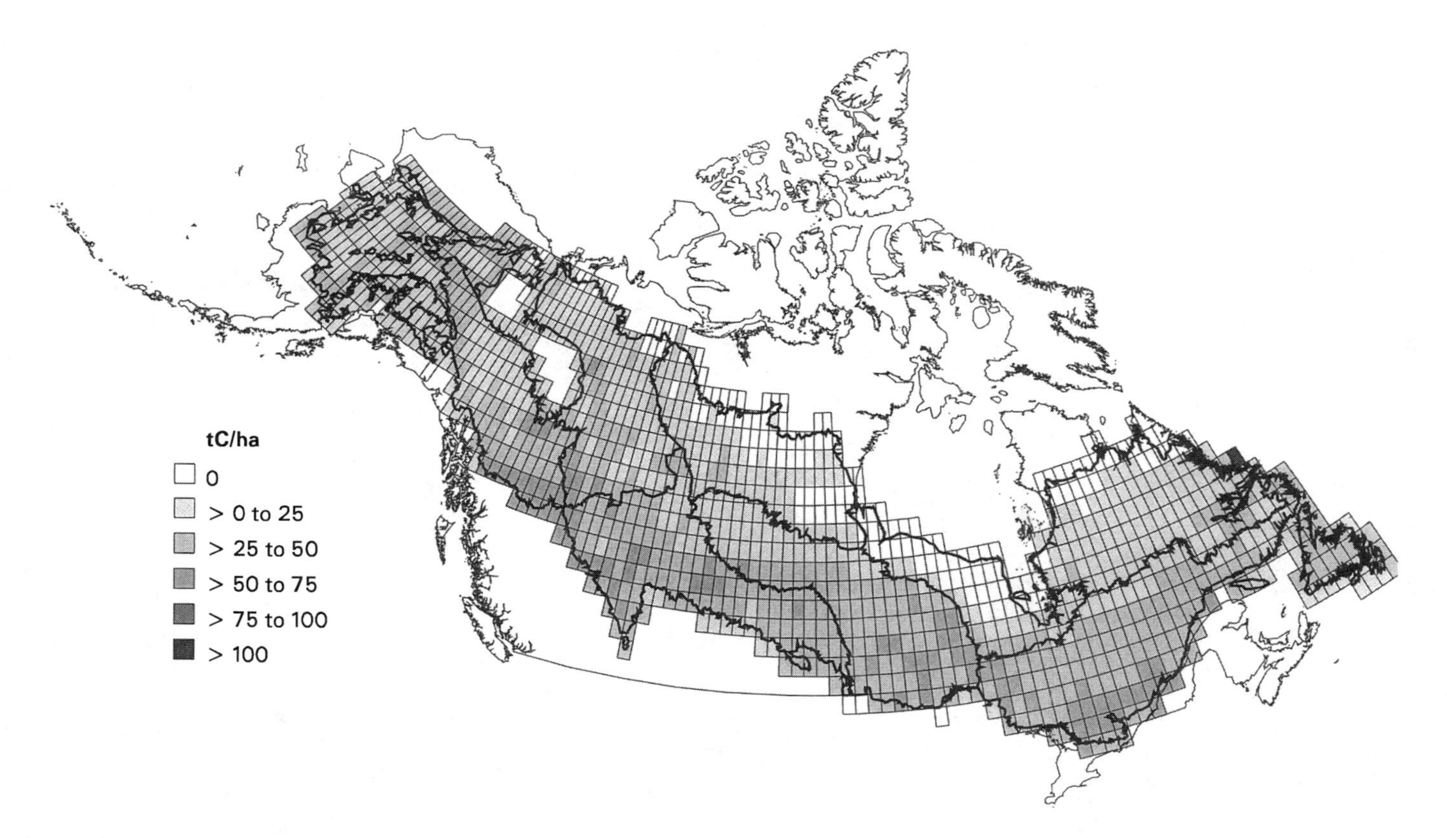

Figure 14.2. Distribution of aboveground biomass in the North American boreal forest ecozones. (Based on data from Lowe et al. 1996 and Kasischke 1995.)

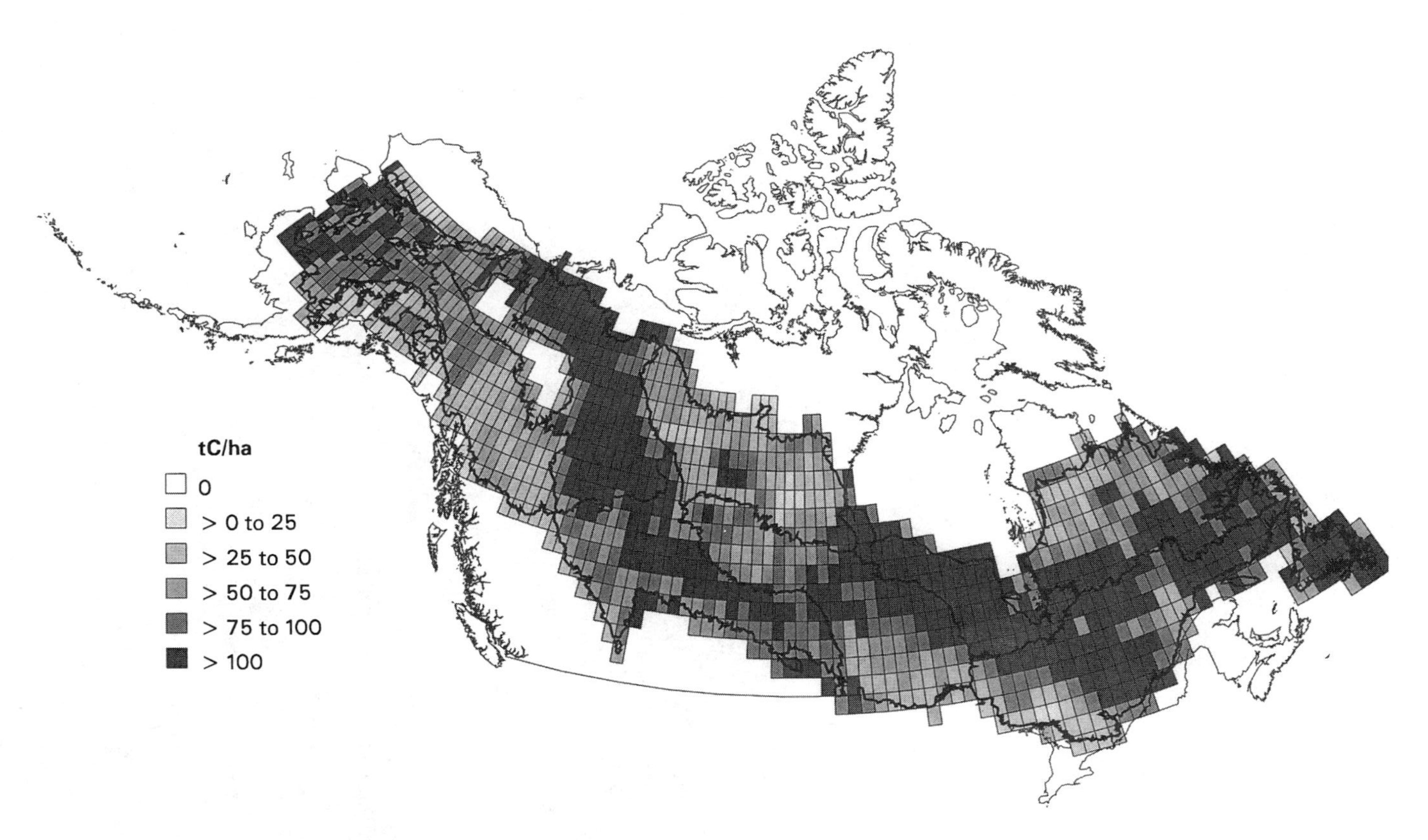

Figure 14.3. Distribution of carbon present in the top 30 cm of the soil horizon in the North American boreal forest ecozones. (Based on data from Tarnocai 1997 and Lacelle 1997.)

available for the state of Alaska results in an fire cycle of 172 years, compared with 253 years if only a 15-year record were used. Therefore, for Alaska the average fire cycle is 30% lower when a longer fire history is used. Table 14.3 shows that fire cycle is highly variable throughout the North American boreal forest, ranging from a low of 91 years for the West Boreal Shield to 905 years for the East Boreal Shield. For the entire boreal forest, the average fire return interval is 215 years. If we adjust this figure based on the bias established for Alaska (30%), then the average fire return interval would be 165 years for the North American boreal forest.

Figures 14.2 and 14.3 present the aboveground biomass and below-ground carbon data sets for the North American boreal forest zone. Table 14.3 summarizes the average carbon levels for each ecozone. The aboveground levels were estimated by using the assumption that vegetation is 50% carbon. The amount of carbon present throughout the region is highly variable. For the entire North American boreal forest zone, there is 15.2 Gt C in the aboveground vegetation and 49.2 Gt C in the ground layer.

Summary

The data sets presented in this chapter provide the basis for developing a better understanding of the dynamics of the carbon budget in the North American boreal forest region. For example, in Chapter 23 these data are used by Kasischke and coworkers to estimate direct carbon release from fires in this region. The data sets are useful for examining the spatial distribution of carbon release and also for understanding how this release varies on a yearly basis. In Chapter 25, these data are used to analyze how changes in future climate will influence carbon storage in this region.

References

Bray, J.R., and E. Gorham. 1964. Litter production in forests of the world. *Adv. Ecol. Res.* 2:101–157.

Ecoregions Working Group. 1989. *Ecoclimatic Regions of Canada, First Approximation.* Canadian Committee on Ecological Land Classification. Ecological Land Classification Series 23. Sustainable Development Branch, Canadian Wildlife Service, Environment Canada, Ottawa, Ontario, Canada.

Elliot-Fisk , D.L. 1988. The boreal forest, pp. 33–62 in M.G. Barbour and W.D. Billings, eds. *North American Terrestrial Vegetation.* Cambridge University Press, Cambridge, UK.

Gallant, A.L., E. F. Binnian, J.M. Omernik, and M.B. Shasby. 1995. *Ecoregions of Alaska.* U.S. Geological Survey Professional Paper 1567. U.S. Government Printing Office, Washington, DC.

Gray, S.L., and K. Power. 1997. *Canada's Forest Inventory 1991: The 1994 Version—Technical Supplement.* Information Report BC-X-363. Pacific Forestry Centre, Canadian Forest Service, Victoria, BC, Canada.

Kasischke, E.S., N.H.F. French, L.L. Bourgeau-Chavez, and N.L. Christensen, Jr. 1995. Estimating release of carbon from 1990 and 1991 forest fires in Alaska. *J. Geophys. Res.* 100:2941–2951.

Lacelle, B. 1997. Canada's soil organic carbon data base, pp. 93–101 in R. Lal, J.M. Kimble, R.F. Follet, and B.A. Stewart, eds. *Soil Processes and the Carbon Cycle.* CRC Press, Boca Raton, FL.

Lacelle, B., C. Tarnocai, S. Waltman, J. Kimble, F. Orozco-Chavez, and B. Jakobsen. 1997. *North American Soil Carbon Map (Provisional).* Unpublished report. Agriculture and Agri-food Canada, USDA, INEGI, Institute of Geography, University of Copenhagen.

Lowe, J.J., K. Power, and S.L. Gray. 1994. *Canada's Forest Inventory 1991.* Information Report PI-X-115. Petawawa National Forestry Institute, Canadian Forest Service, Chalk River, Ontario, Canada.

Lowe, J.J., K. Power, and M.W. Marsan. 1996. *Canada's Forest Inventory 1991: Summary by Terrestrial Ecozones and Ecoregions.* Information Report BC-X-364E. Pacific Forestry Centre, Canadian Forest Service, Victoria, BC, Canada.

Penner, M., K. Power, C. Muhairwe, R. Tellier, and Y. Wang. 1997. *Canada's Forest Biomass Resources: Deriving Estimates from Canada's Forest Inventory.* Information Report BC-X-370. Pacific Forestry Centre, Canadian Forest Service, Victoria, BC, Canada.

Rowe, J.S. 1972. *Forest Regions of Canada.* Publication 1300. Department of the Environment, Canadian Forestry Service, Ottawa, Canada.

Slaughter, C.W., and L.A. Viereck. 1986. Climatic characteristics of the taiga in interior Alaska, pp. 9–21 in K. Van Cleve, F.S. Chapin III, P.W. Flanagan, L.A. Viereck, and C.T. Dyrness, eds. *Forest Ecosystems in the Alaskan Taiga.* Springer-Verlag, New York.

Tarnocai, C. 1997. The amount of organic carbon in various soil orders and ecological provinces in Canada, pp. 81–92 in R. Lal, J.M. Kimble, R.F. Follet, and B.A. Stewart, eds. *Soil Processes and the Carbon Cycle.* CRC Press, Boca Raton, FL.

Van Cleve, K., C.T. Dyrness, L.A. Viereck, J. Fox, F.S. Chapin III, and W. Oechel. 1983. Taiga ecosystems in interior Alaska. *BioScience* 33:39–44.

Viereck, L.A. 1970. Forest succession and soil development adjacent to the Chena River in interior Alaska. *Arct. Alp. Res.* 2:1–26.

Viereck, L.A. 1983. The effects of fire in black spruce ecosystems in Alaska and northern Canada, pp. 201–220 in R.W. Wein and D.A. MacLean, eds. *The Role of Fire in Northern Circumpolar Ecosystems.* John Wiley & Sons, New York.

Viereck, L.A., K. Van Cleve, and C.T. Dyrness. 1986. Forest ecosystem distribution in the taiga environment, pp. 23–43 in K. Van Cleve, F.S. Chapin III, P.W. Flanagan, L.A. Viereck, and C.T. Dyrness, eds. *Forest Ecosystems in the Alaskan Taiga.* Springer-Verlag, New York.

Wahrhaftig, C. 1965. *Physiological Divisions of Alaska.* Professional Paper 482. USDI Geological Service, Washington, DC.

Wiken, E. 1986. *Terrestrial Ecozones of Canada.* Ecological Land Classification Series 19. Lands Directorate, Environment Canada, Ottawa.

Yarie, J. 1983. *Forest Community Classification of the Porcupine River Drainage, Interior Alaska, and Its Application to Forest Management.* U.S. Forest Service Technical Report PNW-154. U.S. Department of Agriculture, Portland, OR.

15. Historical Fire Records in the North American Boreal Forest

Peter J. Murphy, James P. Mudd, Brian J. Stocks,
Eric S. Kasischke, Donald Barry, Martin E. Alexander,
and Nancy H.F. French

Introduction

There is currently a great deal of interest in understanding and quantifying the extent of natural and human-caused fire in the different biomes throughout the world. In Chapter 8, Shvidenko and Nilsson examined the fire statistics for the Russian boreal forest region and showed that there is much uncertainty in these data because of the management and monitoring practices in this country. Although a much more accurate record for fire activity exists for the North American boreal forest region, this data set is still incomplete. In this chapter, we discuss the pedigree of the data sets that comprise the North American boreal forest record so that their limitations can be understood. Finally, the recent advances in geographic information systems (GIS) and their adoption by natural resource management agencies have led to the creation of databases that can be used to examine the spatial extent of fires in more detail. In this chapter, we also discuss the use of these spatial data sets for examining fire in the North American boreal forest.

Historical Perspectives on Fire

The North American boreal zone is large and diverse, covering nearly 5.5 million km^2 and extending over 5,000 km from east to west and 1,500 km from south to north. The size, accessibility, and political boundaries throughout this region have contributed to the evolution of different approaches to forest fire management and maintenance of fire records in Canada and Alaska.

Virtually all forested areas in Canada and Alaska have developed under the influence of recurrent disturbances, of which fire has been the most pervasive and influential. Besides natural ignition sources (mostly from lightning), aboriginal people also used fire to enhance their hunting, gathering, and travel opportunities. For example, Col. S.B. Steele who led the Northwest Mounted Police into western Canada reported in 1874 (Steele 1915): "Indians . . . willfully set the prairies on fire [in the autumn] so that the bison would come to their part of the country to get the rich green grass which would follow in the spring." Although the actual amount of fires started by native people has been exaggerated by some authors (Lutz 1959), there is little doubt that they represented a source of fire ignition.

To newly arriving Europeans, forest fire was an alien phenomenon. Early travelers typically referred to fires as dangerous and destructive and to burned areas as dark, dismal, and desolate. This viewpoint was representative of a European mindset that strongly influenced fire control policies. Fires were a real problem to European settlers, and given the continuity of fuels available in Alaskan and Canadian forests and prairies, wildfires could become very large. Textbooks list several of the record U.S. fires, such as the 1825 Miramichi-Maine fire of 1.2 million ha, the 1871 Michigan and Peshtigo fires of 1.0 and 0.5 million ha, respectively, and the 1910 Great Idaho fire of 1.2 million ha. All these fires resulted in significant loss of human life and property. Large fires in Canada were not as well documented but were recorded in an anecdotal manner:

> The Plains are, and have been the several days past, burning in a most dreadful manner. Fires are raging in all directions, and the sun obscured with smoke that covers the whole country, and should the remarkable dry weather which has now continued so long, not change very soon, the Plains must be burned to such an extent as to preclude all hopes of our getting a large supply of dry provisions, for which appearances on our arrival here were very flattering. (*Hudson's Bay Company Journal,* Fort Edmonton, October 12, 1812)

> From beyond the south branch of the Saskatchewan to Red River all the prairies were burned last autumn, a vast conflagration extended for 1,000 miles in length and several hundreds in breadth. The dry season had so withered the grass that the whole country of the Saskatchewan was in flames . . . we traced the fire from the 49th parallel to the 53rd, and from the 98th to the 108th degree of longitude. It extended no doubt to the Rocky Mountains. (Hind 1860)

These comments suggest that the incidence of recurrent and large fires was not unusual. Explorers, trappers, missionaries, and others entering Alaska during this period also made frequent reference to widespread fire activity in diaries and other reports (Lutz 1959). Fires were so extensive that they may have been viewed as part of the scenery. For example, fire was used as a backdrop in many historical photographs dating to the turn of the century. Lake core sediments taken for pollen analysis also show the presence of charcoal and recurrent spikes of greater infusions of charcoal (Chapter 9), again suggesting that fire was a frequent visitor with recurrent spikes of higher-intensity burns. However, with the apparent reduction of fire incidence in some areas (especially in forests in areas of high population density), perceptions have also changed, so that now the green forest is seen as normal and disturbances as unusual.

Evolution of Fire Management and Monitoring in the Twentieth Century

The creation of the databases used to estimate the area burned in the North American boreal forest region is closely tied to fire management activities. The records maintained by the various fire agencies in Canada and Alaska provide the basis for fire statistics for this region. In this section, we discuss the evolution of the various fire management agencies in Canada and Alaska to provide a perspective on how the fire records evolved.

Canada

When the Dominion Forestry Branch took responsibility for forest management on the Dominion lands of Canada in 1899, fire control was one of their major mandates. This centralized agency was in the position to develop and maintain records and allowed MacMillan (1909) to publish the first record of forest fires in Canada. With the objective of controlling forest fires, it was important to establish a baseline against which to measure fire control effectiveness. In addition, this record could be used for public awareness to illustrate the magnitude of the fire problem.

Although the Canadian government by 1918 was creating a more complete record of annual fire activity, it is important to realize the limitations of this record. At that time, fires were fought primarily with hand tools, with the portable power pump just invented. Detection was very limited, with fire patrols typically undertaken by canoe. Access to fires was accomplished by using canoe, foot, and horse. Vehicular travel was available only on a limited number of roads. The Depression of the 1930s and World War II during the 1940s greatly limited the resources applied to effective fire management in Canada. As a consequence, major efforts in fire monitoring and control began in earnest in the 1950s. This increased effort also coincided with the expansion of the use of aircraft for fire monitoring and suppression.

With the expanding use of Canadian forests for both industrial and recreational purposes throughout the twentieth century, forest fire activity also increased dramatically. There is little doubt that in areas of higher populations, there was an increase in human ignitions (Chapter 19). There was also an expanding awareness of the large number of lightning fires in previously unmonitored regions of the country. The fire management capability mobilized to deal with increasing fire activity was expanded accordingly, with improved fire occurrence prediction and detection capabilities and increasingly more effective initial attack strategies heavily dependent on water-bombing aircraft and helicopters.

As provincial and territorial fire management agencies developed across Canada, they began compiling fire report records on all fires within their jurisdiction. Today, all provinces and territories have responsibility for reporting and managing the fires within their boundaries. The first "national" statistics were reported by Dwight (1918), and this led to periodic summaries over the next few decades. By 1940 there were sufficient data that Wright (1940) was able to present a graph showing the annual number of forest fires and total area burned in Canada and to postulate long-term trends. Beginning in 1948, annual fire statistics were summarized in publications produced by the federal forestry organization of the time, culminating in a series of reports by the Canadian Forest Service in the 1980s and early 1990s (Higgins and Ramsey 1992). Since 1992, Statistics Canada has summarized annual fire statistics as part of a larger report on the state of Canada's forests.

Lightning accounts for 35% of Canada's fires, yet these fires result in 85% of the total area burned (Stocks 1991). This is because lightning fires occur randomly and in remote regions; therefore, they present access problems usually not associated with human-caused fires. As a result, lightning fires generally grow larger, as detection and subsequent initial attack are often delayed.

A recent evaluation of Canadian fire statistics (Stocks 1991) also identified some of the reasons why Canadian fire impact varies significantly, both geographically and between years. Sophisticated fire management programs are largely successful at controlling the vast majority of forest fires at an early stage, such that only 2–3% of fires grow larger than 200 ha in size, but these fires account for 97–98% of the area burned across Canada. In addition, the practice of modified or selective protection policy in remote regions of Canada results in many large fires in low-priority areas being allowed to carry out their natural function. Recent studies comparing fire sizes relative to levels of protection indicate that, on average, fires in the largely unprotected regions of the boreal zone are much larger than fires in intensively protected regions (Stocks 1991; Ward and Tithecotte 1993).

Alaska

The transition of Alaska from a territory to a state in 1959 has led to complex history of fire-management jurisdiction in this region. Organized forest fire control in the Alaskan Territory began in July of 1939 when the U.S. Congress appropriated $37,500 to establish the Alaskan Fire Control Service (AFCS) under the General Land Office of the Department of the Interior. In 1946, the General Land Office and the U.S. Grazing Service were combined to form the Bureau of Land Management (BLM). The AFCS was abolished in 1947, and its duties were transferred to the Branch of Timber and Resource Management under the regional administrator of the Alaska BLM. Its budget (which had grown to $170,000 under the AFCS) was cut to $28,000. By 1949, the budget of the new BLM Division of Forestry increased to $140,000. In 1950, the Division of Forestry acquired its first airplane. In the early 1950s, the first use was made of helicopters in Alaska. In 1957, the Division of Forestry designated two operational districts—the Anchorage and Fairbanks districts.

When Alaska was granted statehood on January 3, 1959, the BLM Division of Forestry began work on a protection agreement with the State of Alaska's Department of Natural Resources. Under this agreement, BLM became a contractor providing fire protection for the state. In 1958, the BLM initiated a pilot program dropping chemical retardants on fires in Oregon and Alaska. In 1959, the first smoke-jumper force was established in Alaska. By the early 1960s, the BLM owned an expanding fleet of aircraft—four Cessna 180s, three Gruman Gooses, a DC-3, and a P-51 mustang. In 1960, the BLM Division of Forestry began to provide organized training for fire crews who were hired from Alaska villages.

The BLM fire suppression organization remained within the operations districts in Fairbanks and Anchorage throughout the 1970s. In 1970, the BLM Division of Forestry was replaced by the BLM Division of Fire Control, and in 1977 the name was changed again to the Division of Fire Management. In 1971, the Alaska Native Claims Settlement Act was passed by the U.S. Congress, which allowed for the selection of 52 million acres by Alaska Native Regional and Village Corporations. The State of Alaska began to provide for protection of its own lands in 1976 by establishing its own wildland fire-fighting organization. In 1980, the Alaska National Interest Lands Conservation Act was passed by the Congress, which allocated 104.1 million acres to U.S. National Parks and U.S. Fish and Wildlife Service refuges, in addition to extending the time limit for the State of Alaska to select the 104 million acres of land granted to it under the statehood act.

In 1978, the Alaska Land Managers' Cooperative Task Force created a fire subcommittee to develop and coordinate an interagency fire program and organization for the state. The first fire management plan to be developed was the Forty Mile Area Fire Plan in 1979. By 1980, a Fire Planning Working Group was formed to coordinate fire planning throughout the state. This group incorporated the changing views of the various land management agencies about the role of wildfire.

During the late 1970s and throughout the 1980s, fire suppression activities in Alaska changed greatly. The State of Alaska began to develop its own wildfire suppression infrastructure, beginning with an organization on the Kenai Peninsula in south central Alaska and expanding along the populated roadside areas throughout the state. By 1981, the State provided protection in the Anchorage and Copper River areas and part of the Fairbanks district. In 1982, the Alaska Fire Service was formed to provide for protection for the agencies of the U.S. Department of the Interior—the BLM, the National Park Service, the U.S. Fish and Wildlife Service, and the Bureau of Indian Affairs. Protection of lands selected by the native corporations formed by the Alaska Native Claims Settlement Act and lands under the management of the military were also included under the umbrella of the Alaska Fire Service. Throughout the 1980s and 90s, the State of Alaska and the Alaska Fire Service have entered into cooperative agreements with each other. This arrangement has resulted in reduced costs to each organization through the sharing of tactical resources and the mutual protection of each other's lands. In general, the State of Alaska protects all land south of 64° north latitude and the Alaska Fire Service protects all land to the north of that line. The State of Alaska protects the roadside areas along the major highways, with the exception of the oil pipeline corridor extending from Prudhoe Bay in the extreme north of Alaska to the just south of the Yukon River crossing north of Fairbanks. The Alaska Fire Service protects roadless areas, with the exception of the southwest area of Alaska for which the State assumed protection responsibility in 1985. This pattern of protection responsibilities has remained in a fairly stable form since 1985.

Spatial Fire Databases

Canada

Map Generation

Although some provincial fire history map summaries have been produced (Donnelly and Harrington 1978; Delisle and Hall 1987), by the late 1980s there was a growing need to represent the Canadian forest fire situation in a broader context. This information was required to address emerging issues such as global biomass burning/atmospheric chemistry and climate change and led to recognition of need for a spatial and temporally explicit national database of large Canadian fires. To achieve this end, fire report data and maps for all fires greater than 200 ha were collected from all provincial and territorial fire management agencies and Parks Canada. Although some agencies provided digital databases, in all other cases the fire maps were digitized. All data sets were entered into a GIS database along with appropriate attribute files for each fire. This large-fire database (LFDB) now includes about 5,000 fires covering the 1980–1995 period. It is being updated annually. In addition, the LFDB is being extended back in time, with the goal to map all large fires nationally back to about 1950. Analysis of this database shows that some records are still missing. Efforts are underway to track down these missing data.

A preliminary version of the LFDB was used in an examination of the spatial distribution of all 1980–1989 Canadian fires greater than 200 ha (Stocks et al. 1996). This analysis showed that by far the greatest area burned occurred in the boreal region of west central Canada and attributed this to a combination of fire-prone ecosystems, extreme fire weather, lightning activity, and reduced levels of protection in this region.

Limitations in the Forest Fire Record

Although the Canadian fire record extends back to 1918 and the official Alaskan record begins in 1940, there are many reasons to believe that these records are much more incomplete than originally thought. In Canada, it had been surmised that the fire record was incomplete primarily with respect to the jurisdictions of the Yukon and Northwest Territories, which did not begin contributing to the national database until 1946, and Newfoundland, which was added to the national record in 1949. In reality, there were deficiencies in all records, in large measure because the areas protected were only expanded as resources became available, meaning large areas were only sporadically monitored or not monitored at all. In mid-Canada, for example, Kiil (1979) stated that "fire statistics prior to 1945 are based on less than one-half of the 1,404,000 km^2 being protected today."

In a paper prepared for a 1996 National Workshop on Wildland Fire Activity in Canada (Simard 1997), Murphy and associates (1996) chronicled in great detail the deficiencies in the Canadian record. To emphasize the incompleteness of the official record, they referred specifically to (1) the lack of complete fire records for northern British Columbia prior to 1960 (J. Parminter, personal communication); (2) the sparsity of fire records in northern Alberta prior to 1951 (Delisle and Hall 1987); (3) the lack of data on fires in the northern zone of Saskatchewan and Manitoba before 1950; (4) a map of Ontario fires back to 1920 showing no fires above 52° north latitude prior to 1978 (Donnelly and Harrington 1978; (5) the reporting of fires only around communities in the Northwest Territories prior to 1964; (6) northern Quebec fires not being reported prior to 1980; and (7) the fact that Newfoundland and Labrador fire records were not considered complete until 1948 and 1958, respectively (Wilton and Evans 1974). They also cited a number of documented and significant fires not included in the national statistics. Some examples are (1) a series of fires covering approximately 8 million ha in central Alberta and Saskatchewan in 1919; (2) a 1-million-ha fire in northern British Columbia and the Northwest Territories in 1942 (Holman 1944); and (3) a 1.4-million-ha fire in northern British Columbia and Alberta in 1950 (Murphy and Tymstra 1986).

By the mid-1970s, satellite coverage of Canada was becoming routine, and there is general agreement that the post-1975 fire record is reasonably complete. At this time, however, policy implications were coming to the fore, and fire-fighting budget constraints, coupled with a growing awareness of the ecological role of natural fire in northern forests, led to a policy in which fires were fought based on "values-at-risk." This translates into fires in remote regions being fought only when communities or resources are threatened and is a large contributing factor to the substantial areas burned in Canadian forests since 1980.

Alaska

Map Generation

Permanent fire records in Alaska generally contain a page of statistical information about the fire (e.g., ownership, location, size, weather parameters, suppression data, cause data), a map of the fire, and a narrative of actions taken. There can also be a variety of different reports attached to the permanent record for a particular fire event such as monitoring reports, escaped fire situation analysis reports, and other supporting documents. The permanent fire record is actually derived from an active fire file compiled during the fire season. This active file often contains hand-written notes as well as preliminary maps of the fire boundary at different points of time. At the end of the fire season, the permanent fire record is created. During this time, parts of the fire record are sometimes lost or misplaced, including the fire boundary maps. Thus, many permanent records include only the general location and size of a specific fire event.

Three techniques are used to generate a map of a fire location and boundaries: (1) ground surveys; (2) airborne surveys; and (3) interpretation of aerial or satellite imagery. Although it may seem that ground or foot surveys would produce inaccurate fire boundaries, in fact, maps generated in this fashion can be very accurate. Much of Alaska lies in hilly or mountainous terrain containing distinct drainage basins. In addition, other geographic features such as rivers, streams, lakes, and roads provide easily documented reference points. Using topographical maps, the position of a fire within stream and river basins (including elevation) and relative to other distinct geographic features (ponds, lakes, and mountains) can often be determined from ground surveys. Similar references are used during aerial surveys, with the advantage of having a vantage point with a more synoptical field of view. The accuracy of surveys is now being improved through the use of global positioning system (GPS) technology.

Prior to 1970, most fire perimeters and locations were drawn by hand on plain paper (i.e., a standard baseline map was not used). Since 1970, this practice was discontinued, and the boundaries and locations of fires were drawn on U.S. Geological Survey (USGS) maps at a scale of 1:250,000 or 1:63,360. The current practice is to map fire on foot or by detection aircraft, drawing the fire boundaries directly on USGS quadrangle maps. The use of maps derived from satellite imagery has also begun. In the 1940s, aircraft were not readily available for use in mapping fires. Most fires that were fought at that time were probably mapped on foot, and large remote interior fires were probably not mapped at all.

The fire boundary maps within the Alaskan data archive can be given an overall quality rating of poor, fair, and good. Figure 15.1 presents examples of the different quality of fire boundaries available in this data archive:

- *Poor.* Fire perimeters are drawn using generic shapes (e.g., a circle, an ellipse, a star, and no map scales are provided. Inaccurate coordinates are often recorded, and no named geographic references are provided or generic geographic reference terms (e.g., *river, lake*) are used (Fig. 15.1a).
- *Fair.* The map has a delineated scale and provides at least one named geographic reference. Sometimes the coordinates are in error. The perimeter boundary does conform to the general outline of the fire (Fig. 15.1b).
- *Good.* The map has a delineated scale with a distinctly drawn fire perimeter. It contains a number of geographic reference points, and the reference latitude/longitude coordinates for the fire location are consistent with these features (Fig. 15.1c).

In the older maps in the fair and poor categories, several additional analyses were undertaken prior to digitization of the fire boundary. In the fair category, it was

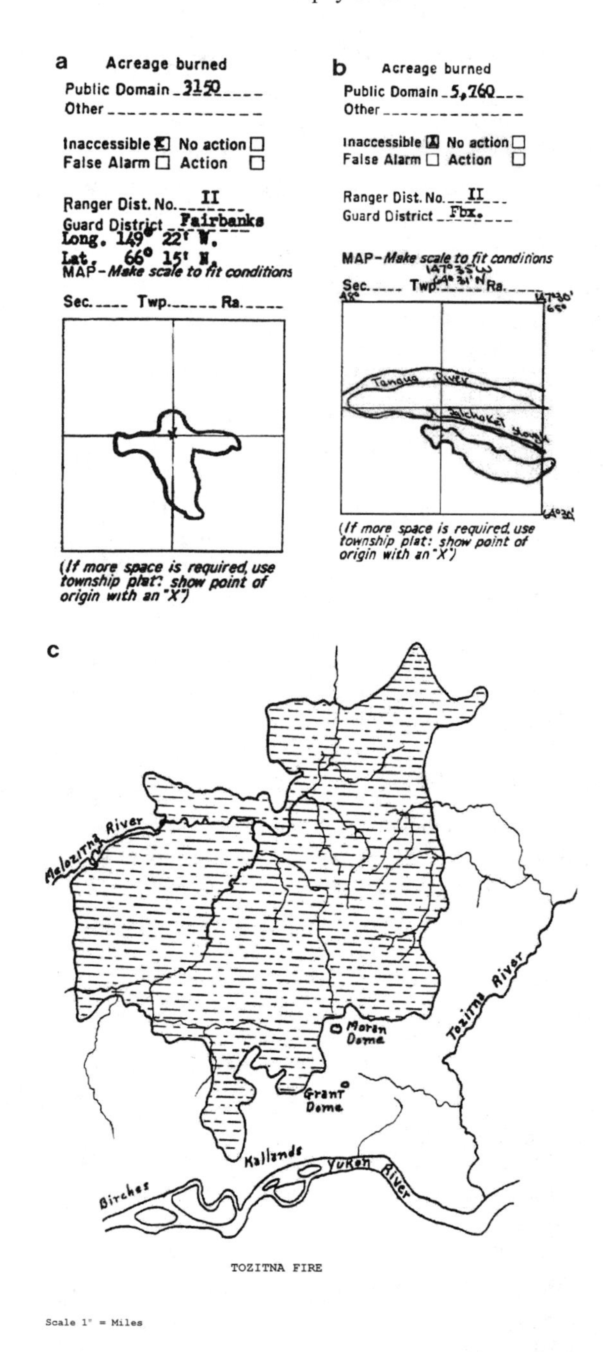

Figure 15.1. Examples of maps of fire boundaries from records of the Alaska Fire Service illustrating (a) poor quality; (b) fair quality; and (c) good quality.

often necessary to correct the location of the map relative to the named geographic features identified in the written portion of the permanent record. In addition, the fire perimeter was redrawn on the actual USGS map of the area so that it was consistent with the named geographic features. When possible, the same procedure was used for the poor-quality maps, but in many cases, a best guess had to be made with respect to the actual location of a fire boundary.

Quality of Data Within the Permanent Fire Records

One of the major problems with the Alaska LFDB is the presence of missing fire maps within the permanent fire records. Missing maps account for 15% of the total fire area summarized in Figure 15.2. Most (80%) of the missing data records are from 5 years—1954, 1966, 1968, 1969, and 1971. The amount of missing data has decreased significantly since 1950. For the 1950s, 14% of the total fire area reported has missing records. During the 1960s, several permanent data records were lost, resulting in 57% of the total fire area reported for this time period having missing data records. In the 1970s, this figure fell to 12%, and in the 1980s it fell even farther, to 8%. Finally, during the 1990s, maintenance of permanent records has improved to the point at which only one fire record (representing 0.2% of the total fire area) has been lost.

In terms of the overall quality of the data within the Alaska LFDB, the following assessment is presented.

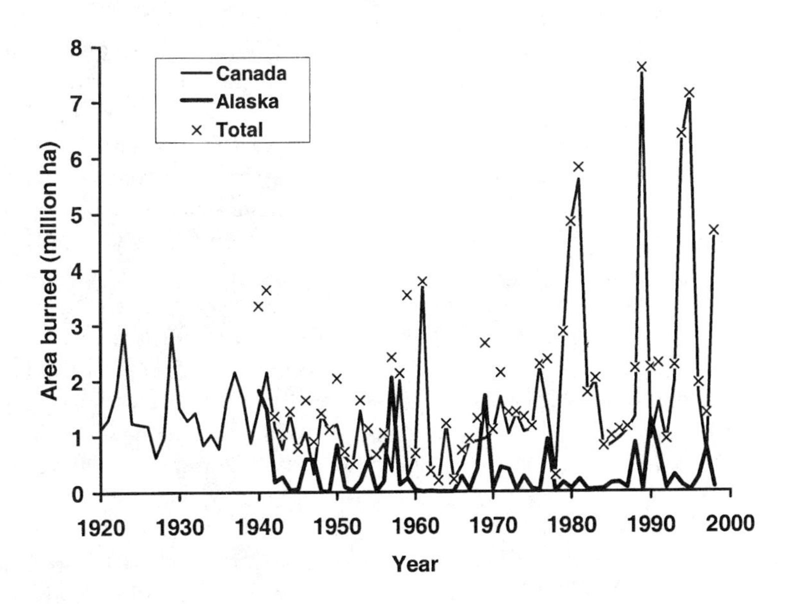

Figure 15.2. Annual patterns of area burned in Canada and Alaska. Analyses have shown that more than 95% of the area burned in Canada occurs in its boreal forest ecozones.

- *1950–1959.* The overall quality of the fire maps is fair, ranging from poor to good in individual years. Most large fires are well documented. There are questions as to whether all fires were actually detected and mapped during this time period.
- *1960–1969.* Although the quality of the maps improves, as noted above, many permanent data records are missing for this time period. During this time period, aerial monitoring activities increased and the likelihood of missed fires decreased.
- *1970–1979.* Although record maintenance procedures improved during this time period, some maps for larger fires are poor or missing.
- *1980–1997.* The quality of data maps for this time period is assessed to be very good, with mapping of fire boundaries being very consistent and the number of missing data records being very low.

Analysis of the Fire Record

The data records that have been compiled for the North American boreal forest region allow for a number of different analyses to be performed. In this section, we examine some of the longer-term trends within the overall record, as well as use the Alaskan and Canadian LFDB to examine the spatial and temporal characteristics of these fires. This latter data set provides the basis for the carbon budget assessments presented in Chapters 21 and 25.

Figure 15.2 presents a plot of annual area burned for Canada since 1920 and for Alaska since 1940 based on the best reconstruction of fire statistics from both regions. For the most part, because of its much larger forest area, the overall trend of fire activity in the North American boreal forest region follows the Canadian fire record. It is interesting to note, however, that in 15% of the years since 1940, the area burned in Alaska was greater than that burned in Canada. One of the emerging patterns in the North American boreal forest fire record since 1970 is the occurrence of episodic fire years. Since 1970, the average area burned during the 6 greatest fire years was 6.2 million ha yr^{-1}, with only 1.5 million ha yr^{-1} burned in the remaining years.

Figure 15.3 presents a plot of average area burned per year for different decades since 1920. This plot clearly shows there has been a trend in fire activity in the North American boreal forest over the past half century, with an apparent dip in area burned during the 1950s and 1960s followed by a steady increase over the past three decades. It is interesting to note that without inclusion of the Alaska data set, there would be a slight decrease in fire activity during the 1990s relative to the 1980s.

Most analyses of longer-term fire data in the North American boreal forest region have focused on Canada. The trends noted by previous researchers (Van Wagner 1988; Stocks 1991; Auclair and Carter 1993; Kayll 1996) are consistent with those presented in Figure 15.3. These researchers have all shown that Canadian fire occurrence remained relatively constant at about 5000 fires annually

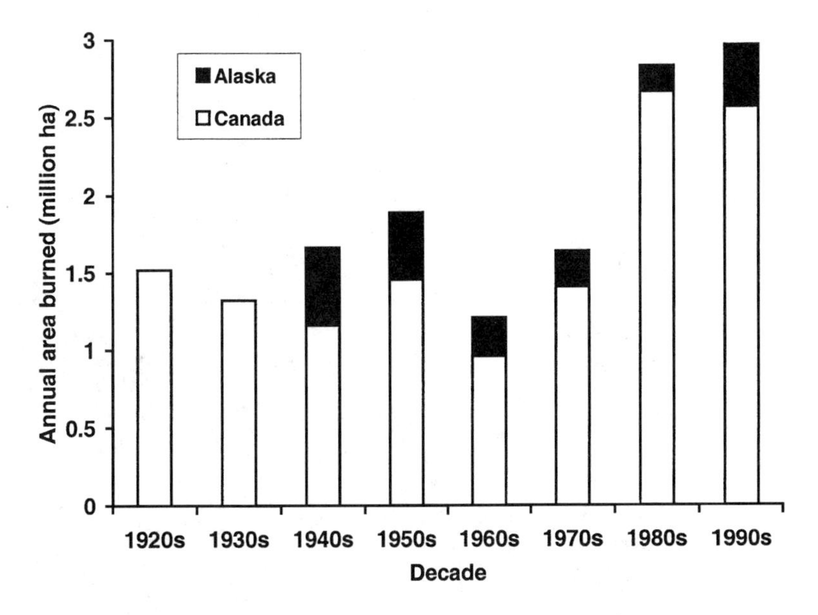

Figure 15.3. Decadal average area burned for Canada and Alaska.

prior to 1960 and has increased to close to 10,000 fires a year in the 1980s and 1990s, with little interannual variation. The apparent dramatic increase in area burned in recent decades has prompted speculation that climate change may be a factor. However, despite the fact that average temperatures across the North American boreal zone have increased sharply over the past 30 years (see Fig. 1.1), the incompleteness of the fire data record prior to 1970 does not permit a correlation of fire occurrence and climate over longer time periods.

Figure 15.4[1] presents the fire boundaries contained within the Alaskan and Canadian LFDBs for the period of 1980–1994. Analysis of these data shows that 97.5% of the total area burned during this time occurred within the boreal forest ecozones of North America (Fig. 14.1). Figure 2.2 presents the fire boundaries for the period of 1950–1998 for Alaska. These figures illustrate how the use of GIS allows for a more in-depth analysis of the spatial and temporal patterns of fire in this region. For example, the fires in Alaska are clearly concentrated in the boreal forest region, bounded on the north by the Brooks Range and on the south by the Alaskan Range.

Combining the LFDB with other geographic layers allows for examination of other characteristics of fire. For example, this database can be spatially partitioned by using the ecozones identified for the North American boreal forest (Fig. 14.1). This allows for the determination of fire characteristics for these regions (Table 14.3), as well as examination of yearly fire totals for each region (Fig. 15.5).

[1]Figure 15.4 will be found in the color insert.

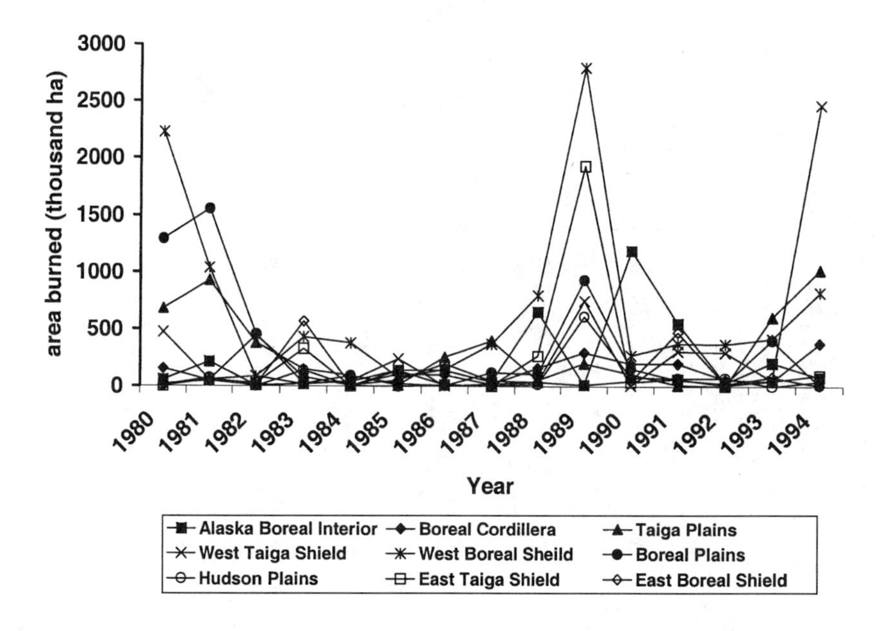

Figure 15.5. Annual area burned for the different ecozones of the North American boreal forest for 1980–1994.

Figure 15.5 reveals that there are distinct regional trends in annual area burned that capture the same theme illustrated in Figure 15.1: a significant portion of specific regions burns over very short time periods (1–2 years) during episodic fire events. For example, in 1980 and 1981, 4.2% of the total forest area of the boreal plains ecozone burned, as did 4.6% of the forest area of the West Boreal Shield, and 2.7% of the Boreal Cordillera. In 1989, 3.9% of the total forest area of the West Boreal Shield and 2.9% of the total land area of the East Taiga Shield burned. In 1990 and 1991, 4.2% of the Alaska Boreal Interior burned. In 1993 and 1994, 2.8% of the Taiga Plains burned. Finally, in 1994, 4.7% of the West Taiga Shield burned.

Summary

The databases discussed in this chapter present the basis for studying and understanding the spatial and temporal patterns of fire in the North American boreal forest. Figures 2.2, 15.4, and 15.5 make it exceedingly obvious that fire is one of the major disturbance forces in this region and has the ability to affect large segments in the landscape over very short periods of time. The further development of the LFDB in Canada as well as the production of fire boundary maps for Russia by using satellite imagery will provide significant new information to conduct additional analyses of fires in the boreal forest on a global scale.

Acknowledgments

The historical information provided above is drawn largely from an unpublished paper, "Organizational History of the Alaska Fire Service," written by Holly Jewkes, a graduate student at the University of Alaska, Fairbanks, in 1997.

The production of the Alaskan LFDB was the result of the efforts of numerous individuals. The authors thank Susan Kasischke, who sorted through the Alaskan data records from 1950 through 1989 and made copies of all fire boundaries for this time period, and Peter Harrell of Duke University, who performed the same function for 1990 and 1991. The authors also thank numerous Duke University graduate students who assisted in digitizing these records. Several Alaska Fire Service employees were involved in producing and verifying the fire perimeters included in this Alaskan database, including Anne Burns, Tami DeFries, Lynn Emerick, Marlene Eno, James Higgens, Kent Slaughter, Roger Stilipec, and Tim Theisen, who did most of the recent digitizing of the 1988–1998, >100-acre fires; Mary Lynch and Pat Houghton, who did most of the verification and sorting through old fire records to find fire perimeters overlooked by the original research done by ERIM International scientists; Tim Hammond, a contract employee; and Gayle Moore, who provided GIS expertise and oversight. The records from which these fire perimeters were retrieved consist of more than 25,400 individual reports and more than 160,000 pages.

References

Auclair, A.N.D., and T.B. Carter. 1993. Forest fires as a recent source of CO_2 at northern latitudes. *Can. J. For. Res.* 23:1528–1536.

Delisle, G.P., and R.J. Hall. 1987. *Forest Fire History Maps of Alberta, 1931 to 1983.* Canadian Forest Service, Northern Forest Center, Edmonton, Alberta, Canada.

Donnelly, R.E., and J.B. Harrington. 1978. *Forest Fire History Maps of Ontario.* Miscellaneous Report FF-Y-6. Environment Canada, Canadian Forest Service, Forest Fire Research Institute, Ottawa, Ontario, Canada.

Dwight, T.W. 1918. *Forest Fires in Canada 1914–15–16.* Forestry Branch Bulletin 64. Department of the Interior, Ottawa, Ontario, Canada.

Higgins, D.G., and G.S. Ramsey. 1992. *Canadian Forest Fire Statistics: 1988–1990.* Information Report PI-X-107E/F. Forestry Canada, Petawawa National Forest Institute, Chalk River, Ontario, Canada.

Hind, H.Y. 1860. *Narrative of the Canadian Exploring Expedition of 1857 and of the Assiniboine and Saskatchewan Expedition of 1858.* 2 vols. Longman, Green, Longman, and Roberts, London.

Holman, H.L. 1944. *Report on Forest Fire Protection in the Mackenzie District,* Vol. 464. File 50050. National Archives of Canada RG 39. Northwest Territory Lands, Parks and Forests Branch, Dominion Forest Service, Ottawa, Ontario, Canada.

Hudson's Bay Company. 1812. *Hudsons Bay Post Record.* Fort Edmonton, October 12, 1812. Hudsons Bay Archives, Winnipeg, Canada.

Kayll, A.J. 1996. *Forest Fire Activity in Canada 1920 to 1995.* Report to A. Simard, Canadian Forest Service, Natural Resources Canada, Ottawa, Ontario, Canada.

Kiil, A.D. 1979. Fire research programs and issues in mid-Canada, pp. 5–14 in D. Quintillo, ed. *Proceedings International Fire Management Workshop.* Information Report NOR-X-215. Canadian Forest Service, Northern Forest Research Center, Edmonton, Alberta, Canada.

Lutz, H.J. 1959. *Aboriginal Man and White Man as Historical Causes of Fires in the Boreal Forest, with Particular Reference to Alaska.* Bulletin 65. Yale School of Forestry, New Haven, CT.

MacMillan, H.R. 1909. *Forest Fires in Canada during 1908.* Forestry Branch Bulletin 7. Department of the Interior, Ottawa, Ontario, Canada.

Murphy, P.J., and C. Tymstra. 1986. The 1950 Chinchaga River fire in the Peace River region of British Columbia/Alberta: preliminary results of simulating forward distances, pp. 20–30 in *Proceedings Third Western Region Fire Weather Committee Scientific and Technical Seminar.* Northern Forestry Center, Canadian Forestry Service, Edmonton, Alberta, Canada.

Murphy, P.J., M.E. Alexander, and B.J. Stocks. 1996. The Canadian forest fire record. Paper presented at National Workshop on Wildland Fire Activity in Canada, April 1–4, 1996. Edmonton, Alberta, Canada.

Simard, A.J. 1997. *National Workshop on Wildland Fire Activity in Canada: Workshop Report.* Information Report ST-X-13. Canadian Forest Service. Ottawa, Ontario, Canada.

Steele, S.B. 1915. *Forty Years in Canada: Reminiscences of the Great Northwest, with Some Account of His Service in South Africa.* Dodd & Mead, New York.

Stocks, B.J. 1991. The extent and impact of forest fires in northern circumpolar countries, pp. 197–202 in J.S. Levine, ed. *Global Biomass Burning: Atmospheric, Climatic, and Biospheric Implications.* MIT Press, Cambridge, MA.

Stocks, B.J., B.S. Lee, and D.L. Martell. 1996. Some potential carbon budget implications of fire management in the boreal forest, pp. 89–96 in M.J. Apps and D.T. Price, eds. *Forest Ecosystems, Forest Management and the Global Carbon Cycle.* NATO ASI Series, Subseries I, Vol. 40, Global Environmental Change. Springer-Verlag, Berlin.

Van Wagner, C.E. 1988. The historical pattern of annual burned area in Canada. *For. Chron.* 64:182–185.

Ward, P.C., and A.G. Tithecott. 1993. *The Impact of Fire Management on the Boreal Landscape of Ontario.* AFFMB Publication 305. Ontario Ministry of Natural Resources, Sault Ste. Marie, Ontario, Canada.

Wilton, W.C., and C.H. Evans. 1974. *Newfoundland Forest Fire History 1619–1960.* Inf. Report N-X-116. Newfoundland Forest Research Centre, Canadian Forest Service, St. John's, Newfoundland, Canada.

Wright, J.G. 1940. Long-term trends in forest-fires in Canada. *For. Chron.* 16:239–244.

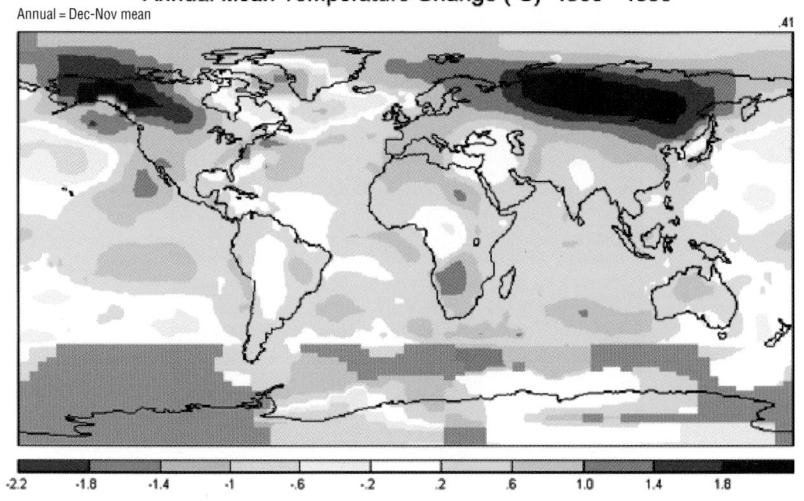

Figure 1.1. Average surface temperature change observed over the 30 years 1965–1995 in the boreal forest region. (Courtesy of Jim Hansen, Goddard Institute for Space Studies.)

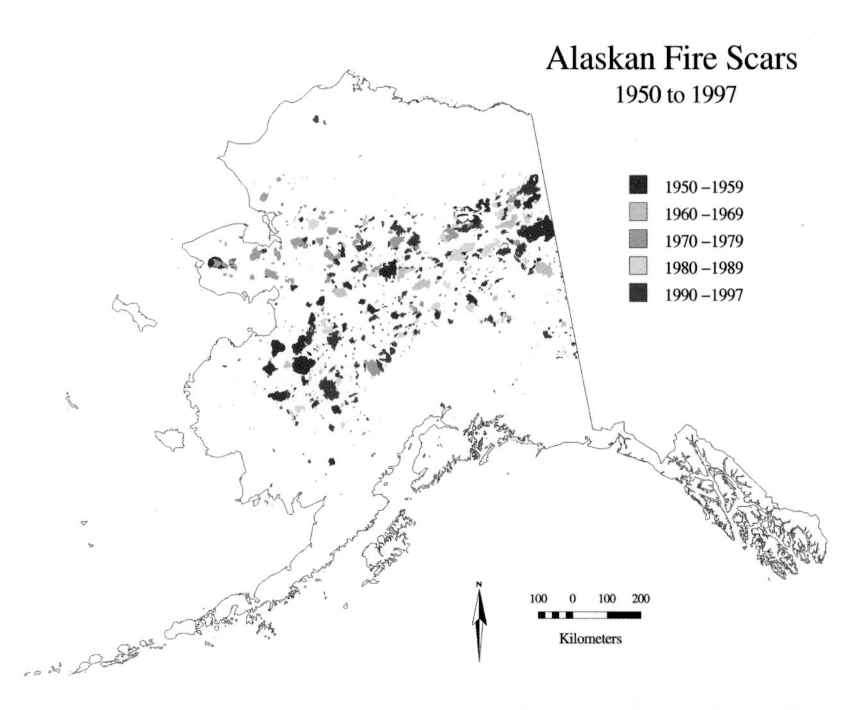

Figure 2.2. Locations of fires in Alaska based on fire records from 1950 to 1997.

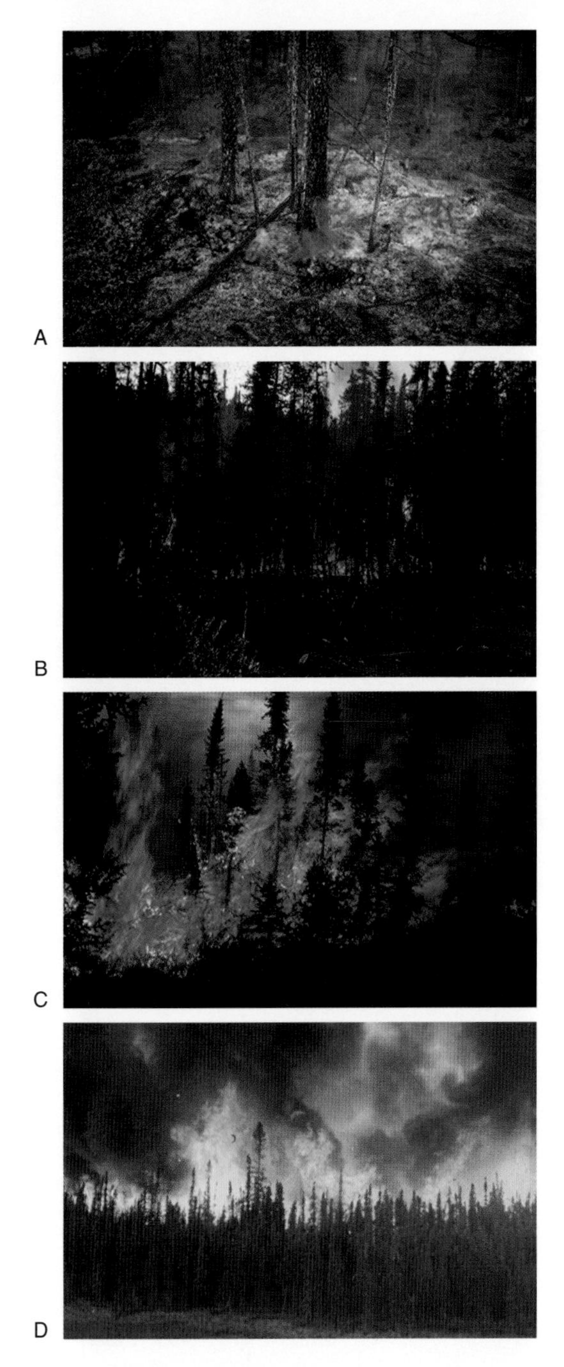

Figure 7.1. Illustration of the variation in fire behavior in the black spruce fuel type of western and northern North America: (a) ground fire; (b) surface fire; (c) intermittent or passive crown fire; and (d) fully developed crown fire. (Photographs courtesy of R.A. Lanoville, Department of Resources, Wildlife and Economic Development, Fort Smith, Northwest Territories.)

Figure 14.1. Ecozones of the North American boreal forest.

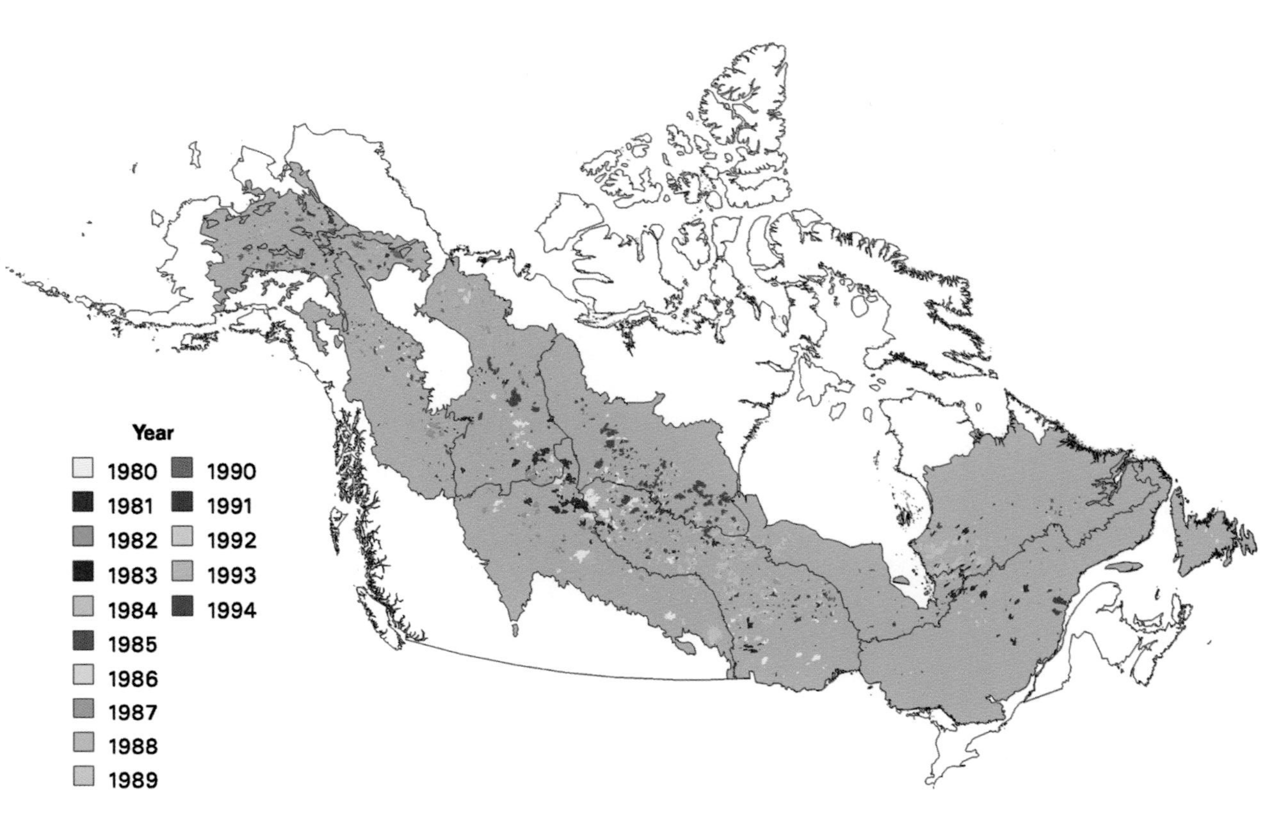

Figure 15.4. Map of large fires occurring in the North American boreal forest for 1980–1994.

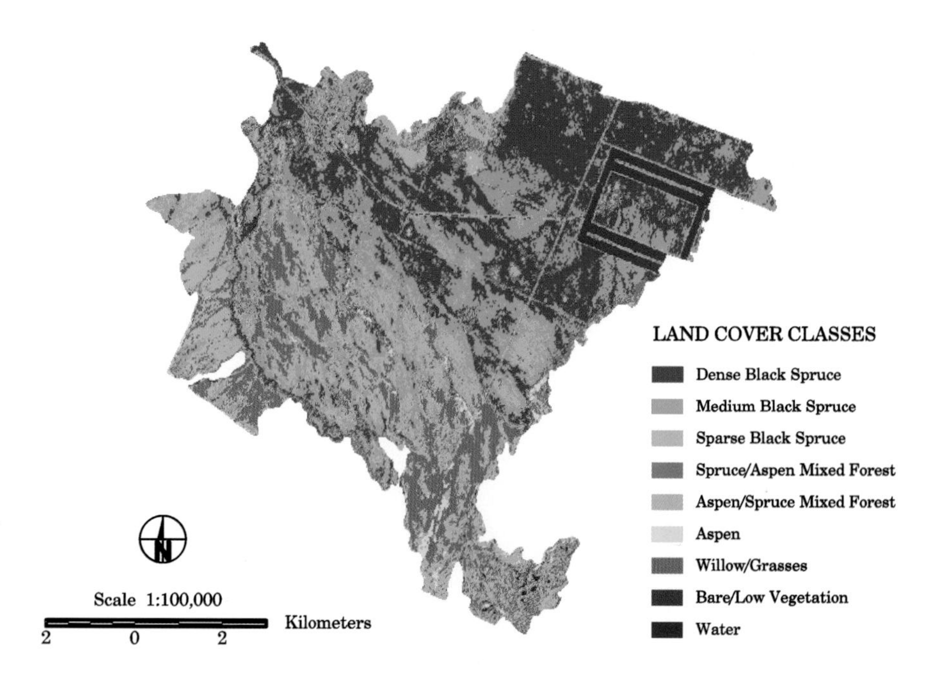

Figure 17.2. Vegetation/forest-cover map over the Gerstle River region of Alaska generated from Landsat TM imagery collected in August 1992.

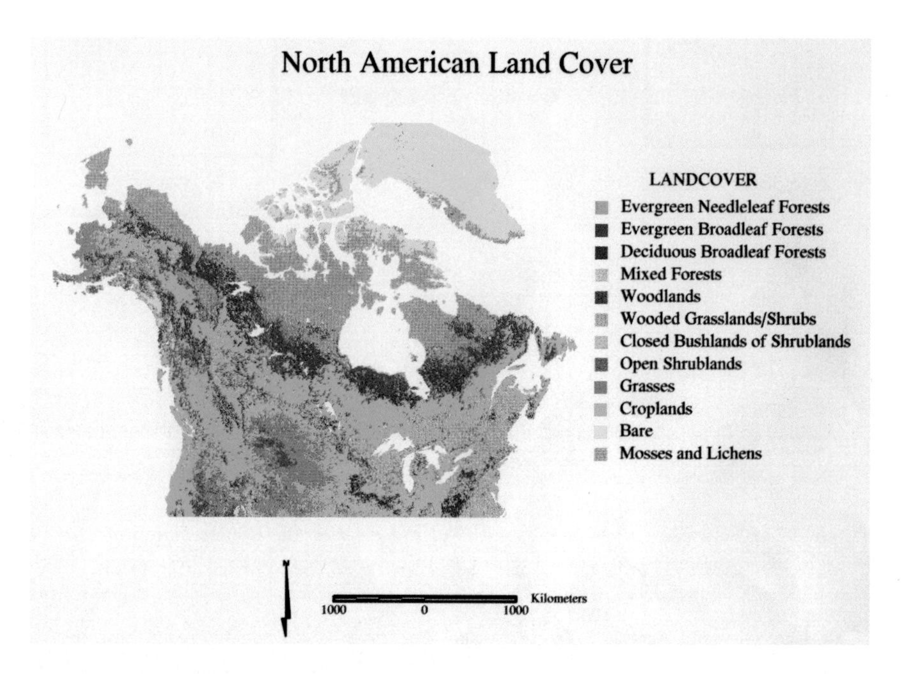

Figure 17.3. Vegetation-cover map at an 8-km resolution of North American derived from AVHRR imagery. (DeFries et al. 1995b.)

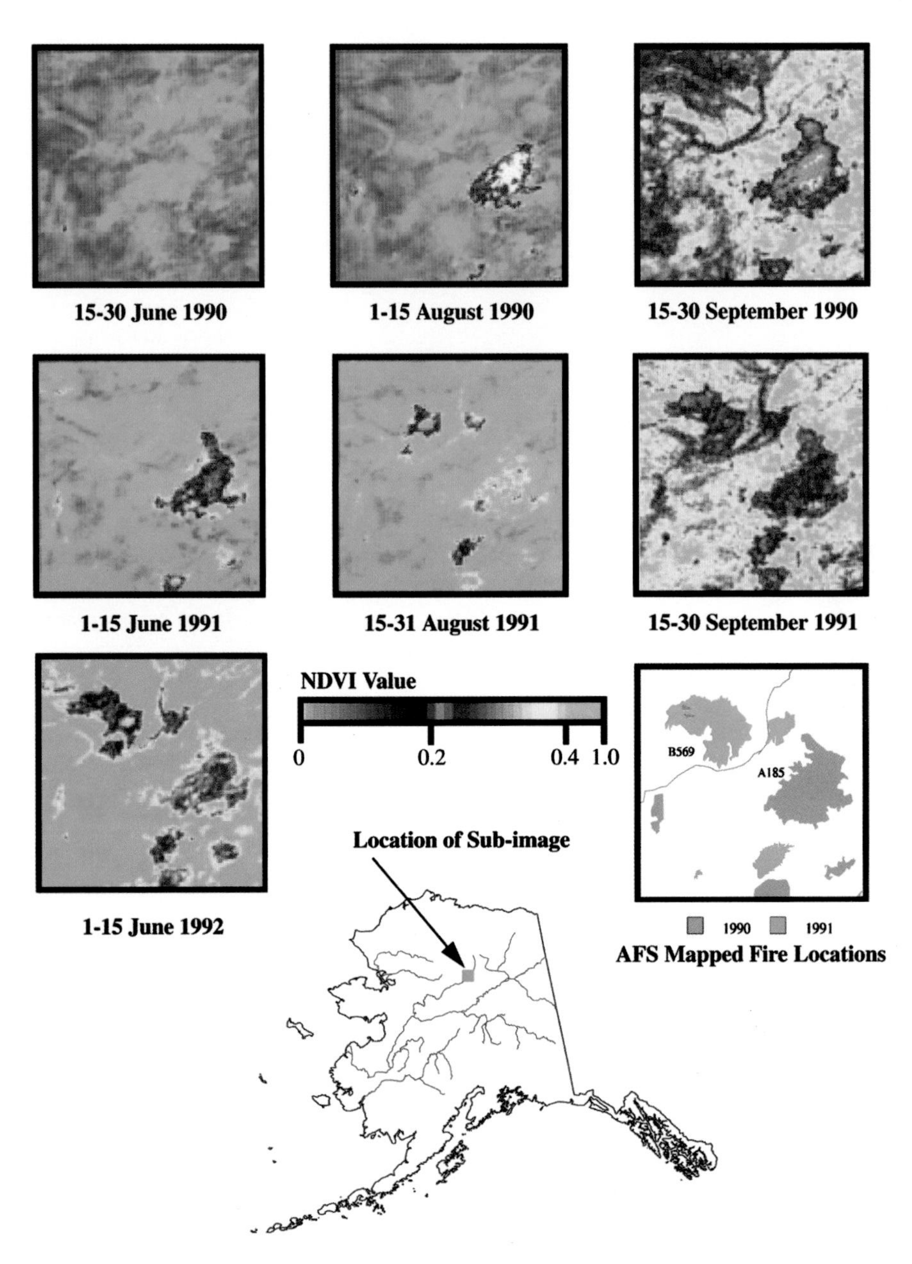

Figure 17.4. Series of AVHRR-derived NDVI images over the Bettles region of interior Alaska, where green represents a high vegetation index and white a low index. (From Kasischke and French 1995, with permission from Elsevier Science.) A series of major fires occurred in this region during the summers of 1991 and 1992. These fires can be detected as a combination of a drop in NDVI between spring and late summer as well as a delaying of green-up during the following spring.

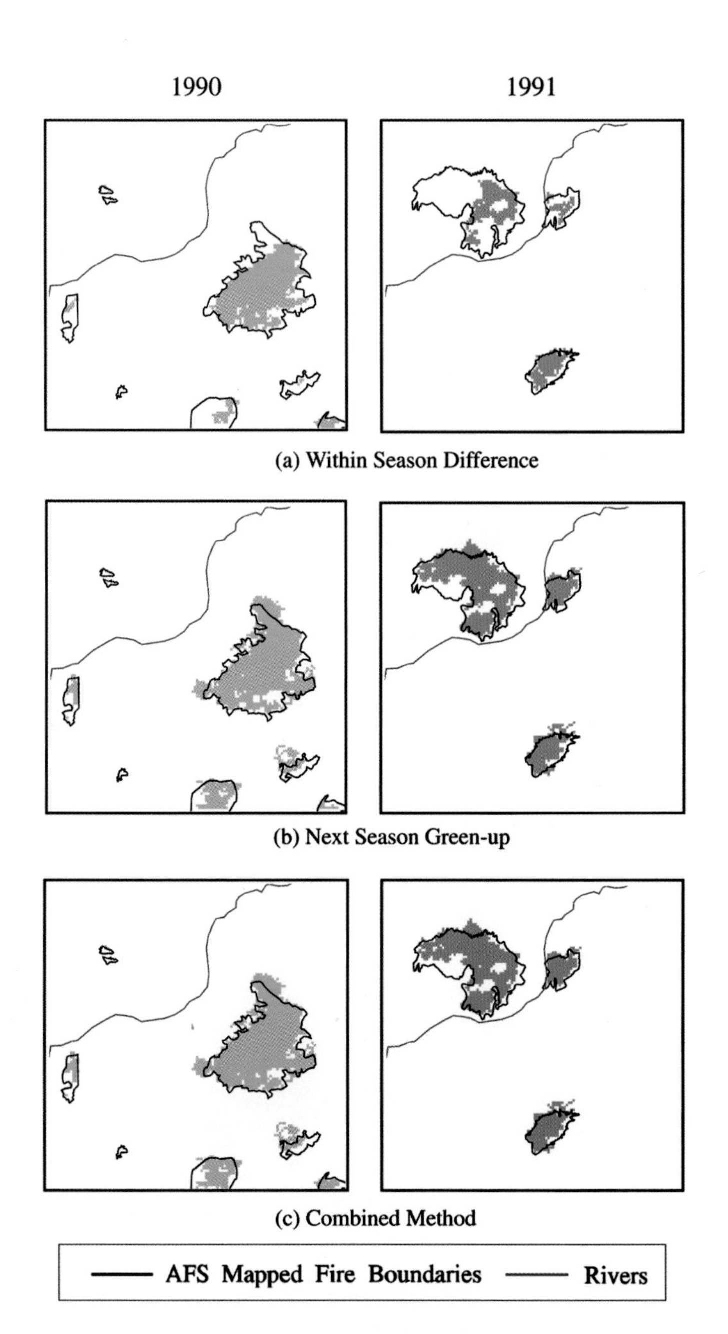

Figure 17.5. Results of using two fire detection algorithms on AVHRR NDVI composite imagery collected over Alaska. (From Kasischke and French 1995, with permission from Elsevier Science.) The first method (a) depends on detection of the drop in NDVI late in the summer associated with the fire, and the second (b) depends on delay of greening-up of the five disturbed regions during the following spring. Optimum results were obtained by combining the fire areas detected by using both methods (c).

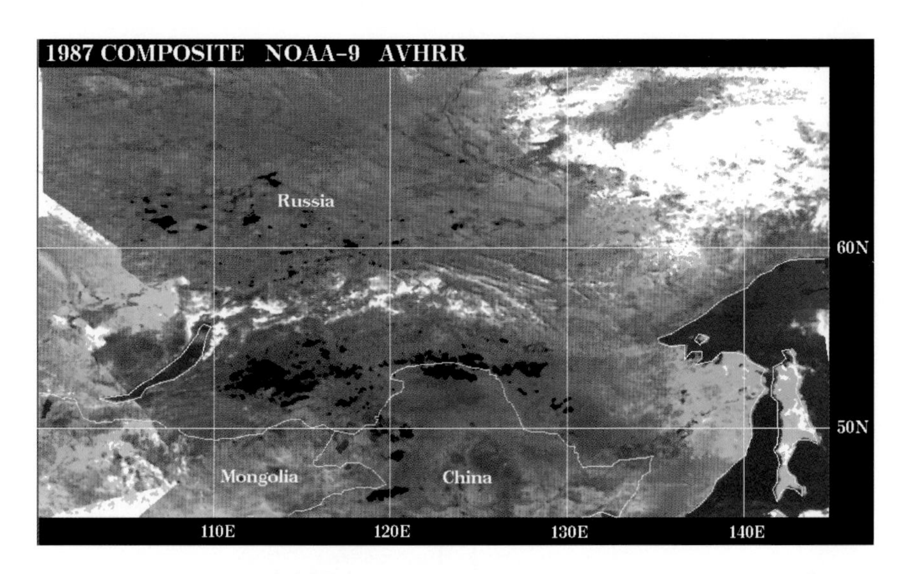

Figure 17.6. Mosaicked AVHRR image of southeastern Russia and northwestern China containing scars from fires detected on AVHRR imagery. (From Cahoon et al. 1994.) The black areas within this image are the locations of the 1987 fires.

Figure 17.7. False-color image generated from Landsat TM data collected over the Northwest Territories of Canada illustrating distinct signatures associated with recent forest fires. This image was created assigning red to TM band 3, blue to TM band 4, and green to TM band 5. The oldest fires appear light green, to orange, to pink, to red, and to dark brown for the most recent fires.

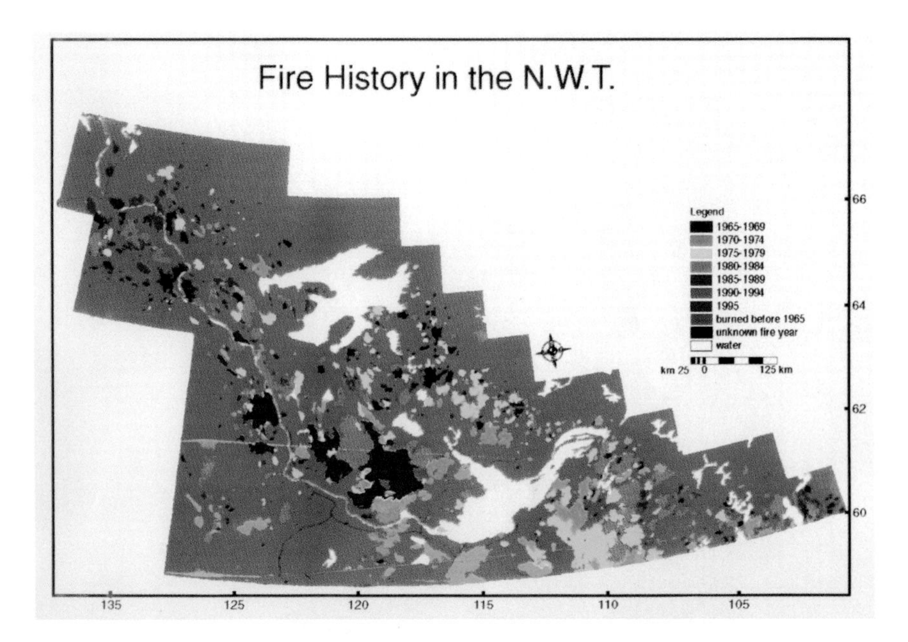

Figure 17.8. GIS-derived map of forest fire boundaries in the Northwest Territories based on an analysis of Landsat TM imagery.

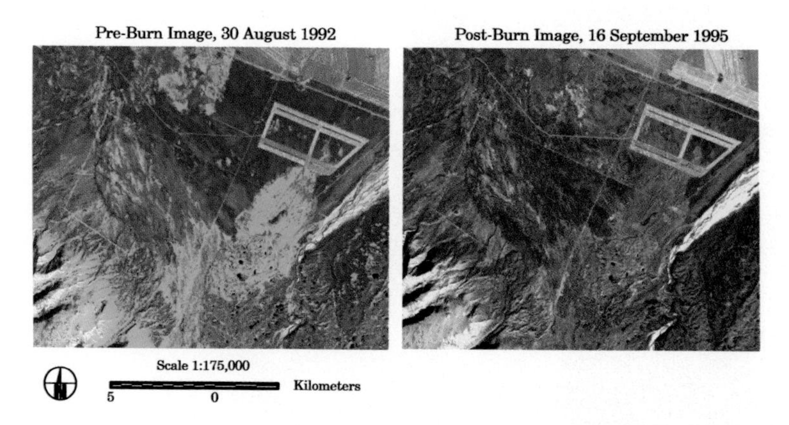

Figure 17.9. False-color, shortwave infrared Landsat TM images of an area burned by a forest fire in 1994. The left image was collected prior to the fire and the right image 1 year after the fire. The Gerstle River is in the right-hand side of both images, which were created by assigning red to TM band 7, blue to TM band 4, and green to TM band 3. The Landsat data from the preburn image were used to generate Figure 17.2.

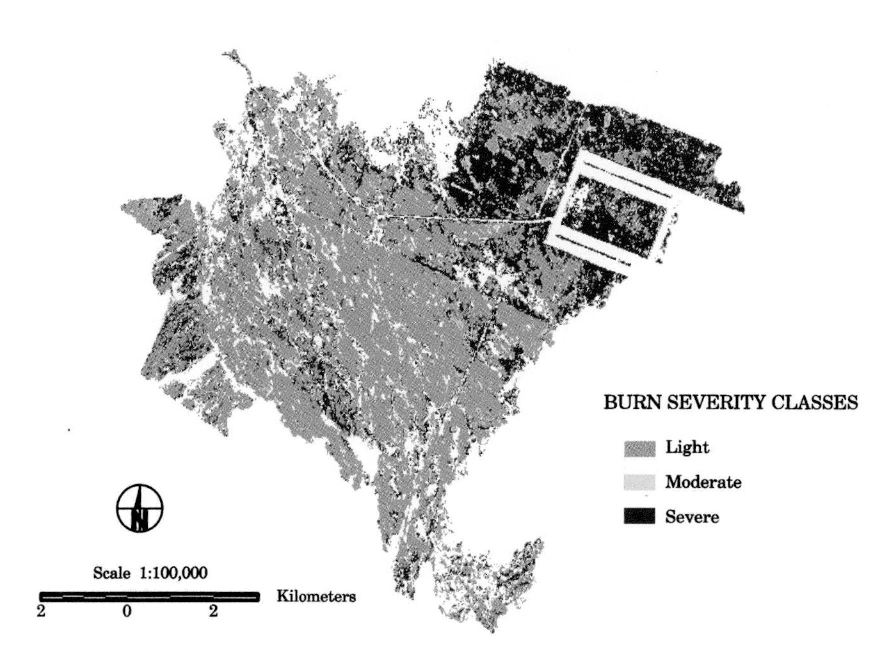

Figure 17.10. Classification of burn severity in black spruce forests based on analysis of Landsat TM imagery collected in the fall of 1995 (see Fig. 17.9).

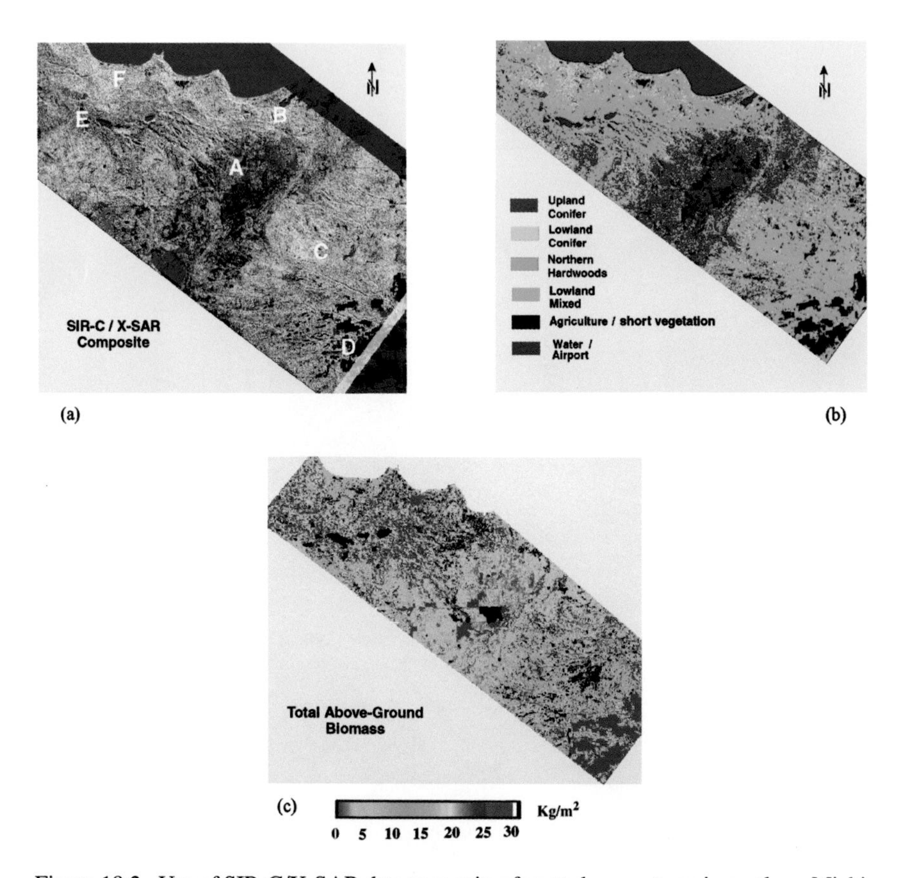

Figure 18.2. Use of SIR-C/X-SAR data to monitor forested ecosystems in northern Michigan: (a) multichannel radar color composite; (b) land-cover map; and (c) forest biomass. (Reprinted from Bergen et al. 1998 with permission from Elsevier Science. Image courtesy of Kathleen Bergen and Craig Dobson of the University of Michigan.)

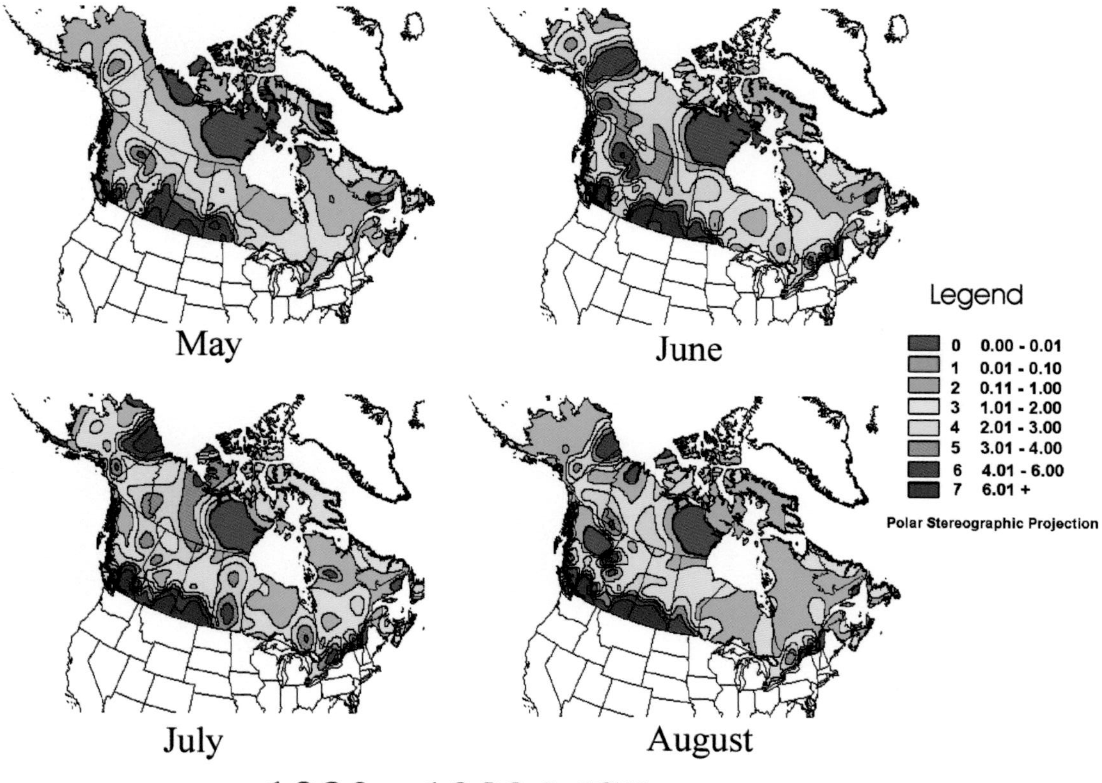

Figure 20.1. Average monthly severity rating maps for Canada and Alaska based on measured 1980–1989 daily weather.

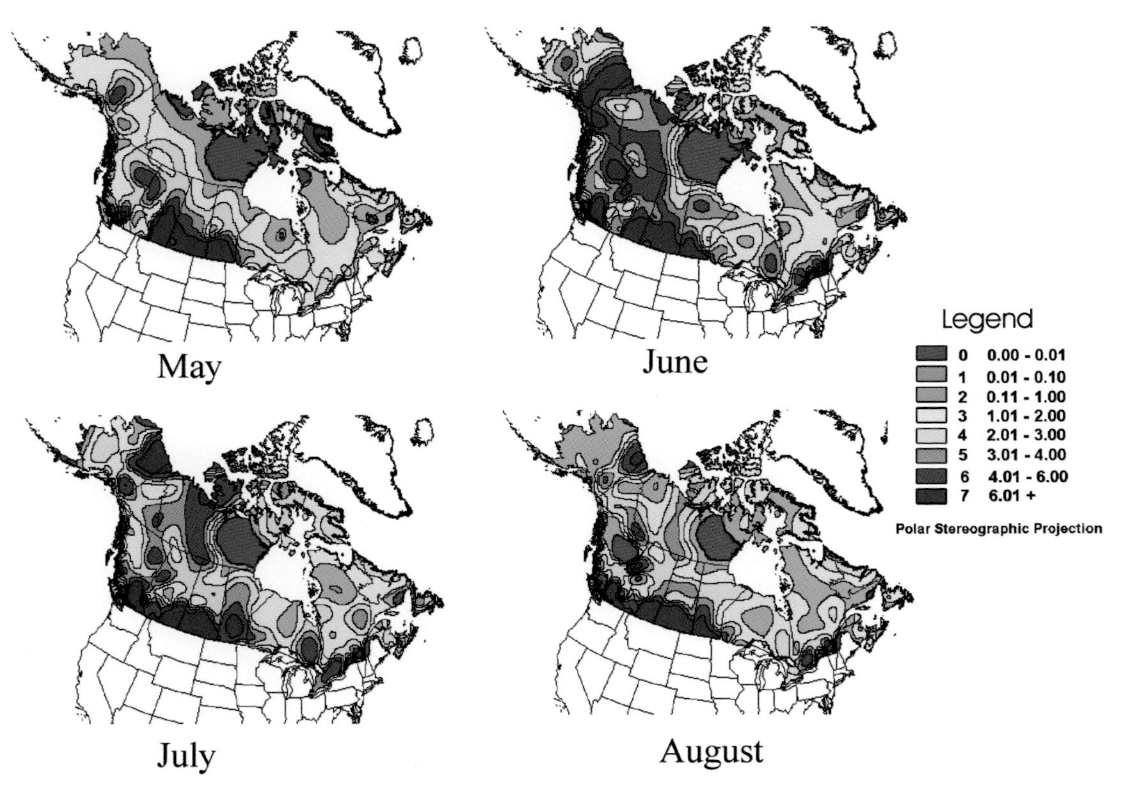

Figure 20.2. Average monthly severity rating maps produced by the Canadian Climate Center for Canada and Alaska under a 2 ↔ CO_2 climate.

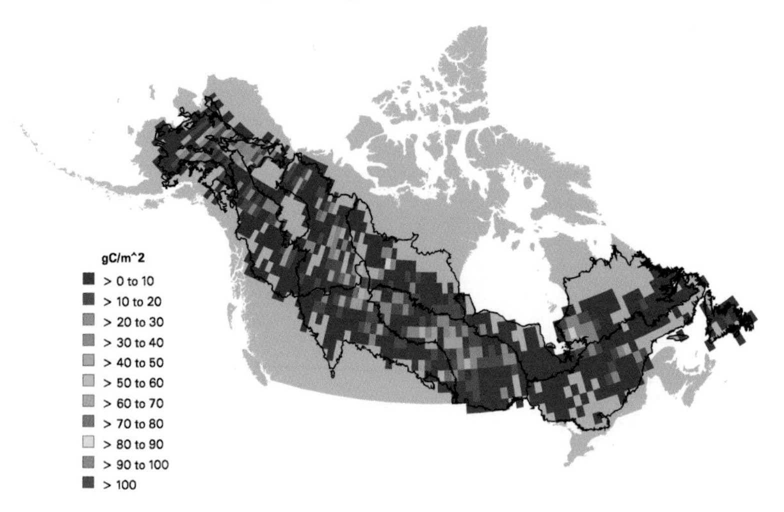

Figure 21.1. Average annual carbon release from the North American boreal forest region, 1980–1994.

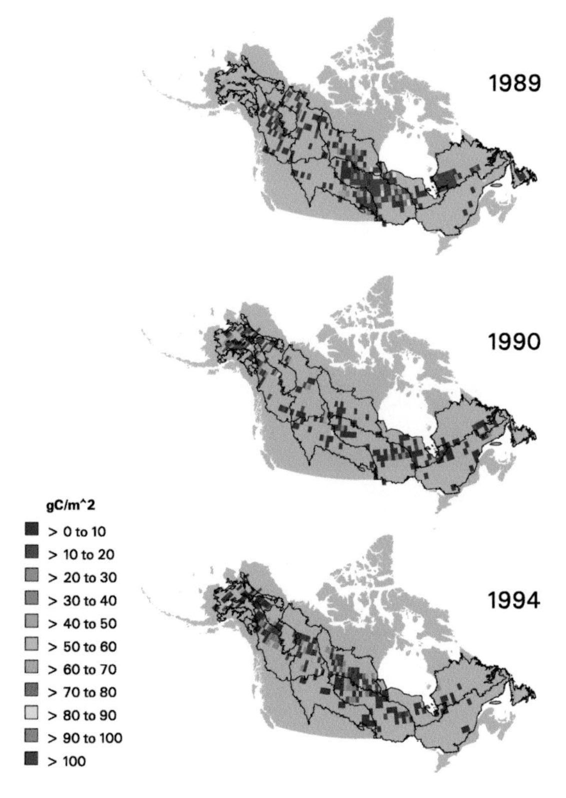

Figure 21.2. Carbon release from the North American boreal forest region for select years.

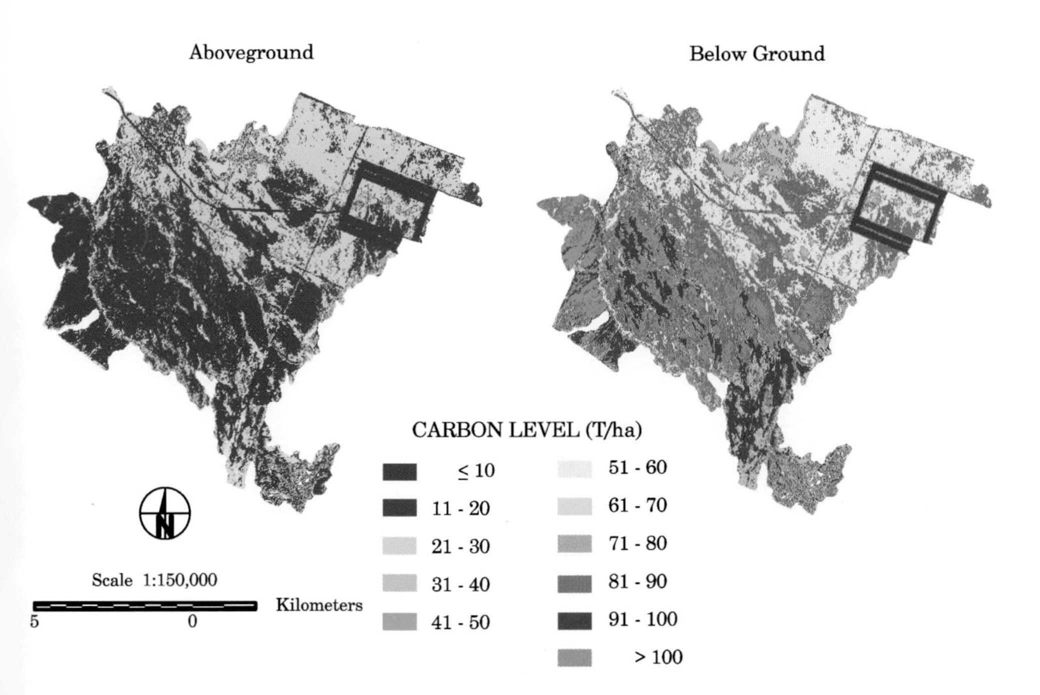

Figure 23.4. Patterns of above-ground and ground-layer carbon storage in an area burned in 1994 based on combining a land-cover classification of a 1992 Landsat TM image (Fig. 17.2) and field data.

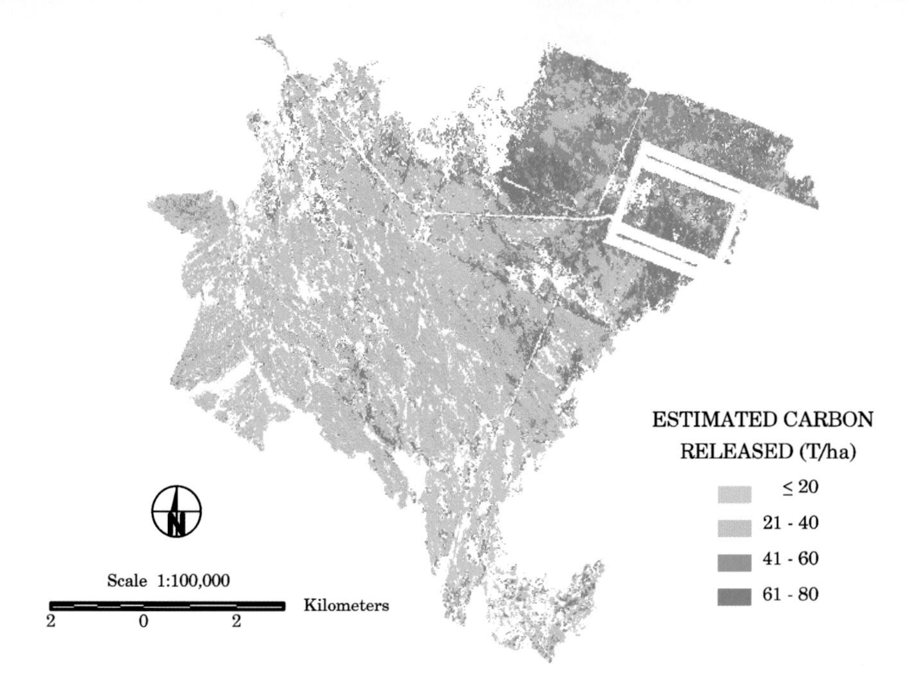

Figure 23.5. Patterns of percentage carbon consumption during fires based on fire severity estimated from Landsat TM imagery (Fig. 17.10) and ground measurements.

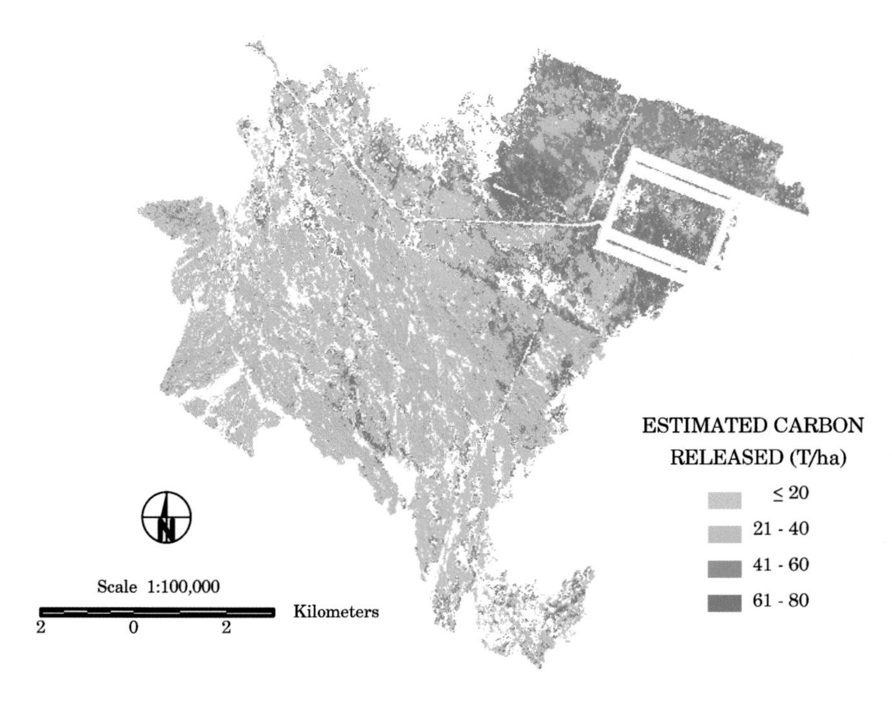

Figure 23.6. Patterns of carbon release from fires based on combining carbon levels (Fig. 23.4) and percentage carbon consumption (Fig. 23.5).

16. Fire and the Carbon Budget of Russian Forests

Anatoly Z. Shvidenko and Sten Nilsson

Introduction

This chapter reports on a study to estimate carbon emissions from fires in the Russian boreal forest (excluding postfire forest regeneration). These emissions result from the effects of fire on the dynamics of the primary carbon pools of terrestrial ecosystems: phytomass, coarse woody debris, and soils. Fire-related carbon flux is divided into two parts: direct fire emissions and postfire emissions. Direct fire emission is the carbon released from biomass burning during the year in which the fire occurred. Postfire emissions are the result of both unburned residuals of forest combustibles (FC), postfire dieback (mortality), and changes in soil organic matter. Due to the significant interseasonal variation of the extent of fire and hence carbon emissions, the results are presented as the annual average for the period 1988–1992.

The area evaluated in this study includes the Russian Forest Fund and state land reserves (all territories situated north of the forest zone). These categories represent 75% of the total land of Russia. The basic features of the fire regime in Russian forests are presented in Chapter 8.

Models for Estimating Fire-Related Carbon Fluxes

For a specific area, the total carbon flux generated by fire, TCF_{t_1}, during a year t_1, can be expressed as:

$$TCF_{t_1} = DF_{t_1} + PFF_{t<t_1} \qquad (16.1)$$

where DF_{t_1} is the direct fire emission during the year t_1 and $PFF_{t<t_1}$ is the postfire biogenic flux generated by fires occurring during previous years $t < t_1$. The values of DF_{t_1} and $PFF_{t<t_1}$, as well as the explicit form of Equation (16.1), depend on the type, extent, and severity of the fire, landscape and weather conditions under which the fires occurred, and the type and characteristics of the vegetation cover.

The direct flux (direct fire emission) is defined as

$$DF_{t_1} = \sum_{ilkq} [S_{ilkq} (FC)_{ilkq} C_{ilkq}] \gamma_q \qquad (16.2)$$

where S_{ilkq} is the estimate of burned vegetation areas (ha), $(FC)_{ilkq}$ is the amount of combustable forest material (t ha^{-1} of of dry matter),C_{ilkq} is the fraction of FCs consumed during the fire, and γ_q is the fraction of the material that is carbon.

The indexes in Equation (16.2) are as follows: (a) i is the territorial unit for which calculations were performed. We used a zonal aggregation of ecoregions (Kurnaev 1973; Shvidenko et al. 1996), in which i = 1–7: (1) subarctic desert and tundra; (2) forest tundra, sparse taiga, and meadow forests; (3) northern taiga; (4) middle taiga; (5) southern taiga; (6) mixed forests, deciduous forests, and forest steppe; and (7) steppe, semidesert, and desert. (b) l is the aggregated land-use class (l = 1–6): (1) forested area dominated by coniferous species; (2) forested areas dominated by deciduous species; (3) unforested areas; (4) peatlands; (5) vegetative nonforest land; and (6) unproductive nonforest lands. (c) k is the type of forest fire (k = 1–5): (1) crown fire, (2) superficial on-ground fire; (3) stable on-ground fire; (4) peat fire; and (5) underground fire. (d) q is the FC category (q = 1–14), including three aboveground phytomass pools: (1) "large wood" (i.e., stem-wood and large branches with a top diameter > 8 cm under the bark excluding the stump); (2) branches with a diameter less than 8 cm: and (3) foliage and herbaceous plant; three classes of roots with a diameter (d) (4) greater than 8 cm; (5) 1–8 cm, and (6) less than 1 cm; two classes of standing dry wood including wood of dry standing trees, of dry branches of live trees, and fallen wood, (7) large (d > 8 cm) and (8) small wood (d < 8 cm); two classes of dry wood laying on the ground, (9) large (d > 8 cm) and (10) small wood (d < 8 cm); two classes of dead roots, (11) large (d > 1 cm) and small (d < 1 cm); and three classes of organic soils, (12) litter, (13) labile soil (fibric), and (14) stabile humus.

Postfire flux depends on the rate of decomposition of biomass residuals from the fire and material resulting from postfire dieback (mortality), as well as on changes in the structure and content of soil organic matter. Let O_{ij} (t) (Fig. 16.1) be a function that describes the amount of dead organic matter coming into a decomposition pool j in year t, TD_{ij} (t*) the decomposition time of this amount, and O_{ij} (t*) the value of this function in year t.* For a simple exponential model, the process of decomposition of organic matter of pool j within the ith territory can be described as

$$G_{ij} (t^*,\tau) = O_{ij} (t^*) \exp (-\alpha_{ij}\tau) \qquad (16.3)$$

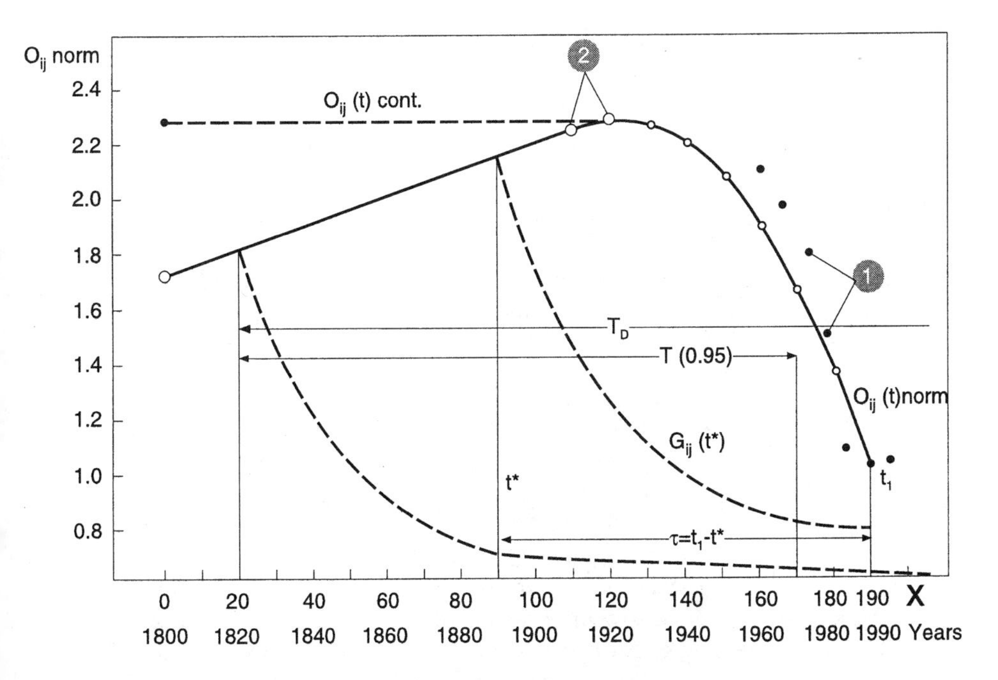

Figure 16.1. Model for evaluation of postfire biogenic flux: 1, data from the State Forest Account; 2, historical reconstructions.

where G_{ij} (t^*,τ) is the mass of organic matter that has not been decomposed during the period τ (where σ is the number of years between the year of fire and the year of the postfire flux estimation, e.g., $\tau = t^* - t_1$), and α_{ij} is the decomposition constant.

In Equation (16.3), the i subscript refers to territorial units, and the j subscript refers to the category of decomposition pool. Three decomposition pools were used and defined as a function of different FC classes: (1) fast (the green parts of vegetation and fine roots); (2) medium (woody material with small diameters, $8 < d < 1$ cm); and (3) slow (woody material with large diameters, $d > 8$ cm).

To calculate the postfire biogenic flux during year t_1 caused by fires that occurred during previous years, we have to sum up the input of G_{ij} (t^*,τ) for all years of which $G_{ij}(t^*,\tau) > 0$ in the year t_1, and

$$G_{ij} = O_{ij}\ (t - \tau)\ \exp\ [-\ \alpha_{ij}(\tau - 1)] - O_{ij}\ (t - \tau)\ \exp\ [-\alpha_{ij}\tau] \tag{16.4}$$

is the amount of organic matter decomposed for the period. We limit our consideration by the time $T_{0.95}$ that is needed for decomposition of 95% of the initial value of organic matter. For Equation (16.3), $T_{0.95}$ depends only on α_{ij} ($T_{0.95} = \ln 20 / \alpha_{ij}$).

The postfire biogenic flux during year t_1 used by fires that occurred during previous years can be estimated by

$$(PFF)_{ij}(t_1) = 1.05\,\chi\,[\exp(\alpha_{ij}) - 1] \sum_{\tau=0}^{\phi+1} O_{ij}(t-\tau)\exp(-\alpha_{ij}\tau) + \delta SOC \qquad (16.5)$$

where χ is the fraction of carbon from decomposed organic matter that is released to the atmosphere ($0 < \chi < 1$), $\phi = \text{int}\,[T_{0.95}]$ (the integer part of $T_{0.95}$), and ϕSOC is the change of soil organic carbon during year t_1. The correction coefficient 1.05 is used due to the above truncation of the curves $G_{ij}(t^*,\tau)$ by $T_{0.95}$.

Development of Model Parameters: Direct Carbon Release

Average Area Burned Annually in 1988–1992

The estimates of the average annually burned area from 1988 to 1992 are presented in Table 8.3. To take into account differences in carbon release and postfire dieback under different fire regimes and in different forest types, we divided areas of on-ground fires into steady and superficial ones, based on expert estimates and analysis of regional distribution of major fire by seasons. Nonforestlands were divided into two classes (vegetative nonforestlands and low-productive nonforestlands) based on (1) a classification of natural non-forest vegetation into 64 classes using the vegetation map of the Former Soviet Union; and (2) average estimates of phytomass and detritus by the above classes calculated by the IIASA Boreal Forest Study. These classes (which had average phytomass <50% of total average inside a definite vegetation zone) were treated as low-productive nonforestlands. Peat fire areas were distributed by vegetation zones proportionate to areas of wetlands of Asian Russia, which contains about 83% of all Russian bogs. Areas used in the calculations are given in Table 16.1.

Forest Combustibles

The amount of FCs depends heavily on the geographic zone, the type and productivity of the forest, the frequency of fires, and the soil moisture conditions (Table 16.2). Variability in the levels of FC can be high between individual stands and different forest types. For example, Furyaev (1996) reported the average storage of on-ground FCs for pine stands in the middle Siberia plateau ranged between 12.7 t ha^{-1} (pine with lichens) and 23.0 t ha^{-1} (pine with green mosses and red bilberry). Numerous other researchers have reported similar levels of variation (Kasesalu 1972; Usoltsev et al. 1979; Dichenkov 1993).

As a rule, all types of fires start as on-ground fires. Thus, the amount and properties of on-ground combustibles play an extremely important role in the development of wild forest fires. Many field observations are provided for the northern Eurasia boreal zone with respect to combustibles. The analyses presented

Table 16.1. Burnt Areas by Classes Used in the Calculation

Zone	Average annual burned area by types of fire (million ha)							
		On-ground fires by land-cover categories						
		Forested areas		Unforested areas				
	Crown fire	Total	Steady	Total	Steady	Vegetative nonforest lands	Low-productive nonforest lands	Peat
Subarctic and tundra	—[a]	—	—	—	—	0.34	0.09	0.06
Forest tundra and sparse taiga and meadow forests	0.055	0.26	0.20	0.10	0.08	0.14	0.04	0.09
Northern taiga	0.060	0.29	0.20	0.11	0.08	0.14	0.05	0.06
Middle taiga	0.074	0.36	0.23	0.14	0.09	0.17	0.10	0.07
Southern taiga	0.037	0.17	0.10	0.08	0.05	0.08	0.06	0.06
Mixed and deciduous forests and forest steppe	0.014	0.07	0.0	0.03	0.02	0.03	0.04	0.01
Steppe, semidesert, and desert	—	0.01	0.01	0.01	—	—	0.01	—
Total	0.240	1.16	0.78	0.48	0.32	0.90	0.39	0.35

[a]—, No data.

Table 16.2. Phytomass and Carbon of Russian Forests

	Phytomass component (Tg, dry mass)						Carbon
Regions and indicators	Stem wood[a]	Branches[a]	Foliage	Roots	Understory[b]	Total	Density (kg C m^{-2})
Total Russia	38,100.9	5,644.4	11,936.9	2,624.7	4,663.7	64,372.5	4.20
European Russia	9,553.0	1,559.6	3,061.6	980.9	963.9	16,118.7	4.77
Asian Russia	28,547.9	4,084.8	8,875.3	1,643.8	3,699.8	48,253.8	4.05
Asian Russia by zones[c]							
Forest tundra, northern taiga, and sparse taiga	3,603.8	481.6	180.9	1,245.7	820.0	7,010.6	2.04
Middle taiga	14,301.7	1,865.4	801.8	4,478.7	2,013.1	24,124.9	4.29
Southern taiga	7,261.5	1,173.2	438.8	2,143.7	570.2	11,629.3	5.74
Mixed forests	1,938.1	322.9	133.3	609.5	209.9	3,227.1	5.92
Forest steppe	1,442.8	241.7	88.9	397.6	86.6	2,261.9	5.18
Shrubs						1,402.1	

[a]Stem wood and branches are presented over bark; percentage of the bark comprises—for example, for stems in Asian Russia—15.6% of total stem mass over bark.
[b]Understory includes phytomass of bushes and undergrowth and green forest floor (the average ratio of the last two categories is 1:3).
[c]Ecoregions served as initial units of the calculations.

above and many other sources allow us to make the following aggregated conclusions:

1. There is a distinct climatic gradient (defined mainly by longitude, although some latitude and altitude effects exist over large regions) of the total amount of on-ground FCs that is dependent on the level of primary productivity, rate and specifics of accumulation, and decomposition of on-ground organic matter. Polar deserts have a fragmented vegetation cover with the level of on-ground vegetation combustibles (OGVC) from 0.5 to 3.0 t ha^{-1}, of which green parts (mosses and lichens) comprise 45–50% (Alexandrova 1971). For a typical subarctic tundra zone, the amount increases from the north to the south, where the average is 55.0 t ha^{-1} in the European north (of which 22.0 t ha^{-1} is phytomass) and up to 100.0 t ha^{-1} in west Siberia (where phytomass comprises 27.0 t ha^{-1}). The productivity and amount of OGVC in the southern tundra is minimal in the European north and increases in west Siberia (up to 80.0–100.0 t ha^{-1}, of which 7.0–25.0 t ha^{-1} is phytomass), reaches its maximum in middle Siberia (phytomass 40.0 t ha^{-1} and mortmass about 60.0 t ha^{-1}), and decreases to the east (phytomass about 13.0 t ha^{-1}, but the accumulation of mortmass can be very high). The amount of phytomass and the OGVC increase toward forest tundra (Vilchek 1987; Bazilevich 1993a). Regularities of the OGVC's distribution within forest zones are much more complicated to estimate or identify due to anthropogenic land-cover transformations. The range of natural variability is high, from 1.0 to 5.0 t ha^{-1} in dry pine stands (with very high frequency of fire) to 50 to more than 90 t ha^{-1} in wet sites.
2. At the landscape level, there is an evident gradient of soil moisture for all taiga zones and the storage of OGVC increases from dry to humid and wet sites. This gradient is superimposed over the effects of relief. There are higher amounts in the valleys and at the bottom of gorges than on upper slopes.
3. The above two factors are often masked by the nature of the fire regime. The amount of the OGFC and the severity of fires are inversely proportional to the frequency of the fires. The fire regimes are related to the amount of litter (Academy of Sciences of the USSR 1983). Measurements by Sheshukov (1978) of the content of FCs in mature (150–180 years) spruce stands in the northern taiga, which were generated after large fires, identified about 40 t ha^{-1} of combustible content in the moss and upper layers of litter. In larch stands with marsh tea and sphagnum layers, the amount was estimated at 132.0 t ha^{-1}. The amount of litter in wet pine stands affected by fire 54 years ago in west Siberia was 26.0 t ha^{-1}, and in pine stands on fresh sites with the last fire 105 years ago, it was 46.0 t ha^{-1} (Furyaev 1996).
4. Large amounts of FCs are left on clear-cut areas. A study of spruce-fir uneven-aged forests harvested by heavy machinery in the southern taiga zone of Krasnoyarsk Kray identified 40.0–49.0 t ha^{-1} of stem remains and 8.0–14.0 t ha^{-1} of slash (Zvetkov and Ivanova 1985). Even more storage (up to 90.0–110.0 t ha^{-1}) was reported for middle taiga permafrost areas (Peshkov 1991).

5. High levels of FCs are found in stands disturbed by insects and diseases. For instance, the stock of dry wood can reach 75–90% of the initial growing stock in dead spruce-fir forests in the Far East (up to 300–350 m^3 ha^{-1}) (Zuranov 1973).

In our calculations, we used average data for on-ground combustibles by vegetation zones. Such averaging is very rough because of the aggregation of data for different forest types. Table 16.3 contains data on the OGVC used in the study. These data are based on (1) available results of measurements of on-ground FCs such as in Table 16.2; and (2) transformed data of phytomass and detritus estimations made as a part of numerous productivity evaluations. The total number of measurements used was about 1,800. Unforested areas included natural sparse forests, burns and dead stands, unregenerated harvested areas, and grassy glades, which explains the high average values of OGVC.

Inventory data on phytomass, detritus, and organic soils in forested areas are useful for estimating the amount of available combustibles for different types of fire, especially crown and soil fires. There have been many study reports on biomass inventories during the past decade (Isaev et al. 1995; Alexeyev and Birdsey 1998; Alexeyev et al., this volume). The IIASA Forest Study provided a new inventory of phytomass of forest ecosystems based on using smaller geographic regions and examination of additional data sorces (Shvidenko and Nilsson, in press). The final result of this inventory was in line with previous studies. For example, for total phytomass of Russian forests, the Alexeyev and Birdsey (1998) study estimates were 14% less and the study estimates of Isaev and colleagues (1995) were 14% higher compared with the Shvidenko and Nilsson (in press) study. In addition, differences for individual regions were significant.

In Table 16.4, we present average data on phytomass for Russia by the above vegetation zones and some details by major coniferous forest types (dark coniferous and light coniferous forest as forests with the highest crown fire risk). The total amount of carbon in phytomass of all Russian forests is estimated to be 32.9 Gt or 42.4 t ha^{-1}, with the regional variability from about 15.0 to 115.0 t ha^{-1}.

Coarse woody debris (CWD) (defined as dry wood > 1 cm in top diameter that has not lost its structure, e.g., dry stems, dry branches of living trees, wind-fallen wood) was estimated based on regional data from the forest inventory, different publications, and other data archives. Total amount of CWD was defined at 6.0 Gt C, of which above-ground carbon comprises 5.1 Gt (or 85.7% of total). Average data for European Russia are half of the extent for Asian Russia (3.8 and 7.5 t C ha^{-1}, respectively, for aboveground carbon). Litter has been estimated separately based on digitized maps developed by the IIASA Forest Study. Total amount of carbon accumulated in litter (on forested areas) is at 8.72 Pg C, or 11.4 t C ha^{-1}. Variability of carbon density of both CWD and litter is very high and calculated as an average by zonal aggregations of ecoregions. The total amount of carbon in peat was estimated at 117.0 Pg on an area (the depth of the organ soil horizons is > 5 cm) of 328.0 million ha (Rojkov et al. 1997).

Table 16.3. Average On-Ground Vegetation Combustibles (OGVC) (dry matter, kg m^{-2})

	Forested area		Unforested area		Nonforest natural vegetation		Nonforest unproductive land	
Vegetation zone	Average ± SD	n	Average ± SD	n	Average ± SD	n	Average ± SD	n
Subarctic and tundra	—[a]	—	—	—	1.79 ± 0.07	197	0.47 ± 0.11	26
Forest tundra and sparse taiga	2.70 ± 0.11	123	3.17 ± 0.45	24	2.07 ± 0.23	64	0.51 ± 0.19	21
Northern taiga	3.10 ± 0.14	94	4.03 ± 0.61	35	1.48 ± 0.20	56	0.36 ± 0.09	17
Middle taiga	3.27 ± 0.15	116	4.41 ± 0.55	48	1.06 ± 0.17	94	0.31 ± 0.07	54
Southern taiga	3.31 ± 0.12	199	4.98 ± 0.92	67	0.98 ± 0.11	78	0.41 ± 0.11	61
Mixed forests, deciduous forests, and forest steppe	1.73 ± 0.05	157	1.90 ± 0.26	71	0.90 ± 0.27	76	0.44 ± 0.15	23
Steppe	0.64 ± 0.03	83	0.67 ± 0.17	9	0.71 ± 0.09	73	0.33 ± 0.06	41

[a]—, No data.

Table 16.4. Phytomass of Russian Forests by Vegetation Zones and Basic Aggregated Forest Types of Coniferous (t ha^{-1} of dry matter)

	Phytomass component								
Zone	Stem[a]	Bark	Branches[a]	Foliage	Roots	Understory[b]	Green forest floor	Total	Shrubs
Forest tundra, sparse taiga, and meadow forests	28.06	4.49	4.77	2.13	8.84	1.68	4.07	49.55	15.8
Northern taiga	38.05	5.53	5.71	2.66	11.22	1.73	3.97	63.34	24.6
Middle taiga	57.00	8.73	7.47	3.44	15.44	2.54	5.54	91.43	38.3
Southern taiga	72.97	9.97	11.61	4.86	21.72	2.23	3.78	117.16	42.7
Mixed forest and forest steppe	73.23	10.29	12.55	4.74	25.36	1.90	2.40	120.18	
Deciduous forests	77.13	10.64	14.14	4.48	30.94	1.87	1.66	130.22	—[c]
Average	56.91	8.26	8.35	3.66	16.56	2.20	4.32	92.00	28.56
By basic forest types of coniferous—Asian part									
DCM[d]	74.89	9.86	10.58	7.42	20.37	1.41	6.22	120.89	—
LCM	57.05	9.06	5.59	2.09	14.48	3.08	6.49	88.78	—
DCP	68.93	9.17	9.89	6.69	18.91	1.50	6.56	112.48	—
LCP	51.28	7.25	5.51	2.27	13.61	2.44	5.33	80.44	—

[a]Stem wood and branches are presented over bark; percentage of the bark comprises—for example for stems in Asian Russia—15.6% of total stem mass over bark.
[b]Understory in Table 16.4 includes phytomass of bushes and undergrowth.
[c]—, No data.
[d]DCM, LCM, DCP, LCP, forests with dominance of dark (spruce, fir, Russian cedar) and light (pine, larch) coniferous species in mountains and on plains, respectively.

Based on our estimates of phytomass, CWD, and litter, we present a hypothetical model of forest fuel for areas in Siberian forests for the main canopy species. The hypothetical stand contains 59.2 t C ha^{-1} in above- and on-ground FCs, of which phytomass comprises 40.7 t C ha^{-1}, litter 11.4 t C ha^{-1}, and CWD 7.1 t C ha^{-1}. Carbon of peat and mineral layers of soil is not included in this amount.

We can compare this result with a model developed by Grishin (1992) for analyses of the burning efficiency, which identifies a maximum amount of burnt matter of 117.0 t ha^{-1} (of dry matter) in coniferous forests. This can be considered as a theoretical maximum of matter that could burn during a crown fire. The total amount of aboveground FCs is estimated to be greatest in the Far Eastern mature dark coniferous (spruce-fire) forest, where the living phytomass reaches 350–400 t ha^{-1}. For mixed cedar forests, the amount is estimated to be 300–330 t ha^{-1}, and for larch with green mosses, 350.0–370.0 t ha^{-1} (Dukarev 1989).

The above data present the maximum actual amount of living and dead vegetation available for crown fires. These data are consistent with other studies on the storage of aerial FCs (small branches and needles) (Usoltsev 1988; Gabeev 1990; Shepashenko et al. 1998). Usoltsev (1988) estimated the dry phytomass of canopy combustibles in stocked pine plantations in northern Kazakhstan to be in the range of 9.0–15.0 t ha^{-1} depending on site indexes and age of stand. For middle-aged and immature pine stands in the southern Krasnoyarskiy kray, Semetchkina (1977) estimated a content of dry branches of 0.7–4.8 t ha^{-1}, with the content of branches of a diameter of less than 1 cm to range between 0.4 and 0.9 t ha^{-1}. Usenja (1997) estimated the amount of FCs in canopy layers of pine stands of 15–80 years for western Europe to range from 2.1 to 4.1 t ha^{-1} (pine with lichens) to 5.9 to 9.5 t ha^{-1} (pine with *Oxalis*). For all five forest types investigated, the maximum amount of FC was reported in stands of 40–50 years of age.

The variability of the aerial FCs is very high. For instance, the storage of needles (dry weight) in pine stands with lichens in middle taiga of the European north was estimated to range between 2.18 and 5.05 t ha^{-1} (or 3.5–14.2% of total aboveground phytomass) in 40- to 51-year-old stands, and up to 7.12 t ha^{-1} (11.5%) in stands 80 years of age (Rusanova and Sloboda 1974; Kazimirov et al. 1977).

The data of Table 16.4, as well as average zonal data of CWD, were used in calculations of the carbon emission caused by crown fire and for estimation of postfire dieback on forested areas.

Fraction of Biomass Consumed

The fraction of forest fuels consumed during a fire depends on the type and intensity of the fire, forest type and site characteristics, previous and current weather conditions, timing of the fire during the growing season, and composition and humidity of the FCs. Many scientists have observed that it requires a minimum level of basic burning conductors (BBC) to create a forest fire, estimated to be 1.5–250.0 t ha^{-1}, depending on the weather conditions and the spatial distribution of FCs (Ivanova 1978; Furayev 1996). The critical moisture content under which burning occurs depends on the types of FCs. Generally, ground fuels (mosses, lichens, herbaceous material, and litter) become flammable with a water content of 25–40% (Sofronov 1967; Zhukovskaja 1970). Once a fire is ignited, the resultant heat can burn fuels with higher moisture contents. For example, Vonsky and associates (1974) showed that the total consumption of litter with mosses and lichens was observed with a water content of 43% and for litter and sphagnum from 85 to 122%. Similar data for lichens and dry herbs were obtained in experiments by Valendik and Gavel (1975).

The amount of fuel consumed has been estimated from data collected in many field measurements, usually in controlled burns. In 64 experimental fires in pine stands in southern taiga, Furyaev (1996) estimated that the amount of biomass consumption was 73–96% for on-ground fuels ranging from 6.4 to 34.6 t ha^{-1}. Where the fuels were greater than 21.0 t ha^{-1}, the consumption was more than 92%. Similar data have been reported by Kasischke and co-workers (1995) for Alaska and by Valendik and Isakov (1978) in other forests of Russia.

According to Vasilenko (1976), the average consumption coefficient in larch forests in Sakhalin Island and in the Lake Baikal region, where the average FC was 14.0–16.0 t ha^{-1}, depends on the season, forest type, and specifics of sites. Consumption levels were 50–80% in spring and 100% in in summer and autumn. Consumption of the undergrowth in broad-leaved coniferous forests of Far East is estimated to be 62–64% (Kozlov 1973). An extremely high consumption is observed in forests damaged by winds and insect outbreaks. For instance, in dark coniferous stands of west Siberia killed by an insect infestation, the average amount of combustibles was about 130.0 t ha^{-1} and consumption 90–100% (Furyaev 1970). Our measurements in wind-fallen dark coniferous forests in 1964 on the Sakhalin Island indicated a maximum level of FCs of 160.0 t ha^{-1} and the consumption during a severe fire between 67 and 93%.

Consumption of litter is extremely variable. Surface fires in spring to early summer consume only a thin layer of fresh litterfall, 10–15% of on-ground combustibles. In on-ground fires of light to medium intensity, the amount of litter burned is estimated to be some 30% in the Ural Mountain forests (Firsova 1960) and up to 50% in central Siberia (Popova 1978). Average estimates for forests of the Far East are 20–25%. According to Furyaev (1996), consumption of litter by fire of low and medium intensity in west and middle Siberia forests is about 30%.

Based on numerous results of measurements, we developed models that allow us to estimate the average percentage of consumed on-ground FCs as a function of their total amount. The models are presented in a form of regression equations:

$$C_{st} = 103.5 - 28.00/x^1 + 1.4583/x^2 \tag{16.6a}$$

and

$$C_{su} = 31.25 - 6.26/x^1 \tag{16.6b}$$

where C_{st} and C_{su} are percentages of consumed fuel under steady and superficial on-ground fires, and x is the amount of on-ground combustibles (kg/m^2, $0.2 \leq x \leq 5.0$).

During crown fires, trunk wood and branches of trees (>7–10 mm in diameter) usually do not burn, but needles and dry branches burn completely. The amount of burned biomass and the strength of the fire increase significantly in forests with a readily combustible shrub layer or in areas with slash from forest cutting. Kurbatsky (1970) has estimated the maximum amount of aboveground phytomass that can be burned in a crown fire to be 20–30% of the total aboveground phytomass.

In our calculations, we used the following assumptions for the consumption of FC under crown fires: (1) bark of stems and branches, 30%; (2) needles, 100%; (3) twigs, 70% of mass of needles; (4) detritus (dry stems, dry fall-down wood, and dry branches of living trees), 90%; (5) on-ground combustibles, as in steady on-ground fires.

Consumption at peat and underground fires (and to some extent under stable humus-litter fire) depends mostly on the water table level and the water content of peat. The consumption rate is usually very high and can reach 50–90% of the total organic matter in organogenic horizons of burnt areas. In our calculations, we used a conservative assumption that the average consumption of peat is 50%. Based on many indirect evidence, the average depth of burned peat layer has been estimated at 15 cm, average densities of peat was estimated based on regional ratios of areas of different types of peat areas (Sabo et al. 1981; two categories were used—high moor peat with a density of 0.08–0.13 g m^{-3}, and fen peat, density 0.10–0.30 g m^{-3}). The average density of peat comprised 0.155 g m^{-3}. For an underground fire, all assumptions used were the same, excluding the depth of burnt-out layer, which has been estimated at 0.6 m.

Carbon Coefficent (γ)

In this study, we used $\gamma_q = 0.5$ for woody FCs, 0.45 for the rest of vegetation, and 0.58 for peat (Filippov 1968; Telizin 1973).

Development of Model Parameters: Postfire Carbon Release

Fraction of Carbon Released (χ)

Generally, χ depends on many factors, and very few values have been reported in the literature. These values range from 0.77 to 0.92 (Chagina 1970; Kurz et al. 1992; Vedrova 1995). In this study, we used an average value of 0.88.

Decomposition Rates for Organic Matter (α_{ij} and $T_{0.95}$)

The rate of decomposition of organic matter (α_{ij}) is highly variable, depending on geographic zones, ecosystem and site characteristics, tree species and size of residuals, and other factors (Academy of Sciences 1983; Kobak 1988; Bazilevich 1993b). Stemwood of larch in the forest-tundra on permafrost is estimated to have a decomposition time of 200–300 years (Ivshin 1993). Glasov (1989) estimates the average decomposition time for spruce stem wood in the southern taiga to be

25–30 years. Storoshenko (1990) showed a complete decomposition cycle for uneven-aged spruce stands in the southern taiga of up to 70 years. Krankina and Harmon (1995) reported the decomposition rates of pine and spruce stems over the first 70 years of decomposition to be 0.034 yr^{-1} and 0.045 yr^{-1} for birch in Russian northwestern boreal forests. Our estimates of the decomposition rate for 95% of the stem wood of aspen and white birch is up to 0.037 yr^{-1} in the southern Far East and 0.016 yr^{-1} for stone and yellow birch. Even within a climatic zone, the variation may be very large. Bogatiriev and Fless (1983) showed that for the northern taiga in the Komi Republic, the period for litter decomposition was 100–300 years for wetlands and 2–10 years in productive stands. Taking into account the uncertainties of initial data and the limited knowledge of the rate of the decomposition of the woody pools, we used average zonal estimates for the organic matter decomposition rate shown in Table 16.5, which have been estimated based on many publications. In Table 16.5, α_{ij} is a coefficient from Equation (20.5). We used data from Table 16.5 for both above- and below-ground decomposition pools. The values within the parentheses identify the reported range in years.

Amount of Organic Matter (O (t))

As can be seen from Table 16.5, the time it takes for tree materials to decompose can be nearly 200 years in the case of larch stems in the forest tundra site. This means that ideally O_{ij} (t) should be quantified for the period from t_1 (i.e., 1990) back to the 18th century. Because official statistical data before 1988 are not accurate (Chapter 8), O_{ij} (t) was estimated indirectly.

For the period from 1961 to 1988, regional areas of different forest stand classes estimated by the Russian forest inventory are thought to be relatively reliable for

Table 16.5. Rate of Organic Matter Decomposition by Different Decomposition Pools

	Fast (litter) pool		Medium-fast pool		Slow pool	
Zone	α	$T_{0.95}$	α	$T_{0.95}$	α	$T_{0.95}$
Subarctic and tundra	0.038	78.8 (50–110)	0.03	99.9	—	—
Forest tundra, sparse taiga, and meadow forest	0.072	41.6 (25–60)	0.043	69.7	0.017	176
Northern taiga	0.16	18.7 (15–35)	0.075	39.9	0.027	111
Middle taiga	0.32	9.4 (5–20)	0.097	30.9	0.03	100
Southern taiga	0.75	4.0 (2–8)	0.16	18.7	0.047	64
Mixed forest, deciduous forests, and forest steppe	1.2	2.5 (1–5)	0.27	11.1	0.07	43
Steppe, semidesert, and desert	4.0	0.75 (0.2–1.5)	0.37	8.1	0.13	23

all the Forest Fund lands managed by state forest authorities (including long-term leased areas). These data indicate the following areas of burned and dead forests: 1961, 70.6 million ha; 1966, 68.4 million ha; 1973, 53.6 million ha; 1978, 43.9 million ha; 1983, 36.8 million ha; 1988, 34.9 million ha; and 1993, 35.2 million ha (SNKh SSSR 1962; Goskomles SSSR 1968, 1976, 1986, 1990, 1991; FSFMR 1995).

For this study, O_{ij} (t) [1961–1988] was quantified based on regional estimates of (1) burnt areas and dead stands, which are a consequence of stand-replacing fires; (2) ratios between areas of different types of fires; (3) quantitative characteristics of natural forest regeneration on burned areas; and (4) following the hypothesis that these regularities as well as the average consumption of forest fuel were constant during the period considered. The reconstruction of O_{ij} (t) for the period 1800–1960 was made using mainly regional expert estimates and taking into account (1) some fragmentary data on the fire history available in Russian forest and geographic literature; (2) current distribution of forests by levels of transformation (pristine, natural, and anthropogenic, as well as indigenous and secondary, forests were considered), types of age stand structure (areas of uneven-aged, relatively uneven-aged, graduated uneven-aged, and even-aged forests were estimated), and age classes; (3) an assumption that during 1800–1920, areas of burned forests linearly increased due to human intervention for about 0.27% (of areas of burns by 1800) annually, and before these data burnt areas have been in equilibrium of natural fire cycles, estimated for the boreal zone at the level of 6.7% for the Russian forestlands. Dynamics of organic input for decomposition was described in the form

$$O_{ij}\,(t) = b_{ij} \cdot O_{\mathrm{norm}} \tag{16.7}$$

where O_{norm} is the relative change of O_{ij} (t) compared to 1990, that is, $O_{ij\ \mathrm{norm}}$ (1990) = 1, and b_{ij} is a scaling coefficient (i.e., the corresponding amount of organics by decomposition pools and zones for 1990). Finally,

$$O_{\mathrm{norm}} = 2.000 + 0.005455x, \text{ for 1800–1910} \tag{16.8a}$$

and

$$O_{\mathrm{norm}} = -1.89 + 0.06805x - 0.2781 \cdot 10^{-3}x^2, \text{ for 1910–1990} \tag{16.8b}$$

where $x = 0, 1, \ldots, 190$ for 1800, 1801, . . . , 1990 (Fig. 16.1). Using the above model, the final conclusion was that for all of Russia the total average annual burned area [and, respectively, the proportional value of O_{ij} (t)] increased (compared to 1990) from 1.71 in 1800 to about 2.27 in 1920–1930 and after that decreased to 1.0 in 1990. To check the sensitivity of the approach, the control assumption was that O_{ij} was stable during 1800–1920 and equal to 2.27, as is $O_{ij\ \mathrm{cont}}$ in Figure 16.1.

Postfire Dieback

Based on the analyses of postfire dieback in Chapter 8, we assumed that during 1988–1992, 0.24 million ha of forests died due to peat and steady on-ground fires from previous years. Areas of these stands have been distributed proportionally to zonal areas of crown and peat fires. The zonal percentage of the dieback on the rest of on-ground, steady fire areas have been estimated from 40% (forest tundra and sparse taiga) to 32% in the zones of deciduous forests and forest steppe. Data for superficial fires were from 22 to 10%, respectively. The amount of unburned residuals from crown fires was calculated as the difference between the total amount of phytomass and detritus and the estimates of the parts that had been burned out.

Changes in Organic Soil

The quantification of the postfire dynamics of soil organic carbon is very uncertain, because the processes to be considered are dependent on a variety of site and regional conditions. A usual assumption is that organic soils of undisturbed forest ecosystems are in relative equilibrium over the short term. Only some categories of forests on wet or moist sites (e.g., the major spruce forest types in Siberia and the Far East) can be considered as undisturbed. Direct losses of carbon during fires can be very large, based on the level of burning of organic soils. The changes in the mineral layers of soil are more difficult to estimate.

As a rule, prescribed burning and superficial on-ground fires do not result in any losses of mineral soil organic matter in the boreal forests (Dyrness et al. 1989). In many cases, there is a slight increase of soil organic matter during the years after fires (Johnson 1992). By contrast, steady on-ground litter and turf fires have the potential to result in considerable postfire losses of humus content. For example, Sands (1983) reports up to 40% loss during a 25-year period in the top 60 cm of soil. On average, it is thought that such results are the exception rather than the rule (Dyrness et al. 1989). Changes in the physical and chemical properties of soils (e.g., an increase in the intensity of oxidation of soil organic matter and decreased soil density and air permeability), changes in soil moisture, and even changes in the direction of soil-forming processes are common after forest fires (Snytko 1973; Saposhnikov 1978; Richter et al., this volume).

Severe fires over large areas can change many processes over an entire landscape, particularly influencing ground thermal and hydrological conditions. It has been long recognized that fire in permafrost regions slows the rate of paludification (e.g., it staves off the creation of bogs) (Stepanov 1925; Pjavchenko 1952). Two major conclusions concerning postfire effects in the taiga forests in the Far East, Siberia, and the European north are as follows:

- Site fertilization and heat effects of forest fire are common for taiga forest regions, especially for areas with permafrost.
- There are no essential changes in the amount of organic matter in the mineral horizons after most on-ground fires. The postfire rehabilitation of litter and soil

organic matter is estimated to be between 2 and 7 years in most cases but sometimes can be as long as 10–15 years (Firsova 1960; Popova 1978; Saposhnikov and Kostenkova 1984; Saposhnikov et al. 1993). The exceptions are severe peat and turf fires, from which recovery times are longer.

Based on the above observations, in our calculations we applied a simple approach using aggregated regional estimates of the total loss of the organic matter in the mineral soil layers. Consumption of litter and detritus is calculated separately as the consumed FCs. For both direct and postfire emissions from mineral soils, only losses caused by crown and steady on-ground fire were taken into account. Based on the soil map of Russia, the average content of carbon in the top 30-cm layer of (forest) mineral soils is estimated to be from 40 to 90 t ha^{-1}. Using limited results of measurements and expert estimates, we estimated average losses of carbon from the above layer to be 1.5–4.0% for different regions and types of soils, with an average of about 2.5%. For the first 3 years of postfire successions, our estimates of losses of soil humus from mineral horizons during this period varied from 2 to 11% of its content in the top 0.3 m of soil, with a total average loss of about 5%.

Estimates of Carbon Fire Emissions for 1988–1992

Based on the above methods and data, we estimated direct fire emissions and postfire biogenic fluxes (Tables 16.6 and 16.7). The conclusion is that annually during the 5-year period 1988–1992, fires in territories of the Russian Forest Fund and state land reserve emitted to the atmosphere about 127 Tg C yr^{-1}, of which 58 Tg C yr^{-1} is direct fire emission. Postfire carbon fluxes comprise about 64 Tg C yr^{-1}, of which postfire losses of SOC due to enhanced respiration are estimated to be about 9 Tg C yr^{-1}. Such a ratio is defined by the character of the curve $O_{ij}(t)$. The direct emissions generated 16.6 t ha^{-1}. If we exclude peat fire from this amount, the average is 10.8 t C ha^{-1}.

Conclusion: Uncertainties and Research Needs

Our results on the impact of forest fires in Russia on the carbon budget (emissions of 150 Tg C yr^{-1}) are lower than estimates reported by others. Stocks (1991) estimates that some 34 Tg C yr^{-1} of direct fire emission for the former Soviet Union (FSU) (excluding peat and underground fires). Dixon and Krankina (1993) estimate 47 Tg C yr^{-1} of direct emissions and 117–281 Tg C yr^{-1} of postfire fluxes as an average for the period 1971–1991 in Russia. Kolchugina and Vinson (1993) estimate 199 Tg C yr^{-1} for the FSU (during the past 20–30 years, fires in Russian forests corresponded to about 98% of the forest fires in the FSU). In a recent paper, Kolchugina and Vinson (1995) estimate the direct fire emissions for Russian forests at 78 Tg C yr^{-1} and the postfire fluxes at 42 Tg C yr^{-1}.

Table 16.6. Yearly Average Direct Forest Fire Carbon Emissions Caused by Fires, 1988–1992

		Emissions ($Tg\ C\ yr^{-1}$)		
Subject/type of fire	Area (million ha)	Total	Vegetation	Soils
Forested areas	1.40	18.1	16.6	1.5
Crown fire	0.24	5.1	4.8	0.3
On-ground fire	1.16	13.0	11.8	1.2
Unforested areas	0.48	6.5	6.1	0.4
Vegetative nonforest lands	0.90	5.7	5.4	0.3
Low-productive nonforestlands	0.39	1.0	0.9	0.1
Peat fires	0.35	23.5	3.1	20.4
Below-ground fires	0.012	3.2	0	3.2
Total	3.32	58.0	32.1	25.9

The average estimate on aboveground organic matter consumed during fires in boreal forests is reported to be 11.3 $t\ C\ ha^{-1}$ (Stocks 1991; Cahoon et al. 1994). Kasischke and colleagues (1995) reported significantly higher releases for fires in the Alaskan boreal forests during 1990 and 1991 fire seasons, from 22.5 to 26.1 $t\ C\ ha^{-1}$ for forested and 2.9 to 3.9 $t\ C\ ha^{-1}$ for nonforested areas. Seiler and Crutzen (1980) used an average value for boreal forests of 16.9 $t\ C\ ha^{-1}$. These estimates can be compared with our data for Russia during 1988–1992: a general average total emission of 16.6 $t\ C\ ha^{-1}$, and without peat fires, 10.8 $t\ C\ ha^{-1}$.

To evaluate uncertainties, we used sensitivity analysis (see, e.g., Kendall and Stuart 1966) based on personal (a priori) probabilities. Such an approach does not contribute significantly to the estimation of the real accuracy of the results and confidence intervals. But it estimates the impact of changes of initial data on the results. Our general conclusion is that the method and models used are rather robust for the majority of the calculations. For instance, if we provide calculations for $O_{ij\ \mathrm{cont}}$ (1800–1910), the final result changes to about 6%. The exclusions mostly deal with areas and intensity of peat and underground fires, change of soil respiration after fire on permafrost, and variation of the postfire dieback. Using the relevant expert estimations of accuracy and confidence limits of initial data and basic assumptions, it can be concluded that the total flux generated by fire on the territories defined above at 122 $Tg\ C\ yr^{-1}$ is estimated with the accuracy of about 15% at a confidence (a priori) probability of 0.85. Nevertheless, we must stress that such a numerical conclusion *for a definite year* is only true if the model is completely adequate. Evidently, this is not the case in this chapter (the same could be said about methods applied in all other available publications) because the retrospective part of the model used is deterministic, whereas fire dynamics is a stochastic process with large interseasonal variability. How much the uncertainties are changed if we include this consideration can only be estimated based on a corresponding stochastic model. We can only note that based on rough estimates it cannot change the magnitude of the results.

Table 16.7. Average Annual Postfire Emissions in 1988–1992 Resulting from Decomposition of Dieback Caused by Fires of Previous Years (1800–1988) (Tg C yr^{-1})

	Average input organic matter for decomposition in 1990 by pools				Average total postfire emissions of previous years in 1990 by pools[a]				
Vegetation zone	Fast	Medium	Slow	Total	Fast	Medium	Slow	Soil	Total
Subarctic and tundra	1.6	0.1	—[b]	1.7	2.2	0.1	—	0.7	3.0
Forest tundra and sparse taiga	2.3	0.3	1.8	4.4	2.9	0.4	2.8	2.0	8.1
Northern taiga	3.2	1.0	5.4	9.6	4.1	1.2	8.1	2.1	15.5
Middle taiga	4.6	1.6	7.4	13.6	6.8	2.0	10.9	2.4	22.1
Southern taiga	3.2	0.7	4.6	8.5	7.0	1.0	6.1	1.1	15.2
Mixed forests, deciduous forests, and forest steppe	0.6	0.3	1.7	2.6	2.1	0.4	2.1	0.2	4.8
Steppe, steppe desert, and desert	0.3	—	—	0.3	0.3	—	—	0.1	0.4
Total	15.8	3.9	20.9	40.6	25.4	5.1	30.0	8.6	69.1

[a]To get the flux from Equation (16.5), data for vegetation pools must be multiplied for the corresponding coefficients 1.05 and 0.88 [i.e., the total postfire emission is $(25.4 + 5.1 + 30.0) \times 1.05 \times 0.88 + 8.6 = 64.5$ Tg C yr^{-1}].
[b]—, No data.

References

Academy of Sciences of the USSR. 1983. *Role of Litter in Forest Biogeocenosis.* Short reports of the All-Union Meeting. Nauka, Moscow. (in Russian)

Alexandrova, V.D. 1971. Experiences from estimating above- and below-ground phytomass in the polar desert of Franz-Jozef Land, pp. 33–37 in *Biological Productivity and Turnover of Chemical Elements in Vegetational Communities.* Nauka, Leningrad. (in Russian)

Alexeyev, V.A., and R.A. Birdsey, eds. 1998. *Carbon Storage in Forests and Peatlands of Russia.* General Technical Report NE-244. U.S. Department of Agriculture, Forest Service, Northeastern Forest Experiment Station, Radnor, PA.

Bazilevich, N.I. 1993a. *Biological Productivity of Ecosystems of the Northern Eurasia.* Nauka, Moscow. (in Russian)

Bazilevich, N.I. 1993b. Geographical productivity of soil-vegetational formations in northern Eurasia. *Soil Sci.* 10:10–18. (in Russian)

Bogatirev, L.G., and A.D. Fless. 1983. Litter's structure and classification in the northern taiga forest biogeocenoses, pp. 22–23 in *Role of Litter in Forest Biogeocenoses.* Nauka, Moscow. (in Russian)

Cahoon, D.R., Jr., B.J. Stocks, J.S. Levine, W.R. Cofer III, and J.M. Pierson. 1994. Satellite analysis of the severe 1987 forest fires in northern China and southeastern Siberia. *J. Geophys. Res.* 99:18,627–18,638.

Chagina, E.G. 1970. Carbon balance under litter's decomposition in cedar forests of West Sajan Mountains, pp. 246–252 in A.B. Shukov, ed. *Problems of Forestry,* Vol. 1. Institute of Forest and Timber, Russian Academy of Sciences, Krasnoyarsk, Russia. (in Russian)

Dichenkov, N.A. 1993. *Ways to Increase the Effectiveness of Forest Fire Risk Evaluation.* Institute of Forests, Belarus Academy of Sciences, Gomel, Belarus. (in Russian)

Dixon, R.K., and O.N. Krankina. 1993. Forest fires in Russia: carbon dioxide emission to the atmosphere. *Can. J. For. Res.* 23:700–705.

Dukarev, V.N. 1989. Investigations on biological productivity of the coniferous forests of Sikhote-Alin, pp. 9–10 in *Results of Investigations of Forests in the Far East and Tasks for Intensified Multipurpose Forest Utilization.* Far Eastern Forestry Research Institute, Khabarovsk, Russia. (in Russian)

Dyrness, C.T., K. Van Cleve, and J.D. Leviston. 1989. The effect of wildfire on soil chemistry in four forest types in interior Alaska. *Can. J. For. Res.* 19:1389–1396.

Filippov, A.V. 1968. Some pyrological properties of forest combustibles, pp. 351–358 in *Problems of Burning and Extinguishing.* All-Union Research Institute for Fire Protection, Moscow. (in Russian)

Firsova, V.P. 1960. Influence of forest fires on soils. *Pap. Inst. Biol. Ural Fil., Acad. Sci. USSR* 16:41–52. (in Russian)

FSFMR. 1995. *Forest Fund of Russia* (state by January 1, 1993). Federal Service of Forest Management of Russia, Moscow, Russia. (in Russian)

Furyaev, V.V. 1970. Impact of fires and insect infestations on formation of forests between rivers Ket and Culim, pp. 408–421 in A.B. Shukov, ed. *Problems of Forestry,* Vol. 1. Institute of Forest and Timber, Russian Academy of Sciences, Krasnoyarsk, Russia. (in Russian)

Furyaev, V.V. 1996. *Role of Fires in the Forest Regeneration Process.* Nauka, Novosibirsk, Russia. (in Russian)

Gabeev, V.N. 1990. *Ecology and Productivity of Pine Forests.* Nauka, Novosibirsk, Russia. (in Russian)

Glasov, M.V. 1989. Bioproductivity and pecularities of biological turnover in spruce forest ecosystems, pp. 52–53 in *Ecology of Forests in the North.* Short reports of the All-Union Meeting, October 2–7, 1989, at Siktivkar, Russia. Nauka, Moscow. (in Russian)

Goskomles SSSR (before 1988, Gosleshoz) (Forest Fund of the USSR). 1968 (status as of January 1, 1966, p. 744); 1976 (status as of January 1, 1973, vol. 1, vol. 2, p. 561, vol. 3, p. 800); 1982 (status as of January 1, 1978, vol. 1, p. 601, vol. 2, p. 683); 1986 (status as of January 1, 1983, vol. 1, p. 891, vol. 2, p. 973); 1990 (status as of January 1, 1988, vol. 1, p. 1005); 1991, vol. 2, p. 1021. USSR State Committee of Forests, Moscow. (in Russian)

Grishin, A.M. 1992. *Mathematical Modeling of Forest Fires and New Methods for Fighting Them.* Nauka, Novosibirsk, Russia. (in Russian)

Isaev, A., G. Korovin, D. Zamolodchikov, A. Utkin, and A. Pryashnikov. 1995. Carbon stock and deposition in phytomass of the Russian forests. *Water Air Soil Pollut.* 82:247–256.

Ivanova, G.A. 1978. Impact of herbs on the speed of on-ground fires, pp. 145–147 in N.P. Kurbatsky, ed. *Burning and Fire in Forests.* Institute of Forest and Timber, Russian Academy of Sciences, Krasnoyarsk, Russia. (in Russian)

Ivshin, A.D. 1993. *Influence of Air Pollutants from the Norilsk Metallurgical Enterprise on the State of Spruce-Larch Stands.* Institute of Ecology of Plants and Animals, Ekaterinburg, Russia. (in Russian)

Johnson, D.W. 1992. Effects of forest management on soil carbon storage. *Water Air Soil Pollut.* 64:83–120.

Kasesalu, C.P. 1972. *Interaction Between Forest Vegetation and Soil in Dry Pine Stands of Estonia.* Tartu State University, Tartu, Estonia. (in Russian)

Kasischke, E.S., N.H.F. French, L.L. Bourgeau-Chavez, and N.L. Christensen, Jr. 1995. Estimating release of carbon from 1990 and 1991 forest fires in Alaska. *J. Geophys. Res.* 100:2941–2951.

Kazimirov, N.I., A.D. Volkov, S.S. Zjabchenko, A.A. Ivanchikov, and R.M. Morozova. 1977. *Exchange of Matter and Energy in Pine Forests of European North.* Nauka, Leningrad. (in Russian)

Kendall, M.J., and A. Stuart 1966. *Theory of Probability Distribution.* Nauka, Moscow. (in Russian)

Kobak, K.I. 1988. *Biotic Compounds of the Carbon Cycle.* Hydrometeoizdat Press, Leningrad. (in Russian)

Kolchugina, T.P., and T.S. Vinson. 1993. Comparative analysis of carbon budget components for forest biomes in the Former Soviet Union. *Water Air Soil Pollut.* 70:207–227.

Kolchugina, T.P., and T.S. Vinson. 1995. Role of Russian forests in the global carbon balance. *Ambio* 24:258–264.

Kozlov, A.G. 1973. Estimation of above-ground phytomass and amount of forest combustibles of undergrowth of coniferous species in broadleaved-coniferous forests of the Far East, pp. 234–241 in A.S. Ageenko, ed. *Increased Productivity of Far Eastern Forests,* Vol. 13. Far Eastern Forestry Research Institute, Khabarovsk, Russia. (in Russian)

Krankina, O.N., and M.E. Harmon. 1995. Dynamics of the dead wood carbon pool in northwestern Russian boreal forests. *Water Air Soil Pollut.* 82:227–238

Kurbatsky, N.P. 1970. Classification of forest fires, pp. 384–407 in A.B. Shukov, ed. *Problems of Forestry,* Vol. 1. Institute of Forest and Timber, Russian Academy of Science, Krasnoyarsk, Russia. (in Russian)

Kurnaev, S.F. 1973. *Forest Growth Division of the USSR.* Nauka, Moscow. (in Russian)

Kurz, W.A., M.J. Apps, T.M. Webb, and J.P. McNamee. 1992. *The Carbon Budget of the Canadian Forest Sector: Phase I.* Internal Report NOR-X-32y. Forestry Canada, Northwest Region, Northern Forestry Centre, Edmonton, BC, Canada.

Peshkov, V.V. 1991. Impact of cleaning of clear cut areas on fire danger and intensity in northern forest of Amur oblast, pp. 89–96 in D.F. Efremov, ed. *Scientific Backgrounds of Forestry in Far East,* Issue 33. Far Eastern Forestry Research Institute, Khabarovsk, Russia. (in Russian)

Pjavchenko, N.I. 1952. Reasons for paludification of pine burns. *Forestry* 12:56–64. (in Russian)

Popova, E.D. 1978. Duration of pyrogenic influence on the properties of forest soils, pp. 185–186 in *Burning and Fires in Forests.* Russian Academy of Sciences, Krasnoyarsk, Russia. (in Russian)

Rojkov, V., V. Vagner, S. Nilsson, and A. Shvidenko. 1997. Carbon of Russian wetlands, pp. 112–113 in *Fifth International Carbon Dioxide Conference.* Cairns, Queensland, Australia, September 8–12, 1997. Extended Abstracts. CSIRO Division of Atmospheric Research, Aspendale, Australia.

Rusanova, G.V., and A.V. Sloboda. 1974. Biological productivity of pine stand with lichens in middle taiga of Komi Republik. *Bot. J.* 59:1827–1838. (in Russian)

Sabo, E.D., Yu.N. Ivanov, and D.A. Shatilo. 1981. *Reference Book of Hydro-Amelioration.* Forest Industry, Moscow. (in Russian)

Sands, R. 1983. Physical changes to sandy soils planted by *Pinus radiata,* pp. 146–152 in R. Ballard and S.P. Gessel, eds. *IUFRO Symposium on Forest Site and Continuous Productivity.* Report GTR PNW-163. U.S. Department of Agriculture, Forest Service, Washington, D.C.

Saposhnikov, A.P. 1978. Role of fire in forest soil formation, *Ecology* 1:43–46. (in Russian)

Saposhnikov, A.P., and A.F. Kostenkova. 1984. Influence of forest fire on litter in cedar forests of the southern Sikhote-Alin and peculiarities by the regeneration, pp. 139–146 in *Dynamic Processes in Forests of the Far East.* Far Eastern Scientific Center, Russian Academy of Sciences, Vladivostok, Russia. (in Russian)

Saposhnikov, A.P., G.A. Selivanova, and T.M. Iljna. 1993. *Soil Generation and Peculiarities of Turnover in Mountain Forests of Southern Sikhote-Alin.* Far Eastern Forestry Research Institute, Khabarovsk, Russia. (in Russian)

Seiler, W., and P.J. Crutzen. 1980. Estimates of gross and net fluxes of carbon between the biosphere and atmosphere. *Clim. Change* 2:207–247.

Semetchkina, M.G. 1977. Structure of Minnsinsk pine forests with regard to phytomass components, pp. 76–90 in *Inventory Investigations of Siberian Forests.* Russian Academy of Sciences, Krasnoyarsk, Russia. (in Russian)

Shepashenko, D., A. Shvidenko, and S. Nilsson. 1998. Phytomass and carbon of Siberian forests. *Biomass Bioenergy* 14:21–32.

Sheshukov, M.A. 1978. Impact of fires on development of taiga biogeocenoses, pp. 166–167 in N.P. Kurbatsky, ed. *Burning and Fire in Forests.* Institute of Forests and Timber, Russian Academy of Sciences, Krasnoyarsk, Russia. (in Russian)

Shvidenko, A., and S. Nilsson. In press. Phytomass, increment, mortality and carbon budget of Russian forests. *Clim. Change.*

Shvidenko, A., S. Nilsson, V. Roshkov, and V. Strakhov. 1996. Carbon budget of the Russian boreal forests: a systems approach to uncertainty, pp. 145–162 in M. Apps and D. Price, eds. *Forest Ecosystems, Forest Management and the Global Carbon Cycle,* NATO ASI Series, Series I, Vol. 40, Springer-Verlag, Heidelberg, Germany.

SNKh SSSR. 1962. *Forest Fund of the USSR.* Council of the National Economy, Moscow. (in Russian)

Snytko, V.A. 1973. Geochemical research on topogeosystems, pp. 3–10 in *Topological Aspects of Behavior of Matter in Geosystems.* Institute of Geography, Academy of Sciences of the USSR, Irkutsk, Russia. (in Russian)

Sofronov, M.A. 1967. *Forest Fires in the Mountains of Southern Siberia.* Nauka, Moscow. (in Russian)

Stepanov, N.N. 1925. Physical and chemical peculiarities of forest burns, pp. 85–95 in *Reports on Forest Experimental Activities,* Vol. 2. Forest Management and Forest Industry, Leningrad. (in Russian)

Stocks, B.J. 1991. The extent and impact of forest fires in northern circumpolar countries, pp. 197–202 in J.S. Levine, ed. *Global Biomass Burning: Atmospheric, Climatic, and Biospheric Implications.* MIT Press, Cambridge, MA.

Storoshenko, V.G. 1990. Some criteria on stability of forest ecosystems, pp. 298–299 in *Problems of Stability of Biological Systems.* Short reports of the All-Union School, October 15–20, 1990. Russian Academy of Sciences, Moscow. (in Russian)

Telizin, G.P. 1973. Elementary composition of forest combustibles in the Far East, pp. 351–358 in *Utilization and Regeneration of Forest Resources in the Far East.* Reports of the Far Eastern Forestry Research Institute 15. Far Eastern Forestry Research Institute, Khabarovsk, Russia. (in Russian)

Usenja, V.V. 1997. Investigation of the amount of forest combustibles of different groups in pine stands, pp. 124–127 in V.F. Baginsky, ed. *Problems of Forest Science and Forestry.* Institute of Forests, Belarus National Academy of Sciences, Gomel, Russia. (in Russian)

Usoltsev, V.A. 1988. Tables of measurements of forest combustibles under crown fires, pp. 148–155 in *Forest Inventory and Forest Management.* Lithuanian Agricultural Academy, Kaunas, Lithuania. (in Russian)

Usoltsev, V.A., A.A. Makarenko, and A.S. Atkin. 1979. Regularities of aboveground phytomass formation in northern Kazakhstan. *For. Sci.* 5:5–12. (in Russian)

Valendik, E.N., and N.F. Gavel. 1975. The share of different burned forest materials, pp. 127–137 in *Problems of Forest Pyrology.* Institute of Forests and Timber, Russian Academy of Sciences, Krasnoyarsk, Russia. (in Russian)

Valendik, E.N., and R.V. Isakov. 1978. Intensity of forest fires, pp. 40–55 in *Prediction of Forest Fires.* Institute of Forests and Timber, Russian Academy of Sciences, Krasnoyarsk, Russia. (in Russian)

Vasilenko, A.V. 1976. Role of fire in forestry, pp. 98–102 in V.G. Chertovsky, ed. *Current Research in Forest Typology and Pyrology.* Arkhangelsk Institute of Forests and Forest Chemistry, Arkhangelsk, Russia. (in Russian)

Vedrova, E.F. 1995. Carbon pools and fluxes of 25-year-old coniferous and deciduous stands in middle Siberia. *Water Air Soil Pollut.* 82:230–246.

Vedrova, E.F., L.S. Shugalej, and I.N. Beskorovajnaja. 1989. Development and properties of litter in planted forest biogeocenoses, pp. 77–78 in D.F. Efremov, ed. *Results of Studies of the Far Eastern Forests and Problems Connected with Intensified Multiple Forest Use.* Far Eastern Forestry Research Institute, Khabarovsk, Russia. (in Russian)

Vilchek, G.E. 1987. Productivity of typical tundra in Taimir. *Ecology* 5:38–43. (in Russian)

Vonsky, S.M., V.A. Shdanko, and L.V. Tetjusheva. 1974. Impact of precipitation on change of humidity and burnability of forest green floor and litter, pp. 66–72 in *Forest Fires and Technical Means of Fighting Them.* Reports of the Leningrad Forestry Research Institute, Vol.19. Leningrad Forestry Research Institute, Leningrad. (in Russian)

Zhukovskaja, V.I. 1970. Moistening and drying up of hygroscopic forest combustibles, pp. 105–141 in *Questions of Forest Pyrology.* Institute of Forests and Timber, Siberian Division, Academy of Sciences of the USSR, Krasnoyarsk, Russia. (in Russian)

Zuranov, V.P. 1973. Dynamics of dryness in spruce stands and specifics of forest regeneration in the north of Sikhote-Alin, pp. 241–244 in A.S. Ageenko, ed. *Increase of Productivity of Forests in the Far East.* Reports of the Far Eastern Forestry Research Institute 13. Far Eastern Forestry Research Institute, Khabarovsk, Russia. (in Russian)

Zvetkov, P.A., and V.V. Ivanova. 1985. Amount of slush after utilization of aggregate harvest techniques, pp. 124–132 in N.P. Kurbatsky, ed. *Forest Fires and Their Consequences.* Institute of Forests and Timber, Krasnoyarsk, Russia. (in Russian)

17. Using Visible and Near-Infrared Satellite Imagery to Monitor Boreal Forests

Frank J. Ahern, Helmut Epp, Donald R. Cahoon, Jr.,
Nancy H.F. French, Eric S. Kasischke, and
Jeffery L. Michalek

Introduction

During the 1970s, two satellite remote sensing systems launched by the United States (Landsat and AVHRR) fostered the beginning of the modern remote sensing era. These systems spurred an entire generation of scientists to develop new approaches for the use of satellite imagery to monitor the earth's surface, leading to the operational use of remote sensing satellites for a broad range of applications.

The Landsat and AVHRR satellite sensors epitomize an entire class of earth-observing systems operating in the visible and infrared regions of the electromagnetic spectrum. They detect two categories of signatures originating from the earth's surface, those resulting from (1) reflected solar radiation and (2) emitted thermal energy. This chapter deals with satellite remote sensing systems that record the signal origination from reflected solar radiation in the visible and near-infrared region of the electromagnetic spectrum.

Satellite remote systems operating in the visible and near-infrared region of the electromagnetic spectrum use optics or filters to split the electromagnetic spectrum into different wavelength regions while arrays of sensors sensitive to electrical impulses detect the signals in these regions. The electrical impulses detected by the sensors are converted to digital values. The optics of these systems are mechanically rotated to scan the earth's surface in the cross-track direction, with the satellite movement providing the capability to create the continuous advancement of scans required to generate an image. The digital signals are transmitted to a ground-receiving station, where they are recorded. The signals can then be computer processed to re-create an image of the earth's surface of the areas that were scanned. This class of sensor is typically referred to as a multispectral scanner (MSS).

The original MSS onboard the Landsat satellite had four different channels, a ground resolution of 80 m, and collected imagery with a swath width of 185 km. Since the launch of the original Landsat satellite in 1972, an additional three systems with the same characteristics were deployed. In 1982, the United States launched an improved version of the Landsat MSS, the Thematic Mapper (TM), which contained seven channels of data and an improved resolution (Table 17.1). Finally, in 1979 the United States deployed a coarser-resolution MSS system called the Advanced Very High-Resolution Radiometer (AVHRR) onboard its weather satellites, which had a 1.1-km ground resolution (at nadir) but a much larger swath width (2,500 km). This allowed for imaging the earth's surface once per day.

In this chapter, we first review the basic characteristics and operation of these systems, as well as one additional system, the SPOT system launched by the French government in 1987. Although other MSS systems have been deployed, the data collected by these systems are typical of this class of sensor. We then discuss different types of information on surface characteristics that can be produced through analysis of signatures recorded by these sensors. Sections are then presented that discuss the use of these data for (1) mapping of forest and vegetation cover, (2) detection and mapping of burn scars, (3) estimation of burn severity, and (4) monitoring of seasonal growth characteristics.

Characteristics and Capabilities of Operational Satellites

NOAA AVHRR

The U.S. National Oceanographic and Atmospheric Administration (NOAA), has launched a series of meteorological satellites commonly referred to by the name of this organization. The primary earth-imaging sensor on the NOAA satellites is called the AVHRR (Kidwell 1991). The sensor was not originally intended for land-surface measurements but for monitoring cloud and sea-surface temperatures. Its name is rather misleading inasmuch as it is the lowest-resolution sensor used regularly by earth scientists (see Table 17.1). Beginning with NOAA 6 (launched in 1979), the spectral bands of the AVHRR were modified to provide imagery in the red and near-infrared spectral regions, making its data very attractive for low-spatial-resolution, high-temporal-resolution monitoring of large areas of the earth's land surface.

The AVHRR system uses a spin-scan radiometer with a single stable detector for each spectral band. Although this results in negligible within-scene radiometric variations, because the radiometer was not intended for long-term quan-

Table 17.1. Summary of Sensor Characteristics for Different Satellite Systems

	Satellite					
	NOAA	Landsat 1–3	Landsat 4–5	SPOT 1–3	SPOT 1–3	
Sensor	AVHRR	MSS	TM	HRV[a]	HRV	
Spatial resolution (m)	1,100	80	30	20	10	
Swath (km)	2,940	185	185	60	60	
Repeat (days)	1	18	16	3–5	3–5	
Spectral bands	Wavelength bands (μm)					Band features
Panchromatic					0.51–0.73	Covers green, red, and near-infrared
Blue			0.45–0.52			Atmospheric aerosols, water solids
Green			0.52–0.60	0.50–0.59		Vegetation chlorophyll absorption
Red	0.58–0.68		0.63–0.69	0.61–0.68		Vegetation chlorophyll absorption
Near-infrared	0.73–1.10		0.76–0.90	0.79–0.89		Foliage, high reflection
Shortwave infrared			1.55–1.75			Water absorption, vegetation structure
Shortwave infrared			2.08–2.35			Mineral and water absorption, vegetation structure
Middle infrared	3.55–3.93					Fires, volcanoes
Thermal infrared	10.3–11.3		10.4–12.5			Earth surface temperature, atmospheric water vapor
Thermal infrared	11.5–12.5					Earth surface temperature, atmospheric water vapor

[a]HRV, haute resolution visible.

titative monitoring of the earth, there is no in-flight calibration capability. Considerable effort has been expended on postlaunch radiometric calibration using terrestrial targets such as White Sands, New Mexico (Teillet and Holben 1994). Despite these efforts, the accuracy of the absolute calibration of data from the AVHRR series is not suitable for many applications. There are several sources of variations in the illumination and viewing geometry in the AVHRR sensor. First, there is the wide field of view of the sensor necessary to achieve daily global coverage. Second, there is the fact that many NOAA satellites do not cross a given latitude at the same time each day (because of orbital degradation of the older satellites). These cause large variations in the radiance received from any given ground-cover type. The wide field of view also subjects AVHRR data to widely varying atmospheric conditions. In clear areas, the most serious of these is varying aerosol haze. There is no on-board sensor to measure this haze and no fully satisfactory way to remove it from data. Finally, subresolution cumulus clouds and their shadows are another source of radiometric contamination.

Several groups have devoted considerable effort toward dealing with these problems, and progress has been reported (Cihlar et al. 1994, 1996, 1997). In addition, some mathematical operations, such as computing the normalized difference vegetation index (NDVI), can decrease the effects of variable illumination, viewing geometry, and spatially varying aerosols in the atmosphere.

Geometric correction of AVHRR data is also difficult. The large field of view results in severe panoramic distortion, with the spatial resolution increasing from 1.1×1.1 km at nadir to 4×6 km near the edge of the swath. Because the satellite transmits imprecise ephemeris information, correction of systematic geometric distortions based on accurate knowledge of the satellite attitude and position cannot be performed. Geometric correction requires acquisition of numerous ground control points per scene. The critical points near the edge of the scene are imaged with low resolution and are thus imprecise. The difficulty of carrying out accurate geometric corrections creates substantial registration errors in multitemporal images. Depending on the processing used, these can appear as blurring or as radiometric anomalies in multitemporal images. Cihlar and co-workers (1996) have estimated the effective resolution of multitemporal composite images to be no better than about 8 km. The large ground resolution also means that many pixels consist of mixtures of two or more ground-cover classes.

Despite these shortcomings, the AVHRR has been used successfully for a number of forest monitoring applications. One of the earliest was in monitoring extensive deforestation of tropical rainforests. Indeed, AVHRR was probably the most effective data source in bringing world attention to the rapid deforestation in parts of the Brazilian Amazon in the 1980s (Nelson and Holben 1986; Malingreau and Tucker 1988). AVHRR has provided approximate estimates of tropical deforestation, but accurate figures require data from higher-resolution satellites.

At present, researchers are beginning to overcome many of the difficulties inherent in AVHRR data and extract biophysical information, such as absorbed photosynthetically active radiation, from the data. This enables scientists to refine data analysis and modeling capabilities and prepare for the higher-quality data expected from NASA's moderate-resolution imaging spectroradiometer (MODIS) sensor, which will have slightly greater resolution (500 × 500 m) and a greatly increased number of spectral channels, 36 in all.

Landsat and SPOT

The U.S. Landsat and French Système pour l'Observation de la Terre (SPOT) were the first remote sensing satellite systems designed primarily for routine operational use for a large number and variety of natural resource management applications. Landsat-1 was launched in 1972 and intended to serve as a proof of concept. Anticipating positive results, the Landsat program managers included follow-on satellites to ensure data continuity, a data archiving and distribution system to address the needs of operational users, and a vigorous applications development and technology transfer program to develop the market for this revolutionary form of data.

The Landsat and SPOT satellites were placed into sun-synchronous orbits, which means that they cross a given latitude in midmorning each day. This minimizes variations in the solar illumination and viewing geometry. The midmorning overpass time allows for good solar illumination and precedes the afternoon buildup of cumulus clouds at temperate latitudes. The Landsat orbit is also designed to provide repetitive coverage of a fixed pattern of scene centers on the earth's surface. This greatly simplifies predicting, requesting, and ordering data. Landsat has monitored much of the earth more or less continually since 1972, providing a very valuable archive for studies of land-cover change in many regions of the world.

Landsats 1 to 3 carried a MSS with four spectral bands and a spatial resolution of 80 m. Landsats 4 and 5 carried an improved MSS, called the TM, with seven spectral bands and a spatial resolution of 30 m (120-m resolution in the thermal infrared band) (see Table 17.1). The technical improvements incorporated into the TM made its data much more usable for a wide range of applications in forested environments.

The SPOT system was designed as a commercial system from the outset. Each satellite carries two haute resolution visible instruments, each observing a swath 60 km in width. A movable mirror allows SPOT to collect data anywhere within an accessibility swath of 950 km under the satellite. Thus it is designed to collect data for customers who have ordered particular scenes. Routine observation of the earth is a second priority. Depending on latitude, SPOT has an opportunity to collect an image of a given location on earth at least every 5 days. SPOT data can be collected in a panchromatic mode with 10-m resolution and a multispectral mode with 20-m resolution. The SPOT multispectral bands include the visible and near-infrared regions of the electromagnetic spectrum, which the early Landsat satellites showed to be very useful for land-cover discrimination. The TM contains two bands in the shortwave infrared. This region has been found to be particularly effective for mapping vegetation damage, including forest fire damage (Ahern and Archibald 1986; Vogelmann and Rock 1986; Ahern et al. 1991; Michalek et al., in press).

As Landsat and SPOT are operated as part of commercial systems, data from these satellites are available through numerous distributors, and assistance for users is readily available. However, the high cost of the data has restricted its use, particularly for research purposes.

Radiometric corrections for the Landsat and SPOT sensors can be conveniently divided into relative correction and absolute calibration. Relative corrections remove within-scene radiometric variations caused by the unequal response functions of the multiple detector elements used for each band in these sensors. Relative corrections are usually derived from statistical information from the data to be corrected and have been routinely and effectively applied by most data suppliers since about 1975. Absolute calibration refers to the establishment of a known and accurate relationship between the pixel intensity (almost always on an 8-bit or 256 gray-level scale) for each band and the corresponding scene radiance. Both Landsat and SPOT sensors have included on-board calibration lamp systems to facilitate absolute calibration.

There has been minimal demand for absolute calibration of Landsat data, and not all Landsat data suppliers have included the calibration coefficients in their data products. If global change investigators require well-calibrated data, they can, in principle, perform the calibration themselves if calibration lamp data are included with their image data products. In practice, this can be difficult, especially if the data supplier cannot furnish accurate and complete documentation concerning the calibration procedure used to produce their data products.

The absolute calibration of the SPOT system has been more rigorously controlled, and users of SPOT data can usually be confident in the absolute calibration of original data products delivered by suppliers.

Landsat and SPOT data products are normally supplied with several different levels of geometric corrections. Level 0, or raw data, has no geometric corrections. User correction of such products requires very specialized software. Level 1 corrections (referred to as system corrections) remove all within-scene distortions by using an accurate model of sensor distortions in which the data are provided with the scan lines oriented perpendicular to the satellite ground track. Level 2 corrections (referred to as precision corrections) provide the data in a more conventional map projection (e.g., Universal Transverse Mercator) with north at the top. Even with level 2 corrections, the data are not truly orthographical because the effects of topographical elevation have not been removed. Topographical distortions are small near nadir but can be appreciable at near the edge of TM scenes (7.5° off nadir). With oblique SPOT scenes (up to 27° off nadir), using a digital elevation model in conjunction with specialized software can produce an orthographical image. Demand for such corrections is growing as user requirements for geometric accuracy increases. This trend has accelerated as more and more users acquire geographic information systems to manage and manipulate maplike data.

Forest Cover Mapping

Image Classification Approaches

One of the unique capabilities promised with the collection of digital multispectral data from satellite systems was automated land-cover classifications that were automatic, repeatable, and reliable. Indeed, the term *spectral signature* lingers in the remote sensing vocabulary as a relic of the dream that a vast number of land-cover types could be distinguished automatically, each through its unique spectral signature. Unfortunately, many land-cover types can produce the same (or nearly the same) spectral radiance measured by satellites. In addition, a given cover type can produce many different spectral radiance measurements, depending on factors such as growth phenology, homogeneity versus heterogeneity, terrain slope and aspect, and atmospheric conditions.

Nonetheless, the concept of mathematical classification of remotely sensed data into land-cover types remains valid. Numerous techniques have been devised, tested, and used to accomplish scene classifications for both research and operational purposes. There are three fundamental types of single-scene classification: supervised classification, unsupervised classification, and visual interpretation. Despite occasional claims to the contrary, all are highly subjective, difficult to repeat in different areas with different data, and not very automated. All can be reliable when performed by experienced technicians, assuming the information required to separate the desired classes is actually contained in the remotely sensed data set. The three approaches are well described in all modern textbooks (e.g., Lillesand and Kiefer 1987; Jensen 1996) and will not be reviewed in any detail here.

Supervised classification approaches have gradually become more accurate and capable of distinguishing more classes as researchers have included ancillary information such as slope, aspect, elevation, and other factors that affect land-cover types or their spectral reflectance properties. In some cases, texture features can be added to the data set to improve both supervised and unsupervised classification. Postclassification filters can be used to create a generalized land-cover map. Such generalizations can be more intelligently implemented by using decision rules such as "class A is likely to occur adjacent to class B, but not likely to occur adjacent to Class C." All these capabilities are often present in advanced image analysis systems, but using them effectively requires such theoretical knowledge, skill, and experience that only skilled practitioners use them routinely in extensive mapping projects.

When shape, context, and more subtle characteristics of a scene become important for successful classification, visual interpretation of enhanced color images is often the only satisfactory approach. However, choosing the best spectral features, applying optimum radiometric transformations to the original data, and assigning output information to color features in a way that renders the information of interest readily visible also requires much theoretical knowledge, skill, and experience.

A limitation in using AVHRR data for land-cover mapping is the small number of available channels (two). Because of this limitation, a different approach has been developed for these data that uses seasonal variation in vegetation cover and greenness as a basis for discrimination of different land-cover categories. A common data product generated by using AVHRR data is the periodic map of vegetation greenness. A vegetation index, called the NDVI, is generated for AVHRR as $(B2 - B1) / (B2 + B1)$, where $B2$ is the measured radiance in the near-infrared AVHRR band (0.73–1.10 μm) and $B1$ is the measured radiance in the red AVHRR band (0.58–0.68 μm). This index is used because the ratio of these two bands not only minimizes errors in the data introduced by sensor (scanning and atmospheric artifacts) but also minimizes information present by nonvegetated surfaces in the scene (e.g., variations in soil background). To eliminate the effects of cloud cover, a composite NDVI image is generated for a given time period (typically 10–15 days) by selecting the maximum value for the NDVI signatures from all daily AVHRR data collected over the time period.

Researchers have found that the shape and characteristics of the seasonal NDVI curve for different forest- or vegetation-cover types differ significantly. Figure 17.1 presents a seasonal NDVI curve for three different forest covers found in interior Alaska: aspen, white spruce, and black spruce. Techniques have been

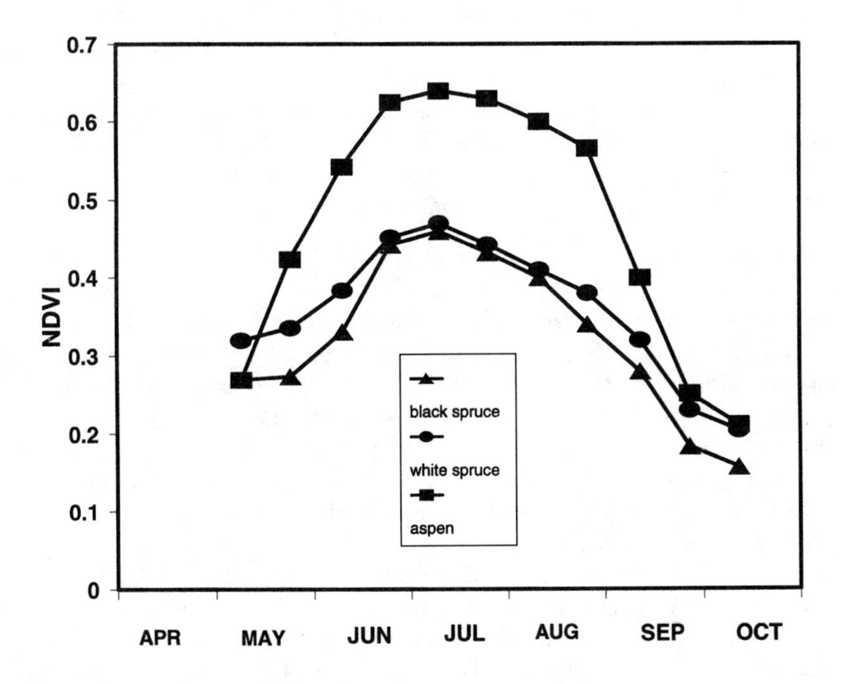

Figure 17.1. Seasonal patterns of the normalized difference vegetation index (NDVI) derived from AVHRR data collected during 1992 over three different forest types in interior Alaska.

developed to use different characteristics of the seasonal NDVI curves as a basis for classification of different vegetation types. Among the more common discriminants used on seasonal NDVI curves are maximum NDVI, mean NDVI, NDVI amplitude, NDVI threshold (the greenness of the vegetation at the beginning of the growing season), length of growing season, rate of greenup, and rate of senescence (DeFries et al. 1995a; Markon et al. 1995).

Forest Classifications Using Landsat Data

Most Canadian provinces have used Landsat TM data to produce general vegetation classifications of much of or all their territories. In addition, several federal and state agencies within the state of Alaska use Landsat data to generate forest- and vegetation-cover maps of the land under their jurisdictions. The maps have been produced with a variety of themes (generally closely related to land cover and land use) at a variety of scales for a variety of purposes. Some governments, such as the Province of Ontario, have produced digital files rather than hard-copy maps.

An example of a forest-cover classification derived for a region of interior Alaska is presented in Figure 17.2.[1] This image was generated by using a supervised classification approach, in which analysis of aerial photographs and field surveys provided the locations of specific sites used to derive the spectral signatures for the nine different land-cover categories derived for this region. The purpose of this classification was to estimate how much biomass was present in this region prior to a fire (that occurred in 1994) to assess how much carbon was released by biomass burning (Michalek et al., in press). The use of this classified image is further discussed in Chapter 23.

Forest Classifications with AVHRR Data

Several recent efforts have focused on using AVHRR data to map the forested lands of the United States and Canada. The earliest work was in Canada, where a mosaic of single-date AVHRR images was classified into a small number of broad classes (e.g., broadleaf forest, needleleaf forest, agricultural land, tundra, rangeland, and pasture) by the Manitoba Centre for Remote Sensing (Palko 1993). Several digital classification techniques were used, with water and agricultural areas superimposed onto the map using a priori knowledge. Test maps were sent for evaluation to all the provinces and territories, and the classification was extensively edited to remove errors reported by the evaluators.

A vegetation-cover classification for the conterminous United States was prepared by using AVHRR data (Loveland et al. 1991; Brown et al. 1993). Unsupervised clustering was used, with the input data being 17 biweekly (14-day) maximum NDVI images composited from AVHRR data, digital elevation data, climate data, and ecoregion maps. The clusters were labeled by using extensive

[1]Figure 17.2 will be found in the color insert.

reference to air photos and other data sources. Many clusters were split, and others were merged in the process. The final result was a cover-type map of the United States containing 70 classes. A similar approach was used to map vegetation-cover characteristics in Alaska by Markon and associates (1995).

More recently in Canada, Beaubien and Semard (1993), Beaubien and co-workers (1996), and Cihlar and colleagues (1996) have developed a technique that uses multitemporal composite AVHRR, band 1, band 2, band 4, and NDVI as inputs to image classifiers. These researchers developed a classification method in which a color enhancement is used to train and guide a minimum distance classification algorithm to produce a map of Canada with 31 classes. Gaston and colleagues (1994) used a global, coarser-resolution (16 km) global vegetation index (GVI) product set to produce a forest-cover map for the Former Soviet Union with 42 different classes. This approach used an average monthly GVI signature from 4 years to minimize the effects of interannual variations in vegetation greenness.

Because of the vast land area of the United States, Canada, and Russia, the AVHRR-derived maps are nearly impossible to validate rigorously. Having been produced with different methodologies and different cover-type labels, they cannot be intercompared with each other. In addition, a recent analysis of the use of AHVRR data for mapping forest-cover type in boreal regions by Kasischke and French (1997) revealed some shortcomings of present AVHRR classification approaches. These include the effects of clouds and haze within the imagery (even using the 10- to 15-day composite approach), as well as the lack of accounting for large-scale disturbances (e.g., forest fires) in the classifications. For many studies, a simpler land-cover map with a limited number of categories may be more appropriate. One such map with 11 different vegetation categories is presented in Figure 17.3[2] (DeFries et al. 1995b).

Burn Scar Mapping Using AVHRR Imagery

Broad-scale detection of area burned and mapping of fires in boreal regions is most easily accomplished by using AVHRR imagery (Flannigan and Vonder Haar 1986; Cahoon et al. 1991, 1996; Kasischke et al. 1993; French et al. 1995). For detection of burned areas in the boreal forest, the resolution of this system is adequate because of the large size of fires in this region. Because the NOAA satellites are in sun-synchronous polar orbits and the AVHRR instrument has a wide swath, as the orbits converge near the poles there is opportunity for several passes over the same boreal forest region each day.

Several different approaches have been demonstrated to use AVHRR imagery to map fire boundaries. The first approach uses a composite (twice-monthly) NDVI image data set produced for Alaska. The second was conducted in northern China and Siberia and used a multichannel classification approach on a set of

[2]Figure 17.3 will be found in the color insert.

single-date images. The approaches, results, and conclusions from these two studies are presented as examples of ways that the AVHRR imagery can be used for fire studies in boreal forests.

Alaska Statewide Fire Mapping Using AVHRR Composite NDVI Imagery

Kasischke and French (1995) developed a technique for detecting wildfire scars and estimating the area burned by wildfire in the state of Alaska that uses NDVI images generated from AVHRR data collected over 15- or 16-day intervals throughout the growing season. The composite data set was generated by the EROS Data Center's Alaska Field Office in Anchorage, using 1-km-resolution, local area coverage data. The composite NDVI images were generated by choosing the maximum NDVI value for each pixel from the set of imagery from the 15 or 16 days of the composite period. Each composite image represents what is expected to be a nearly cloud-free image of the state of Alaska over a 15- to 16-day period. A complete description of this data set and calculation of the NDVI is presented by Eidenshink (1992).

The NDVI images of Alaska were analyzed by comparing the vegetation index late in the growing season (after wildfires had burned) with the vegetation index from data collected early in the growing season (before most wildfires had begun). When the vegetation index was found to drop substantially from early in the season, it was assumed that a fire had occurred. Additional information on greenup time from NDVI data collected early in the following growing season was added to the results from the within-season difference analysis to obtain estimates of the amount of area burned. Sites disturbed by fire were found to green up later during the following spring compared with unburned areas, so the boundaries of fire scars were easily seen in the early June images. Figure 17.4 illustrates these effects on a region in interior Alaska where large fires occurred in 1990 and 1991. Figure 17.5 presents the results of the classifiers based on within-season NDVI differences and delay in next-season greenup.[3]

The information on fire location and area derived from this analysis was compared with records collected and archived by the Alaska Fire Service (AFS). This analysis showed that the AVHRR technique developed by Kasischke and French (1995) detected more than 83% of the fires that occurred in the state of Alaska in 1990 and 1991 and greater than 78% of the area burned. Larger fires were detected with higher accuracy. For example, 100% of the fires greater than 10,000 ha were detected. Of the area burned in fires larger than 10,000 ha, 88% of the area mapped by the AFS was mapped using the AVHRR imagery.

Errors using NDVI composite imagery generated from AVHRR data results from several sources. In a comparison with high-resolution satellite imagery from the SPOT system (20-m resolution) over a fire that occurred in 1991, it was found that the AVHRR area estimate was 16% higher than an estimate made using the

[3]Figures 17.4 and 17.5 will be found in the color insert.

SPOT image (French et al. 1996). Two reasons for the overestimate was the inclusion of unburned "islands" in the AVHRR estimate and a poor mapping of an intricate outer boundary. The SPOT image was able to detect the unburned areas within the burn and did a more precise job of mapping the outer boundary. It is unrealistic, however, to rely on high-resolution systems for regional fire mapping. A more realistic approach is to use coarse-resolution AVHRR for fire detection and initial size estimates and then move to fine resolution systems if more precision or detailed information is required.

Fire Scar Mapping in China and Siberia Using AVHRR Imagery

Cahoon and co-workers (1992, 1994, 1996) used AVHRR multispectral imagery to map fire scars in Asia for 2 years, 1987 and 1992. The 1987 fire season in eastern Asia was particularly eventful due to unusually dry conditions. It is estimated that in 1987, some 14 million ha burned in northeastern China and southeastern Siberia (Cahoon et al. 1994). The area burned by fires in 1987 was assessed by using predominately AVHRR global area coverage imagery with a nadir resolution of approximately 4 km (Kidwell 1991). Each available image was mosaicked into a composite clear-sky scene from which the burned areas were delineated (Fig. 17.6[4]).

For the 1992 study (Cahoon et al. 1996), AVHRR 1-km images were used to map the burned areas. The images were grouped into geographic blocks, mapped into clear-sky mosaics, and total area burned estimated. The 1992 estimate for the total area burned in the Russian boreal forest is 1.5 million ha. The great difference in the total area burned between 1987 and 1992 demonstrates the tremendous interannual variability of boreal fire activity. A brief description of the methodologies used in these studies and the expected error of estimating the burned area is described in the remainder of this section.

Due to a high percentage of total cloud cover over the boreal region (Warren et al. 1986), to map burned areas over the large geographic extent of the 1987 and 1992 fire activity, the use of image mosaicking techniques was necessary. The first step in the mosaicking technique involved selection of cloud-free scenes and radiometric calibration of the data (Cahoon et al. 1992). The method for creation of the clear-sky mosaic involved the rectification of each calibrated AVHRR scene to be used in the mosaic and then filling each channel of the mosaicked scene with pixel values that retain spectral information about the burned area (Cahoon et al. 1994). To determine burned area, an unsupervised, minimum-distance classification was performed on channels 1, 2, and 4 to define burned versus unburned image pixels (Cahoon et al. 1992). Channels 1 and 2 provide spectral information about the burned scar and smoke aerosol, and channel 4 (a thermal infrared channel with a center wavelength of 3.7 μm) is used to help differentiate water from burned areas. The geometric area of the burn is calculated and integrated over the entire image but can also be subdivided for political or

[4]Figure 17.6 will be found in the color insert.

geographic regions. When smoke attenuates the upwelling radiation in channels 1 and 2 and obscures the ground, band 4 is used for the mapping of burned area. The thermal contrast between burned and unburned forests in band 4 is sufficient to identify burned areas. This method has also been used to monitor fire growth during active burning of a widespread fire episode in northeastern Siberia (Stocks et al. 1996).

An error analysis for surface area estimation has been accomplished for both the AVHRR 1-km and 4-km data by evaluating the surface area of lakes of known size (Cahoon et al. 1992, 1996). Lakes provide an ideal surrogate for burns, because they are high-contrast targets, have irregular boundaries, and often contain islands, mimicking unburned tree islands surrounded by burned forest. Most of the error in area estimation can be attributed to edge pixels, both at the perimeter and around islands. The error analysis showed that for both the 1-km and 4-km data, the area error increased as the target size decreased. For lakes down to 10,000 ha, the mean area error was 2% and 4% for the 1-km and 4-km products, respectively (Cahoon et al. 1992). The surface area estimation errors showed more variability at 10,000 ha, but 90% of all cases had an area error less than about 6% for the 4-km imagery, whereas for the 1-km imagery, at 10,000 ha, 90% of all the observations were less than a 4% error. As the surface area increases, the errors and error variance of both the 1-km and 4-km imagery decrease rapidly up to 100,000 ha and then slowly level off and converge to less than 2%. Because most of the area burned results from large fires, these errors are acceptable for estimation of area burned per year in the North American boreal forest.

Burn Mapping in the Northwest Territories with Landsat Thematic Mapper Data

The Northwest Territories contain one-eighth of the forested area of Canada. These forests are fire-dependent ecosystems and support periodic, high-intensity, stand replacement fires. About 80% of the fires are ignited by lightning and the remainder by human carelessness. The fire management policy adopted by the government of the Northwest Territories in 1990 recognizes the natural role of fire in the northern forests (Chapter 5). It also recognizes the need to suppress or otherwise control wildfires that have the potential to threaten human life, property, renewable resource values, or areas having cultural values.

The Northwest Territories Centre for Remote Sensing has produced a fire history database going back over 30 years using information from a variety of sources (Epp and Lanoville 1996). The major source of information has been Landsat TM imagery, with additional information provided by fire observation reports (from aerial reconnaissance data) and AVHRR satellite imagery.

This fire database is updated on a yearly basis using Landsat TM imagery. The date of the imagery is based on the end of the fire season, usually from September 1 to October 30 or before the first snowfall (bearing in mind cloud cover and smoke if any fires were still burning). In some cases, no imagery was available due to cloud cover, and imagery has to be obtained the following year. Visual analysis of the TM data started in 1990 with imagery collected in 1988 and 1989.

The initial examination of the Landsat TM imagery showed that it was possible to visually detect the boundaries of most of the fires that had been mapped by using historical records based on aerial reconnaissance. It was determined, however, that the fire scars did not have unique spectral signatures that could be used for supervised classification of digital data. Instead, the most appropriate means to develop fire boundary maps was through visual interpretation that allowed combining the spectral, spatial, and context information present in the imagery.

Film transparencies were produced from Landsat TM data to detect and map fire boundaries in the Northwest Territories. The three bands chosen to generate the false-color composite were those that resulted in the highest contrast between burned and unburned areas, while also exhibiting different stages of regeneration and roads, tracks, and cut lines. It was found that two combinations were most effective: (1) band 3 (0.63–0.69 μm), band 4 (0.76–0.90 μm), and band 5 (1.55–1.75 μm) were assigned to blue, green, and red, respectively. In some cases, band 7 (2.08–2.35 μm) was used instead of band 5, because it has greater haze penetration and also highlights burn areas. To enhance the image products, histogram equalization approaches were applied to the digital data before generating image products.

Figure 17.7[5] presents a band 3-4-5 false-color composite image over the Northwest Territories. More recent fire scars appear red and brown in this image, and smoke from active fires is visible along the right-hand border of this image. Boundaries of older fire scars can also be detected on this image as areas of green, whereas older forest stands appear blue on the image.

A PROCOM II visual analysis system was used to interpret the film transparencies produced from Landsat TM imagery. This instrument was used to project the Landsat image onto a 1:250,000 national topographical system (NTS) map sheet. The fire boundaries detected through visual interpretation were then drawn onto the map sheet, including islands within a fire that did not burn. Each fire was then given its own unique fire number. Initially, a fire history from 1967 to 1989 was developed by using 1988 and 1989 TM imagery. With the aid of the aerial reconnaissance database, the imagery was visually interpreted by outlining polygons based on the color of successive fires inwhich the oldest fires appeared light green, to orange, to pink, to red, and to dark brown for the most recent fire.

The polygons drawn on each of the 1:250,000 NTS map sheets were then digitized and imported into geographic information system. A mosaic of all the map sheets was then created showing the fire history for the whole area (Fig. 17.8[6]). Attribute data associated with each fire consists of location of each fire by latitude and longitude, size, start date and time of fire, proximity to other fires, resources used to fight the fire, and date and time the fire was out.

[5]Figure 17.7 will be found in the color insert.
[6]Figure 17.8 will be found in the color insert.

The fire boundary database aids in natural resource management in two ways: by aiding in management of fire suppression resources and by aiding communities in more efficient use of the forest's resources.

The success in controlling and suppressing forest fires depends on early detection of the fires, potential fire behavior, the initial attack force, and the lapse time in traveling from the attack base to the fire. Given that the fire environment is continually changing, in the case of the Northwest Territories, over a very large area, responding to it in an effective way by preparing for fire occurrences with sufficient initial attack resources positioned to take effective action is considered a challenge.

The digital fire history database, in the form of a map, provides any manager information from the detailed level of 1:50,000 to the reconnaissance level of 1:2.5 million scale. The fire manager can determine what resources to use and when to deploy them based on the fire history of any area where a new fire has started. For example, if the old fires surrounding the new fire are from 10 to 15 years old, there is not enough fuel to allow the fire to spread, allowing the manager to deploy resources elsewhere. Yearly averages of number of hectares burned, compared with resources used in fire suppression, can also be obtained from the database.

Most communities in the Northwest Territories still use wood as a primary fuel source for heating and cooking. It is, therefore, important for the community to know where the nearest wood source is and whether it is accessible. Hunting and trapping are also important economic activities. Old burns eventually become good habitat for fur-bearing mammals as forests regenerate and mature. These burn areas are also good hunting areas for species such as moose as the area develops new growth. Maps of the size and location of burns therefore provide valuable information for members of the local community

Estimating Burn Severity Using Landsat TM Imagery

One of the key information needs in estimating the amounts of greenhouse gas release during fires in boreal forests is determining how much ground-layer biomass (e.g., litter, moss, lichen, organic soil) is consumed during biomass burning. Recent studies by Michalek and associates (in press) showed that in some cases, Landsat TM imagery can be used to estimate this severity.

Figure 17.9[7] presents a Landsat TM image collected 1 year after a fire had burned a 10,000-ha area near the Gerstle River, Alaska. The prefire vegetation map derived from a 1992 Landsat TM image of the same region is presented in Figure 17.2. The false-color image in Figure 17.9 was generated by assigning bands 7, 5, and 3 to the colors red, green, and blue, respectively. Field surveys showed that the pink areas in Figure 17.9 corresponded to areas where severe burning of the organic soil had occurred, and the red areas corresponded to areas where a moderate level of burning had occurred. The causes of the spectral

[7]Figure 17.9 will be found in the color insert.

differences are the result of two factors: (1) the exposure of mineral soil in the severely burned areas and (2) the absence of shadows in the severely burned areas where no trees remained standing because their roots had been consumed by fire.

Michalek and co-workers (in press) developed a supervised classification approach (using information from ground surveys and interpretation of aerial photographs) to estimate burn severity in the black spruce stands of the area where a fire had occurred in 1994. Three levels of fire severity were mapped: severe, moderate, and light. Figure 17.10[8] presents these burn categories.

Monitoring Seasonal Growth Characteristics Using AVHRR Imagery

Even though AVHRR data have a limited number of channels in the visible/near-infrared region of the electromagnetic spectrum, the ability of this system to collect daily data on a global basis has proved to be extremely valuable in terms of monitoring seasonal patterns of vegetation growth. This value was demonstrated by Tucker and associates (1986), who showed that interannual and latitudinal variations in the atmospheric concentration of CO_2 was strongly correlated to changes in the vegetation index estimated from AVHRR data. This correlation is expected because the seasonal changes in greenness detected by AVHRR and other remote sensing systems are, in fact, driven by changes in the amount of chlorophyll present on the land surface. This, in turn, is correlated with the rates of photosynthesis, which drives variations in the atmospheric concentration of CO_2.

Several recent studies have shown the potential of using AVHRR for monitoring seasonal and longer-term variations in vegetation growth in the boreal forest. A study by Braswell and colleagues (1997) illustrated that the interannual variations in vegetation growth in higher-latitude ecosystems associated with El Nino events can be monitored by using AVHRR data. Based on a study of a 10-year global AVHRR data set, Myneni and co-workers (1997) concluded that vegetation greenness in the higher northern latitudes (45–70°) had increased significantly and that the length of the growing season had increased by 12 days (± 4 days) over this time period. Although provocative, these results are somewhat controversial because of issues associated with calibration of the AVHRR data set.

Implications for Monitoring Carbon Cycles

Remote sensing systems operating in the visible and infrared regions of the electromagnetic spectrum represent an important source of information for monitoring the carbon cycle in boreal forests. A strong argument can be made that without these systems, quantifying the relative source-sink strength of the boreal

[8]Figure 17.10 will be found in the color insert.

forest carbon pool is not possible. In terms of monitoring the carbon cycle, satellite remote sensing systems such as Landsat and AVHRR provide four types of information. First, they can provide accurate and up-to-date maps of forest cover, which, when combined with ground measurements of biomass and carbon density, can provide information on the distribution of carbon throughout the boreal region. Second, remote sensing data can reliably monitor patterns of disturbance (both extent and severity), including that from human activities (logging and land clearing), fire, insects, and disease. Third, these systems can be used to monitor the patterns and rates of forest regrowth after disturbance. Finally, they can be used to monitor the intra- and interannual patterns of net primary production.

Summary

In this chapter, we have discussed a number of methods to monitor the boreal forest by using satellite systems that collect data from the visible and infrared regions of the electromagnetic spectrum. Although we have focused our attention on data collected by the Landsat and AVHRR systems, data from other satellite systems are available as well.

Of particular note is the Earth Observing System that will be launched by the United States and several international partners in 1999. The MODIS will contain a total of 36 channels, 19 of which will be in the visible and infrared regions of the electromagnetic spectrum. Of particular importance are the seven bands or channels designed for direct observation of the earth's land surface. Although MODIS is designed to collect a global data set on a daily basis, its resolution is much better than AVHRR, with a 500-meter pixel spacing. This system should be particularly useful for monitoring the boreal forest.

References

Ahern, F.J., and P.D. Archibald. 1986. Thematic Mapper information about Canadian forests: early results from across the country, pp. 683–697 in *Proceedings of the Tenth Canadian Symposium on Remote Sensing.* Canadian Aeronautics and Space Institute, Ottawa, Ontario, Canada.

Ahern, F.J., T. Erdle, D.A. MacLean, and I. Kneppeck. 1991. A quantitative relationship between forest growth rates and Thematic Mapper reflectance measurements. *Int. J. Remote Sens.* 12:387–400.

Beaubien, J., and G. Simard. 1993. Methologie de classification des données AVHRR pour la surveillance du couvert vegetal, pp. 597–603 in *Proceedings of the 16th Canadian Remote Sensing Symposium.* Canadian Aeronautics and Space Institute, Sherbrooke, Quebec, Canada.

Beaubien, J., J. Cihlar, Q. Xiao, and G. Simard. 1996. La cartographie du territoire à partir de compositions de données AVHRR multitemporelles. *Actes, 9e Congrès de l'Association Québécoise de Télédétection.* Association Québécoise de Télédétection, Montreal, Quebec, Canada.

Braswell, B.H., D.S. Schimel, E. Linder, and B. Moore III. 1997. The response of global terrestrial ecosystems to interannual temperature variability. *Science* 278:870–872.

Brown, J.F., T.R. Loveland, J.W. Merchant, B.C. Reed, and D.O. Ohlen. 1993. Using multisource data in global land-cover characterization: concepts, requirements, and methods. *Photogram. Eng. Remote Sens.* 59:977–987.

Cahoon, D.R., Jr., J.S. Levine, W.R. Cofer III, J.E. Miller, P. Minnis, G.M. Tennille, T.W. Yip, B.J. Stocks, and P.W. Heck. 1991. The great Chinese fire of 1987: a view from space, pp. 61–65 in J.S. Levine, ed. *Global Biomass Burning: Atmospheric, Climatic and Biospheric Implications.* MIT Press, Cambridge, MA.

Cahoon, D.R., Jr., B.J. Stocks, J.S. Levine, W.R. Cofer III, and C.C. Chung. 1992. Evaluation of a technique for satellite-derived area estimation of forest fires. *J. Geophys. Res.* 97:3805–3814.

Cahoon, D.R., Jr., B.J. Stocks, J.S. Levine, W.R. Cofer III, and J.M. Pierson. 1994. Satellite analysis of the severe 1987 forest fires in northern China and southeastern Siberia. *J. Geophys. Res.* 99:18,627–18,638.

Cahoon, D.R., Jr., B.J. Stocks, J.S. Levine, W.R. Cofer III, and J.A. Barber. 1996. Monitoring the 1992 forest fires in the boreal ecosystem using NOAA AVHRR satellite imagery, pp. 795–802 in J.S.. Levine, ed. *Biomass Burning and Climate Change,* Vol. 2: *Biomass Burning in South America, Southeast Asia, and Temperate and Boreal Ecosystems, and the Oil Fires of Kuwait.* MIT Press, Cambridge, MA.

Cihlar, J., D. Manak, and N. Voisin. 1994. AVHRR bidirectional reflectance effects and compositing. *Remote Sens. Environ.* 48:77–88.

Cihlar, J., H. Ly, and Q. Xiao. 1996. Land cover classification with AVHRR multichannel composites in northern environments. *Remote Sens. Environ.* 58:36–51.

Cihlar, J., H. Ly, Z Li., J. Chen, H. Pokrant, and F. Huang. 1997. Multitemporal, multichannel AVHRR data sets for land biosphere studies—artifacts and corrections. *Remote Sens. Environ.* 60:35–57.

DeFries, R., M. Hansen, and J. Townshend. 1995a. Global discrimination of land cover types from metrics derived from AVHRR pathfinder data. *Remote Sens. Environ.* 54:209–222.

DeFries, R.S., C.B. Field, I. Fung, C.O. Justice, S. Los, P.A. Matson, E. Matthews, H.A. Mooney, C.S. Potter, K. Prentice, P.J. Sellers, J.R.G. Townshend, C.J. Tucker, S.L. Ustin, and P.M. Vitousek. 1995b. Mapping the land surface for global atmosphere-biosphere models: toward continuous distributions of vegetation's functional properties. *J. Geophys. Res.* 100:20,867–20,822.

Eidenshank, J.C. 1992. The 1990 conterminous U.S. AVHRR data set. *Photogram. Eng. Remote Sens.* 57:809–815.

Epp, H., and R.A. Lanoville. 1996. Satellite data and geographic information systems for fire and resource management in the Canadian Arctic. *Geocarto Int.* 11:97–104.

Flannigan, M.D., and T.H. Vonder Haar. 1986. Forest fire monitoring using NOAA satellite AVHRR. *Can. J. For. Res.* 16:975–982.

French, N.H.F., E.S. Kasischke, L.L. Bourgeau-Chavez, and D. Barry. 1995. Mapping the location of wildfires in Alaska using AVHRR data. *Int. J. Wildlands Fire* 5:55–61.

French, N.H.F., E.S. Kasischke, R. D. Johnson, L.L. Bourgeau-Chavez, A.L. Frick, and S.L. Ustin. 1996. Using multi-sensor satellite data to monitor carbon flux in Alaskan boreal forests, pp. 808–826 in J.S. Levine, ed. *Biomass Burning and Climate Change,* Vol. 2: *Biomass Burning in South America, Southeast Asia, and Temperate and Boreal Ecosystems, and the Oil Fires of Kuwait.* MIT Press, Cambridge, MA.

Gaston, G.G., P.L. Jackson, T.S. Vinson, T.P. Kolchugina, M. Botch, and K. Kobak. 1994. Identification of carbon quantifiable regions in the former Soviet Union using unsupervised classification of AVHRR global vegetation index images. *Int. J. Remote Sens.* 15:3199–3221.

Jensen, J.R. 1996. *Introductory Digital Image Processing—A Remote Sensing Perspective.* Prentice Hall, Upper Saddle River, NJ.

Kasischke, E.S., and N.H.F. French. 1995. Locating and estimating the areal extent of wildfires in Alaskan boreal forests using multiple season AVHRR NDVI composite data. *Remote Sens. Environ.* 51:263–275.

Kasischke, E.S., and N.H.F. French. 1997. Natural limits on using AVHRR imagery to map patterns of vegetation cover in boreal forest regions. *Int. J. Remote Sens.* 18:2403–2426.

Kasischke, E.S., N.H.F. French, P. Harrell, N.L. Christensen, S.L. Ustin, and D. Barry. 1993. Monitoring of wildfires in boreal forests using large area AVHRR NDVI composite data. *Remote Sens. Environ.* 44:61–71.

Kidwell, K.B. 1991. *NOAA Polar Orbiter Data (TIROS-N, NOAA-6, NOAA-7, NOAA-8, NOAA-9, NOAA-lO, NOAA-11) Users Guide.* National Environmental Satellite Data and Information Service, Washington, DC.

Lillesand, T.M., and R.W. Kiefer. 1987. *Remote Sensing and Image Interpretation,* 2nd ed. John Wiley & Sons, New York.

Loveland, T.R., J.W. Merchant, D.O. Ohlen, and J.F. Brown. 1991. Development of a land-cover characteristics database for the conterminous U.S. *Photogram. Eng. Remote Sens.* 57:1453–1463.

Malingreau, J.P., and C.J. Tucker. 1988. Large scale deforestation in the southeastern Amazon basin of Brazil. *Ambio* 17:49–55.

Markon, C.J., M.D. Fleming, and E.F. Binnian. 1995. Characteristics of vegetation phenology over the Alaskan landscape using AVHRR time-series data. *Polar Rec.* 31:179–190.

Michalek, J.L., N.H.F. French, E.S. Kasischke, R.D. Johnson, and J.E. Colwell. In press. Using Landsat TM data to estimate carbon release from burned biomass in an Alaskan spruce forest complex. *Int. J. Remote Sens.*

Myneni, R.B., C.D. Keeling, C.J. Tucker, G. Asrar, and R.R. Nemani. 1997. Increased plant growth in the northern high latitudes for 1981 to 1991. *Nature* 386:698–702.

Nelson, R., and B. Holben. 1986. Identifying deforestation in Brazil using multiresolution satellite data. *Int. J. Remote Sens.* 7:429–448.

Palko, S. 1993. Canada's new data set from NOAA satellite imagery, pp. 457–463 in *Proceedings of the Canadian Conference on GIS,* March 23–25, 1993. Canadian Institute of Geometrics, Ottawa, Ontario, Canada.

Stocks, B.J., D.R. Cahoon, Jr., W.R. Cofer III, and J.S. Levine. 1996. Monitoring large-scale forest fire behavior in northwestern Siberia using NOAA-AVHRR satellitle imagery, pp. 802–807 in J.S. Levine, ed. *Biomass Burning and Climate Change,* Vol. 2: *Biomass Burning in South America, Southeast Asia, and Temperate and Boreal Ecosystems, and the Oil Fires of Kuwait.* MIT Press, Cambridge, MA.

Teillet, P.M., and B.N. Holben. 1994. Towards operational radiometric calibration of NOAA AVHRR imagery in the visible and near-infrared channels. *Can. J. Remote Sens.* 20:1–10.

Tucker, C.J., I.Y. Fung, C.D. Keeling, and R.H. Gammon. 1986. Relationship between atmospheric CO_2 variation and a satellite-derived vegetation index. *Nature* 319:195–199.

Vogelmann, J.E., and B.N. Rock. 1986. Assessing forest decline in coniferous forests of Vermont using NS-001 Thematic Mapper Simulator data. *Int. J. Remote Sens.* 7:1303–1321.

Warren, S.G., C.J. Hahn, J. London, R.M. Chervin, and R.L. Jenne. 1986. *Global Distribution of Total Cloud Cover and Cloud Type Amounts over Land.* NCAR/TN-273+STR. National Center for Atmospheric Research, Boulder, CO.

18. Monitoring Boreal Forests by Using Imaging Radars

Eric S. Kasischke, Laura L. Bourgeau-Chavez,
Nancy H.F. French, and Peter A. Harrell

Introduction

The past decade has seen the deployment of a new class of space-borne remote sensors that use imaging radars to monitor the earth's surface. Several operational radar systems have been launched on satellites by the European Space Agency (ESA), Japan, and Canada during this decade for monitoring the earth's polar oceans and geological exploration. In addition, the U.S. National Aeronautics and Space Administration (NASA), in conjunction with the space agencies of Germany and Italy, deployed a multichannel imaging radar system on NASA's space shuttle. This system collected a multifrequency, multipolarization data set over a large portion of the earth's surface between the latitudes of 60° north and 60° south during two separate missions in April and October 1994.

The data sets collected by these systems have spurred much research on using imaging radars to monitor terrestrial ecosystems (Kasischke et al. 1997a). In particular, ESA's Earth Resources Satellite (ERS) system has spurred research on the use of imaging radars to monitor processes in boreal ecosystems (Kasischke et al. 1995a). In this chapter, we first review the basic operational characteristics of imaging radar systems, including those that distinguish them from systems operating in the visible and infrared regions of the electromagnetic (EM) spectrum (as discussed in Chapter 17). We then illustrate the different ways these systems can be used to monitor characteristics of boreal ecosystems.

Imaging Radars: Systems and Principles of Monitoring Vegetated Landscapes

Imaging radars differ from other types of remote sensing systems in several important ways. First, they operate in the microwave region of the EM spectrum at wavelengths that are hundreds and thousands times longer than in the visible and infrared region. Second, they transmit and receive their own EM energy, which allows these systems to control the polarization of the transmitted and received EM energy along with the wavelength of operation. These two factors give radars the ability to collect imagery regardless of cloud cover and solar illumination conditions, a distinct advantage over visible spectrum systems that can only collect data during daylight hours and cloud-free conditions.

The heritage of today's imaging radar systems began in the 1930s, when engineers began developing electronic target detection methods. The word *radar* originated from this early heritage, standing for *ra*dio *d*etection *a*nd *r*anging system. The earliest systems initially transmitted a microwave pulse and detected its return. These radars were designed to detect pulses reflected by airplanes and ships. By measuring the time between the transmission and reception of the pulse, the distance could easily be calculated because the pulse travels at the speed of light, which is known. The sophistication of radar systems continued to grow throughout the remainder of the century; today, scanning radars are used to create maps of the earth's surface. The systems used in satellites today are called synthetic aperture radars (SAR). A complete description of the operational methodologies of these systems is beyond the scope of this chapter. Those interested in these details are referred to Brown and Porcello (1969), Jensen and associates (1977), or Elachi (1987).

Scientists and engineers using imaging radars have adopted a reference terminology developed during the 1930s and 1940s. At this time, the operational characteristics of these systems were military secrets, and a code was adopted in which a letter referred to the wavelength regions or bands. Thus, systems operating with a wavelength of 3 cm were referred to as X-band radars; with a 6-cm wavelength, C-band radars; with a 24-cm wavelength, L-band radars; and with a 65-cm wavelength, P-band radars. Because of their convenience, these letter designations are still used today. In addition, imaging radar systems typically transmit and receive horizontally or vertically polarized EM energy and receive energy reflected from the earth's surface in either of these polarizations. The most sophisticated imaging radar systems in operation today transmit and receive four different polarization combinations—horizontally transmit, horizontal receive (HH); horizontally transmit, vertically receive (HV); vertically transmit, horizontally receive (VH); and vertically transmit, vertically receive (VV). It is common in the literature to use a shorthand designation whereby a radar system is described by its band and polarization transmission/reception characteristics. For example, the ESA's ERS SAR system is a C-VV system, meaning it operates with a 6-cm wavelength and transmits and receives vertically polarized EM energy.

Those interested in using imaging radar data for earth resource monitoring have many choices. Data are currently available from two different operating spaceborne imaging radar systems: ERS-2 and Radarsat. Extensive archives of radar imagery collected during the 1990s are available for these systems, as well as for the JERS-1, ERS-1, and SIR-C/X-SAR systems. The operational characteristics of these systems and their capabilities for monitoring terrestrial ecosystems are discussed in the study by Kasischke and co-workers (1997a).

The fact that imaging radar systems operate with wavelengths between 1 and 65 cm has important ramifications on how these systems detect characteristics of the earth's surface. The wavelengths of the EM energy transmitted by radars are long enough that they can actually pass through or penetrate many components of the earth's surface, including vegetation. The degree of transmission/reflection is dependent primarily on the electrical conductivity of the material that makes up an object. Materials with high conductivity have poor transmission and high reflectivity, and materials with low conductivity have high transmission and low reflection.

Because of this sensitivity, imaging radar systems are extremely sensitive to variations in moisture conditions of the earth's surface and the vegetation canopy that covers it. The degree of transmission is also dependent on the length of the radar wavelength relative to the size of an object. For example, the leaves/needles and branches in the canopies of trees in the boreal forest are on the same dimensional scale as the wavelengths of X- and C-band radars (3 and 6 cm, respectively) but are typically smaller than the wavelengths of L- and P-band radars (24 and 65 cm). Thus, the overstory canopy of these trees tends to reflect a major portion of the EM energy from X- and C-band systems, and these same canopies tend to transmit most of the energy transmitted by L- and P-band systems. However, if disease or fire has recently killed the forests so the canopies consist of dead material (which has a low electrical conductivity due to lack of moisture), then EM energy will be transmitted regardless of wavelength.

A final factor that influences the degree of transmission by EM energy at microwave wavelengths is the orientation of the surface objects, which is important when considering the polarization of a radar system. Vertically oriented objects (e.g., tree trunks) tend to reflect more energy when vertically transmitted radar systems are used, whereas horizontally oriented objects (e.g., some tree branches) tend to reflect more energy when horizontally-transmitted radar systems are used. Because of this sensitivity to the orientation of the surface scatterers, imaging radar systems with multiple polarizations can be used to infer information on the structure of vegetation canopies.

The returns recorded from imaging radars typically originate from several sources within the scene being illuminated. This is because of the sensitivity to the conductivity and orientation of objects on the earth's surface and the fact that when covered with vegetation, most of the earth's surface consists of numerous objects with complex orientations. This is particularly true with imaging systems whose return comes not from a single point source (e.g., a building or other manufactured structure) but from a large area equivalent to the resolution of the system. For current space-borne imaging radar systems, this resolution is on the order of tens of meters.

Understanding the capabilities of imaging radars to monitor and map boreal ecosystems requires a basic knowledge of microwave scattering from vegetated surfaces. A radar image contains spatial information on variations in the radar scattering coefficient, $\sigma°$. When interpreting radar imagery from complex vegetation, it is necessary to think in terms of the different canopy layers that contribute to the detected signature. For vegetation canopies containing shrubs and trees, there are three distinct layers to consider (Fig. 18.1a): (1) a canopy layer that consists of smaller branches and foliage; (2) a trunk layer that consists of larger branches and trunks or boles; and (3) a surface layer that may be covered by bare soil, litter, mosses, or water. For vegetated surfaces not containing live trees (e.g.,

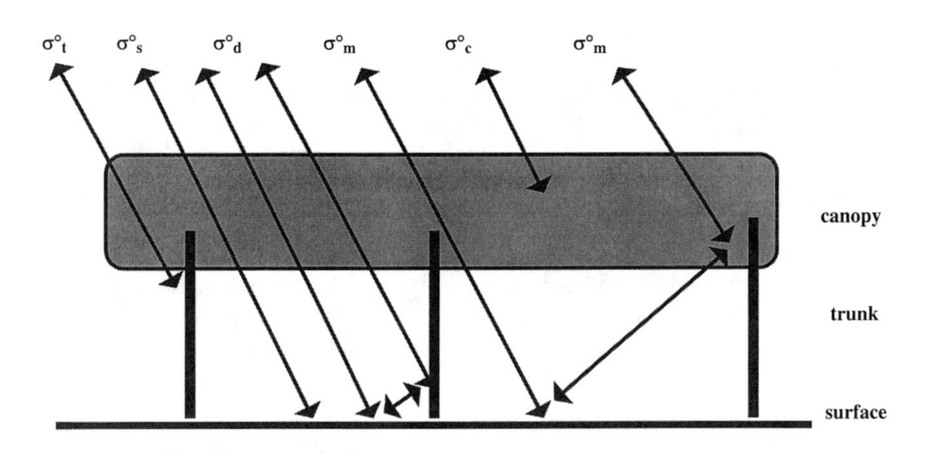

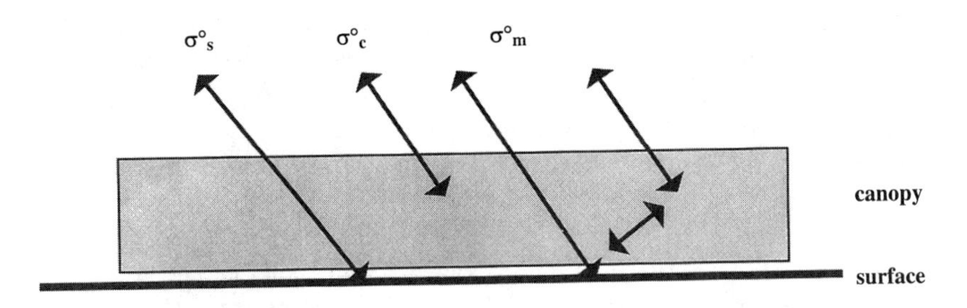

Figure 18.1. Models of microwave scattering from vegetated surfaces: (a) vegetation canopies dominated by trees and (b) herbaceous vegetation. (Reproduced with permission, the American Society for Photogrammetry and Remote Sensing. Kasischke and Bourgeau-Chavez 1997.)

a burned forest or a nonforested peatland or wetland), a simpler two-layer model can be used (Fig. 18.1b): (1) a canopy layer consisting of herbaceous vegetation, short shrubs, and sometimes, standing-dead tree boles; and (2) a surface layer covered by bare soil, litter, mosses, or water.

In the interaction of microwave energy with vegetated surfaces, the canopy and trunk layers function in two distinctly different ways. First, they are direct sources of scattering of microwave energy. Second, the components of these layers absorb or attenuate microwave energy.

The total radar scattering coefficient from woody vegetation, σ°_{t-w}, can be expressed (Wang et al. 1994, 1995; Dobson et al. 1995) as

$$\sigma^{\circ}_{t-w} = \sigma^{\circ}_{c} + \alpha^{2}_{c}\,\alpha^{2}_{t}\,(\sigma^{\circ}_{m} + \sigma^{\circ}_{t} + \sigma^{\circ}_{s} + \sigma^{\circ}_{d}) \tag{18.1}$$

where

σ°_{c} is backscatter coefficient of the crown layer of smaller woody branches and foliage,
α_{c} is the transmission coefficient of the vegetation canopy,
α_{t} is the transmission coefficient of the trunk layer,
σ°_{m} is multiple-path scattering between the ground and canopy layer,
σ°_{t} is direct scattering from the tree trunks,
σ°_{s} is direct surface backscatter from the ground, and
σ°_{d} is double-bounce scattering between the trunks and ground.

In terms of the total radar scattering coefficient from vegetation with herbaceous vegetation, σ°_{t-h}, by eliminating all terms pertaining to the trunk layer, Equation (18.1) can be simplified to

$$\sigma^{\circ}_{t-h} = \sigma^{\circ}_{c} + \alpha^{2}_{c}\,(\sigma^{\circ}_{s} + \sigma^{\circ}_{m}) \tag{18.2}$$

The various terms in Equations (18.1) and (18.2) are not only dependent on the types of vegetation present in a wetland but also on the wavelength and polarizations of the incident microwave radiation and the dielectric constant of the vegetation and the ground surface. The scattering and attenuation terms in Equations (18.1) and (18.2) are all directly proportional to the dielectric constant. Live vegetation, with a higher water content, has a higher dielectric constant than drier dead vegetation. The presence of dew or moisture from precipitation acts to increase the dielectric constant of vegetation surfaces. Whether the vegetation is frozen also strongly influences the radar signature. Studies have shown that frozen vegetation has a much lower dielectric constant than unfrozen vegetation (Way et al. 1991); hence, frozen vegetation has a lower radar backscatter than unfrozen vegetation (Way et al. 1990, 1994; Rignot and Way 1994; Rignot et al. 1994a).

The condition of the ground layer is very important in microwave scattering from vegetated surfaces. Two properties of this layer are important: the root-mean-square (RMS) surface roughness (relative to the radar wavelength) and the reflection coefficient. In general, an increasing RMS surface roughness (1) increases the amount of microwave energy backscattered (increasing σ°_{s}) and

(2) decreases the amount of energy scattered in the forward direction (decreasing σ_m° and σ_d°). The reflection coefficient is dependent on the dielectric constant (or conductivity) of the ground layer. A dry ground layer has a low dielectric constant and therefore has a low reflection coefficient. As soil moisture increases, so does the dielectric constant, and hence the reflection coefficient. Given a constant RMS surface roughness, as the soil dielectric constant increases, so does both the amount of backscattered and forward-scattered microwave energy (resulting in increases in σ_s°, σ_m°, and σ_d°).

Finally, the presence of a layer of water over the ground surface of a vegetated landscape (e.g., in many wetland ecosystems) has two results: (1) it can significantly reduce, even eliminate, any RMS surface roughness; and (2) it can significantly increase the reflection coefficient of the surface layer. In terms of microwave scattering, (1) the elimination of any RMS surface roughness means that all the energy is forward scattered, eliminating the surface backscattering term (σ_s°) in Equations (18.1) and (18.2); and (2) the increased forward scattering and higher reflection coefficient lead to significant increases in the ground–trunk and ground–canopy interaction terms (σ_d° and σ_m°, respectively).

Monitoring Boreal Ecosystems with Imaging Radars

During the past 10 years, an increasing body of research has focused on using imaging radar data to map and monitor ecosystems in high northern latitudes. Studies have not only been conducted using radar systems to map vegetation cover but also to estimate aboveground biomass, study the effects of fires in boreal forests, monitor effects from changes in temperature, and detect flooding. Examples of these applications are presented in this section.

Mapping of Forest Type and Vegetation Cover

Multichannel radar imagery can be used in the same way as imagery collected by multichannel systems that operate in the visible/infrared region of the EM spectrum to generate images depicting variations in land or vegetation cover. These image classification approaches distinguish different land covers using spectral, spatial, or temporal information. For most applications, two approaches are commonly applied to radar imagery: (1) maximum-likelihood classification, including supervised and unsupervised clustering approaches (Rignot et al. 1994a); and (2) knowledge-based approaches such as hierarchical decision trees (Dobson et al. 1996) or those based on the surface scattering properties inferred from the radar imagery (van Zyl 1989).

Figure 18.2[1] presents a multichannel radar image composite generated from NASA SIR-C/X-SAR data collected over northern Michigan. Figure 18.2 also

[1]Figure 18.2 will be found in the color insert.

contains a land-cover map generated from these data by using a hierarchical decision tree approach (Dobson et al. 1996; Bergen et. al. 1998). If this map were compared to one generated from visible/near-infrared spectrum data (e.g., that collected by Landsat), it is highly likely that there would be significant differences in the depictions of land cover. This naturally leads one to ask the question: Which map is more accurate? To answer this question, one needs to consider the application for which the classification is being used. For instance, if the application requires estimating to what extent the area within the imaged scene is contributing to net primary productivity based on the amount of photosynthetically active plant material present, then the Landsat-derived map would provide the best representation of the scene. This is because reflection of solar energy is proportional to the chlorophyll present in the vegetation. However, if one were studying the effects of aerodynamic roughness on the exchange of energy and water between the land surface and the atmosphere, the radar imagery might provide a better representation because the signatures recorded in this data source are due to variations in the canopy structure.

In summary, there is probably no right or wrong answer in terms of the information content of land-cover maps derived from visible/near-infrared imagery versus those derived from radar imagery. However, although large data sets of visible/infrared imagery of the entire earth's surface exist, there is only a limited set of multichannel radar imagery available, with very little from boreal regions. Therefore, visible/near-infrared imagery may represent the better data source for generation of land-cover maps in this region.

Estimating Aboveground Biomass

Because of SAR's sensitivity to variations in forest canopy structure, remote sensing scientists have been evaluating the use of imaging radars for estimating aboveground woody biomass. The consensus on this use is that for a wide range of forest types, imaging radars can be used to estimate biomass up to levels between 150 and 200 t ha^{-1}, depending on forest type (see, e.g., Kasischke et al. 1994a, 1995b, 1997a). Studies have also shown that there is not a generic biomass-estimation algorithm that can be applied to any forest type, but rather, specific algorithms based on allometric relationships have to be developed for different forest types based on test stands within the regions of interest (Kasischke et al. 1997a).

Several studies have been conducted on using imaging radar data to estimate biomass in the boreal forests of interior Alaska (Rignot et al. 1994b; Harrell et al. 1995). These studies have shown that biomass variations in boreal forests are proportional to variations in radar backscatter, as illustrated in Figure 18.3. Although significant relationships exist between C-band radar signatures and aboveground biomass in areas where lower biomass levels are present (Kasischke et al. 1994a; Harrell et al. 1995), it is generally agreed that L-band radar systems produce optimum results. Studies by Rignot and colleagues (1994b) estimate that the errors associated with using radar imagery for biomass estimation in boreal

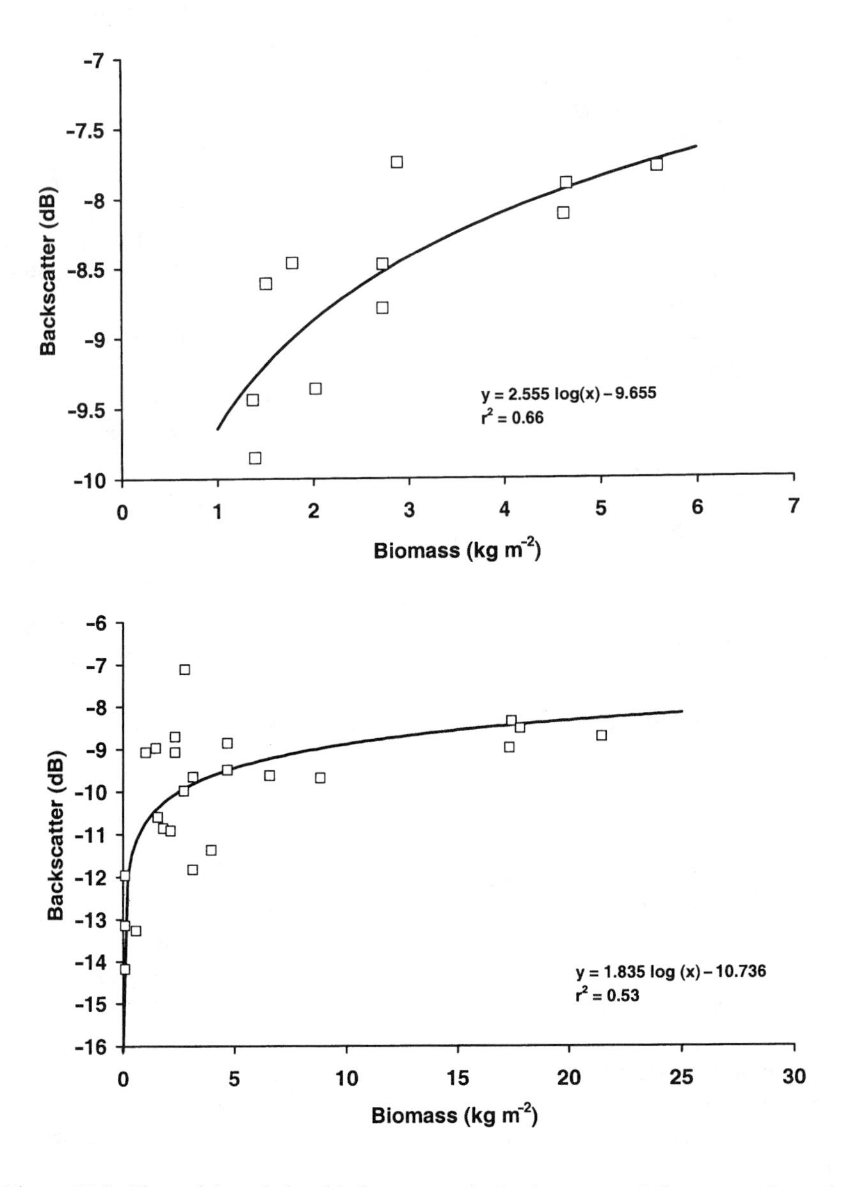

Figure 18.3. Plots of the relationship between radar backscatter and aboveground woody plant biomass in Alaskan black spruce forests: (top) for ERS-1 SAR data and (bottom) for JERS-1 SAR data. (Reprinted from Harrell et al. 1995 with permission from Elsevier Science.)

forests are on the order of 20%, which is similar to the uncertainties obtained in other regions (Kasischke et al. 1995b; Harrell et al. 1997).

One factor that must be considered in biomass estimation in boreal regions is the influence of the moisture in the ground surface. This is important because many boreal forest types have open canopies with an exposed understory, including the ground covered by mosses and lichens. Moisture variations in these exposed ground surfaces have a greater influence on the radar signature than do moisture variations of ground surfaces in forests with closed canopies (Wang et al. 1994). Thus, the relationship between biomass and radar backscatter can vary significantly during the growing season in boreal forests. In general, lower errors are found by using data collected during drier conditions (Rignot et al. 1994b; Harrell et al. 1995, 1997).

Studies have shown that radar imagery can be used to develop spatial maps of forest biomass. Figure 18.2 also contains one such map generated from SIR-C/X-SAR imagery over northern Michigan (Dobson et al. 1996; Bergen et al. 1998).

Monitoring Fire-Disturbed Boreal Forests

One of the most striking discoveries on the initial sets of ERS C-band radar imagery collected over boreal forests was the extremely bright signatures over forested regions recently disturbed by fires (Kasischke et al. 1992). Studies showed that fires in boreal forests result in signatures that are at certain times during the growing season two to four times brighter than the adjacent undisturbed forests (Kasischke et al. 1994b). These characteristic signatures not only occurred in recently disturbed forests but also in forest fire scars that were greater than 10 years old (Bourgeau-Chavez et al. 1997). Recent studies have shown that these characteristic signatures are present in fire-disturbed boreal forests across Canada (Kasischke et al. 1997b).

Figure 18.4 presents a series of C-band ERS radar images collected during the summer of 1995 over a fire that burned in the summer of 1994. Landsat Thematic Mapper imagery of this region is presented in Figure 17.9. Studies involving the collection soil moisture measurements in other forests of this region show a strong correlation between volumetric moisture in the upper 10 to 15 cm of soil and radar image intensity (French et al. 1996a), as illustrated in Figure 18.5. Studies in fire-disturbed boreal forests show that surface moisture is typically highest during the spring immediately after the disappearance of the winter snow pack. The absence of a deep thaw layer in combination with little topography results in a nearly saturated upper soil layer at this time. As the spring progresses and the depth of thaw increases, in the absence of precipitation the upper soil layer quickly dries out, resulting in a decreasing radar signature. This situation is illustrated in the May and June images in Figure 18.4. Bright radar signatures during the growing season are the result of precipitation events, which are illustrated in the July 30 and August 31 images in Figure 18.4.

One factor that allows for the detection of soil moisture–related signatures in fire-disturbed boreal forests is the slow rates of regrowth on these sites. For example, during 1995, vegetation was harvested from many different sites located in the 1990 Tok fire. The aboveground biomass present in these sites ranged between 0.1 and 1 t ha^{-1}, levels that would result in virtually no appreciable attenuation of microwave energy. However, as the fire-disturbed sites become

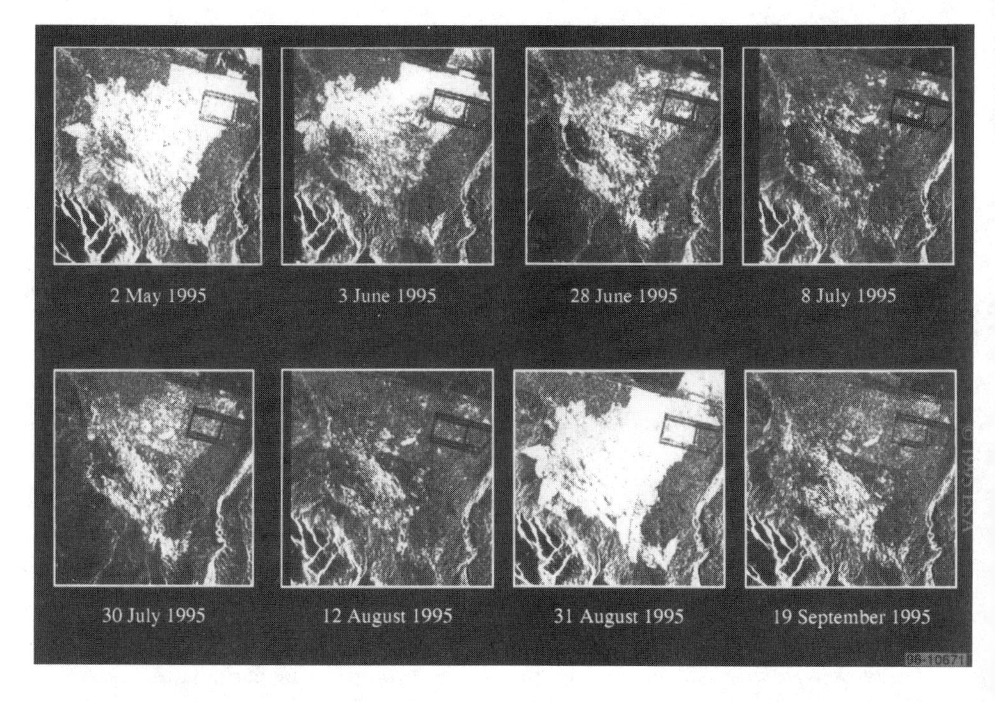

Figure 18.4. ERS SAR images collected over the 1995 Gerstle River/Hajdukovich Creek fire in Alaska, illustrating the spatial and temporal variations in image intensity associated with differences in soil moisture.

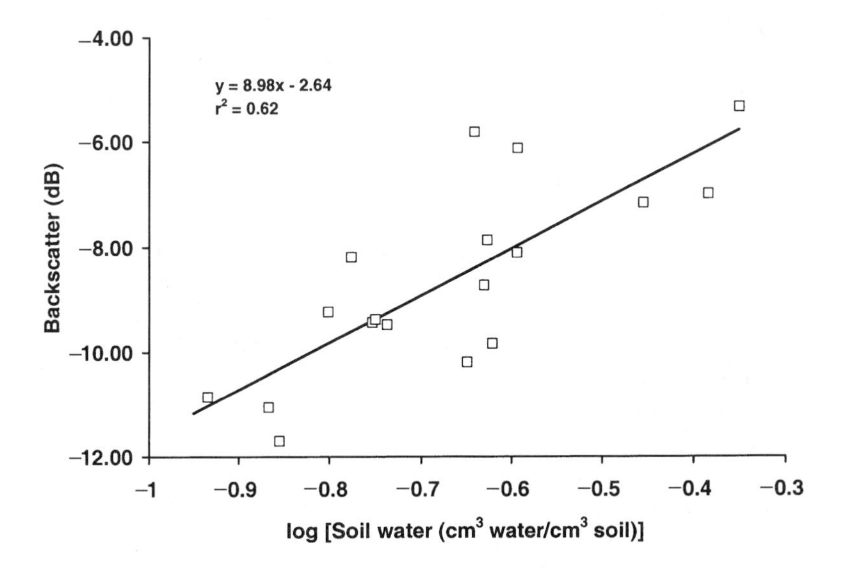

Figure 18.5. Relationship between variations in soil moisture in fire-disturbed boreal forests and radar image intensity on ERS SAR imagery.

older and more biomass accumulates, then attenuation by vegetation becomes more significant. The lower image intensities in the 1987 burned area in Figure 18.5 are due in part to attenuation by the vegetation canopy.

Studies have shown that the bright radar signatures observed on ERS SAR imagery over fire-disturbed forests are also present on Radarsat imagery (French et al. 1999). However, these studies also showed that the intensity of the radar signature decreases as the incidence angle of the system increases. Radarsat data collected at incidence angles greater than 30° did not exhibit a bright signature from fire-disturbed forests.

Finally, because of their sensitivity to variations in surface moisture conditions, imaging radar systems also have the potential for assessing fuel moisture conditions. Bourgeau-Chavez and colleagues (in press) showed that seasonal variations in ERS SAR signatures from burned and unburned black spruce forests were significantly correlated with the drought index code produced by the Canadian Fire Weather Index system.

Temperature-Related Phenomena

During the late winter (early March) of 1988, NASA conducted a series of data collection missions over the Bonanza Creek Experimental Forest located 35 km west-southwest of Fairbanks, Alaska, with their airborne multifrequency, multipolarization imaging radar. Some of these missions took place during a record warming period when the high temperatures during the day reached +10°C, whereas others took place during days when the high temperature was near normal, around −10°C. Examination of the imagery collected during these two sets of conditions revealed that the radar signatures of forested areas were two to four times brighter on the warm days than on the cold days. Analysis of ground data collected coincident with the radar imagery showed that the electrical conductivity of the tree boles in these forests was temperature dependent, being low when the air temperatures were below freezing and high when the temperatures were above freezing (Way et al. 1990; Kwok et al. 1994).

Subsequent analysis of ERS radar imagery collected during the early 1990s has shown that seasonal variations in radar image intensity related to changes in temperature are present in many different boreal ecosystems (Rignot and Way 1994; Rignot et al. 1994c; Way et al. 1994; Kasischke et al. 1995a; Morrissey et al. 1996). These studies have shown that the same general pattern of higher radar image intensities associated with warmer temperatures found in the initial aircraft studies were also present on the ERS-1 imagery.

Based on these observations, Way and co-workers (1994) suggested that radar imagery represents a unique means to detect the frozen/thawed status of vegetation in boreal forests. Way and colleagues (1994) argue that because coniferous trees begin respiring almost immediately after spring thaw begins, radar imagery presents a unique means to provide inputs into biophysical models that estimate total seasonal net primary productivity for these ecosystems. However, recent studies of Boehnke and Wismann (1997) have shown that radar backscatter measurements from the ERS scatterometer system were sensitive to spring thawing events in the boreal forest region of Siberia but relatively insensitive to fall freezing events. The same studies showed that although the thawing signatures were evident on coarse-resolution ERS scatterometer data, these thawing signatures were much more difficult to detect in individual landscape units using fine-resolution ERS SAR imagery collected over the same region. Based on these studies, it appears that using ERS scatterometer data may provide the most useful means to monitor vegetation temperature-related events.

Monitoring Flooding in Wetland Ecosystems

One of the most useful applications of imaging radar systems is the detection of flooding in wetland ecosystems (Hess et al. 1990; Morrissey et al. 1994, 1996; Kasischke and Bourgeau-Chavez 1997). This capability exists because of the strong influence that the presence or absence of water has on forward scattering of microwave energy (e.g., the σ_s°, σ_m°, and σ_d° terms in Equations [18.1] and [18.2]). The presence of standing water under a plant canopy has two results relative to the unflooded case. First, it eliminates any direct surface scattering (σ_s°). Second, it enhances multiple-bounce scattering between the ground and canopy layer (σ_m°) and the ground and trunk layer (σ_d°) when present. In terms of different wetland types, relative to an unflooded canopy, the presence of standing water typically results in increases in radar backscatter in flooded forests at longer wavelengths (L-band) and at shorter wavelengths (C-band) when leaves are absent (Hess et al. 1990). In nonforested wetlands, it has been demonstrated that shorter-wavelength radar systems (C-band) are best for detection of flooding. In areas where grass hummocks are present, such as tundra, flooding serves to increase radar image intensity due to increased multiple-bounce scattering (Morrissey et al. 1994, 1996). In areas where no hummocks are present, the presence of water may result in little or no change in image brightness (in areas with dry soils) or result in decreases in backscatter in areas with high soil moisture (Kasischke and Bourgeau-Chavez 1997).

Implications for Monitoring Carbon Cycles

As discussed in Chapter 17, one of the more common applications for remote sensing imagery in the visible and near-infrared region is the generation of land/vegetation-cover maps. These maps provide a strong basis for developing estimates of carbon budgets based on assigning different carbon levels to the different vegetation categories. Although similar maps can be generated by using imaging radar data, there has not been a demonstration of radar-derived maps having a higher degree of accuracy than maps based on Landsat or AVHRR imagery. One justification for using radar imagery is the potential for creating a map of forest stand characteristics, such as density or height (Dobson et al. 1995). Such maps may provide an increase in information in terms of estimating the biomass present in different forest stands, thereby improving the ability to estimate the levels of carbon. However, currently there are no operational multichannel imaging radar systems (which are required to generate accurate land-cover maps), thus the logical choice for creating land-cover maps is existing systems such as Landsat.

There are other capabilities of imaging radar systems that can be used to monitor processes related to carbon cycling in boreal forests. The sensitivity of longer-wavelength radars, such as the L-band JERS-1 system, to changes in aboveground biomass makes this a tool that can be used to monitor rates of forest regrowth in boreal forests. Because large areas of boreal forests are disturbed by fires annually (Chapters 15 and 16), the rates of regrowth are an extremely important component of the carbon budget.

Because fires are so widespread in boreal forests, mapping their spatial extent is necessary to determine how much carbon is released from biomass burning (Kasischke et al. 1995c). Although AVHRR data provide a very good means to detect and map the boundaries of forest fires (Chapter 17), radar imagery can improve the estimates of area burned in two ways (Bourgeau-Chavez et al. 1997). The spatial resolution of imaging radars (25 m) is very much greater than AVHRR data (1,000 m at best); thus, the estimates derived from imaging radars are going to give a greater degree of precision. Although this improvement in precision can also be obtained by using fine-resolution imagery from visible/near-infrared systems, radar imagery has two important advantages. The first is cost and the second is data availability. Radar imagery is currently much less costly than data provided by Landsat and SPOT. In addition, the collection of Landsat and SPOT data requires cloud-free conditions. Because radars operate independently of cloud cover and solar illumination, collection of data is not restricted by weather conditions.

The ability of imaging radars to monitor variations in soil moisture in fire-disturbed boreal forests has extremely strong implications for carbon balance studies in boreal forests (French et al. 1996b). It has been hypothesized that climate change will result in large losses of carbon from the soils of boreal forests because of three factors (Kasischke 1995d): (1) increases in annual area burned; (2) increases in fire severity; and (3) increases in the rates of soil respiration. It is this latter process for which SAR data provide useful information, because the patterns of soil respiration are dependent on soil moisture (see Chapter 11).

Finally, monitoring levels of flooding in boreal ecosystems is very important in carbon cycling studies. First, because methanogenisis only occurs in flooded areas under anaerobic conditions, radar imagery can be used to indirectly estimate the source-sink strength of methane and carbon dioxide fluxes in boreal wetlands (Morrissey et al. 1996). Second, recent studies showed that boreal tundra areas that are drying out actually become important sources of atmospheric CO_2 (Oechel et al. 1995).

Summary

Radar imagery is an important source of data for monitoring specific processes and surface characteristics in boreal forests. As with other sources of remotely sensed data, radar imagery can efficiently provide certain types of information, whereas other information may best be provided by other means, either through ground surveys or other remote sensing systems. Of particular importance in fire and carbon cycling in boreal forests are imaging radars' unique capabilities to (1) detect differences in aboveground woody biomass (for monitoring forest regrowth); (2) detect the levels of inundation in wetland ecosystems (for improving estimates of rates of methane/carbon dioxide fluxes); and (3) monitor variations in soil moisture in fire-disturbed regions (for improving mapping of fire boundaries as well as estimating seasonal/spatial patterns of soil respiration).

References

Bergen, K.M., M.C. Dobson, L.E. Pierce, and F.T. Ulaby. 1998. Characterizing carbon in a northern forest by using SIR-C/X-SAR imagery. *Remote Sens. Environ.* 63:24–39.

Boehnke, K., and V.R. Wismann. 1997. Detecting soil thawing in Siberia with ERS scatterometer and SAR, pp. 35–40 in *Proceedings of the 3rd ERS Symposium on Space at the Service of Our Environment,* Florence, Italy, March 17–21, 1997. Publication SP-414. ESA, Noodwijk, The Netherlands.

Bourgeau-Chavez, L.L., P.A. Harrell, E.S. Kasischke, and N.H.F. French. 1997. The detection and mapping of Alaskan wildfires using a spaceborne imaging radar system. *Int. J. Remote Sens.* 18:355–373.

Bourgeau-Chavez, L.L., E.S. Kasischke, and M.D. Rutherford. In press. Evaluation of ERS SAR data for prediction of fire danger in a boreal region. *Int. J. Wildland Fire.*

Brown, W.M., and L.J. Porcello. 1969. An introduction to synthetic aperture radar. *IEEE Spectrum* 6:52–66.

Dobson, M.C., F.T. Ulaby, L.E. Pierce, T.L Sharik, K.M. Bergen, J. Kellndorfer, J.R. Kendra, E. Li, Y.C. Lin, A. Nashashibi, K.L. Sarabandi, and P. Siqueira, 1995. Estimation of forest biomass characteristics in northern Michigan with SIR-C/XSAR data. *IEEE Trans. Geosci. Remote Sens.* 33:877–894.

Dobson, M.C., L.E. Pierce, and F.T. Ulaby 1996. Knowledge-based land-cover classification using ERS-1/JERS-1 SAR composites. *IEEE Trans. Geosci. Remote Sens.* 34:83–99.

Elachi, C. 1987. *Introduction to the Physics and Techniques of Remote Sensing.* John Wiley & Sons, New York.

French, N.H.F., E.S. Kasischke, L.L. Bourgeau-Chavez, P. Harrell, and N.L. Christensen, Jr. 1996a. Monitoring variations in soil moisture on fire disturbed sites in Alaska using ERS-1 SAR imagery. *Int. J. Remote Sens.* 17:3037–3053.

French, N.H.F., E.S. Kasischke, R. D. Johnson, L.L. Bourgeau-Chavez, A.L. Frick, and S.L. Ustin. 1996b. Using multi-sensor satellite data to monitor carbon flux in Alaskan boreal forests, pp. 808–826 in J.S. Levine, ed. *Biomass Burning and Climate Change,* Vol. 2: *Biomass Burning in South America, Southeast Asia, and Temperate and Boreal Ecosystems, and the Oil Fires of Kuwait.* MIT Press, Cambridge, MA.

French, N.H.F., L.L. Bourgeau-Chavez, Y. Wang, and E.S. Kasischke. 1999. Initial observations of Radarsat imagery at fire-disturbed sites in interior Alaska. *Remote Sens. Environ.* 68:89–94.

Harrell, P., L.L. Bourgeau-Chavez, E.S. Kasischke, N.H.F. French, and N.L. Christensen. 1995. Sensitivity of ERS-1 and JERS-1 radar data to biomass and stand structure in Alaskan boreal forest. *Remote Sens. Environ.* 54:247–260.

Harrell, P.A, E.S. Kasischke, L.L. Bourgeau-Chavez, E. Haney, and N.L. Christensen. 1997. Evaluation of approaches to estimating of aboveground biomass in southern pine forests using SIR-C imagery. *Remote Sens. Environ.* 59:223–233.

Hess, L.L., J.M. Melack, and D.S. Simmonett. 1990. Radar detection of flooding beneath the forest canopy: a review. *Int. J. Remote Sens.* 11:1313–1325.

Jensen, H., L.C. Graham, L.J. Porcello, and E. Leith. 1977. Side-looking airborne radar. *Sci. Am.* 237:84–95.

Kasischke, E.S., and L.L. Bourgeau-Chavez. 1997. Monitoring south Florida wetlands using ERS-1 SAR imagery.. *Photogram. Eng. Remote Sens.* 33:281–291.

Kasischke, E.S., L.L. Bourgeau-Chavez, N.H.F. French, P. Harrell, and N.L. Christensen, Jr. 1992. Initial observations on the use of SAR imagery to monitor wildfires in boreal forests. *Int. J. Remote Sens.* 13:3495–3501.

Kasischke, E.S., L.L. Bourgeau-Chavez, N.L. Christensen, Jr., and E. Haney. 1994a. Observations on the sensitivity of ERS-1 SAR image intensity to changes in aboveground biomass in young loblolly pine forests. *Int. J. Remote Sens.* 15:3–16.

Kasischke, E.S., L.L. Bourgeau-Chavez, and N.H.F. French. 1994b. Observations of variations in ERS-1 SAR image intensity associated with forest fires in Alaska. *IEEE Trans. Geosci. Remote Sens.* 32:206–210.

Kasischke, E.S., L. Morrissey, J.B. Way, N.H.F. French, L.L. Bourgeau-Chavez, E. Rignot, J.A. Stearn, and G.P. Livingston. 1995a. Monitoring seasonal variations in boreal ecosystems using multitemporal spaceborne SAR data. *Can. J. Remote Sens.* 21:96–109.

Kasischke, E.S., N.L. Christensen, Jr., and L.L. Bourgeau-Chavez. 1995b. Correlating radar backscatter with components of biomass in loblolly pine forests. *IEEE Trans. Geosci. Remote Sens.* 33:643–659.

Kasischke, E.S., N.H.F. French, L.L. Bourgeau-Chavez, and N.L. Christensen, Jr. 1995c. Estimating release of carbon from 1990 and 1991 forest fires in Alaska. *J. Geophys. Res.* 100:2941–2951.

Kasischke, E.S., N.L. Christensen, Jr., and B.J. Stocks. 1995d. Fire, global warming and the mass balance of carbon in boreal forests. *Ecol. Appl.* 5:437–451.

Kasischke, E.S., J.M. Melack, and M.C. Dobson. 1997a. The use of imaging radars for ecological applications—a review, *Remote Sens. Environ.* 59:141–156.

Kasischke, E.S., N.H.F. French, and L.L. Bourgeau-Chavez. 1997b. Monitoring of the effects of fire in North American boreal forests using ERS SAR imagery, pp. 363–368 in *Proceedings of the 3rd ERS Symposium on Space at the Service of Our Environment,* Florence, Italy, March 17–21, 1997. Publication SP-414. ESA, Noodwijk, The Netherlands.

Kwok, R., E. Rignot, J.B. Way, A. Freeman, and J. Holt. 1994. Polarization signatures of frozen and thawed forests of varying environmental state. *IEEE Trans. Geosci. Remote Sens.* 32:371–381.

Morrissey, L.A., G.P. Livingston, and S.L. Durden. 1994. Use of SAR in regional methane exchange studies. *Int. J. Remote Sens.* 15:1337–1342.

Morrissey, S.L. Durden, L.A., G.P. Livingston, J.A. Stearn, and L.S. Guild. 1996. Differentiating methane source areas in arctic environments with multitemporal ERS-1 SAR data. *IEEE Trans. Geosci. Remote Sens.* 34:667–673.

Oechel, W.C., G.L. Vourlitis, S.J. Hastings, and S.A. Bochkarev. 1995. Change in arctic CO_2 flux over two decades of climate change at Barrow, Alaska. *Ecol. Appl.* 5:846–855.

Rignot, E., and J.B. Way. 1994. Monitoring freeze-thaw cycles along north-south Alaskan transects using ERS-1 SAR. *Remote Sens. Environ.* 49:131–137.

Rignot, E., C. Williams, J.B. Way, and L. Viereck. 1994a. Mapping of forest types in Alaskan boreal forests using SAR imagery. *IEEE Trans. Geosci. Remote Sens.* 32:1051–1059

Rignot, E., J.B. Way, C. Williams, and L. Viereck.1994b. Radar estimates of aboveground biomass in boreal forests of interior Alaska. *IEEE Trans. Geosci. Remote Sens.* 32:1117–1124.

Rignot, E., J.B. Way, K. McDonald, L. Viereck, C. Williams, P. Adams, C. Payne, W. Wood, and J. Shi. 1994c. Monitoring of environmental conditions in taiga forests using ERS-1 SAR. *Remote Sens. Environ.* 49:145–154.

Van Zyl, J.J. 1989. Unsupervised classification of scattering behavior using radar polarimetry data. *IEEE Trans. Geosci. Remote Sens.* 27:36–45.

Wang, Y., E.S. Kasischke, F.W. Davis, J.M. Melack, and N.L. Christensen, Jr. 1994. The effects of changes in loblolly pine biomass and soil moisture variations on ERS-1 SAR backscatter—a comparison of observations with theory. *Remote Sens. Environ.* 49:25–31.

Wang, Y., F.W. Davis, J.M. Melack, E.S. Kasischke, and N.L. Christensen, Jr. 1995. The effects of changes in forest biomass on radar backscatter from tree canopies. *Int. J. Remote Sens.* 16:503–513.

Way, J., J. Paris, E.S. Kasischke, C. Slaughter, L. Viereck, N. Christensen, M.C. Dobson, F. Ulaby, J. Richards, A. Milne, A. Sieber, F.J. Ahern, D. Simonett, R. Hoffer, M. Imhoff, and J. Weber. 1990. The effect of changing environmental conditions on microwave signatures of forest ecosystems: preliminary results of the March 1988 Alaska aircraft SAR experiment. *Int. J. Remote Sens.* 11:1119–1144.

Way, J.B., E. Rignot, K.C. McDonald, R. Oren, R. Kwok G. Bonan, M.C. Dobson, L.A. Viereck, and J.A. Roth. 1994. Evaluating the type and state of Alaska taiga forests with imaging radar for use in ecosystem models. *IEEE Trans. Geosci. Remote Sens.* 32:353–370.

IV. MODELING OF FIRE AND ECOSYSTEM PROCESSES AND THE EFFECTS OF CLIMATE CHANGE ON CARBON CYCLING IN BOREAL FORESTS

Eric S. Kasischke and Brian J. Stocks

Introduction

As discussed in Chapter 2, one of the key societal challenges facing the scientific community is identifying the sources of the inter- and intra-annual variations in the atmospheric concentration of CO_2 illustrated in Figures I.1 and I.2. Although Figure II.1 presents a simplified diagram of the sources and sinks of carbon in the boreal forest and the different chapters of Section II discuss ecosystem processes that control the exchange of carbon, extrapolating these observations to explain the variations in the atmospheric concentration of CO_2 presents many additional challenges.

It is important to remember that except over very small spatial and temporal scales, it is impossible to directly measure the carbon present in a specific region of the boreal forest as well as that being exchanged with the atmosphere. In almost all cases, these values are estimated through the use of models. To understand the data presented in Figures I.1 and I.2, it is necessary to study and integrate many different aspects of the carbon budget, including

1. Quantifying how much carbon is present in the different compartments of the boreal forest.
2. Determining the net rates of carbon exchange between these different compartments and the atmosphere.
3. Understanding the factors that control the rate of exchange between the boreal forest and the atmosphere.

4. In the light of human activities and these natural processes, developing approaches to predict how the rates of carbon exchange in the boreal forest will change in the future.

A very simple model of the amount of carbon entering or leaving the atmosphere from the boreal forest on an annual basis (net biome flux [NBF]) can be expressed as

$$\mathrm{NBF} = F_e + R_p + R_h + O_f - P_s \tag{IV.1}$$

where F_e is the amount of carbon entering the atmosphere through the burning of fossil fuels, R_p is the carbon emitted to the atmosphere through plant respiration, R_h is the amount of carbon emitted through heterotrophic respiration from microbes and fungi, O_f is the carbon released during fires, and P_s is the amount of carbon being removed from the atmosphere through plant photosynthesis.

A variety of modeling approaches has been developed to estimate the amounts of carbon present in the different carbon reservoirs represented as ellipses in Figure II.1, as well as for the processes controlling exchanges between carbon pools (the boxes in Figure II.1). These ellipses and boxes illustrate two broad categories of boreal forest carbon models: carbon inventory and carbon flux models. Another important class of models are those that link climate and human activities to the processes controlling carbon fluxes.

Research over the past half century by the scientific community in Russia, Scandinavia, and North America has led to a very strong understanding of the sizes of the different carbon pools as well as the fundamental processes controlling the rates of exchange of carbon between pools. However, for most regions of the boreal forest, we do not have a capability to accurately estimate the rates of carbon exchange between the boreal land surface and the atmosphere with any degree of precision. The reasons for this lack of precision are complex and related to issues of scaling. First, the rates of carbon exchange between different pools occur over different spatial and temporal scales requiring different modeling approaches. Second, estimating carbon fluxes over large regions requires aggregation of estimates not only from different carbon flux models but also for different geographic units. The methods of carbon modeling often vary for different geographic units because they are being carried out independently by scientists conducting research in different countries.

Table IV.1 summarizes three broad categories of models (based on spatial scales) used to study the boreal forest carbon budget: global-scale models; continental/regional-scale models; and subregional/local-scale models. These models are inherently hierarchical in nature—the understanding of carbon processes developed through the smaller-scale models are typically incorporated into the larger-scale models. Therefore, the ultimate understanding of global-scale processes of carbon exchange reflected in Figures I.1 and I.2 depends not only on the correct formulation of local-scale models but on an adequate integration of this information into the larger-scale models.

The coarsest-resolution models deal with understanding the seasonal and intra-annual variations in the atmospheric CO_2 data records (Fig. I.1). Studies using these models develop approaches to examine the spatial and temporal trends present in these data records and then use global-scale models of oceanic and terrestrial processes to examine the sources of these variations (Tans et al. 1990; Randerson et al. 1997; Fan et al. 1998). A variety of global-scale models has been developed to examine atmospheric sources and sinks of carbon (Seiler and Crutzen 1980; Running and Hunt 1993; Houghton 1995; Raich and Potter 1995; Haxeltine and Prentice 1996; Randerson et al. 1996; McGuire et al. 1997; Field et al. 1998), to examine changes in plant formations as a function of climate change (Prentice et al. 1992; Smith et al. 1992a; Neilson and Marks 1995, 1992b; Woodward et al. 1995) and to examine how fire and climate change will influence carbon budgets in the boreal forest biome (Kasischke et al. 1995a).

Several models have been developed to estimate the amounts of carbon stored and net primary productivity in different boreal forest regions at continental scales (Kolchugina and Vinson 1993; Kurz and Apps 1993; Gaston et al. 1994; Krankina and Dixon 1994; Isaev et al. 1995; Shvidenko et al. 1996; Alexeyev and Birdsey 1998; Izrael and Avdjushin 1998). Several modeling efforts have estimated the release of carbon from different regions from fire (Dixon and Krankina 1993; Shvidenko et al. 1995; Shvidenko and Nilsson, this volume, Chapter 16; French et al., this volume), and others have investigated the effects of long-term land-cover change on the Russian boreal forest carbon budget (Mellilo et al. 1988).

Finally, several models have been developed to examine carbon storage and fluxes at regional and local (plot) scales. Cahoon and colleagues (1994) and Kasischke and associates (1995b) estimate regional-scale releases of carbon from fires. Several researchers have developed approaches to model net primary productivity at 1-km grid spacings at regional scales (Running and Coughlin 1988; Field et al. 1995; Liu et al. 1997). Models of longer-term patterns of changes in forest composition and carbon storage and release in response to fire and climate change have been developed by Bonan (1988), Bonan and co-workers (1990), Bonan and Van Cleve (1992), Harden and colleagues (1992, 1997), Hunt and Running (1992), and Peng and associates (1998). Finally, researchers have begun to develop models that use satellite remote sensing data to model patterns of carbon storage and release at landscape scales (Bergen et al. 1998; Michalek et al. 1999; Kasischke et al., this volume, Chapter 10).

Section Overview

The chapters of this section present a review of different approaches to modeling processes that control carbon exchange in the boreal forest region, as well as modeling of carbon fluxes themselves. In Chapter 19, Anderson and co-workers explore the various factors that influence fire ignitions in the boreal forest region. This understanding is important because without ignitions, there would be no fire regardless of fuel levels and fuel conditions. Modeling of ignition sources is

Table IV.1. Summary of Models Used to Estimate Carbon Storage and Flux in the Boreal Forest

Reference	Model purpose	Model outputs	Spatial units
Global models			
Tans et al. 1990; Randerson et al. 1997; Fan et al. 1998	Link terrestrial surface processes (fossil fuel burning, biomass burning, ecosystem net primary production) to seasonal and latitudinal variations in atmospheric CO_2	Atmospheric CO_2 concentration for different latitudes	Global
Field et al. 1998	Examine total (terrestrial and oceanic) CO_2 fluxes from net primary productivity	Atmospheric CO_2 concentration for different latitudes	Global
Running and Hunt 1993; Haxeltine and Prentice 1996; McGuire et al. 1997	Biogeochemistry models that simulate carbon, nutrient, and water cycles in terrestrial ecosystems as a function of climate and atmospheric CO_2 concentration (BIOME3, TEM, and BIOME-BGC)	Stored carbon and atmospheric CO_2 concentration for different latitudes	0.5° grid, Latitude regions
Prentice et al. 1992; Smith et al. 1992a, b; Neilson and Marks 1995; Woodward et al. 1995	Biogeographical models that predict dominant plant life forms under varying conditions, ecophysical constraints, and resource limitations (BIOME2, DOLY, MAPSS)	Vegetation cover, gross primary productivity	0.5° grid, major biomes
Raich and Potter 1995; Randerson et al. 1996	Seasonal soil respiration	CO_2 emissions	0.5° grid, major biomes
Seiler and Crutzen 1980	Carbon emissions from biomass burning	Carbon release	Major biomes
Houghton 1995	Global carbon release from land-cover change using statistical estimates of carbon storage and patterns of land-cover change	Annual changes in carbon levels	Major biomes
Kasischke et al. 1995a	Long-term changes in carbon storage based on changes in fire regime	Changes in boreal forest carbon balance	Global boreal forest

Continental/regional models			
Kolchugina and Vinson 1993; Kurz and Apps 1993; Krankina & Dixon 1994; Isaev et al. 1995; Shvidenko et al. 1996; Alexeyev and Birdsey 1998; Izrael and Avdjushin 1998	Standing carbon stocks in Russia and Canada based on extrapolation of plot and survey data	Above- and below-ground carbon level	Country-wide, regional subunits (ecoregions, economic regions)
Gaston et al. 1994	Standing carbon stocks in Russia based on remote sensing maps of land cover and biomass	Above- and below-ground carbon level	1-km grid
Melillo et al. 1988	Changes in Russian carbon stock from land-cover changes based on statistical analyses	Changes in carbon balance	Country-wide
Kolchugina and Vinson 1993; Krankina and Dixon 1994; Isaev et al. 1995; Shvidenko et al. 1996; Izrael and Avdjushin 1998	Annual net accumulation of carbon in Russia based on analysis of statistics	Annual net carbon flux	
Dixon and Krankina 1994; Shvidenko et al. 1995; Shvidenko and Nilsson, this volume, Chapter 16	Annual carbon release in Russia from fires based on national statistics	Annual carbon flux	Country-wide, regional subunits (ecoregions)
French et al., this volume	Annual carbon release in North American boreal forest based on fire location maps	Annual carbon flux	Country-wide, 1.0° grid
Subregional/local models			
Running and Coughlin 1988; Field et al. 1995; Liu et al. 1997	Biogeochemistry models of photosynthesis, respiration, and NPP	Daily variations in CO_2 flux	1-km grid
Bergen et al. 1998	Standing carbon stocks based on remote sensing maps of land cover and biomass	Above- and below-ground carbon level	50-m grid

continued

Table IV.1. Summary of Models Used to Estimate Carbon Storage and Flux in the Boreal Forest (*continued*)

Reference	Model purpose	Model outputs	Spatial units
Bonan 1989; Bonan et al. 1990	Changes in stand structure/biomass levels in response to climate change	Aboveground biomass levels, stand characteristics	Individual forest stands
Bonan and Van Cleve 1992; Hunt and Running 1992; Peng et al. 1998	Annual changes in stand carbon level in response to climate change	Above- and below-ground carbon level	Individual forest stands
Harden et al. 1992, 1997	Long-term soil carbon accumulation	Soil carbon level	Landscape level, individual forest stands
Cahoon et al. 1994; Kasischke et al. 1995b	Estimates of carbon release based on fire locations/fire maps	Carbon release	1- to 8-km grids
Michalek et al., in press; Kasischke et al., this volume, Chapter 10	Carbon release from individual fires based on field data and satellite imagery	Carbon release	50-m grids

particularly important in areas where humans are present in forested regions. In Chapter 20, Stocks and associates explore the relationship between historical and predicted climate patterns and indices of fire probability throughout the North American boreal forest. This model shows that if the average air temperatures in this region continue to rise, seasonal fire severity ratings (which indicate the probability of fire given an ignition source) will rise as well.

In Chapter 21, French and associates produce estimates of the carbon released from the North American boreal forest through fire. Producing these estimates depended on the spatial data sets of biomass and carbon distribution presented in Chapter 14, the fire distribution data sets presented in Chapter 15, and an understanding of the patterns of biomass consumed during fires discussed in Chapter 10 and elsewhere.

In Chapter 22, Shugart and co-workers discuss approaches to modeling patterns of succession and tree growth in boreal forests at landscape and subregional scales. Such models are necessary to determine how changes in climate will influence forest growth at landscape scales. In their chapter, Shugart and colleagues address two critical issues in terms of successional modeling: (1) scaling models from plot scale to landscape and subregional scales; and (2) incorporating patterns of disturbance (from fire and insects) into these models.

In Chapter 23, Kasischke and associates review different approaches for using observations made from satellite imagery to estimate carbon fluxes from boreal forests. Just as eddy tower flux correlation measurements have enabled scientists to approach measurement of net ecosystem production in a new and innovative way, the observational capabilities provided by satellite remote sensing systems also present scientists with new ways of monitoring boreal landscapes. The challenge lies in developing new modeling approaches that can exploit these observations.

For the most part, the chapters in this book have focused on forest ecosystems in the boreal region. The nonforested peatlands and wetlands of this region store large amounts of carbon in organic soils and represent the single largest carbon reservoir in this area (Table 2.1). In Chapter 24, Morrissey and colleagues present a review of this carbon pool and present estimates of how much carbon is released from these ecosystems through fire.

In the final chapter of this section and book (Chapter 25), Kasischke presents a model that explores how future carbon storage in the North American boreal forest will likely be influenced by changes in climate. This model considers the effects of climate on plant growth, heterotrophic respiration, and the fire regime of this region. It shows that the increases in net primary production [$P_s - R_p$ in Equation (IV.1)] will be offset by increases in heterotrophic respiration (R_h) and carbon release during fires (O_f), resulting in a net loss of carbon from this region.

References

Alexeyev, V.A., and R.A. Birdsey. 1998. *Carbon Storage of Ecosystems of Forests and Peatlands of Russia.* General Technical Report NE-244. U.S. Department of Agriculture, Forest Service, Northeastern Forest Experiment Station, Radnor, PA.

Bergen, K.M., M.C. Dobson, L.E. Pierce, and F.T. Ulaby. 1998. Characterizing carbon dynamics in a northern forest using SIR-C/X-SAR imagery. *Remote Sens. Environ.* 63:24–39.

Bonan, G.B. 1988. Environmental processes and vegetation patterns in boreal forests. PhD dissertation. University of Virginia, Charlottesville, VA.

Bonan, G.B., and K. Van Cleve. 1992. Soil temperature, nitrogen mineralization, and carbon source-sink relationships in boreal forests. *Can. J. For. Res.* 22:629–639.

Bonan, G.B., H.H. Shugart, and D.L. Urban. 1990. The sensitivity of some high-latitude boreal forests to climate parameters. *Clim. Change* 16:9–29.

Cahoon, D.R., Jr., B.J. Stocks, J.S. Levine, W.R. Cofer III, and J.M. Pierson. 1994. Satellite analysis of the severe 1987 forest fires in northern China and southeastern Siberia. *J. Geophys. Res.* 99:18,627–18,638.

Dixon, R.K., and O.N. Krankina. 1993. Forest fires in Russia: carbon dioxide emissions to the atmosphere. *Can. J. For. Res.* 23:700–705.

Fan, S., M. Gloor, J. Hahlman, S. Pacala, J. Sarmiento, T. Takahashi, and P. Tans. 1998. A large terrestrial carbon sink in North America implied by atmospheric and oceanic carbon dioxide data and models. *Science* 282:456–458.

Field, C.B., J.T. Randerson, and C.M. Malmstrom. 1995. Ecosystem net primary production: combining ecology and remote sensing. *Remote Sens. Environ.* 51: 74–88.

Field, C.B., M.J. Behrenfeld, J.T. Randerson, and P. Falkowski. 1998. Primary production of the biosphere: integrating terrestrial and oceanic components. *Science* 281:237–240.

Gaston, G.G., P.L. Jackson, T.S. Vinson, T.P. Kolchugina, M. Botch, and K. Kobak. 1994. Identification of carbon quantifiable regions in the former Soviet Union using unsupervised classification of AVHRR global vegetation index images. *Int. J. Remote Sens.* 15:3199–3221.

Harden, J.W., E. Sundquist, R. Stallard, and R. Mark. 1992. Dynamics of soil carbon during deglaciation of the Laurentian ice sheet. *Science* 258:1921–1924.

Harden, J.W., K.P. O'Neill, S.E. Trumbore, H. Veldhuis, and B.J. Stocks. 1997. Moss and soil contributions to the annual net carbon flux of a maturing boreal forest. *J. Geophys. Res.* 102:28,805–28,816.

Haxeltine, A., and I.C. Prentice. 1996. BIOME3: an equilibrium terrestrial biosphere model based on ecophysiological constraints, resource availability, and competition among plant functional types. *Global Biogeochem. Cycles* 10:693–709.

Houghton, R.A. 1995. Land-use change and the carbon cycle. *Global Change Biol.* 1:275–287.

Hunt, E.R., Jr., and S.W. Running. 1992. Simulated dry matter yields for aspen and spruce stands in the North American boreal forest. *Can. J. Remote Sens.* 18:126–133.

Isaev, A., G. Korovin, D. Zamolodchikov, A. Utkin, and A. Pryaznikov. 1995. Carbon stock and deposition in phytomass in Russian forests. *Water Air Soil Pollut.* 82:247–256.

Izrael, Y.A., and S.I. Avdjushin. 1998. *Russian Federation Climate Change Country Study: Preparation of the Climate Change Action Plan Report.* Progress Report 5 (January–March 1998). U.S. Department of Energy, Washington, DC.

Kasischke, E.S., N.L. Christensen, Jr., and B.J. Stocks. 1995a. Fire, global warming and the mass balance of carbon in boreal forests. *Ecol. Appl.* 5:437–451.

Kasischke, E.S., N.H.F. French, L.L. Bourgeau-Chavez, and N.L. Christensen, Jr. 1995b. Estimating release of carbon from 1990 and 1991 forest fires in Alaska. *J. Geophys. Res.* 100:2941–2951.

Kolchugina, T.P., and T.S. Vinson. 1993. Equilibrium analysis of carbon pools and fluxes of forest biomes in the Former Soviet Union. *Can. J. For. Res.* 23:81–88.

Krankina, O.N., and R.K. Dixon. 1994. Forest management options to conserve and sequester terrestrial carbon in the Russian Federation. *World Res. Rev.* 6:88–101.

Kurz, W.A., and M.J. Apps. 1993. Contribution of northern forests to the global C cycle: Canada as a case study. *Water Air Soil Pollut.* 70:163–176.

Liu, J., J.M. Chen, J. Cihlar, and W.M. Park. 1997. A process-based boreal ecosystem productivity simulator using remote sensing inputs. *Remote Sens. Environ.* 62:158–175.
McGuire, A.D., J.M. Melillo, D.W. Kicklighter, Y. Pan, X. Xiao, J. Helfrich, B. Moore III, C.J. Vorosmarty, and A.L. Schloss. 1997. Equilibrium responses of global net primary production and carbon storage to doubled atmospheric carbon dioxide: sensitivity to changes in vegetation nitrogen concentration. *Global Biogeochem. Cycles* 11:173–189.
Melillo, J.M., J.R. Fruci, R.A. Houghton, B. Moore III, and D.L. Skole. 1988. Land-use change in the Soviet Union between 1850 and 1980: causes of a net release of CO_2 to the atmosphere. *Tellus* 40B:116–128.
Michalek, J.L., N.H.F. French, E.S. Kasischke, R.D. Johnson, and J.E. Colwell. In press. Using Landsat TM data to estimate carbon release from burned biomass in an Alaskan spruce forest complex. *Int. J. Remote Sens.*
Neilson, R.P., and D. Marks. 1995. A global perspective of regional vegetation and hydrologic sensitivities for climate change. *J. Veg. Sci.* 5:715–730.
Peng, C., M.J. Apps, D.T. Price, I.A. Nalder, and D.H. Halliswell. 1998. Simulating carbon dynamics along the boreal forest transect case study (BFTCS) in central Canada. *Global Biogeochem. Cycles* 12:381–392.
Prentice, I.C., W. Cramer, S.P. Harrison, R. Leemans, R.A. Monserud, and A.M. Solomon. 1992. A global biome model based on plant physiology and dominance, soil properties, and climate. *J. Biogeogr.* 19:117–134.
Raich, J.W., and C.S. Potter. 1995. Global patterns of carbon dioxide emissions from soils. *Global Biogeochem. Cycles* 9:23–36.
Randerson, J.T., M.V. Thompson, C.M. Malmstrom, C.B. Field, and I.Y. Fung. 1996. Substrate limitations for heterotrophs: implications for models that estimate the seasonal cycle of atmospheric CO_2. *Global Biogeochem. Cycles* 10:585–602.
Randerson, J.T., M.V. Thompson, T.J. Conway, I.Y. Fung, and C.B. Field. 1997. The contribution of terrestrial sources and sinks to trends in seasonal cycles of atmospheric carbon dioxide. *Global Biogeochem. Cycles* 11:535–560.
Running, S.W., and J.C. Coughlin. 1988. A general model of forest ecosystem processes for regional applications I: hydrological balance, canopy gas exchange and primary production processes. *Ecol. Mod.* 42:125–154.
Running, S.W., and E.R. Hunt, Jr. 1993. Generalization of a forest ecosystem process model for other biomes, BIOME-BGC, and an application for global scale models, pp. 141–158 in J.R. Ehleringer and C. Field, eds. *Scaling Processes between Leaf and Landscape Levels.* Academic Press, San Diego, CA.
Seiler, W., and P.J. Crutzen. 1980. Estimates of gross and net fluxes of carbon between the biosphere and atmosphere from biomass burning. *Clim. Change* 2:207–247.
Shvidenko, A., S. Nilsson, R. Dixon, and V.A. Rojkov. 1995. Burning biomass in the territories of the former Soviet Eurasia: impact on the carbon budget. *Q. J. Hung. Meteor. Serv.* 99:235–255.
Shvidenko, A., S. Nilsson, V.A. Rojkov, and V.V. Strakhov. 1996. Carbon budget of the Russian boreal forest: a systems analysis approach to uncertainty, pp. 145–162 in M.J. Apps and D.T. Price, eds. *Forest Management and the Global Carbon Cycle.* NATO ASI Series, Subseries I, Vol. 40, Global Environmental Change. Springer-Verlag, Berlin.
Smith, T.M., H.H. Shugart, G.B. Bonan, and J.B. Smith. 1992a. Modeling the potential response of vegetation to global climate change. *Adv. Ecol. Res.* 22:93–116.
Smith, T.M., R. Leemans, and H.H. Shugart. 1992b. Sensitivity of terrestrial carbon storage to CO_2-induced climate change: comparison of four scenarios based on general circulation models. *Clim. Change* 21:367–384.
Tans, P.P., I.Y. Fung, and T. Takahashi. 1990. Observational constraints on the global atmospheric CO_2 budget. *Science* 247:1431–1438.
Woodward, F.E., T.M. Smith, and W.R. Emanuel. 1995. A global land primary productivity and phytogeography model. *Global Biogeochem. Cycles* 9:471–490.

19. Modeling of Fire Occurrence in the Boreal Forest Region of Canada

Kerry Anderson, David L. Martell, Michael D. Flannigan, and Dongmei Wang

Introduction

Fire is a significant component of most boreal forest ecosystems. It is important to understand its occurrence and spread to assess the potential impact of global climate change on boreal forest ecosystems. This chapter presents an overview of our understanding of the processes and models that have been developed and used to predict both people-caused and lightning-caused fire occurrences in the boreal forest. We draw heavily on our experience with fire occurrence in the boreal forest region of Canada, but some of our observations may be applicable to other parts of the circumpolar boreal forest as well as other biomes.

We begin by describing the fire occurrence process and the terminology that is used throughout the chapter. We note the importance of weather and provide a very brief description of the Canadian forest fire danger rating system used to predict fire occurrence in Canada. We then describe fire occurrence prediction systems that have been developed and illustrate how they could be used to predict changes in fire occurrence processes that might result from climate change and conclude with a discussion of future research needs.

The factors that influence fire occurrence in the boreal forest include the properties of the forest vegetation, weather, and ignition agents. Fire and forest managers often view forest vegetation as a fuel complex that includes both live and dead vegetation. Common boreal forest fuels include organic soils, leaves, needles, lichen, mosses, twigs, cones, bark, and branches. The chemical composition and physical structure of the fuel and its moisture content all play significant roles in determining whether ignition will occur given exposure to a specified temperature for a specified length of time.

Weather plays a significant role in fire occurrence as it determines the moisture content of the forest fuel complex, and lightning ignites many fires in the boreal forest. Fire danger rating systems have been developed to simplify the use of weather data to estimate the moisture content of selected components of a forest fuel complex each day (Stocks et al. 1989).

In Canada, fire danger is modeled by using the Canadian Forest Fire Weather Index (FWI) System (Van Wagner 1987). Three codes represent the moisture contents of (1) fine fuels (fine fuel moisture code [FFMC]); (2) loosely compacted organic matter (duff moisture code); and (3) the deep layer of compact organic matter (drought code). These moisture codes are derived from daily noon observations of temperature, relative humidity, wind speed, and 24-hour precipitation.

In addition to surface weather variables, the circulation patterns of the upper atmosphere influence fire activity. Sunny skies, warm temperatures, and low relative humidities often accompany the high atmospheric pressure systems. The presence of an upper atmospheric ridge with these high atmospheric pressure systems has been shown to be associated with high fire activity (Newark 1975; Flannigan and Harrington 1988).

Fire Occurrence Process and Terminology

Fire ignition agents are generally separated into two categories—natural and anthropogenic. Natural ignition sources include lightning, which is the most important natural ignition source in the boreal forest. Lightning started 38% of the annual average of 11,000 fires during the 1980s in Canada (Higgins and Ramsey 1992). The remaining fires are mostly human caused and can be accidental or deliberate. A small percentage (3%) of fires is started by unknown causes.

The life cycle of a forest fire can be partitioned into several distinct stages as shown in Figure 19.1. The process begins with the ignition event that occurs when some external heat source such as a discarded flaming match or a lightning strike comes into contact with the forest fuel complex and heats it up to its ignition temperature. The moisture content and other physical properties of the fuel, as well as the ambient weather, will determine the subsequent behavior of the fire. If the fuel moisture content is high but less than the moisture content of extinction (the moisture content above which the fire cannot continue to burn; Rothermel 1972), the fire will continue to smolder and spread very slowly in the duff layer with little or no visible flame. Fires that smolder in the duff layer generally emit very small amounts of energy and smoke and are commonly referred to as smoldering or holdover fires. A holdover fire may be extinguished if the moisture content rises above the moisture content of extinction or it consumes all its fuel.

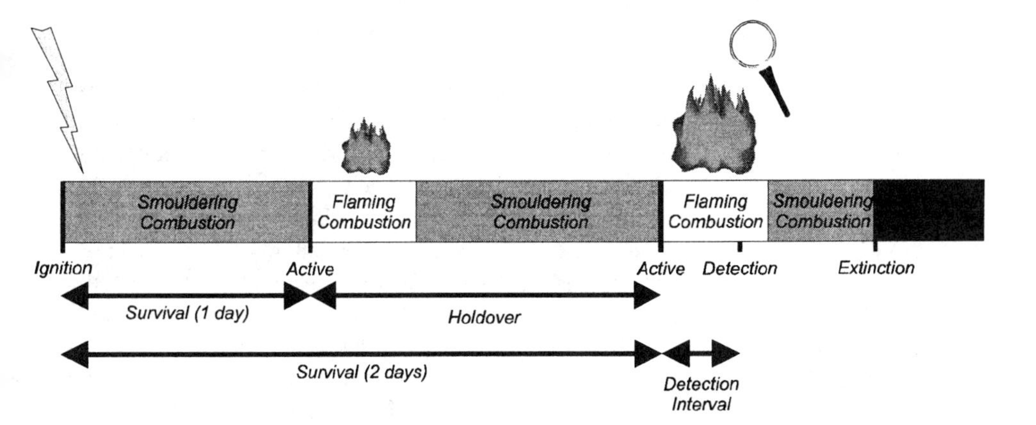

Figure 19.1. Life cycle of a forest fire. The process begins with an ignition from some external heat source, such as a lightning flash or a campfire. The ignition may become active as a flaming combustion or survive smoldering in the forest floor. During periods of peak burning conditions, the smoldering fire may become active by bursting into flaming combustion. The active fire may or may not be detected. In the figure shown, an ignition triggers a smoldering fire. This fire becomes active during the first day but goes undetected, eventually slipping back into a smoldering state. On the second day, the fire becomes active again and this time is detected. It is not actioned but slips back into a smoldering state, ultimately extinguishing itself.

If the holdover fire manages to survive until the weather and fuel moisture conditions reach the point at which sustained spread can occur, the fire will begin to spread through the forest fuel complex as a surface fire or a crown fire and emit significant amounts of smoke and energy. Fires that reach this stage are considered active. Active and smoldering fires that are spotted by people are classed as detections. Fires that are detected and reported to a forest fire management agency are classed as fire arrivals (Cunningham and Martell 1976).[1] The term *fire occurrence* is commonly used to refer to fire ignition and/or arrival, but its meaning is usually clear from the context in which it is used.

Clearly, not all fires will be detected and reported. The probability that an ignition event will lead to a smoldering fire and the probability that a smoldering fire will continue to smolder or begin to spread as an active fire will depend on the condition of the forest fuel complex and the weather. The probability that an ignition event will lead to a detection and arrival will depend on those factors and the presence and behavior of people in the surrounding area. The analysis of fire occurrence processes is complicated by the fact that only reported fires or arrivals are recorded in the records of fire agencies.

[1]Some forest fire management agencies use the term *fire arrival* to refer to what we describe as an active fire (B. Todd, personal communication).

Modeling Human-Caused Fire Occurrence Processes

People ignite forest fires both intentionally and by accident; therefore, human-caused fire occurrence can be viewed as a random or chance process with parameters determined by demographic, weather, and biophysical factors. Cunningham and Martell (1973) described the theoretical rationale for using the Poisson process to model people-caused fire occurrence. Suppose a large number of persons (n) pass through a forest on a particular day and the probability that any one of them will ignite a fire (p) is very small. The number of fires ignited that day will have a binomial distribution. It can be shown that in the limit as n becomes very large, p becomes very small, and the product of n and p remains constant, the binomial distribution converges to the Poisson distribution with a single parameter $\lambda = np$. The probability that x fires will occur is then given by the following formula for the Poisson distribution (Blake 1979):

$$P(x) = \lambda^x e^{-\lambda} / x! \tag{19.1}$$

for $x = 0,1,2, \ldots$.

Cunningham and Martell (1973) presented statistical test results for the Sioux Lookout district of northwestern Ontario that suggest it is reasonable to use the Poisson distribution to model daily people-caused fire occurrence.

Given the Poisson distribution, the challenge is to estimate λ, the expected number of fires per day. It is difficult to incorporate demographic and biophysical variables (see, e.g., Poulin-Costello 1993; Garcia et al. 1995) into fire occurrence estimation procedures. One pragmatic approach for developing estimates of λ is to control for demographic and biophysical parameters by delineating relatively small geographic areas that are reasonably homogeneous with respect to vegetation and land-use patterns. It is then possible to relate daily fire occurrence within that area to fire danger-rating indices observed at a weather station in or near the designated area. Even with such simplifications, human-caused fire occurrence still poses significant parameter estimation challenges due to the small number of fires that occur each day and because the number of fires that occur is not normally distributed, which precludes use of simple linear regression analysis techniques. Several procedures have been developed to circumvent such difficulties.

Cunningham and Martell (1973) used historical observations of the FFMC and daily people-caused fire occurrence to develop an empirical relationship between λ and the FFMC. The FFMC was designed to be representative of the moisture content of fine fuels, which often play a significant role in people-caused fire occurrence processes. They developed a step function that relates the average number of fires per day to the FFMC by partitioning the FFMC into a finite number of subintervals and using historical data to estimate the average number of fires per day for each FFMC category.

The use of step functions to relate daily fire occurrence to fire danger-rating indices is simple to implement but has many limitations. The most obvious is its failure to capture the smooth nature of changes in the rate at which the average number of fires per day increases as the fire danger increases. Simple linear regression techniques are not suitable because daily fire occurrence is not normally distributed, and very low fire occurrence rates further complicate fire occurrence parameter estimation procedures.

Martell and associates (1987) addressed this problem by using logistic (logarithmic) regression techniques. If one partitions the FFMC scale into a finite number of categories and plots the fraction of days on which fires occur as a function of the FFMC, one often sees a logistic relationship. This observation is consistent with empirical studies of basic fire ignitions processes (see Blackmarr's 1972 work described by Bradshaw et al. 1984). They defined a people-caused fire day as a day during which one or more people-caused fires occur and used a logistic model (Lee 1980; Harrell 1986) to relate the probability of a fire day to the FFMC as follows:

$$P(\text{fire day}) = e^{(a+b\ \text{FFMC})}/[1 + {}^{(a+b\ \text{FFMC})}] \quad (19.2)$$

where a and b are empirically derived coefficients.

Forest fire management agencies typically classify human-caused fires into several categories that reflect the land-use activities resulting in fires. The Ontario Ministry of Natural Resources uses eight people-caused fire categories (recreation, resident, miscellaneous, railway, industrial [forestry], industrial [other], incendiary, and unknown). Fire occurrence also varies seasonally because of changes in plant phenology and land-use patterns. For example, the likelihood that people will cause accidental fires decreases as the season progresses and dead vegetation from the previous year becomes dominated by lush new vegetation. Martell and colleagues (1987) therefore partitioned the fire season into three subseasons (spring, early summer, and summer) and derived logistic models for each fire cause and subseason. Once the probability of a fire day has been predicted, the formula for the Poisson distribution can be used to transform the predicted probability into a prediction concerning λ, the average number of people-caused fires per day.

It is difficult to partition the fire season into a finite number of intervals as it is not clear how many categories should be used or where the boundaries between the subseasons should fall, and significant changes in model parameters can result when one moves across those boundaries. Walker and co-workers (1979) describe how harmonic or periodic regression analysis techniques were used to incorporate seasonal variables in urban fire occurrence prediction models. Martell and associates (1989) adapted these procedures to incorporate seasonal variables in people-caused fire occurrence prediction models. They also aggregated the eight basic fire causes into two broad categories. The approach had the added benefit that one can readily include other variables such as the buildup index, which is thought to influence recreation fire occurrence (e.g., fires that result from campfires). Their periodic logistic fire occurrence prediction system is the basis of the operational people-caused fire occurrence prediction systems used in Ontario, and variants of the technique have been adopted in other parts of Canada.

The basic Poisson model with seasonal and fire danger-rating variables can readily be enhanced. Poulin-Costello (1993) used both logistic and Poisson regression techniques to model human-caused fire occurrence in the Kamloops area of British Columbia. She found that Poisson regression methods produced better predictions than logistic regression methods in relatively dry areas with high fire occurrence rates. Given the widespread availability of statistical software that can be used to carry out Poisson regression analysis, Poisson regression methods should replace logistic regression techniques for fire occurrence prediction models in the future.

Lightning-Caused Fire Occurrences

Lightning is a major cause of fires in the boreal forest. Although lightning fires cause only 38% of the fires, they account for 82% (approximately 2 million ha) of the total area burned nationwide each year. The reason for the disparity in proportions is that most lightning-caused fires occur in remote areas. This results in longer detection times, so that when fire-fighting resources finally arrive, the fires are large. These large fires are difficult to contain, increasing the likelihood of escape. Also, dispatched resources must be transported by air, significantly increasing the costs to contain these fires.

Lightning-caused forest fires are initiated by a cloud-to-ground lightning flash. Not all lightning flashes that strike the forest ignite fires. The likelihood of a lightning flash triggering an ignition is determined by the characteristics of the flash, fuel conditions, and precipitation.

It is generally accepted that long-continuing currents (LCC) present within return strokes of lightning flashes are the cause of ignitions in forest fuel types (Fuquay et al. 1967, 1972, 1979). Unfortunately, little is known about the long-continuing current other than simple observations (Uman 1987) due to the complexities involved in distinguishing strokes with an LCC from those without (Shindo and Uman 1989). Approximately 20% of negative lightning flashes and 85% of positive flashes have a continuing current (Uman 1987).

Latham and Schlieter (1989) used an electric arc generator that simulated lightning discharges (Latham 1987) to study lightning-caused fire ignitions. Simulated discharges were sent through various fuel samples to observe occurrence of ignition. The study showed that the most important predictor of ignition was the duration of the current. The study showed that moisture contents and fuel depth were important factors, too.

The survival or smoldering phase is the time between the ignition of a fuel and the time in which flaming combustion begins. Between these two times, the lightning-caused ignition remains smoldering in the fuel (possibly for several days) until it either dies out or bursts out into active flaming combustion under the right weather conditions.

Hartford (1990) determined a probability of the survival of a smoldering fire within commercial peat moss by using logistic regression with the bulk density, the moisture ratio, and the inorganic ratio as predictors. The inorganic ratio is the mass of dry inorganics of added soil plus the dry mass of the inherent minerals in the peat over the dry mass of the mineral-free organics.

Under the conditions of the Hartford (1990) study, the probability is that a fire will survive by burning the entire sample (a $9 \times 9 \times 5$-cm block of peat moss) when ignited on one side. In this study, there were only a few cases of partial burns in which substantial amounts but not the entire sample were burned (W. Frandsen, personal communication). This indicates that extinction occurred within minutes of the withdrawal of the ignition coil. Therefore, Hartford's probability physically represents the probability that a smoldering fire will accelerate to steady state. Once the fire reaches steady state (which typically occurs in the first few minutes), it will continue to burn regardless of time or size, assuming the predictors remain constant.

The final stage of a lightning-caused fire is the active stage at which a smoldering fire bursts into full combustion on the surface. Once a fire reaches this stage, it becomes governed by the three fire behavior components: weather, fuel, and terrain. These are the principal predictors used in the Canadian Forest Fire Behavior Prediction System (Forestry Canada Fire Danger Working Group 1992).

In an operational setting, the active stage alone is not enough to predict a fire occurrence, as the detection ability of forest protection agencies must be considered. Minimum conditions required to sustain combustion are not enough, as a fire may begin flaming combustion, but if it is not spotted by detection efforts, it will simply lapse back into the smoldering stage overnight.

A preliminary survey of lightning-caused fire detections in Saskatchewan's primary protection zone reveals that, on average, lightning-caused fires are detected with head fire intensities of 1,200 kW m^{-1} and at 1.7 m min^{-1} rate of spread. These numbers will vary depending on the detection efforts of protection agencies, but they are indicative of the behavior and intensities required to spot a fire in remote areas.

A probabilistic model can be constructed by calculating the probabilities of each stage of a lightning-caused fire occurrence. Given an individual lightning flash, one can calculate the probability that a lightning-caused fire will occur at a given time t as follows:

$$P(t) = P_{LCC} P_{ign} P_{sur}(t) P_{det} \qquad (19.3)$$

where P_{LCC} is the probability that the flash has a long-continuing current, P_{ign} is the probability of ignition assuming a long-continuing current, $P_{sur}(t)$ is the probability of a smoldering fire surviving until time t, and P_{det} is the probability of detecting a fire in the active stage.

There are a few operational lightning-caused fires occurrence prediction models generally following the probabilistic model described above. Fuquay and assoicates (1979) developed a model of the lightning ignition environment. This model has been generally accepted and has been applied (with variations) by some agencies to predict lightning-caused fire occurrences (Latham 1983; Kourtz and Todd 1991).

Climate Change and Fire Occurrence

Human activities have caused an increase in carbon dioxide, methane, water vapor, and human-made gases such as chlorofluorocarbons and hydrofluorocarbons. These increases in greenhouse gas concentrations are responsible, in part, for the warming of our climate. Global mean surface air temperature has increased by 0.3–0.6°C since the late 19th century (Houghton et al. 1990), with a 1.0–1.5°C rise in temperature observed for the boreal region over the past 30 years (Fig. 1.1).

Fire occurrence should be affected by any climate change that takes place in the boreal forest. Some studies have suggested dramatic increases in fire incidence and area burned in a simulated 2 × CO_2 climate (Overpeck et al. 1990; Flannigan and Van Wagner 1991). However, more recent studies have suggested that there might be a great deal of spatial variability of fire behavior in response to a warmer climate, including areas of decreased fire frequency (Bergeron and Flannigan 1995; Flannigan et al. 1998). The reason for decreasing fire frequency in some regions is the increased precipitation and, more important, the increased precipitation frequency that will be accompanying the increasing temperatures. Other regions may show a dramatic increase in fire frequency if the climate changes according to the general circulation models (GCM) (Stocks et al. 1998).

The amount of ignitions may be directly influenced by changes in the climate. Williams (1992) found a positive correlation between surface temperature and lightning flash rate. Price and Rind (1994) suggest a 30% increase in cloud-to-ground lightning activity for a 2 × CO_2 climate using the Goddard Institute for Space Studies GCM. They estimate that there would be a 44% increase in lightning fires with an area burned increase of 78% for the United States in the 2 × CO_2 environment. Another consideration is the length of the fire season. In Canada, Wotton and Flannigan (1993) predicted that the length of the fire season would increase by 22% (30 days) across Canada using 10 years of daily data from the 2 × CO_2 GCM from the Canadian Climate Centre. Both the increase in lightning activity and the length of the fire season will greatly influence fire occurrence. However, the change in the day-to-day weather associated with a change in climate also needs to be considered. For example, increased lightning activity may be offset due to increased precipitation amount and frequency in some areas. There are indications that some regions will experience an increase in the persistence of blocking ridges in the upper atmosphere (Lupo et al. 1997), which are conducive to increased fire activity.

The weather will also influence the fuel moisture, directly affecting ignition, survival, and arrival of fires. A key factor in determining the fire occurrence in the future will be the precipitation regime. Unfortunately, confidence in the precipitation estimates from the GCMs is low relative to the temperature estimates, although the GCMs are constantly being improved and updated. In a warmer climate, it is likely that overall fire occurrence will increase as a result of increased lightning activity and the increased length of the fire season. Finally, as the climate changes, vegetation cover will also change in response to the changing environment, including disturbance regime (fire, pests, diseases, and windthrow).

A change in the vegetation cover would result in changes in ease of ignition, which, in turn, would influence fire occurrence (Chapter 9).

Research Needs

If global climate change has a significant impact on boreal forest ecosystems, it will affect fire weather, vegetation and forest cover, and land-use patterns. It will also affect fire behavior, which in turn, will influence forest vegetation and land use (Weber and Flannigan 1997). The fire occurrence prediction models that have been developed for fire management planning purposes satisfy such needs reasonably well, but they are not well suited for assessing the potential long-term affects of climate change on boreal forest ecosystems.

There is a need to develop comprehensive models that relate daily fire occurrence to forest vegetation, weather, land-use patterns, and fire management efforts such as prevention. It is clear that the result will be models in which fire danger-rating indices are significant, so it is essential that GCM results are used to generate realistic fire weather scenarios.

One important aspect of fire management is the need to model day-to-day changes in fire weather conditions. Fire management agencies carry out initial attack on new fires as they occur and mount extended attack operations on large fires that result when fires escape initial attack. Initial attack success tends to exhibit serial correlation because fires become more likely to escape after several days of severe burning weather when suppression resources become depleted and the fire danger grows in severity. Martell (1990) has indicated that it is reasonable to model day-to-day changes in the FWI, one of the components of the Canadian FWI system, as a Markov chain. To do this, a GCM-based sequence of days will be required to assess the joint impact of fire management, forest vegetation, and fire weather on fire regimes. Ultimately, higher-resolution regional climate models (Caya et al. 1995) may be required to model fire occurrence under climate change.

If climate change does alter the composition of the boreal forest, it may do so gradually over time. Static models may provide some insight concerning climate change impacts on the boreal forest, but there is an urgent need for dynamic models that can be used to model fire occurrence over long periods of time over which the climate, boreal forest ecosystems, and land-use patterns are changing. There will be a need for fire occurrence prediction models that are compatible with dynamic climate and ecosystem models.

References

Bergeron, Y., and M.D. Flannigan. 1995. Predicting the effects of climate change on fire frequency in the southeastern Canadian boreal forest. *Water Air Soil Pollut.* 82:437–444.

Blackmarr, W.H. 1972. *Moisture Content Influences Ignitability of Slash Pine Litter.* Research Note SE-173. Southeastern Forest Experiment Station, U.S. Department of Agriculture, Forest Service, Asheville, NC.

Blake, I.F. 1979. *An Introduction to Applied Probability.* John Wiley & Sons, New York.
Bradshaw, L.S., J.E. Deeming, R.E. Burgan, and J.D. Cohen. 1984. *The 1978 National Fire-Danger Rating System: Technical Documentation.* General Technical Report INT-169. Intermountain Forest and Range Experiment Station, U.S. Department of Agriculture, Forest Service, Ogden, UT.
Caya, D., R. Laprise, M. Giguère, G. Bergeron, J.P. Blanchet, B.J. Stocks, J.G. Boer, and N.A. McFarlane. 1995. Description of the Canadian regional climate model. *Water Air Soil Pollut.* 82:477–482.
Cunningham, A.A., and D.L. Martell. 1973. A stochastic model for the occurrence of man-caused forest fires. *Can. J. For. Res.* 3:282–287.
Cunningham, A.A., and D.L. Martell. 1976. The use of subjective probability assessments concerning forest fire occurrence. *Can. J. For. Res.* 6:348–356.
Flannigan, M.D., and J.B. Harrington.1988. A study of the relation of meteorological variables to monthly provincial area burned by wildfire in Canada (1953–80). *J. Appl. Meteorol.* 27:441–452.
Flannigan, M.D., and C.E. Van Wagner. 1991. Climate change and wildfire in Canada. *Can. J. For. Res.* 21:66–72.
Flannigan, M.D., Y. Bergeron, O. Engelmark, and B.M. Wotton. 1998. Future wildfire in northern forests: less than global warming would suggest? *J. Veg. Sci.* 9:469–476.
Forestry Canada Fire Danger Group. 1992. *Development and Structure of the Canadian Forest Fire Behaviour Prediction System.* Informal Report ST-X-3. Forestry Canada, Science and Sustainable Development Directorate, Ottawa, Ontario, Canada.
Fuquay, D.M., R.G. Baughman, A.R. Taylor, and R.G. Hawe. 1967. Characteristics of seven lightning discharges that caused forest fires. *J. Geophys. Res.* 72:6371–6373.
Fuquay, D.M., A.R. Taylor, R.G. Hawe, and C.W. Schmid, Jr. 1972. Lightning discharges that caused forest fires. *J. Geophys. Res.* 77:2156–2158.
Fuquay, D.M., R.G. Baughman, and D.J. Latham. 1979. *A Model for Predicting Lightning Fire Ignitions in Wildlands Fuels.* Research Paper INT-217. USDA Forest Service, Intermountain Forest and Range Experiment Station, Ogden, Utah.
Garcia, C., C. Vega, P.M. Woddard, S.J. Titus, W.L.Adamowicz, and B.S.Lee. 1995. A logistical model for predicting the daily occurrence of human caused forest fires. *Int. J. Wildland Fire* 5:101–111.
Harrell, F.E., Jr. 1986. The LOGIST procedure, pp. 269–293 in R.P. Hastings, ed. *SUGI Supplemental Library User's Guide.* SAS Institute Inc., Cary, NC.
Hartford, R.A. 1990. Smoldering combustion limits in peat as influenced by moisture, mineral content, and organic bulk density, pp. 282–286 in D.C. MacIver, H. Auld, and R. Whitewood, eds. *Proceedings of the 10th Conference on Fire and Forest Meteorology,* April 17–21, 1989, Ottawa, Ontario. AES, Downsview, Ontario, Canada.
Higgins, D.G., and G.S. Ramsey. 1992. *Canadian Forest Fire Statistics.* Information Report PI-X-107. Forestry Canada, Petawawa National Forestry Institute.
Houghton, J.T., G.J. Jenkins, and J.J. Ephraums, eds. 1990. *Climate Change—The IPCC Scientific Assessment.* Cambridge University Press, Cambridge, UK.
Kourtz, P., and B. Todd. 1991. *Predicting the Daily Occurrence of Lightning-Caused Forest Fires.* Informal Report PI-X-112. Forestry Canada, Petawawa National Forestry Institute, Chalk River, Ontario, Canada.
Latham, D.J. 1983. *LLAFFS—A Lightning Locating and Fire-Forecasting System.* Research Paper INT-315. USDA Forest Service Intermountain Forest and Range Experiment Station, Ogden, UT.
Latham, D.J. 1987. *Design and Construction of an Electric Arc Generator for Fuel Ignition Studies.* Research Paper INT-366. USDA Forest Service, Intermountain Forest and Range Experiment Station, Ogden, UT.
Latham, D.J., and J.A. Schlieter. 1989. *Ignition Probabilities of Wildland Fuels Based on Simulated Lightning Discharges.* Research Paper INT-411. USDA Forest Service, Intermountain Forest and Range Experiment Station, Ogden, UT.

Lee, E.T. 1980. *Statistical Methods for Survival Data Analysis.* Lifetime Learning Publications, Belmont, CA.
Lupo, A.R., R.J. Oglesby, and I.I. Mokhov. 1997. Climatological features of blocking anticyclones: a study of Northern Hemisphere CCM1 model blocking events in present-day and double CO_2 concentration atmospheres. *Clim. Dyn.* 13:181–195.
Martell, D.L. 1990. A Markov chain model of a forest fire danger rating index, pp. 225–231 in D.C. MacIver, H. Auld, and R. Whitewood, eds. *Proceedings of the 10th Conference on Fire and Forest Meteorology,* April 17–21, 1989, Ottawa, Ontario. AES, Downsview, Ontario, Canada.
Martell, D.L., S. Otukol, and B.J. Stocks. 1987. A logistic model for predicting daily people-caused forest fire occurrence in Ontario. *Can. J. For. Res.* 17:394–401.
Martell, D.L., E. Bevilacqua, and B.J. Stocks. 1989. Modelling seasonal variation in daily people-caused forest fire occurrence. *Can. J. For. Res.* 19:1555–1563.
Newark, M.J. 1975. The relationship between forest fire occurrence and 500 mb longwave ridging. *Atmosphere* 13:26–33.
Overpeck, J.T., D. Rind, and R. Goldberg. 1990. Climate-induced changes in forest disturbance and vegetation. *Nature* 343:51–53.
Poulin-Costello, M. 1993. People-caused forest fire prediction using Poisson and logistic regression. MSc thesis, Department of Mathematics and Statistics, University of Victoria, Victoria, British Columbia, Canada.
Price, C., and D. Rind. 1994. The impact of a $2 \times CO_2$ climate on lightning-caused fires. *J. Clim.* 7:1484–1494.
Rothermel, R.C. 1972. *A Mathematical Model for Predicting Fire Spread in Wildland Fuels.* Research Paper INT-115. USDA Forest Service, Intermountain Forest and Range Experiment Station, Ogden, UT.
Shindo, T., and M.A. Uman. 1989. Continuing current in negative cloud-to-ground lightning. *J. Geophys. Res.* 94:5189–5198.
Stocks, B.J., B.D. Lawson, M.E. Alexander, C.E. Van Wagner, R.S. McAlpine, T.J. Lynham, and D.E. Dube. 1989. The Canadian forest fire danger rating system: an overview. *For. Chron.* 65:258–265.
Stocks, B.J., M.A. Fosberg, T.J. Lynham, L. Mearns, B.M. Wotton, Q. Yang, J.-Z. Jin, K. Lawrence, G.R. Hartley, J.A. Mason, and D.W. McKenney. 1998. Climate change and forest fire potential in Russian and Canadian boreal forests. *Clim. Change* 38:1–13.
Uman, M.A. 1987. *The Lightning Discharge.* Academic Press, San Diego, CA.
Van Wagner, C.E. 1987. *Development and Structure of the Canadian Forest Fire Weather Index System.* Forest Technical Report 35. Canadian Forestry Service, Ottawa, Ontario, Canada.
Walker, W.E., J.M. Chaiken, and E.J. Ignall, eds. 1979. *Fire Department Deployment Analysis: A Public Policy Analysis Case Study—The Rand Fire Project.* North-Holland, New York.
Weber, M.G., and M.D. Flannigan. 1997. Canadian boreal forest ecosystem structure and function in a changing climate: impact on fire regimes. *Environ. Rev.* 5:145–166.
Williams, E.R. 1992. The Schumann resonance—a global tropical thermometer. *Science* 256:1184–1187.
Wotton, B.M., and M.D. Flannigan. 1993. Length of the fire season in a changing climate. *For. Chron.* 69:187–192.

20. Climate Change and Forest Fire Activity in North American Boreal Forests

Brian J. Stocks, Michael A. Fosberg, Michael B.Wotton, Timothy J. Lynham, and Kevin C. Ryan

Introduction

After a decade of speculation and debate, there is now a general scientific consensus that rising greenhouse gas levels in the earth's atmosphere will result in significant climate change over the next century. The recent statement by the Intergovernmental Panel on Climate Change (Watson et al. 1995) that "the observed increase in global mean temperature over the last century (0.3–0.6°C) is unlikely to be entirely due to natural causes, and that a pattern of climate response to human activities is identifiable in the climatological record" is a strong endorsement of this conclusion. The recently negotiated Kyoto Protocol to the United Nations Framework Convention on Climate Change recognizes the influence of greenhouse gas concentrations on global warming and requires signatory countries to commit to significant reductions in emissions in the near future, further evidence of a growing acknowledgment that climate change is a reality.

There is also evidence of an emerging pattern of climate response to forcings by greenhouse gases and sulfate aerosols, as evidenced by geographic, seasonal, and vertical temperature patterns. In North America and Russia, this pattern of observed changes has taken the form of major winter and spring warming in west central and northwestern Canada and Alaska and virtually all Siberia over the past three decades, resulting in temperature increases of 2–3°C over this period (Environment Canada 1995; Hansen et al. 1996)

General circulation models (GCM) developed independently in many countries are in agreement in projecting a global mean temperature increase of 0.8–3.5°C by A.D. 2100. This change is much more rapid than any experienced through natural climate variability since the Ice Age 10,000 years ago. In addition, the change is also one to two orders of magnitude faster than changes in the boreal zone over the past 100,000–200,000 years (Kirschbaum and Fishlin 1996). Most significant temperature changes are projected at higher latitudes and over land, particularly in boreal forest and tundra regions. The greatest warming is expected to occur in winter and spring (similar to the trends measured recently), although warming is projected for all seasons. Although GCM projections vary, with a doubling of atmospheric CO_2, winter temperatures are expected to rise 6–10° C and summer temperatures 4–6° C over much of Canada, Alaska, and Russia. Global precipitation forecasts under a $2 \times CO_2$ climate are more variable among GCMs, but indications are that large increases in evaporation over land due to rising air temperatures will more than offset minor changes (about 20%) in precipitation amount, resulting in drier soil conditions during the summer. In addition, changes in the regional and temporal patterns and intensity of precipitation are expected, increasing the tendency for extreme droughts and floods.

As awareness of global warming and climate change has grown, a concurrent concern over the potential impacts of climate change has arisen. In northern circumpolar countries, this translates into the need to project changes to terrestrial and aquatic ecosystems that are vital to economic and recreational well-being. Impacts on boreal forest ecosystem structure and function with changing climate are a virtual certainty, and it is anticipated that changes in disturbance regimes, particularly fire and insects, will drive change in the boreal forest. Over the past decade, research into understanding and modeling fundamental forest ecosystem processes has been accelerated to better anticipate the impact of global warming, and these research activities are expanding (Chapter 22). Among these studies are several investigations aimed at projecting future boreal fire activity and severity under various climate change scenarios. This chapter summarizes these research activities and uses outputs from the Canadian GCM to predict fire danger levels and potential fire impacts in Canadian and Alaskan boreal forests.

Projecting Future Boreal Fire Regimes

Despite their coarse spatial and temporal resolution, GCMs provide the best means currently available to project future climate on a broad scale. However, GCMs cannot be expected to provide the regional-scale higher-resolution projections necessary to estimate future conditions across the North American boreal zone, where large variability in climate is reflected in both forest composition and fire activity. Regional climate models (RCMs) with much higher resolution currently under development (e.g., Caya et al. 1995) and validation (Wotton et al. 1998) will permit more accurate regional-scale climate projections. As boreal fire activity and climate are closely coupled, it follows that changes in the fire regime will accelerate changes in the boreal landscape with increasing climate change. Indeed, fire will be the likely agent for most vegetation shifting in the boreal zone under a changing climate (Stocks 1993). Weber and Flannigan (1997) expanded on this prediction, concluding that "fire regime as an ecosystem process is highly sensitive to climate change because fire behavior responds immediately to fuel moisture" and that "interaction between climate change and fire regime has the potential to overshadow the direct effects of global warming on species distribution, migration, substitution, and extinction." Fosberg and associates (1996) suggested three agents of climate change that will affect future fire regime: the changed vegetation or fuels complex, the changed potential for fire occurrence (both lightning and human caused), and the change in fire severity and behavior due to changes in fire weather.

A climatically induced shifting of vegetation northward is expected with climate change, driven by altered species distribution, increased local mortality, and replacement of whole ecosystems as species are unable to migrate quickly enough to keep up with the rate of changing climate (Smith and Shugart 1993). The boreal zone is projected to shrink, as grasslands and temperate forests expand northward into the southern boreal, whereas expansion into tundra regions is limited by poor soils and permafrost (Rizzo and Wiken 1992). It has been speculated that stand mortality and breakup associated with climate change would result in increased fuel accumulation that would, in turn, increase fire severity and accelerate ecosystem conversion or replacement (Martin 1993).

An increase in lightning frequency across North America is also expected under a $2 \times CO_2$ climate, due to increased convective activity associated with a warmer atmosphere (Fosberg et al. 1990, 1996; Price and Rind 1994). Lightning fire occurrence would increase accordingly due to increased lightning and cloud-to-ground lightning discharges (Price and Rind 1994), in combination with higher ignition probabilities associated with more frequent, drier fuel conditions. This is particularly important in the North American boreal zone where lightning accounts for only about 35% of fire starts but about 85% of the area burned (Stocks 1991). Lightning often results in multiple ignitions simultaneously. Lightning fires can occur in remote regions, both contributing factors to the preponderance of area burned in the boreal zone resulting from lightning-caused fires. Human-caused fires are largely preventable and tend to occur in more populated areas, making their management under a changing climate more possible.

In recent years, GCM outputs have been used by researchers to estimate the level of fire danger and potential fire severity in the boreal zone under a changing climate. Street (1989) compared historical and GCM-derived monthly temperature and precipitation data for two doubled CO_2 scenarios and determined that Ontario fire seasons would be longer and more severe. Flannigan and Van Wagner (1991) used results from three early GCMs to compare seasonal fire weather severity under a $2 \times CO_2$ climate with historical climate records and determined that fire danger would increase by nearly 50% across Canada with climate warming. Wotton and Flannigan (1993) used the Canadian GCM to predict that fire season length across Canada would increase by 30 days in a $2 \times CO_2$ climate. In two recent studies, Fosberg and associates (1996) used the Canadian GCM and Stocks and colleagues (1998) used four current GCMs, along with recent weather data, to evaluate the relative occurrence of extreme fire danger across Canada and Russia and showed a significant increase in both fire weather/danger levels and the geographic expanse of the worst fire danger conditions in both countries under a warming climate. Flannigan and co-workers (1998) report results using the Canadian GCM that indicate a projected increase in fire danger does not appear to be uniform across Canada, as increased precipitation over eastern Canada could result in decreased levels of fire danger and fire activity in that region. Fosberg (1998) also examined the needs of fire management policy makers under a changing climate, determining that a significant increase in the frequency and areal extent of extreme fire danger, which drives fire activity, is likely under current climate change projections.

Projecting Future Fire Danger Conditions in Canada and Alaska: Methodology

In this analysis, daily weather data from the 1980s for 220 stations across Canada and Alaska were used, along with outputs from the Canadian GCM (Boer et al. 1992; McFarlane et al. 1992) to project fire danger conditions under a 2 × CO_2 climate. Local noon measurements of temperature, relative humidity, wind speed, and 24-hour precipitation were used to calculate component codes and indices of the Canadian Forest Fire Weather Index (FWI) System (Van Wagner 1987) for each weather station. Daily FWI values were then converted to daily severity rating (DSR) values by using a technique developed by Williams (1959) and modified by Van Wagner (1970). Fire severity rating can be integrated over periods of varying lengths by using this technique, from DSR through monthly (MSR) to seasonal severity rating values. In this analysis, MSR values are used to provide an assessment of relative fire potential based solely on weather, independent of forest vegetation and fuel conditions. For each grid point of the Canadian GCM, average monthly temperature, relative humidity, wind speed, and precipitation anomalies (differences between the 1 × CO_2 control and 2 × CO_2 GCM runs) were determined. Although significant anomalies were observed for both temperature and precipitation, relative humidity and wind speed showed minimal change between runs and were considered to remain unchanged. The average monthly temperature anomaly for each grid point was then added to the observed daily temperature (from the 1980s data) at the nearest weather station, whereas the monthly precipitation anomaly was factored in as a percentage (positive or negative) of each rainfall event that occurred during that particular month. Two data sets were thus created: the 1980s observed baseline data, and a data set augmented with temperature and precipitation anomalies, which serves as a surrogate for the 2 × CO_2 climate. MSR outputs were then mapped for both scenarios by using a geographic information system, and the areal distribution of fire danger levels across Canada and Alaska were determined.

Distinct fire danger classes cannot been established for MSR values, but a study of fire weather climatology in Russia and Canada in the 1980s (Stocks and Lynham 1996) produced frequency distributions of MSR values, indicating that MSR values of less than 1 to 3 represent low-to-moderate fire potential, values between 3 and 6 represent high-to-very high fire potential, and values greater than 6 constitute conditions for extreme fire potential. In general, MSR values greater than 6 occurred with a frequency of 4–5%.

Results and Discussion

MSR maps for Canada and Alaska, for the fire season months of May through August, based on the 1980s observed weather data, are presented in Figure 20.1.[1] Fire danger conditions are highest in northern and west central Canada, including Alaska, and there is a strong dichotomy between this large region, with its continental climate, and eastern Canada, which generally experiences lower fire danger conditions due to a more maritime climate. In general, fire danger progresses from south to north, with high-to-extreme fire danger being limited to southern Canada in May but expanding quickly north and west in June to include much of west central Canada and Alaska. These fire danger levels are maintained in July but are moderating by August as the fire season begins to wane.

The $2 \times CO_2$ MSR maps (based on Canadian GCM outputs) are presented in Figure 20.2.[2] The monthly progression of fire danger under a $2 \times CO_2$ climate shows a significant increase in the geographic extent of extreme fire danger in May, which indicates an earlier start to the fire season. During June, virtually all west central Canada and interior Alaska are projected to experience extreme fire danger, whereas a slightly more modest increase in fire danger is forecast for July and August. In general, eastern Canada will experience an increase to moderate fire danger levels.

The areal extent of fire danger classes serves as an indication of fire potential, and Figure 20.3 illustrates the significant increases in the monthly geographic expanse of high (rating of 3–6) to extreme (rating >6) fire danger in both Canada and Alaska under a $2 \times CO_2$ climate. These increases, covering approximately 3 million km^2, generally offset a decrease in the areal distribution of low-to-moderate fire danger. Increases in extreme fire danger are common in all months and average about 0.5 million km^2. This is significant because it is commonly accepted that the most severe fires occur under extreme fire danger conditions and that the most severe 10% of fires cause 90% of resource damage. Fire management agencies are generally well prepared (and financed) to handle all but the most severe fire situations, so identifying how this threshold may change with climate change is important.

[1]Figure 20.1 will be found in the color insert.
[2]Figure 20.2 will be found in the color insert.

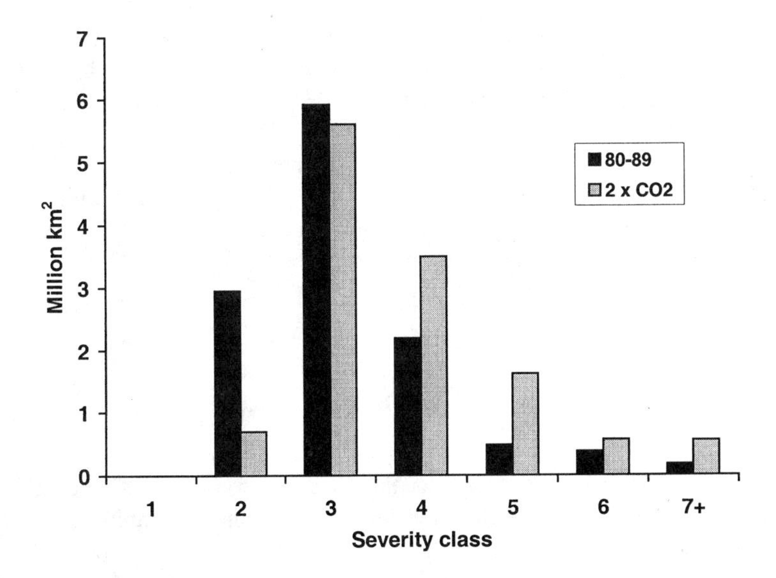

Figure 20.3. Comparison of the areal extent of monthly severity rating classes in Canada and Alaska (by month) for both the 1980–1989 baseline data and the $2 \times CO_2$ climate.

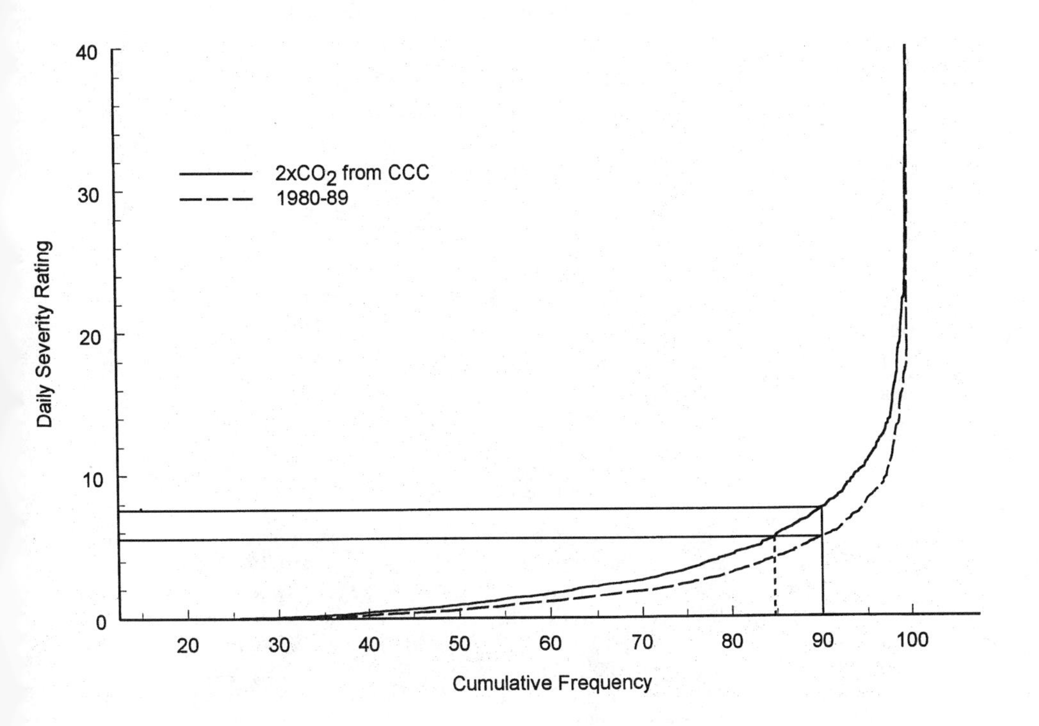

Figure 20.4. Cumulative daily severity rating probability distribution for the 1980–1989 baseline data and the $2 \times CO_2$ climate.

Figure 20.4 shows the cumulative probability of the DSR for both the 1980–89 and 2 × CO_2 scenarios for Thompson, Manitoba, located in the heart of the Canadian boreal zone. The 90th percentile level of the DSR is identified for the 1980–1989 climate, whereas a different level of DSR (an increase of 2, from about 6 to about 8) is defined for the 2 × CO_2 climate. A fire management agency faced with this development would have to increase its capability to match the increase in DSR or accept the risk that the current DSR level will be exceeded 15% of the time in the future rather than the current 10% of the time.

Conclusions

Even state-of-the-art GCMs, such as the Canadian version, suffer from very coarse spatial and temporal resolution levels that hamper our ability to project regional-scale impacts of fire danger, and this will only be rectified with the development and application of higher-resolution RCMs. However, as this study shows, current GCM outputs, when combined with recent observed weather, permit a reasonable assumption of future fire danger levels under a changing climate, which translates into longer and more severe fire seasons in both Canada and Alaska. Increases in fire danger of this magnitude would translate into significant increases in fire activity and area burned, particularly if coupled with budget constraints that limit the ability of fire management agencies to increase the effectiveness of current programs. The result would be more frequent and severe fires, shorter fire return intervals, a skewing of forest age distribution toward younger stands, and a resultant decrease in carbon storage in northern forests (Kasischke et al. 1995; Kasischke, this volume, Chapter 25). This represents a likely positive feedback loop between northern fires and climate change (Kurz et al. 1995). Fire management planning in the future will require a constant balancing act, and future fire climates must be modeled, using higher-resolution models as they become available, and future fire regimes predicted, to assist fire management agencies in adapting to climate change.

References

Boer, G.J., N. McFarlane, and M. Lazare. 1992. Greenhouse gas induced climate change simulations with the CCC second generation GCM. *J. Clim.* 5:1045–1077.

Caya, D., R. Laprise, M. Giguere, G. Bergeron, J.P. Blanchet, B.J. Stocks, G.J. Boer, and N.A. McFarlane. 1995. Description of the Canadian regional climate model, pp. 477–482 in M.J. Apps, D.T. Price, and J. Wisniewski, eds. *Boreal Forests and Global Change*. Kluwer Academic Publishers, Dordecht, The Netherlands.

Environment Canada. 1995. *The State of Canada's Climate: Monitoring Change and Variability*. SOE Report 95-1. Ottawa, Canada.

Flannigan, M.D., and C.E. Van Wagner. 1991. Climate change and wildfire in Canada. *Can. J. For. Res.*21:66–72.

Flannigan, M.D., Y. Bergeron, O. Engelmark, and B.M. Wotton. 1998. Future wildfire in circumboreal forests in relation to global warming. *J. Veg. Sci.* 9:469–476.

Fosberg, M.A. 1998. Global change and fire risk—what do policy makers need to know? pp. 3–21 in D.X. Viegas, ed. *Proceedings 3rd International Conference on Forest Fire Research and 14th Conference on Fire and Forest Meteorology,* November 16–20, 1998, Luso, Portugal. University of Coimbra, Coimbra, Portugal.

Fosberg, M.A., J.G. Goldammer, D. Rind, and C. Price. 1990. Global change: effects on forest ecosystems and wildfire severity, pp. 483–486 in J.G. Goldammer, ed. *Fire in the Tropical Biota: Ecosystem Processes and Global Challenges.* Springer-Verlag, Berlin.

Fosberg, M.A., B.J. Stocks, and T.J. Lynham. 1996. Risk analysis in strategic planning: fire and climate change in the boreal forest, pp. 495–505 in J.G. Goldammer and V.V. Furyaev, eds. *Fire in Ecosystems of Boreal Eurasia.* Kluwer Academic Publishers, Dordecht, The Netherlands.

Hansen, J., R. Ruedy, M. Sato, and R. Reynolds. 1996. Global surface air temperature in 1995: return to pre-Pinatubo level. *Geophys. Res. Lett.* 23:1665–1668.

Kasischke, E.S., N.L. Christensen, and B.J. Stocks. 1995. Fire, global warming, and the carbon balance of boreal forests. *Ecol. Appl.* 5:437–451.

Kirschbaum, M.U.F., and A. Fishlin. 1996. Climate change impacts on forests, pp. 93–129 in R. Watson, M.C. Zinyowera, and R.H. Moss, eds. *Climate Change 1995: Contributions of Working Group II to the Second Assessment Report of the Intergovernmental Panel on Climate Change.* Cambridge University Press, Cambridge, UK.

Kurz, W.A., M.J. Apps, B.J. Stocks, and W.J.A. Volney. 1995. Global climate change: disturbance regimes and biospheric feedbacks of temperate and boreal forests, pp. 119–133 in G.M. Woodwell and F. Mackenzie, eds. *Biotic Feedbacks in the Global Climate System: Will the Warming Speed the Warming?* Oxford University Press, Oxford, UK.

Martin, P. 1993. Vegetation responses and feedbacks to climate: a review of models and processes. *Clim. Dyn.* 8:201–210.

McFarlane, N., G.J. Boer, J.P. Blanchet, and M. Lazare. 1992. The CCC second generation GCM and its equilibrium climate. *J. Clim.* 5:1013–1044.

Price, C., and D. Rind. 1994. Possible implications of global climate change on global lightning distributions and frequencies. *J. Geophys. Res.* 99:10,823.

Rizzo, B., and E. Wiken. 1992. Assessing the sensitivity of Canada's forests to climatic change. *Clim. Change* 21:37–55.

Smith, T.M., and H.H. Shugart. 1993. The transient response of carbon storage to a perturbed climate. *Nature* 361:523–526.

Stocks, B.J. 1991. The extent and impact of forest fires in northern circumpolar countries, pp. 197–202 in J.S. Levine, ed. *Global Biomass Burning: Atmospheric, Climatic, and Biospheric Implications.* MIT Press, Cambridge, MA.

Stocks, B.J. 1993. Global warming and forest fires in Canada. *For. Chron.* 69:290–293.

Stocks, B.J., and T.J. Lynham. 1996. Fire weather climatology in Canada and Russia, pp. 481–487 in J.G. Goldammer and V.V. Furyaev, eds. *Fire in Ecosystems of Boreal Eurasia.* Kluwer Academic Publishers, Dordecht, The Netherlands.

Stocks, B.J., M.A. Fosberg, T.J. Lynham, L. Means, B.M. Wotton, Q.Yang, J.-Z. In, K. Lawrence, G.J. Hartley, J.G. Mason, and D.T. McKenney. 1998. Climate change and forest fire potential in Russian and Canadian boreal forests. *Clim. Change* 38:1–13.

Street, R.T. 1989. Climate change and forest fires in Ontario, pp. 177–182 in D.C. MacIver, H. Auld, and R. Whitewood, eds. *Proceedings 10th Conference on Fire and Forest Meteorology,* April 17–21, 1989, Ottawa, Ontario. AES, Downsview, Ontario, Canada.

Van Wagner, C.E. 1970. *Conversion of William's Severity Rating for Use with the Forest Fire Weather Index.* Informal Report PS-X-21. Canadian Forest Service, Petawawa Forest Experiment Station, Chalk River, Ontario, Canada.

Van Wagner, C.E. 1987. *Development and Structure of the Canadian Forest Fire Weather Index System.* Forest Technical Report 35. Canadian Forest Service, Ottawa, Ontario, Canada.

Watson, R.T., M.C. Zinyowera, and R.H. Moss, eds. 1995. *Climate Change 1995: Impacts, Adaptations and Mitigation of Climate Change: Scientific-Technical Analysis.* Cambridge University Press, Cambridge, UK.

Weber, M.G., and M.D. Flannigan. 1997. Canadian boreal forest ecosystem structure and function in a changing climate: impact on fire regimes. *Environ. Rev.* 5:145–166.

Williams, D.T. 1959. *Fire Season Severity Rating.* Technical Note 73. Canadian Department of Northern Affairs, Natural Resources, Forest Research Division, Ottawa, Ontario, Canada.

Wotton, B.M., and M.D. Flannigan. 1993. Length of the fire season in a changing climate. *For. Chron.* 69:187–192

Wotton, B.M., B.J. Stocks, M.D. Flannigan, R. Laprise, and J.-P. Blanchet. 1998. Estimating current and future fire climates in the boreal forest of Canada using a regional climate model, pp. 1207–1221 in D.X. Viegas, ed. *Proceedings 3rd International Conference on Forest Fire Research and 14th Conference on Fire and Forest Meteorology,* November 16–20, 1998, Luso, Portugal. University of Coimbra, Coimbra, Portugal.

21. Carbon Release from Fires in the North American Boreal Forest

Nancy H.F. French, Eric S. Kasischke, Brian J. Stocks, James P. Mudd, David L. Martell, and Bryan S. Lee

Introduction

Using current estimates of exchanges of carbon between the atmosphere, ocean, and biosphere, researchers have determined that a movement of as much as 2 Gt C yr^{-1} out of the atmosphere pool is unaccounted for (Schlesinger 1991). It has been hypothesized that this missing carbon sink may be found in the terrestrial biosphere and most likely in northern latitudes (Tans et al. 1990; Ciais et al. 1995; Fan et al. 1998; Kasischke, this volume, Chapter 2). Research is underway to model and quantify the carbon cycle in northern (temperate and boreal) forest ecosystems—a biogeochemical cycle that is poorly understood and very difficult to quantify because of the high degree of spatial and temporal variability and complexity in distribution of different forest types and ecosystem processes.

In forest ecosystems, net carbon exchange is controlled by fixation by vegetation photosynthesis (also called gross primary production), respiration (of plants, animals, and soil microbes), soil erosion, and disturbance by fire and other natural events and anthropogenic impacts such as logging and urban development. A permanent land-cover change, such as conversion of forest land to agriculture, will have a long-term impact on carbon storage and exchange, whereas disturbance without conversion is a more short-term mechanism for carbon exchange between the boreal forest and the atmosphere.

Many studies and models of biosphere carbon exchange have been aimed at quantifying net ecosystem exchange of carbon or net ecosystem production (NEP)

(Aber and Melillo 1991; Wofsy et al. 1993; Thompson et al. 1996; Ryan et al. 1997). Although these models present an understanding of mature, steady-state forest systems and are important for tracking terrestrial carbon storage and exchange over the short term, they do not fully take into account the impact of a very important factor in carbon release to the atmosphere—disturbance. In the boreal forest region in particular, fire is very common and widespread. It has an immediate effect on carbon storage by converting long-held carbon reserves in the forest trees and organic soils into atmospheric carbon. These older carbon reserves can take many hundreds to thousands of years to recover to predisturbance levels and may never recover if site conditions are altered sufficiently. Fire also influences gross and net primary production by creating warmer site conditions where atmospheric carbon can be converted to forest biomass more rapidly than in older, established forest sites.

Predictions of global warming due to increased greenhouse gases have shown a higher impact in northern latitudes (Hansen et al. 1996). With regard to carbon cycling, warmer temperatures are expected to (1) increase soil carbon efflux due to more favorable conditions for microbial respiration in soils (Goulden et al. 1998; Richter et al., this volume, Chapter 11); (2) increase the length of the fire season and frequency of fires due to warmer, drier conditions (Flannigan and VanWagner 1991; Stocks 1993; Wotton and Flannigan 1993; Stocks et al., this volume, Chapter 20); (3) create conditions for more rapid carbon uptake by primary production due to a larger proportion of early successional forest tree species (Zimov et al. 1999); and (4) encourage forest ecosystem conversion from forest types adapted to colder conditions to forests found in warmer sites (e.g., conversion from black spruce to white spruce types in marginal sites; see Kasischke et al., this volume, Chapter 12).

Disturbance by wildfire is common in the boreal forest region of North America, particularly in the drier areas of the western continental interior (Chapter 15). Many of the dominant ecosystems of the region rely on fire as an important ecological factor influencing ecosystem development and helping to define the pattern of vegetation type and forest stand age (Chapter 7). Additionally, fire is very important in energy, water, nutrient, and carbon cycling in the boreal region and is an important source to the atmosphere of radiatively significant greenhouse gases. Studies of fire and fire effects are important, therefore, for a complete understanding of the carbon balance of the earth and important for studies of global climate change (Rowe and Scotter 1973; Viereck 1973; Van Cleve et al. 1983; Van Cleve et al. 1986).

Carbon release from boreal biomass burning has been estimated for the global boreal zone (Seiler and Crutzen 1980; Stocks 1991) as well as the various subregions (Russia, Scandinavia, Canada, and Alaska) (Stocks 1991; Dixon and Krankina 1993; Kasischke et al. 1995; Shvidenko et al. 1995; Conard and Ivanova 1997; Shvidenko and Nilsson, this volume, Chapter 10). Seiler and Crutzen (1980) first estimated that 20.7 Tg C or 16 t C ha^{-1} per area burned was released annually from the boreal zone. Further estimates by Stocks (1991) and others (Cahoon et al. 1994) have shown this initial estimate to be substantially lower than is likely due to underestimation of annual area burned and neglecting to account for a large amount of carbon lost from combustion of ground-layer biomass. Most agree that estimates are likely to be improved by doing two things: (1) modifying the basic model put forth by Seiler and Crutzen (1980) to account for the characteristics of the boreal zone; and (2) using refined model inputs. The purpose of this chapter is to describe a methodology for estimating carbon release from biomass burning in the boreal region of North America using an improved model and model inputs.

Approach and Methods

A spatial approach was adopted for this study that divides the North American boreal forest zone into ecozones with homogeneous ecological characteristics (Chapter 14). Using a geographic information system (GIS)–based modeling approach, we made a spatial estimate of biomass burned for the 15-year period from 1980 to 1994. Estimates of carbon released were determined for each fire event, summarized by ecozone, and then combined to come up with the total carbon released from the North American boreal forest for the 15-year period.

The approach used to estimate total carbon released from fire (C_t) is based on the basic model of Sieler and Crutzen (1980):

$$C_t = ABf_c\beta \tag{21.1}$$

where A is the area affected by burning (in hectares) B is the biomass density (t ha^{-1}) in the area burned, f_c is the carbon fraction of the biomass, and β is the fraction of biomass consumed during the fire.

Earlier studies of carbon release from biomass burning only considered consumption of aboveground biomass. Biomass burning in the boreal forest is more complex because the deep layers of mosses and organic soils common in this biome can be consumed during fires (Chapter 10). Our approach is to estimate the aboveground and ground-layer components of biomass and fraction consumed separately, resulting in modification of Equation (21.1) to

$$C_t = A\ [(B_a\, f_{ca}\beta_a) + (C_g\beta_g)] \tag{21.2}$$

where the a and g subscripts refer to the aboveground and ground-layer (mosses, lichen, litter, and the organic soil layers) biomass components, respectively, and C_g is the average carbon density of the ground-layer components (used in place of separate biomass and carbon fraction estimates).

Our model has been implemented in a spatial context to more accurately represent the biomass and burning conditions based on regional variation and to investigate the geographic impact of fire across the boreal zone. Each parameter in Equation (21.2) was determined and entered into a GIS to implement this model. The parameters in Equation (21.2) were obtained in the following manner and are summarized by ecozone in Table 21.1.

Table 21.1. Summary Statistics for Each Model Input Variable by Ecozone

Ecozone	Total area (ha)	Average annual area burned (ha) (A)[a]	Carbon density[b] ($t\ ha^{-1}$)		Carbon fraction consumed (average)	
			Aboveground ($B_a f_c$)	Ground layer (C_g)	Aboveground (β_a)	Ground layer (β_g)
Alaska Boreal Interior	41,629,889	225,775	23	90	0.23	0.36
Boreal Cordillera	60,621,339	153,220	29	61	0.13	0.38
Taiga Plain	58,415,228	325,803	20	160	0.25	0.06
West Taiga Shield	52,611,862	327,630	8	57	0.25	0.05
East Taiga Shield	67,021,473	202,089	10	105	0.25	0.05
West Boreal Shield	71,720,221	788,584	22	79	0.26	0.06
East Boreal Shield	98,846,881	109,811	29	109	0.22	0.06
Boreal Plain	67,653,222	376,575	29	109	0.24	0.11
Hudson Plain	36,655,039	75,916	11	116	0.24	0.05
Total	555,175,154	2,585,405				

[a]Variable from Equation (21.2).
[b]Average carbon density within areas burned. These numbers differ from those in Table 14.3, which lists average carbon density for the entire region.

Area burned (A) is estimated by using records maintained by the Alaskan and Canadian fire management agencies as discussed in Chapter 15. The records consist of mapped fire boundaries drawn from aerial surveys and analyses of satellite imagery as well as fire location and area burned data collected by various governmental agencies (Fig. 15.4). For Alaska, maps for all fires greater than 200 ha have been collected by the Alaska Fire Service, digitized, and entered into a GIS database. For Canada, fire maps for all fires greater than 400 ha for most regions were used. Fire maps were incomplete for Saskatchewan in 1981, so we used fire point locations and recorded area burned for the missing records. Although the Alaskan and Canadian large fire databases are still being assessed, we believe that the accuracy of the digitized data from 1980–1994 is high. For example, the average annual area burned for 1980 to 1994 within the official statistics maintained by the governments of Canada and the United States is 2.585 million ha, whereas the average value within the large fire database is 2.707 million ha, a variance of less than 5%.

Aboveground biomass density (B_a) is based on the published data for Canada and Alaska discussed in Chapter 14 (Fig. 14.2). Biomass density refers to forest biomass and is estimated by using forest inventory data for Canada (Penner et al. 1997) and field data for Alaska (Kasischke 1995). The scale of the data varies by region. Our analysis covers forest areas burned, and we have discounted water and treeless areas within the boreal forest region in the analysis. Carbon fraction of aboveground vegetation (f_{ca}) was assumed to be 0.50. Ground-layer carbon density (C_g) was also determined by using published data (Lacelle et al. 1997). The carbon density of the surface layer (the top 30 cm) was used, because this is the carbon subject to fire (Fig. 14.3).

Two estimates of fraction of biomass consumed (β), for both the aboveground and ground-layer variables, were used in our analysis. We use average values (β_a and β_g) estimated for each ecozone and weighted values estimated for each year and ecozone based on estimated fire severity (β_a' and β_g'). Average fraction of biomass consumed (β_a and β_g) was obtained for each of the ecozones from data generated during controlled fires in Canada (Stocks and Kauffman 1997) and field observations in Alaska (Chapter 10). These studies have clearly shown that the fraction of biomass consumed during fires varies significantly, depending on fuel (forest) type and fuel moisture conditions at time of fire. Higher levels of biomass consumption occur during years with warmer drier conditions. Because permafrost hinders soil drainage, the fraction of consumption of ground-layer biomass tends to be higher when fires occur later in the growing season when the depth to permafrost is greatest.

Because of this variability, we use a simple model to estimate a weighted fraction of biomass consumed (β_a' and β_g') for each year and ecozone based on the annual area burned. The model is based on the fact that both annual area burned and fraction of biomass consumption are closely correlated to interannual variations in climate. Larger areas are burned and higher levels of biomass consumption are experienced during warmer drier years, whereas total area burned and fraction of biomass consumed are lower during cooler wetter years. Based on this basic relationship, it was assumed that fraction of biomass consumed is linearly proportional to the annual area burned. The model uses the average fraction consumed as starting points for consumption estimation. For each region, during the lowest fire year it was assumed that

$$\beta_{a\text{-low}} = 0.5\ \beta_a \quad (21.3a)$$

$$\beta_{g\text{-low}} = 0.5\ \beta_g \quad (21.3b)$$

During the highest fire year in each region, it was assumed that

$$\beta_{a\text{-high}} = \beta_a + 0.1 \quad (21.4a)$$

$$\beta_{g\text{-high}} = \beta_g + 0.2 \quad (21.4b)$$

A regression equation of fraction of biomass consumed as a function of annual area burned was developed for each region based on the low, average, and high values. Using the regression results, expected fraction consumed was determined for each ecozone and year based on annual area burned. Because of the uncertainty for this variable, two estimates of annual carbon release were made: one applying the average biomass consumption values (β_a and β_g) and a second using weighted values (β_a' and β_g') from the biomass consumption model.

Table 21.1 summarizes the basic input parameters used in this study. This table includes the total area for each ecozone, the average annual area burned, the average aboveground and ground-layer carbon densities in the areas that burned within each ecoregion, and the average carbon fraction consumed.

Results

Results computed for each fire event were aggregated by ecozone and annualized to derive an average of 9.6 g C m^{-2} yr^{-1} released, using the weighted fraction of carbon consumed values (Table 21.2), compared with 6.9 g C m^{-2} yr^{-1} released, using the average fraction of carbon consumed values. There is considerable variability in carbon released between the different ecozones. The lowest levels of carbon release were found in the East Boreal Shield (2.4 g C m^{-2} yr^{-1}), whereas the highest level was found in the Alaskan Boreal Interior (22.4 g C m^{-2} yr^{-1}). This latter finding is interesting because the West Boreal Shield had the higher level of fire activity. The higher carbon release levels in the Alaskan Boreal Interior are attributed to the higher ground layer carbon density and fractions of consumption of ground-layer biomass observed for this region (Table 21.1).

Table 21.2 also summarizes the average carbon release during low (1980) and high (1985) fire years for the entire North American boreal forest region. Burning in 1980 (with 5 million ha of fire) produced an average of 19.8 g C m^{-2}. Burning in 1985 (with 0.8 million ha burned) produced 1.7 g C m^{-2}. Again, the data show

Table 21.2. Estimates of Carbon Released from Biomass Burning in the North American Boreal Zone

	Annual carbon released ($g\ C\ m^{-2}\ yr^{-1}$)			
Ecozone	Average fraction consumed (β)	Weighted fraction consumed (β′)	High fire year[a] (1980)	Low fire year[a] (1985)
Alaska Boreal Interior	20.7	22.4	2.6	10.9
Boreal Cordillera	6.9	7.5	5.3	2.9
Taiga Plain	8.5	13.9	32.1	0.1
West Taiga Shield	3.1	5.1	4.5	2.0
East Taiga Shield	2.3	4.6	0.0	1.7
West Boreal Shield	11.5	16.5	54.2	0.4
East Boreal Shield	1.4	2.4	0.1	0.6
Boreal Plain	10.6	15.4	67.2	0.3
Hudson Plain	1.8	3.3	0.2	0
Total	6.9	9.6	19.8	1.7

[a]Fire severity based on weighted fraction consumed value (β′).

that there was great variability between the different ecoregions during these high and low fire years.

An average of 0.053 Gt C yr^{-1} was released from the North American boreal forest region, using the weighted fraction consumed, whereas 0.038 Gt C yr^{-1} was released using the average fraction consumed (Table 21.3). The total amount of carbon released using the weighted fraction consumed ranged between a low of 0.006 Gt C in 1984 to 0.167 Gt C in 1979. The data in Table 21.3 illustrate that the location of the fires does make a difference in total carbon release. For example, nearly the same area burned in 1990 and 1991, but the amount of carbon released in 1990 was 70% greater than the amount released in 1991.

The GIS-based modeling approach used in this study allows examination of the geographic distribution of carbon release throughout the North American boreal forest region. Figure 21.1[1] presents the average carbon release from fires for this region plot on a 1° longitude-by-1° latitude grid. The range of carbon release was from 0.0 to 166.9 g C m^{-2}. These spatial maps provide greater insight into understanding the sources of variation present in the data summarized in Tables 21.2 and 21.3. Figure 21.2[2] presents carbon release maps from 3 different years (1989, 1990, and 1994), in which the total carbon released is not proportional to area burned. For example, although 1994 had a total area burned that was two-thirds of that in 1989, the amount of carbon released in 1994 was less than half of that in 1989. By contrast, 1994 had more than twice as much area burned than in 1990, yet the total carbon released during these 2 years was essentially identical. These variations are easily understood through examination of the patterns of

[1]Figure 21.1 will be found in the color insert.
[2]Figure 21.2 will be found in the color insert.

Table 21.3. Area Burned and Carbon Emissions by Year for the North American Boreal Forest Region

		Carbon emissions (Gt)	
Year	Area burned (ha)	Average fraction consumed (β)	Weighted fraction consumed (β′
1980	4,897,897	0.067	0.109
1981	5,731,446	0.076	0.124
1982	1,497,073	0.029	0.035
1983	1,756,164	0.021	0.028
1984	677,256	0.009	0.006
1985	795,476	0.013	0.010
1986	869,447	0.020	0.017
1987	1,028,940	0.015	0.014
1988	2,137,093	0.042	0.041
1989	7,523,287	0.085	0.167
1990	2,097,540	0.063	0.078
1991	2,120,883	0.044	0.045
1992	890,522	0.010	0.007
1993	1,856,372	0.034	0.035
1994	4,901,677	0.044	0.080
Average	2,585,405	0.038	0.053

carbon release that integrate the three factors controlling this process: (1) area burned; (2) carbon density; and (3) fraction of carbon consumed during fires. The fires in 1994 were distributed along the northern Canadian ecozones (the boreal cordillera, taiga plains, and west taiga shield), whereas the fires in 1989 were concentrated in the boreal plains and the east taiga shield and in 1994 in the Alaskan boreal interior. The reasons for the differences are that the areas burned in 1994 had either a lower carbon density or a lower fraction of carbon consumed.

Discussion

Our estimate of 0.053 Gt C yr^{-1} released from boreal North American biomass burning is higher than previous estimates by nearly an order of magnitude. Seiler and Crutzen (1980) estimated that 0.0215 Gt C yr^{-1} is released from the entire boreal forest. Because North America contains approximately one-third of the world's boreal forests, the Seiler and Crutzen (1980) value scales to 0.007 Gt C yr^{-1} for the North American boreal region. We believe the large difference is found because of improved input data for the model and because of the spatial nature of the model. First, the annual area burned for the North American boreal forest is twice as great as the value used by Seiler and Crutzen for the entire boreal forest, which means that they largely underestimated area burned. Second, by using actual locations of fire events and relatively fine-scale estimates of the input parameters (e.g., aboveground and ground-layer carbon), we are using data that are more accurate than previous approaches. In this way, we are able to quantify what actually burns instead of making estimates based on general information. Rather than using an average value for biomass present in the region, for example, we use a biomass value that is estimated for the place that burned.

The GIS-based modeling approach improves our understanding of carbon release because it also allows us to view the spatial patterns. Patterns in carbon release result from variations in both area burned and biomass density. As shown in the map of the 15-year average (Fig. 21.1), carbon release is highest from the continental region of Canada. This is a result of high incidence of fire (Fig. 15.4) and relatively dense carbon reserves (Figs. 14.2 and 14.3). It is helpful, however, to look at the yearly maps (Fig. 21.2) to understand how the patterns seen in this 15-year average are derived.

Recent studies have developed estimates of NEP for different boreal forest types. Frolking (1997) showed that for black spruce forests, NEP ranges between 15 and 20 g C m^{-2} yr^{-1}. Studies by Goulden and co-workers (1998) showed that during warmer-than-average years, black spruce forests can actually be a net source of atmospheric carbon, with an NEP of −8 g C m^{-2} yr^{-1}. For aspen stands, Black and associates (1996) estimate an NEP of 130 g C m^{-2} yr^{-1}. Although no measurements of NEP exist for jack pine stands, if we assume NEP is proportional to net primary production, we can use the estimates of Gower and colleagues (1997) to estimate an NEP for this forest type of 53 g C m^{-2} yr^{-1}. Based on these values, in the western ecoregions of the North American boreal forest where black spruce is dominant, we estimate that NEP is about 40 g C m^{-2} yr^{-1}.

To determine whether a region is an atmospheric carbon source or sink, one must subtract the carbon released during fires from the estimates of NEP. Given our estimate of NEP of 40 g C m^{-2} yr^{-1}, this means that those regions in the western portion of the North American boreal forest where fire carbon release is greater than 40 g C m^{-2} yr^{-1} (Fig. 21.1) served as a net source of atmospheric carbon over the past 15 years.

Summary and Conclusions

In this chapter, we presented the results of a study that produced a refined estimate of carbon released through biomass burning in the North American boreal zone. Our estimate of 0.053 Gt C yr^{-1} released is substantially higher than previous estimates but is considered more accurate because of the fine-scale spatial information used in the model. Besides being an improved estimate due to more accurate input data, the model presented also reveals spatial patterns of biomass burning. This view of the spatial impact of fire on carbon release is useful for understanding the impact of fire at finer than global scales and for visualizing how a local phenomenon can have global effects during extreme fire years.

We have made more accurate estimates of carbon release with our spatial approach than in previous studies because we were able to quantify the input parameters related to location of fire and distribution of carbon throughout the study region. For a regional or global-scale model, the scale of the input data we have used is reasonable without being overly complex. However, further improvements to our estimates are possible and would result from further improvement in the input variables. Possible improvements are as follows.

Area burned for the period of our study has been recorded with fairly good accuracy; however, the database we used still needs to be checked. For example, we know that there are still some missing fire boundaries from this data set and that the overall accuracy of many regions could be assessed by using satellite imagery (Chapter 17). Once these steps are complete, we have great confidence that this parameter is well quantified. In addition, more years are presently being added to the Canadian large fire database. Once these data have been added, we will be able to examine the patterns of carbon release over a longer time period.

Biomass density maps for Canada are fairly accurate due to the government's commitment to forest inventory. Alaska and areas of nonproductive forests in Canada are not as well mapped. Improved estimates of vegetation and forest cover by using satellite imagery could be used to improve estimates of biomass distribution. This is particularly true in the light of the fact that new satellite-based sensors will soon be available. These include MODIS (which has improved spectral capabilities from any current sensors) and ASTER and Landsat 7 (which have improved spectral information and spatial resolution). Other ways of improving this parameter is to use an approach such as that of Kurz and co-workers (1992), in which biomass is determined by using process models based on inventory data. Using a process model will improve the input data as long as it is properly implemented. What needs to be considered is that fire is an important part of the process that determines resident carbon. Carbon density, therefore, is dependent on the location and severity of fire, the other parameters in our basic model. A "hybrid" model, therefore, may be best for integrating processes and feedbacks into our basic model (Kasischke et al. 1995).

The most difficult and least understood parameter to quantify is the fraction of carbon consumed during biomass burning. This is the likeliest source of error in the current modeling approach. We have used estimates that were determined at the ecozone scale and attempted to improve them by weighting by fire severity. Our carbon emission estimates using the weighted fraction consumed inputs are higher because fires during larger fire years are given higher values than the average. This results in overall higher estimates because of the weighting given to large years. We expect our weighted average estimate to be more accurate than the average value estimates because years with more fires are generally drier years and higher consumption is expected. Alternative ways to improve these estimates include modeling severity based on forest type and weather factors or by sampling burn areas in various regions to measure actual burn severity. The later method could be accomplished by using relatively fine-scale remote sensing data, as was done by Michalek and associates (in press) using Landsat Thematic Mapper imagery (see Chapter 23). Future models of carbon released should include improved spatial estimates of fraction consumed to improve the model results.

In summary, we have demonstrated a modeling technique to determine accurately the impact of fire on carbon flux in the North American boreal region. Although improved data inputs would improve the output, the current result is more accurate than past attempts and allows us to examine the spatial impact of fire within the region. Using this technique we have shown that fire may result in the boreal region or portions of the boreal region being a net source of carbon over the 15-year period studied. This result has global implications despite the fact that fire is a local phenomenon.

References

Aber, J.D., and J.M. Melillo. 1991. *Terrestrial Ecosystems.* Saunders College Publishing, Philadelphia, PA.

Black, T.A., G. den Hartog, H.H. Neumann, P.D. Blanken, P.C. Yang, C. Russel, Z. Nesic, X. Lee, S.G. Chen, and R Staebler. 1996. Annual cycles of water vapour and carbon dioxide fluxes in and above a boreal aspen forest. *Global Change Biol.* 2: 219–239.

Cahoon, D.R., Jr., B.J. Stocks, J.S. Levine, W.R. Cofer III, and J.M. Pierson. 1994. Satellite analysis of the severe 1987 forest fires in northern China and southeastern Siberia. *J. Geophys. Res.* 99:18,627–18,638.

Ciais, P., P.P. Tans, M. Trolier, J.W.C. White, and R.J. Francey. 1995. A large Northern Hemisphere terrestrial CO_2 sink indicated by the $^{13}C/^{12}C$ ratio of atmospheric CO_2. *Science* 269:1098–1102.

Conard, S.G., and G.A. Ivanova. 1997. Wildfire in Russian boreal forests—potential impacts of fire regime characteristics on emissions and global carbon balance estimates. *Environ. Pollut.* 98:305–313.

Dixon, R.K., and O.N. Krankina. 1993. Forest fires in Russia: carbon dioxide emission to the atmosphere. *Can. J. For. Res.* 23:700–705.

Fan, S., M. Gloor, J. Mahlman, S. Pacala, J. Sarmiento, T. Takahashi, and P. Tans. 1998. A large terrestrial carbon sink in North America implied by atmospheric and oceanic carbon dioxide data and models. *Science* 282:442–446.

Flannigan, M.D., and C.E. Van Wagner. 1991. Climate change and wildfire in Canada. *Can. J. For. Res.* 21:61–72.

Frolking, S. 1997. Sensitivity of spruce/moss boreal forest net ecosystem productivity to seasonal anomalies in weather. *J. Geophys. Res.* 102:29,053–29,064.

Goulden, M.L., S.C. Wofsy, J.W. Harden, S.E. Trumbore, P.M. Crill, S.T. Gower, T. Fries, B.C. Daube, S.-M. Fan, D.J. Sutton, A. Bazzaz, and J.W. Munger. 1998. Sensitivity of boreal forest carbon balance to thaw. *Science* 279:214–217.

Gower, S.T., J.G. Vogel, J.M. Norman, C.J. Kucharik, S.J. Steele, and T.K. Snow. 1997. Carbon distribution and aboveground net primary production in aspen, jack pine, and black spruce stands in Saskatchewan and Manitoba, Canada. *J. Geophys. Res.* 102: 29,029–29,041.

Hansen, J., R. Ruedy, M. Sato, and R. Reynolds. 1996. Global surface air temperature in 1995: return to pre-Pinatubo level. *Geophys. Res. Lett.* 23:1665–1668.

Kasischke, E.S., N.L. Christensen, Jr., and B.J. Stocks. 1995. Fire, global warming, and the carbon balance of boreal forests. *Ecol. Appl.* 5:437–451.

Kurz, W.A., M.J. Apps, T.M. Webb, and P.J. McNamee. 1992. *The Carbon Budget of the Canadian Forest Sector: Phase I.* Information Report NOR-X-326. Forestry Canada, Northwest Region, Northern Forestry Centre, Edmonton, Alberta, Canada.

Lacelle, B., C. Tarnocai, S. Waltman, J. Kimble, F. Orozco-Chavez, and B. Jakobsen. 1997. *North American Soil Carbon Map.* Agriculture and Agri-food Canada, USDA, INEGI, Institute of Geography, University of Copenhagen.

Michalek, J.L., N.H.F. French, E.S. Kasischke, R.D. Johnson, and J.E. Colwell. In press. Using Landsat TM data to estimate carbon release from burned biomass in an Alaskan spruce complex. *Int. J. Remote Sens.*

Penner, M., K. Power, C. Muhairwe, R. Tellier, and Y. Wang. 1997. *Canada's Forest Biomass Resources: Deriving Estimates from Canada's Forest Inventory.* Information Report BC-X-370. Pacific Forestry Centre, Canadian Forest Service, Victoria, BC, Canada.

Rowe, J.S., and G.W. Scotter. 1973. Fire in the boreal forest. *Quat. Res.* 3:444–464.

Ryan, M.G., M.B. Lavigne, and S.T. Gower. 1997. Annual carbon cost of autotrophic respiration in boreal forest ecosystems in relation to species and climate. *J. Geophys. Res.* 102:28,871–28,883.

Schlesinger, W.H. 1991. *Biogeochemistry: An Analysis of Global Change.* Academic Press, San Diego, CA.

Seiler, W., and P.J. Crutzen. 1980. Estimates of gross and net fluxes of carbon between the biosphere and atmosphere. *Clim. Change* 2:207–247.

Shvidenko, A., S. Nilsson, R. Dixon, and V.A. Rojkov. 1995. Burning biomass in the territories of the former Soviet Eurasia: impact on the carbon budget. *Q. J. Hung. Meteor. Serv.* 99:235–255.

Stocks, B.J. 1991. The extent and impact of forest fires in northern circumpolar countries, pp. 198–202 in J.S. Levine, ed. *Global Biomass Burning: Atmospheric, Climatic, and Biospheric Implications.* MIT Press, Cambridge, MA.

Stocks, B.J. 1993. Global warming and forest fires in Canada. *For. Chron.* 69:290–293.

Stocks, B.J., and J.B. Kauffman. 1997. Biomass consumption and behavior of wildland fires in boreal, temperate, and tropical ecosystems: parameters necessary to interpret historic fire regimes and future fire scenarios, pp. 169–188 in J.S. Clark, H. Cachier, J.G. Goldammer, and B.J. Stocks, eds. *Sediment Records of Biomass Burning and Global Change.* NATO ASI Series, Subseries 1, Global Environmental Change, Vol. 51. Springer-Verlag, Berlin.

Tans, P.P., I.Y. Fung, and T. Takahashi. 1990. Observational constraints on the global atmospheric CO_2 budget. *Science* 247:1431–1438.

Thompson, M.V., J.T. Randerson, C.M. Malmstrom, and C.B. Field. 1996. Changes in net primary production and heterotrophic respirationHow much is necessary to sustain the terrestrial carbon sink? *Global Biogeochem. Cycles* 10:711–726.

Van Cleve, K., L. Oliver, R. Schlentner, L.A. Viereck, and C.T. Dyrness. 1983. Productivity and nutrient cycling in taiga forest ecosystems. *Can. J. For. Res.* 13:747–766.

Van Cleve, K., F.S. Chapin III, P.W. Flanagan, L.A. Viereck, and C.T. Dyrness, eds. 1986. *Forest Ecosystems in the Alaskan Taiga.* Ecological Studies 57. Springer-Verlag, New York.

Viereck, L.A. 1973. Wildfire in the taiga of Alaska. *Quat. Res.* 3:465–495.

Wofsy, S.C., M.L. Goulden, J.W. Munger, S-M. Fan, P.S. Bakwin, B.C. Daube, S.L. Bassow, and F.A. Bazzaz. 1993. Net exchange of CO_2 in a mid-latitude forest. *Science* 260:1314–1317.

Wotton, B.M., and M.D. Flannigan. 1993. Length of the fire season in a changing climate. *For. Chron.* 69:187–192.

Zimov, S.A., S.P. Davidov, G.M. Zimova, A.I. Davidova, F.S. Chapin III, M.C. Chapin, and J.F. Reynolds. 1999. Contribution of disturbance to increasing seasonal amplitude of atmospheric CO_2. *Science* 284:1973–1976.

22. Ecological Models of the Dynamics of Boreal Landscapes

Herman H. Shugart, Donald F. Clark, and Amber J. Hill

Introduction

Ecosystems are inherently complex and highly integrative entities that are defined by the assemblages of species, the physical environment, and the processes that regulate them. The implementation of models as a tool for scientific simulation, experimentation, and investigation has provided a major advance in the field of ecology (Dzeroski et al. 1997; Kompare et al. 1994; Patten et al. 1997; Rykiel 1989; Laval 1995; Liu and Ashton 1995). Until recently, the ability to carry out simulation experiments over large temporal and spatial scales was nearly impossible, except for a small group of scientists with access to powerful computer mainframes. Modern desktop computer systems are now capable of carrying out complex simulation and experimentation on ever-increasing scales (McHaney 1991). This has resulted in an immense expansion in the use of computer-based simulation models in ecological research.

Present-day models are playing an ever-increasing role in the development of ecological theory at several scales, from understanding the mechanisms of carbon fixation (Farquhar and Sharkey 1982; Farquhar and von Craemmer 1982) and plant-water balance (Cowan 1982, 1986); to scaling of physiological processes to whole plant function (Reynolds et al. 1986); to exploring how ecosystem processes of carbon and nitrogen cycling operate at continental to global scales (Emanuel et al. 1984, 1985). An important role of modeling is to explore ecosystem dynamics that occur at spatial and temporal scales where direct observation and experimentation are prohibitive if not impossible. Examples include investigating spatial and temporal variation in competition on ecosystem functioning (Sharpe et al. 1985, 1986; Wu et al. 1985; Walker et al. 1989), extrapolating the processes of carbon fixation and water balance to the landscape scale and to link ecosystem models with remotely sensed data (Running and Coughlan 1988; Running et al. 1989), and exploring the implications of the evolution of plant adaptations to varying environmental conditions on patterns of ecosystem structure across environmental gradients (Tilman 1988).

Scaling Processes in the Boreal Forest

The interactive nature of different biophysical processes combine to determine ecosystem responses to disturbance or environmental change (Prentice et al. 1993a). Gradual plant community processes, which depend on species physiology, morphology, and life history characteristics, substantially complicate the consequences of first-level rapid responses of basic photosynthetic and microbial processes (Moore et al. 1990). In forest community dynamics, competition for light, space, nutrients, water, and other resources may be of equal or more importance in determining ecosystem responses on time scales from years to centuries than are those of primary photosynthetic processes (Shugart 1984; Solomon and Shugart 1993).

Changes in the storage of carbon in boreal forests is a consequence of transfers of carbon within the forest and between the atmosphere and the forest. It is convenient to think of the processes that control these transfers at three different spatial scales (Fig. 22.1):

1. The site scale is typically dimensioned in hundreds or thousands of square meters. Carbon fluxes measured at the site scale are sometimes characterized as "ecosystem metabolism" because they are analogous to the metabolic processes of an organism. The principal fluxes involve the balance between photosynthesis and respiration of the plants (usually a net positive flux of carbon into the site) and the soil respiration (a negative flux of carbon from the site). There is also a flux of carbon from living plants to the soil by consequence of the death and deposition of boles, branches, and leaves to the forest floor. Tree death, which is a source of dead carbon to the soil, is very difficult to measure at the site scale due to its pattern of variability (Clark et al. 1998).
2. The stand scale is typically dimensioned in units of multiple hectares and is frequently considered as forests of similar age and composition. Stand dynamics are associated with demographic processes that change the age structure and competitive interactions that, in turn, alter the composition of the stand. In the boreal forest, one of the most important agents of such change is wildfire, which can radically alter the composition and age structure of the stand. Fires also act to produce direct fluxes of carbon to the atmosphere through the burning of trees and forest floor materials. Insects also function as

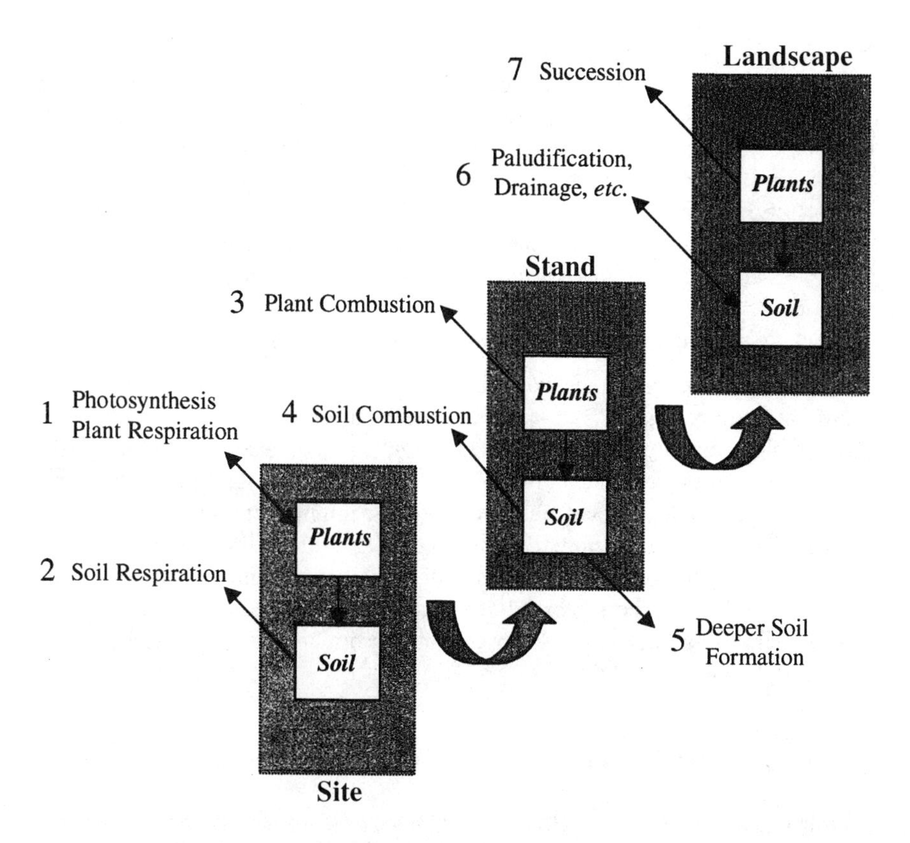

Figure 22.1. Principal carbon fluxes in boreal forest ecosystems. Fluxes of carbon from the forest at different spatial scales are numbered 1 through 7 on the figure and elaborated in Table 22.1. In addition to these fluxes from the forest, there are also significant redistributions of carbon from the living plants to storage in the soil system. These are indicated by the transfers from plant to soil at each of the three scales.

an important agent of disturbance and can affect large areas of forests under severe outbreak conditions. Compared with fire, however, insect disturbances have significantly different temporal effects on the carbon flux; in fires, oxidation of plant carbon acts on the scale of hours and days, whereas this process acts on the scale of months to years after insect outbreaks. Insect-mediated dieback of forests generally leave large amounts of combustible dead plant material that can be a precursor to high-intensity fires, potentially augmenting carbon flux to the atmosphere. Additionally, increasing the fire frequency in many boreal forest ecosystems acts to increase the accumulation of soil carbon and litter, which represents a storage flux for carbon. Soil carbon storage in boreal forest ecosystems is usually tabulated up to a depth of 1 m. Accumulation of soil beyond 1 m in depth represents a missing carbon flux in the light of present-day calculations.

3. The landscape scale is of large dimension, often more than 1 km^2, and is made up of a mosaic of stands. Changes in the landscape mosaic from the processes of ecological succession can result in differences in the amount of carbon stored and in an increased exchange of carbon with the atmosphere. In boreal systems, the soil organic matter accumulation can develop complex and reversible formations of paludified (swamped) forests, along with a variety of patterned bogs, mires, and peatlands. These landscape units can store a prodigious amount of carbon. The changes in landscapes over time and, in particular, the formation of peatlands of various sorts are strongly interactive with the formation and melting of permanent ice (permafrost) in the soil.

These different carbon fluxes and some of the factors that are involved in their control are summarized in Table 22.1 and are be discussed in more detail below. The complex nature of the carbon fluxes from the boreal forest arise in part to the intrinsic patchiness and the multiscale responses of important ecological processes. For example, a local climatic warming might increase both site-level photosynthesis and respiration; thus the net effect might require a detailed evaluation of the flux rate of these processes. Additionally, the same climatic warming might increase the carbon released at the stand level during a wild fire, in turn causing permafrost melting and subsequent bog draining and associated carbon release at the landscape level.

Challenge of Developing Ecological Models of Landscape Dynamics

A basic consideration in modeling the dynamics of landscape systems is whether the landscape can be divided into a number of discrete smaller units (Weinstein and Shugart 1983). For different types of models, the simulated landscapes are assumed to be

1. Mosaic landscapes where one assumes that responses to environmental change are local and relatively independent of landscape configuration and spatial effects. The resultant mosaic landscape models are based on an assumption that the landscape can be divided into discrete noninteractive elements that change dynamically with time.
2. Interactive landscapes where one assumes that responses to environmental change have a pronounced spatial component that either amplifies or attenuates the local responses. Interactive landscape models are constructed recognizing that the individual landscape elements are interactive and can interact mutually, altering each other's dynamics.
3. Homogeneous landscapes where one assumes that the landscape elements respond in a homogeneous fashion and thus local effects are less important or similar for all landscape units. Homogeneous landscape models are based on this assumption that the landscape functions as a uniform dynamic unit.

These three views of landscapes are based on the nature of interactions and dominant processes and each has an associated modeling paradigm. Therefore, consideration of one of these three types of landscape views is relative to the particular phenomena as well as the landscape under consideration (see Shugart 1998).

It is interesting to note that each of these three views is based on the assumption that certain landscape processes (notably the disturbance regime) behave in ways that are exclusive of each other. For example, the spread of wildfire is necessarily modeled as a spatially dependent phenomenon in a fire-spread model but is considered a spatially independent statistical feature of the environment in forest succession models. The implications of this antilogy compel us to consider the development of an inclusive model incorporating all the important boreal forest processes or alternatively developing an exclusive set of models each with different underlying assumptions enabling an intercomparison of how these key features might influence future boreal forest landscape dynamics. The inclusive approach can be expected to have difficult parameter estimation and a high potential for error propagation problems. Additionally, inclusive model formulations may also undermine the importance of recognizing temporal and spatial landscape properties. The latter exclusive approach would look for robust results across a spectrum of models and would encourage extended model testing. We provide an example of an exclusive modeling approach by describing a set of models that could potentially be used to simulate the dynamics of boreal landscapes.

Environmental Change and Scale as Considerations in Understanding Boreal Forests

The issue of understanding space and time scales in ecological systems has been identified as necessary for understanding ecosystems response to large-scale environmental change (O'Neill 1988). Particular attention to scale has been highlighted in the development of hierarchy theory in ecology (Allen and Starr 1982; O'Neill et al. 1986; Shugart and Urban 1988) recognizing that the effects of scale are central to understanding system dynamics. For example, Woodward (1987) postulated that inputs of different frequencies in time or space induce responses from different vegetation processes and can be thought of as producing different patterns. Additionally, Delcourt and colleagues (1983) considered the time and space scale of different disturbance factors, the ecological mechanisms that are excited by these phenomena, and the patterns produced by the interactions between the disturbances and the ecological mechanisms. This is illustrated in a three-part diagram (adapted from Delcourt et al. 1983) in which disturbances, biotic responses, and resultant patterns are indicated at the space and time intervals over which they were typically measured (Fig. 22.2).

Disturbance processes act on an ecosystems at different frequencies such as the slow time and space scales of plate tectonics in separating, moving, and melding the continents over millions of years. A disturbance such as this is relevant to the evolution of the biota and the diversity of the vegetation formations at a conti-

Table 22.1. Major Carbon Fluxes at Different Scales in the Boreal Forest

Carbon flux	Spatial scale	Description	Issues	Controlling factors
1. Photosynthesis and respiration	Site	The balance between fixation of carbon by the photosynthesis and the respiratory energy costs of the plants integrated over 1 year is the net production of the forest.	Net production is a relatively remainder of the difference between two large fluxes. Often, net production is measured by the summation of the incremental change of the size of each tree at a site.	Major environmental factors including light, moisture, nutrients, growing season length, air and soil temperatures, and soil and air growing degree-days. Understanding of the direct effects of CO_2 is a major confounding consideration.
2. Soil respiration	Site	Breakdown by aerobic bacteria, anaerobic bacteria, and fungus of chemical compounds created by production and then shed from the tree.	Decomposition of organic matter in the soil and dead leaf litter layers is usually thought of as the soil respiration, but most methodologies do not, in fact, separate the respiration of living plant roots from the measured CO_2 fluxes from the soil. Methods using chambers to capture gases also are confounded by the respiration of the moss on the soil surface.	Thermal, moisture and nutrient factors tend to control soil respiration. The respiration of the tree roots is under similar but presumably slightly different controls.
3. Plant combustion	Stand	Flux of carbon to atmosphere associated with fires.	In the boreal forest, plant combustion can be a significant source of carbon to the atmosphere. It is variable in its occurrence with both time and space. It is strongly influenced by fuel loads and condition of the vegetation and litter.	Fires are strongly influences by weather conditions with dry, hot weather and high winds representing the worst-case combination. Antecedent conditions are important contributing factors, as are slope and surface conditions.

4. Soil combustion	Stand	Flux of carbon to atmosphere associated with fires.	In the organic upper layers of the boreal soil, burning can be a significant source of carbon to the atmosphere. See comments immediately above.	See comments immediately above.
5. Deep soil formation	Stand	Slow decomposition rates in boreal systems allow for the building of carbon-rich organic soils that deepen with time and bury the older soils below them.	This is largely a data issue in that many computations of boreal soil carbon content only account for the carbon in the upper 1 m, for example, of the soil profile. Soils deepening over time thus can potentially represent carbon "lost" from carbon budget calculations.	Slow decomposition rates and the accumulation of organic matter are a consequence of several of the factors mentioned above.
6. Paludification, bog formation, bog drainage, and so on	Landscape	In boreal landscapes (particularly those that have not burned for some time), there is a tendency for forested systems to convert into complex systems of mires, bogs, and forested swamps. There is also a potential for conversion of these carbon-rich systems back into forests with an associated reduction of carbon storage.	Although the process associated with the dynamic landscape conversions to and from mires and bogs operates fairly slowly, the carbon storage involved is large.	Climate, organic matter accumulation, fires, and permafrost formation all influence the formation and maintenance of bog systems on the boreal landscape. The Russians also report glacial rebound as a factor raising land and draining bogs over large areas in far northern Siberia.
7. Succession	Landscape	The conversion of one type of vegetation with another involves the death of plants, associated growth of the replacement plants, and dynamical changes in the landscape carbon storage.	Successional change is a relatively slow process. Often the pattern of change is inferred from a collection of contemporaneous changes in forests of differing types and development.	Succession is a consequence of climate, soils, the biotic diversity of a given region, and the disturbance regime. It operates under considerable variation according to slope and aspect in high-latitude boreal systems.

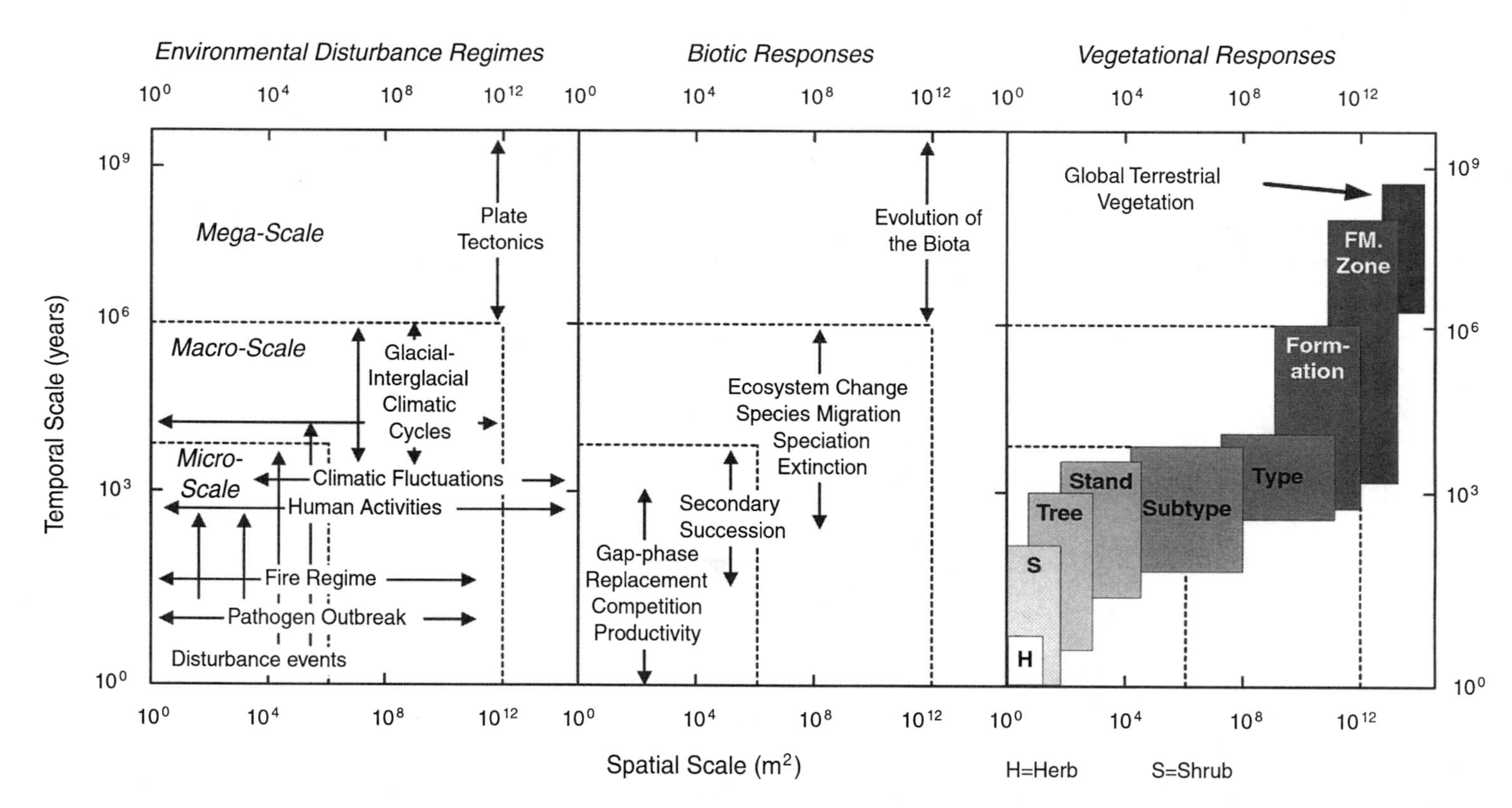

Figure 22.2. Environmental disturbance regimes, biotic responses, and vegetation patterns viewed in the context of space-time domains in which the scale for each process or pattern reflects the sampling intervals required observing it. The time scale for the vegetation patterns is the time interval required to record their dynamics. The vegetation units are mapped as a nested series of vegetation patterns. Examples of disturbance events include wildfire, wind damage, clearcutting, flood, and earthquake. FM, formation. (From Delcourt et al. 1983.)

nental scale (Fig. 22.2). However, plate tectonics are not considered when investigating landscape fire dynamics. Instead, this factor has more relevance to variations in the frequency of disturbance regimes on the scale of decades to century.

Therefore, consideration of the various disturbance processes and how they might vary in both the time and space domains is an essential step in defining what features are relevant. For example, the aim of trying to elucidate what features of a given study region are important constrains the definition of the system as well as the research objectives of the study. This is demonstrated easily by considering the overall goals of several international ecologically oriented research programs. In particular, the stated objectives of the Global Change in Terrestrial Ecosystems (GCTE) program that is a component of the larger International Geosphere and Biosphere Program (IGBP) are

1. To predict the effects of changes in climate, atmospheric composition, and land use on terrestrial ecosystems, including agricultural and forest production systems,
2. To determine how these effects lead to feedbacks to the atmosphere and the physical climate system.

Central to GCTE's strategy to meet this challenge is the development of models to predict change in ecosystem structure and composition, in ecosystem function, and in the performance of agricultural and forest production systems under global change including changes in the disturbance regime. The problem of developing models of the terrestrial surface across a range of spatial and temporal scales is considered an important area of scientific development with regard to the IGBP/GCTE objectives.

In the case of boreal forests, there is a sufficiently large set of ecological models that one can realistically use to discuss the development of interactive multiple-scale models that bridge time and space scales. Such a nested hierarchy of models provides synthesis across disparate data collection scales. What exist at present are a clear demonstration of concept and the existence of the different types of models for the nested hierarchy. The next logical step is to develop and test the hierarchy of models for a given location.

A Nested Set of Boreal Landscape Models

Let us assume that one is interested in the annual to century-scale changes in a boreal landscape in response to changes in fire regime and climatic conditions. The state variables of interest are the standing stock of carbon, the net ecosystem productivity, and the composition of the forest. Several possible sets of hierarchically nested models could be used to attack this generic problem derived from those models outlined in Table 22.2. An example based on work now underway in a project modeling land-use change in the boreal forests along the Baikal-Amur Railroad is provided to illustrate this concept. The Baikal-Amur Mainline (BAM) Railroad runs from Lake Baikal to the eastern shore of Siberia

Table 22.2. Hierarchy of Models Stimulating the Response of Vegetation, Particularly Forests, to Novel Changes in the Environment (including Altered Atmospheric CO_2 Concentrations)[a]

Model type	Attributes	Unique capabilities and limitations
Canopy and tree process models	Mechanistic representations of the rapid response of forest canopies to their environment (canopy models) based on biophysics and tissue physiology. Plant physiological-based productivity and allocation of photosynthate algorithms (tree process models).	Provide mechanistic representation at the detailed process level of essential plant processes. Can have large-parameter demands and are frequently represented in their application to commercial taxa (e.g., *Pinus* sp.). Scaling up to larger temporal or spatial domains remains a challenge.
Individual-based gap models	Individual-based models track the birth, death, and growth of each plant. Stochastic simulators predicting community and successional patterns and dynamics. Fundamental paradigm based on interactions among individual plants and their environment. Spatial resolution of about 0.05 ha (in forests).	Incorporate species-level details of whole plant physiology and demography. Extensively tested against observations of plant communities on gradients and under different past environments. Restricted to gap-scale simulations.
Distribution-based gap models	Size or age distribution functions summarize plant community state.	Tested in plantation forestry with some applications in natural forests. Conceptually independent of individual tree responses. Restricted to gap-scale simulations. Less demographic detail in simulated responses.
Role-based or functional types–based models	Reduced parameter simulators for small tracts of land. Fundamental paradigm is based on realization of likelihood of changes of dominance and subdominant in communities and expected biomass change.	Computationally fast, and able to alter vegetation structure and composition. Need underlying gap models to provide alterations in the likelihood of changes under new environmental conditions.

Static biogeographical models	Relations between climate variables and vegetation features used to determine equilibrium response of vegetation and for vegetation mapping. Used frequently up until now in global carbon pool and land-cover assessments.	Used in conjunction with digitized global climate data sets (typically including some effects of elevation) to produce present potential vegetation patterns (a test on model consistency). Equivalent data sets for altered conditions can be used to produce global change scenarios.
Life form–based models	Models that assemble vegetation based on rules for tolerances of different life forms to ranges of environmental variables. Range from more than 100 plant life forms considered (Box 1981) to tens of life forms considered (Prentice et al. 1993b).	Similar applications to static biogeographical models (above). Capable of producing novel plant assemblages not currently extant.
Global productivity models	Models of ecosystem fluxes of carbon and nutrient elements. Typically systems of differential equations for carbon flux, nitrogen flux, and so on. Models are interfaced with functions representing the performance of plant canopies based on physiology and biophysics. Determine CO_2, H_2O, and energy fluxes for plant canopies under altered environmental conditions (including modified CO_2 conditions).	Can be related to global scale patterns of productivity. Models synthesize current state of understanding of ecosystem fluxes of carbon, nitrogen, and other essential elements. Typically, the models are more empirical than mechanistic. Models assume a given vegetation structure or composition.
Dynamic global vegetation models (DGVM)	Proposed models that would interface directly with general circulation models used by atmospheric scientist to predict climatic change.	Models under development. The ideal DGVM would mechanistically alter structural, compositional, and physiological response of vegetation as a function of the environment (including CO_2 concentration).

[a]From Shugart et al. (1996).

through a previously remote region. Railroads traditionally increase fire regimes and provide access for increased land-use change.

This is an intrinsically multilevel modeling problem. For example, insect outbreaks are strongly a function of the health of individual trees; fires are influenced by stand fuel loads and large-scale weather conditions; vegetation patterns are influenced by successional history, topography, and soil conditions; and so on. It is inappropriate to attempt to meld this mixture of fine-scale and coarse-scale phenomena in a single model. Rather, one can develop a hierarchy by using three modeling approaches. This will involve developing models that are intercompatible and that can pass on and exchange to one another parameters and boundary conditions. The types of models that will be interlinked are

- *Individual-based models.* An important category of individual organism-based tree models that have been widely used in ecological, as opposed to traditional forestry applications, are the gap models (Shugart and West 1980). Several gap models for a wide range of forests (and grasslands and savannas) have been developed for a wide range of locations (Shugart et al. 1992a; Urban and Shugart 1992). More recent models feature the addition of spatial competition among trees, a range of modifications to allow simulation of a wider range of environmental conditions, and additional biological and/or physiological mechanisms. A general review of ecosystems dynamics in boreal forests, including a summarization of the silvics of all the major Russian tree species expressed as parameters for the individual-based model, has already been conducted (Shugart et al. 1992b), which allows for parameterization of boreal forest gap models in this region.

  An individual-based gap model of specific sites in Siberia can be created by extracting concepts, parameters, and mathematical functions from several earlier forest dynamic models including Bonan's model (Bonan 1989a,b), FORSKA (Leemans and Prentice 1987), ROPE (Shao et al. 1996), KOPIDE (Shao 1991), ZELIG (Smith and Urban 1988), and SASMAN. The model by Bonan (1989a,b) focuses strongly on simulating the thermal and moisture balance of the soil. FORSKA models simulated canopy shape (an important feature for conifer-dominated ecosystems). ROPE allows the parameterization of the gap model to be applied at a regional scale. The model SASMAN, based on the aforementioned models, is a gap model that was developed for the NASA-sponsored BOREAS project. This dynamic boreal forest model was tested for two subareas in Saskatchewan and Manitoba provinces of Canada. The model outputs include species-specific number of trees, leaf area index, wood biomass, wood productivity, wood mortality, and vertical distributions of leaf area. This model's output can provide the base parameters for the next model layer.

- *Role-based models.* Models based on the roles of different species in occupying area are collectively referred to as role-based models. A role-based model reduces the species complexity found in an individual-based model with a functional representation of species "roles" defined in terms of their influence on the structure of vegetation and the reciprocal effect of the nature of the structure on the species performance. Acevedo and associates (1996) and Shao and co-workers (1996) have treated the problem of deriving a computationally fast model based on the roles of species from individual-based models. Acevedo and colleagues (1996) were able to develop Markovian simulators directly from a gap model and inspected the behavior of the resultant model in dynamic simulations of forest succession. Shao and co-workers (1996) used a more simulation model–oriented approach to the same question and emphasized the spatial area simulation of the models while solving for the equilibrium vegetation composition. The application of Shao and colleagues (1996) produces regional maps of leaf area and forest type for northern China as a test case.

In principle, the ROPE (Shao et al. 1996) model can be used to produce a new regional simulator for forest cover that will draw its biological rate parameters from the SASMAN-derived model. The parameters for a role-based model can be obtained from an individual-based simulator and the responses of the role-based model can be tested for fidelity with the originating individual-based model. Role-based models can simulate the vegetation structure and composition over large (subcontinental) areas. Shao and associates (1996) developed procedures for estimating parameters for the role-based model from an existing gap model for China, the KOPIDE model (Shao 1991; Shao et al. 1996). Many of these effects were determined by the influence on tree height of two climate parameters (growing degree-days with base temperature of 10°C and the ratio of annual potential evapotranspiration to precipitation). Changes in tree height subsequently affected size of gap formation and the mixture of species. The model was tested by using data from 81 different climate stations in a study area in northeastern China bounded by the borders with Russia and North Korea and covering parts of four provinces: Liaoning, Jilin, Heilongjiang, and Neimenggu. The northern part of this area is covered with larch forests, the southern part is basically covered with deciduous-coniferous mixed forest, and the western vegetation is an oak-grass complex. The pattern of vegetation simulated by the role-based ROPE model corresponds well to the actual pattern of vegetation in this part of China (China's Vegetation Editing Committee 1980).

- *Regional models.* In the case of succession models, one application of differential equations is specifically intended for cases involving the simulation of changes of large areas of land (Shugart et al. 1973). These models are related to the Markov formulations that have been discussed because the sorts of data typically needed to apply either approach are similar. The essential concept in these models is that for a large landscape, the change in the vegetative cover can be thought of as a flow of area from one category to another. The development of these models (e.g., Shugart et al. 1973; Johnson and Sharpe 1976) involves devising a vegetation categorization (cover-states) and determining the rates of change and the transfers of area among these cover-states based either on remeasurement data, on assumptions about the rates of ecological succession, or from model predictions.

The model predictions from the ROPE model can be used to obtain the parameters for a regional forest simulator. The effects of changes in the disturbance rates can also be included directly in this model. The underlying assumptions for developing large-area or regional succession models by using differential equations are

1. The vegetative cover can be divided into a finite number of cover-states or vegetation types (Shugart et al. 1973).
2. The effects of spatial heterogeneity on the dynamics of aerial extent of these cover-states are reasonably constant over the time segments of interest.
3. The dynamics of the amount of area in each cover-state can be thought of as the consequence of the input of land area into the cover-state category and the output of land from the category. This input-output behavior is considered to be determinate. (The amounts of land in each cover-state in the regional successional system provide sufficient information to allow the computation of the future behavior of the system. This is similar to the assumption in the case of the first-order Markov models that the state of the system is sufficient to determine the probabilities of the next system state.) The assumption that a system is determinate does not preclude the consideration of statistical fluctuations or sampling error.
4. If the differential equations used to represent the successional system are linear, then the input-output relations are superposable, implying that the response of a cover-state to two inputs summed together is the same as the sum of the responses taken separately.

It is our intention to develop a spatial and temporal nested set of models, one of each of these approaches for the BAM study area, that will build on our experience in the development and application of each of these sorts of models. A multiple-model approach can be used as a predictive tool to evaluate forest response within the boreal forest, boreal forest interaction with other biomes, and the forest as it interacts with complex environmental fluctuations and land-use change.

Conclusion

Models provide the ability to simulate processes that occur at temporal and spatial scales that are difficult to empirically investigate. We have attempted to provide the framework of a set of nested models that can address multilevel boreal forest dynamics as land-use and climatic conditions shift. Scale has been highlighted as a challenging component of significant importance when considering interactions at the site, stand, or landscape scale within ecosystems. It has been our goal to present modeling as a valuable tool for analyzing boreal system responses to large-scale environmental change and subsequent feedbacks that connect terrestrial ecosystems and atmosphere composition.

Traditionally, boreal forests have not experienced notable land-use change, but in recent history boreal landscapes are being modified. In terms of pollution, land-use change, and altered disturbance regimes, humans exert a strong signal that cannot be ignored. The BAM provides the impetus for land-use change that influences boreal forest fire regime, ecosystem composition, productivity, and the forest ability to store carbon. Boreal forests presently function to store carbon but, because of global change, are predicted to become a source of atmospheric carbon in the future. Recently, the Kyoto conference highlighted the need to reduce atmospheric carbon levels, its anthropogenically related atmospheric increase, primary sources and sinks, and potential associated effects. Considering complete understanding of the basic global carbon cycle is nonexistent, potential effects and strategic planning are difficult to impossible to access. This project invites necessary international scientific cooperation and provides an excellent environment for boreal study. It is our scientific responsibility to address issues of our day, especially those that are effected by global change and confront policy makers. Our ability to study global-scale phenomena has been enhanced by satellite imagery, providing previously unavailable temporal and spatial resolution, and computationally fast computers that can manipulate and store vast amounts of data. The Baikal-Amur project provides a unique opportunity to study boreal forest dynamics by using recently declassified remotely sensed data, state-of-the-art computers, and progressive models within an internationally cooperative scientific community. Our goal is to use models as a tool to extend our predictive temporal and spatial capabilities as global change ensues.

References

Acevedo, M.F., D.L. Urban, and H.H. Shugart. 1996. Models of forest dynamics based on roles of tree species. *Ecol. Mod.* 87:267–284.

Allen, T.F.H., and T.B. Starr. 1982. *Hierarchy: Perspectives for Ecological Complexity.* University of Chicago Press, Chicago.

Bonan, G.B. 1989a. A computer model of the solar radiation, soil moisture, and soil thermal regimes in boreal forests. *Ecol. Mod.* 45:275–306.

Bonan, G.B. 1989b. Environmental factors and ecological processes controlling vegetation patterns in boreal forests. *Land. Ecol.* 3:111–130.

Box, E.O. 1981. *Macroclimate and Plant Forms: An Introduction to Predictive Modeling in Phytogeography.* Dr. W. Junk, The Hague.

China's Vegetation Editing Committee. 1980. *China's Vegetation.* Scientific Publishing House of China, Beijing. (in Chinese)

Clark, D.F., D.D. Kneeshaw, P.J. Burton, and J.A. Antos. 1998. Coarse woody debris in subboreal spruce forests of west-central British Columbia. *Can. J. For. Res.* 28:284–290.

Cowan, I.R. 1982. Regulation of water use in relation to carbon gain in higher plants, pp. 549–587 in O.L. Lange, P.S. Noble, C.B. Osmond, and H. Ziegler, eds. *Physiological Plant Ecology.* Encyclopedia of Plant Physiology (NS), Vol. 12B. Springer-Verlag, Berlin.

Cowan, I.R. 1986. Economics of carbon fixation in higher plants, pp. 133–170 in T.J. Givnish, ed. *On the Economy of Plant Form and Function.* Cambridge University Press, Cambridge, UK.

Delcourt, H.R., P.A. Delcourt, and T. Webb III. 1983. Dynamic plant ecology: the spectrum of vegetation change in time and space. *Q. Sci. Rev.* 1:153–175.

Dzeroski, S., J. Grbovic, W.J. Wally, and B. Kompare. 1997. Using machine learning techniques in the construction of models. II. Data analysis with rule induction. *Ecol. Mod.* 95:95–111.

Emanuel, W.R., G.G. Killough, W.M. Post, and H.H. Shugart. 1984. Modeling terrestrial ecosystems and the global carbon cycle with shifts in carbon storage capacity by land use change. *Ecology* 65:970–983.

Emanuel, W.R., H.H. Shugart, and M.P. Stevenson. 1985. Climate change and the broad-scale distribution of terrestrial ecosystem complexes. *Clim. Change* 7:29–43.

Farquhar, G.D., and S. von Craemmer. 1982. Modeling of photosynthetic response to environmental conditions, pp. 549–587 in O.L. Lange, P.S. Noble, C.B. Osmond, and H. Ziegler, eds. *Physiological Plant Ecology.* Encyclopedia of Plant Physiology (NS), Vol. 12B. Springer-Verlag, Berlin.

Farquhar, G.D., and T.D. Sharkey. 1982. Stomatal conductance and photosynthesis. *Annu. Rev. Plant Physiol.* 33:317–345.

Johnson, W.C., and D.M. Sharpe. 1976. Forest dynamics in the northern Georgia piedmont. *For. Sci.* 22:307–322.

Kompare, B., I. Bratko, F. Steinman, and S. Dzeroski. 1994. Using machine learning techniques in the construction of models. *Ecol. Mod.* 75:617–628.

Laval, Ph. 1995. Hierarchical object-oriented design of a concurrent, individual-based, model of a pelagic Tunicate bloom. *Ecol. Mod.* 82:265–276.

Leemans, R., and I.C. Prentice. 1987. Description and simulation of tree-layer composition and size distributions in a primaeval Picea-Pinus forest. *Vegetatio* 69:147–156.

Liu, J., and P.S. Ashton. 1995. Individual-based simulation models for forest succession and management. *Ecol. and Mgmt.* 73:157–175.

McHaney, R. 1991. *Computer Simulation: A Practical Perspective.* Academic Press, San Diego, CA.

Moore, B., J. Aber, G. Brasseur, R. Dickinson, W. Emanuel, and J. Melillo. 1990. Integrated modeling of the earth system, pp. 16–66 in *Research Strategies for the U.S. Global Change Research Program.* National Academy Press, Washington, DC.

O'Neill, R.V. 1988. Hierarchy theory and global change, pp. 29–45 in T. Rosswall, R.G. Woodmansee, and P.G. Risser, eds. *Scales and Global Change.* SCOPE 35. John Wiley, Chichester, UK.

O'Neill, R.V., D.L. DeAngelis, J.B. Waide, and T.F.H. Allen. 1986. *A Hierarchical Concept of Ecosystems.* Princeton University Press, Princeton.

Patten, B.C., M. Straskraba, and S.E. Jorgensen. 1997. Ecosystems emerging. 1. Conservation. *Ecol. Mod.* 96:221–284.

Prentice, I.C., R.A. Monserud, T.M. Smith, and W.R. Emanuel. 1993a. Modeling large-scale vegetation dynamics, pp. 235—250 in A.M. Solomon and H.H. Shugart, eds. *Vegetation Dynamics and Global Change.* Chapman & Hall, New York.

Prentice, I.C., W.P. Cramer, S.P. Harrison, R. Leemans, R.A. Monserud, and A.M. Solomon. 1993b. A global biome model based on plant physiology and dominance, soil properties and climate. *J. Biogeog.* 19:117–134.

Running, S.W., and J.C. Coughlan. 1988. A general model of forest ecosystem processes for regional applications. I. Hydrological balance, canopy gas exchange and primary production processes. *Ecol. Mod.* 42:125–154.

Running, S.W., R.R. Nemani, D.L. Peterson, L.E. Band, D.F. Potts, L.L. Pierce, and M.A. Spanner. 1989. Mapping regional forest evapotranspiration and photosynthesis by coupling satellite data with ecosystem simulation. *Ecology* 70:1091–1101.

Rykiel, E.J., Jr. 1989. Artificial intelligence and expert systems in ecology and natural resource management. *Ecol. Mod.* 46:3–8.

Shao, G. 1991. Moisture-therm indices and optimum-growth modeling for the main species in Korean pine/deciduous mixed forests. *Sci. Silvae Sinicae* 21:21–27.

Shao, G., H.H. Shugart, and T.M. Smith. 1996. A role-type model (ROPE) and its application in assessing climate change impacts on forest landscapes. *Vegetatio* 121:135–146.

Sharpe, P.J.H., J. Walker, L.K. Penridge, and H. Wu. 1985. A physiologically-based continuous-time Markov approach to plant growth modeling in semi-arid woodlands. *Ecol. Mod.* 29:189–213.

Sharpe, P.J.H., J. Walker, L.K. Penridge, H. Wu, and E.J. Rykiel. 1986. Spatial considerations in physiological models of tree growth. *Tree Physiol.* 2:403–421.

Shugart, H.H. 1984. *A Theory of Forest Dynamics: The Ecological Implications of Forest Succession Models.* Springer-Verlag, New York.

Shugart, H.H. 1998. *Terrestrial Ecosystems in Changing Environments.* Cambridge University Press, Cambridge, UK.

Shugart, H.H., and D.L. Urban. 1988. Scale, synthesis and ecosystem dynamics, pp. 279–290 in L.R. Pomeroy and J.J. Alberts, eds. *Essays in Ecosystems Research: A Comparative Review.* Springer-Verlag, New York.

Shugart, H.H., and D.C. West. 1980. Forest succession models. *BioScience* 30:308–313.

Shugart, H.H., T.R. Crow, and J.M. Hett. 1973. Forest succession models: a rationale and methodology for modeling forest succession over large regions. *For. Sci.* 19:203–212.

Shugart, H.H., T.M. Smith, and W.M. Post. 1992a. The application of individual-based simulation models for assessing the effects of global change. *Annu. Rev. Ecol. Syst.* 23:15–38.

Shugart, H.H., R. Leemans, and G.B. Bonan, ed. 1992b. *A Systems Analysis of the Global Boreal Forest.* Cambridge University Press, Cambridge, UK.

Shugart, H.H., W.R. Emanuel, and G. Shao. 1996. Models of forest structure for conditions of climatic change. *Commonw. For. Rev.* 75:51–64.

Smith, T.M., and D.L. Urban. 1988. Scale and the resolution of forest structural pattern. *Vegetatio* 74:143–150.

Solomon, A.M., and H.H. Shugart, ed. 1993. *Vegetation Dynamics and Global Change.* Chapman & Hall, New York.

Tilman, D. 1988. *Plant Strategies and the Dynamics and Structure of Plant Communities.* Princeton University Press, Princeton.

Urban, D.L., and H.H. Shugart. 1992. Individual-based models of forest succession, pp. 249–293 in D.C. Glenn-Lewin, R.K. Peet, and T.T. Veblen, eds. *Plant Succession: Theory and Prediction.* Chapman & Hall, London, UK.

Walker, J., P.J. Sharpe, L.K. Pendridge, and H. Wu. 1989. Ecological field theory: the concept and field tests. *Vegetatio* 83:81–95.

Weinstein, D.A., and H.H. Shugart. 1983. Ecological modeling of landscape dynamics, pp. 29–45 in H.A. Mooney and M. Godron, eds. *Disturbance and Ecosystems.* Springer-Verlag, New York.

Woodward, F.I. 1987. *Climate and Plant Distribution.* Cambridge University Press, Cambridge, UK.

Wu, H.-I., P.J.H. Sharpe, J. Walker, and L.K. Penridge. 1985. Ecological field theory: a spatial analysis of resource interference among plants. *Ecol. Mod.* 29:215–243.

23. Using Satellite Data to Monitor Fire-Related Processes in Boreal Forests

Eric S. Kasischke, Nancy H.F. French,
Laura L. Bourgeau-Chavez, and Jeffery L. Michalek

Introduction

It is apparent that to monitor the effects of fire on processes related to carbon cycling in boreal forests requires the use of satellite-based remote sensing systems. The information presented in Chapters 17 and 18 illustrated the types of surface characteristics of the boreal forest that can be monitored by using satellite remote sensing systems. The science of remote sensing has made great advances over the past several decades, to the point at which the overall approaches needed to use this technology are now well developed. The actual exploitation of remotely sensed data collected over boreal forests to study specific processes requires three steps: (1) a basic understanding of the physical and biological mechanisms resulting in a signature detected by a remote sensing system must be developed; (2) based on this understanding, approaches or algorithms need to be created and validated to use the remote sensing data to estimate specific surface characteristics; and (3) approaches to incorporate the information derived from the remotely sensed observations into specific processes models must be developed.

The first two steps encompass much of the research that remote sensing scientists or specialists pursue. Remote sensing scientists must collaborate closely with a wide range of earth system scientists to understand and quantify the surface characteristics responsible for the generation of the signatures being detected by the satellite systems. The information presented in Chapters 17 and 18 is based primarily on results of research on the science of remote sensing.

Fully capturing the potential of the information available from a satellite remote sensing system in ecosystem process studies has been and continues to be a very challenging endeavor. First and foremost, satellite remote sensing systems represent a new measurement or observation tool for ecologists and terrestrial scientists, a tool that makes systematical observations of surface characteristics at spatial and temporal scales not previously available. These measurements require the development of new scientific paradigms and approaches to exploit this unique information source. As with any new measurement technology, there is a time lag between when the technology becomes available and when that technology is fully exploited.

The purpose of this chapter is to summarize approaches developed to exploit satellite remote sensing data to monitor fire-related processes in the boreal forest. In this chapter, we discuss five different processes in boreal forests in which remote sensing systems provide unique information to (1) monitor patterns of seasonal net primary productivity; (2) monitor postdisturbance patterns of forest succession; (3) estimate greenhouse gas fluxes during fires; (4) estimate postfire biogenic fluxes of carbon dioxide; and (5) estimate energy exchange between the land surface and atmosphere in fire-disturbed forests. The relationship between these different processes and their importance in carbon cycling are discussed in the introduction to Section II. In this chapter, we outline the basic models or approaches used to address each of these processes and identify the types of remote sensing data that can provide information for these models/approaches.

Estimating Seasonal Patterns of Net Primary Production

Since the discovery that the AVHRR satellite system could be used to monitor seasonal signatures associated with different vegetation cover (see, e.g., Fig. 17.1), much effort has been directed toward development of approaches to exploit this information. Researchers first showed that satellite-based seasonal-growth patterns observed in semiarid regions were related to seasonal plant production (Tucker et al. 1983) as well as being closely correlated with seasonal variations in the atmospheric concentration of carbon dioxide (Tucker et al. 1986).

Based on these initial results, a theory was developed that the seasonal vegetation patterns observed on visible satellite imagery were directly related to the amount of photosynthetically active radiation (PAR) being intercepted by the plant canopy covering the earth's surface. Because the amount of PAR present in a scene can be related to the overall productivity of the plants occupying the land surface, it is thought that the seasonal satellite vegetation patterns are representative of the seasonal net primary production in terrestrial ecosystems (Prince 1991; Goward and Huemmrich 1992; Dye and Goward 1993; Prince and Goward 1996). This has resulted in the development of models to use satellite observations to estimate net primary production from terrestrial biomes (Running and Nemani

1988; Foley et al. 1997) as well as estimate broad-scale patterns of soil respiration (Thompson et al. 1997).

In addition, AVHRR data have been used to study intra-annual variations in vegetation characteristics. Based on a study of 10 years of data, Myneni and co-workers (1997) concluded that the overall net primary production in the boreal region had increased by greater than 10% based on increases in the vegetation index. Using a similar data set, Braswell and colleagues (1997) concluded that land temperature increases in boreal ecosystems associated with El Nino events were correlated with increases in the satellite-observed vegetation index recorded by the AVHRR satellite.

Although the community of remote sensing and global change scientists who are using vegetation index–based models to estimate net primary production is growing, there are some limitations for these models for boreal regions. First, a significant portion of net primary production in boreal forests occupying colder wetter sites originates from mosses (Oechel and Van Cleve 1986). Although some moss cover is visible in open-canopied forests, most is not visible to satellite sensors; therefore, a significant portion of the net primary production in boreal forests is not encompassed within the satellite-based vegetation index. In addition, photosynthesis in bryophytes is very dissimilar to that in vascular plants because of different water transport mechanisms. Thus, it is not possible to directly correlate moss primary production to the signals detected by the satellite sensors. In all likelihood, the current generation of satellite-based models of net primary productivity does not properly account for moss production and hence underestimates ecosystem productivity for the boreal forest.

Monitoring Postdisturbance Patterns of Forest Succession

One of the projected effects of climate change is a major shift of the location and boundaries of the earth's vegetated biomes (Smith et al. 1992). Because the most significant model-predicted temperature rises will occur in high northern latitudes, changes in forest cover within the boreal region can be expected to be particularly dramatic. In addition, fire will act to accelerate shifts in forest types based on its effects on the ground-layer thermal and moisture regimes, as discussed in Chapter 12. Thus, monitoring changes in the patterns of forest succession provides a means of monitoring the effects of climate change on the boreal forest.

The challenge that the remote sensing community faces is to develop approaches to use a satellite systems to detect these changes. For example, in interior Alaska, the increase in temperature of between 1.5° and 2.0°C observed over the past 30 years (Fig. 1.1) would be expected to favor an increase in the white spruce forest type that inhabits warmer drier sites. Because this forest type has deciduous tree species (birch and aspen) early in its successional chronosequence, an increase in fire severity would result in a shift from conifer species to deciduous species. This is especially true for sites where severe fires remove much of the organic soil from the ground surface, conditions that are favorable for the establishment of pioneer species such as aspen and birch.

Traditional approaches that use changes in forest-cover categories to analyze changes in time-series remote sensing data (Hall et al. 1991a) may be limiting, because the difficulty in obtaining the ground-truth data required for classification introduces a high degree of error when using older data sets. A more appropriate approach may be to analyze the changes in the radiometric signatures associated with the different forest types (Johnson and Kasischke 1998).

This approach is illustrated in Figure 23.1. This figure was created by using data from a Landsat Thematic Mapper (TM) image collected over an area in interior Alaska in September 1995 where several forest fires have occurred over the past 40 years. Field surveys conducted during the summer of 1997 showed that some areas within the different burns were dominated by aspen regrowth, whereas others were dominated by black spruce regrowth. Plotting the band 4 (0.76–0.90 μm) reflectance versus the band 5 (1.55–1.75 μm) reflectance in Figure 23.1 for the different-aged forest stands shows that these two forest types follow distinct spectral-temporal trajectories. The differences in trajectories between the two forest types and different-aged forest stands result from changes in the structure and composition of the forest stands during succession.

The spectral trajectories in Figure 23.1 suggest a different paradigm for the analysis of time-series remote sensing data. When examining two or more Landsat scenes collected over the same area at different times, one would expect that differences in the radiometric signatures to be present where different patterns of forest regrowth after disturbances have occurred. The differences in the radiometric signatures between the two dates represent a vector, whose direction and magnitude are diagnostic of the types of changes that have occurred. For example, in Figure 23.2, the vector between a light burn and a heavy burn are clearly different, as are the vectors represented by an area where aspen regrowth is occurring (on the heavy-burn site) versus where the spruce regrowth is occurring (on the light-burn site). Change-vector analysis of time-series remote sensing data (Johnson and Kasischke 1998) represents a methodology to analyze and classify time-series remote sensing data based on the dependent analysis of two different scenes of data, rather than relying on the independent categorization of the two scenes data.

Estimating Greenhouse Gas Fluxes from Fires in Boreal Forests

To estimate the amount of carbon released from a boreal forest fire (C_r in tons) requires the use of a simple equation of the form first introduced by Seiler and Crutzen (1980):

$$C_r = ABf_c\beta \tag{23.1}$$

where A is the total area burned (ha), B is the biomass density in the area burned (ha^{-1}), f_c is the fraction of biomass that is carbon, and β is the fraction of biomass consumed during fires.

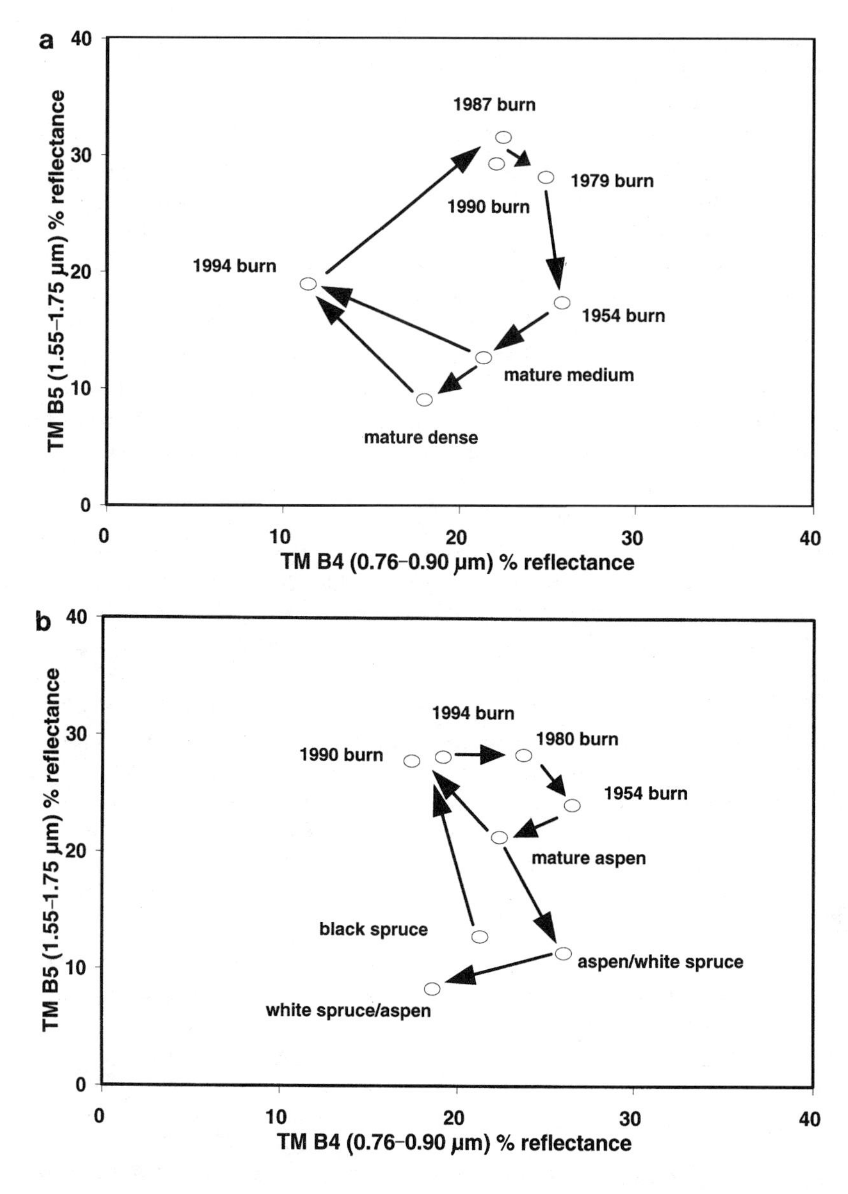

Figure 23.1. Plot of reflectance from Landsat TM bands 4 and 5 from different-aged stands of (a) black spruce and (b) aspen-white spruce. These plots illustrate that forest succession after a fire results in distinct changes in canopy structure and composition that are reflected in the different bands of the Landsat TM system.

Satellite observations are ideally suited to provide many of the parameters required to estimate the amount of carbon consumed and greenhouse gases released during fires in boreal forests. The overall approach combining satellite and ground observations is outlined in Figure 23.3. In using satellite observations,

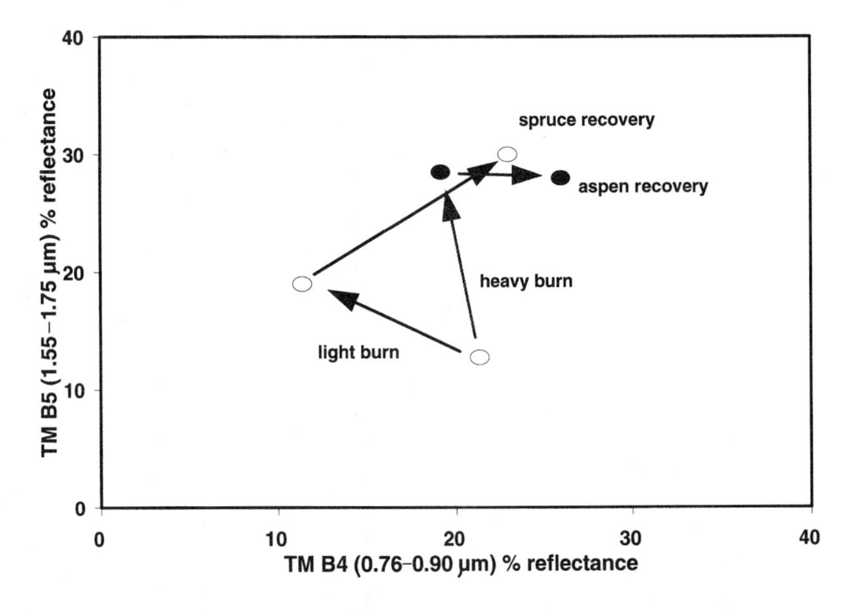

Figure 23.2. Comparison of reflectances from Landsat TM bands 4 and 5 showing that the trajectory of spectral changes are different based on severity of burning as well as patterns of forest regrowth after burning.

most efforts have been fairly simple, in which total area burned (A) observed from satellite data (see Figures 17.4–17.7) has been combined with ground-based models. The ground-based models have been developed by using a single value of biomass density and carbon release for an entire region (Cahoon et al. 1994, 1996) or dividing a region into specific landscape units based on physiography and expected forest types (Kasischke et al. 1995).

Using remotely sensed observations allows for the measurement of additional parameters in Equation (23.1):

- *Biomass/carbon density per unit area (Bf_c).* Remote sensing imagery offers several approaches for estimating biomass/carbon density. The most common approach is to use visible/near-infrared imagery to map forest-cover type in a specific region and then to use ground data to assign a biomass/carbon density value for each cover type (both aboveground and below-ground) (Fig. 23.4[1]). Another approach would be to use radar imagery to directly estimate the aboveground biomass levels for the different forest types in a region (Fig. 18.2).
- *Fraction of biomass consumed (β).* Remote sensing systems cannot directly detect how much biomass is consumed during fires, but important inferences concerning this parameter can be obtained. The patterns of fire and biomass burning (which are strongly controlled by forest type as well as the timing of the

[1]Figure 23.4 will be found in the color insert.

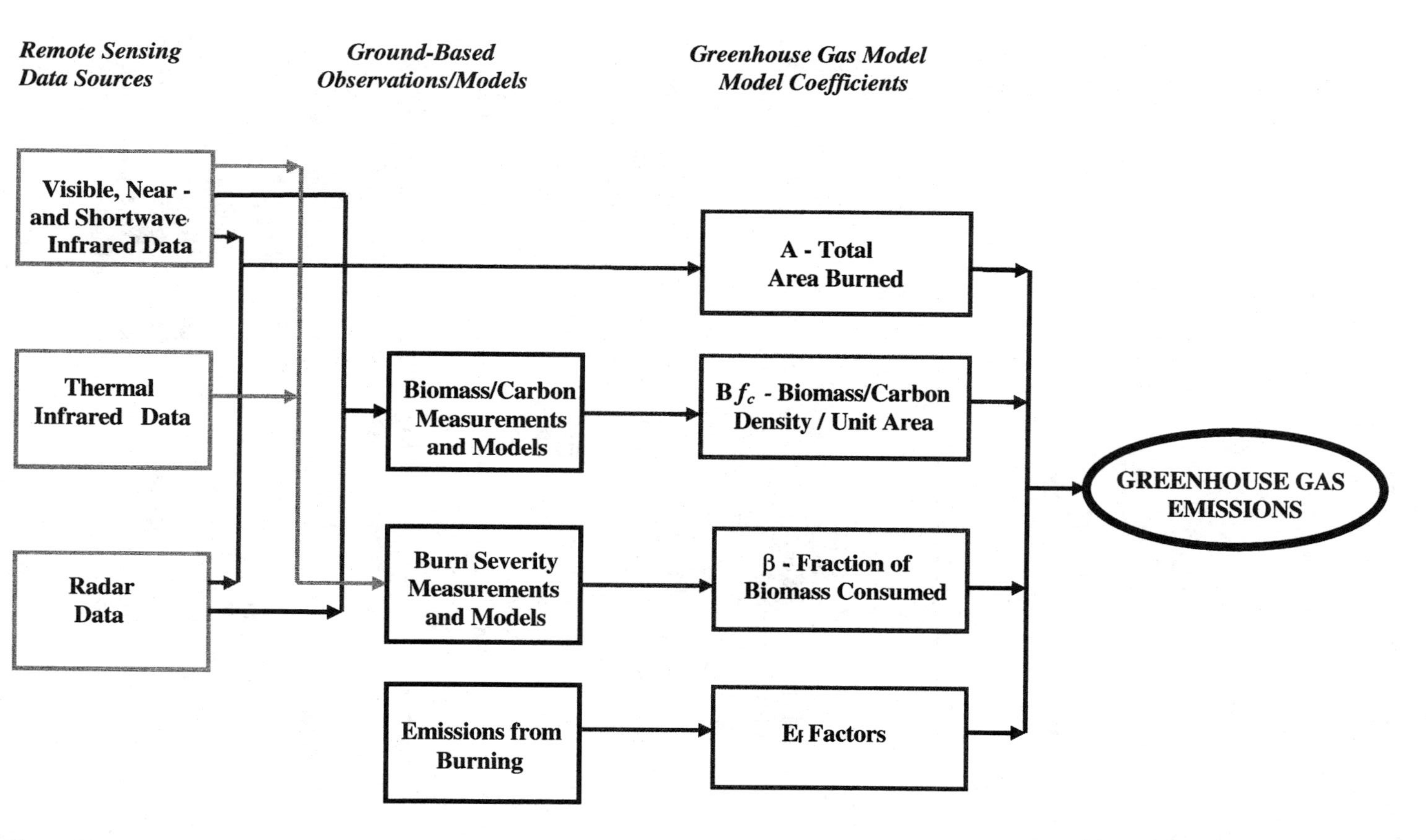

Figure 23.3. Approaches to use information derived from satellite remote sensing systems to estimate greenhouse gas emissions from fires in boreal forests.

fire) have been extensively studied and documented for all forest types in the boreal region. Thus, information derived from remotely sensed data on the timing of the fire and the types of forests where the fires occurred can be used as an inputs to ground-based models of biomass consumption. For example, in forest types where permafrost is present, fires taking place early in the growing seasons typically have low levels of ground-layer burning because the soils have not thawed. Later-season fires typically have a higher level of ground-layer biomass burning because the depth of thaw has increased (see Chapter 10).

Recent research has shown that more direct information on fire severity in black spruce forests of interior Alaska can be derived in some cases by using visible and near/shortwave infrared imagery. Significant burning of organic soil results in two distinct spectral signature variations in these data. The first is the result of the exposure of mineral soil, which has a distinctly different spectral signature from burned or exposed organic soil. The second has to do with variations in the tree shadows present within the scene. Areas with deep organic soil burns often have no standing trees (resulting in little or no shadowing), whereas areas with less severe burns have standing trees (shadows). When combined with ground-truth data, maps of burn severity (Fig. 17.10) can be used to generate maps of percentage biomass or carbon consumed, as illustrated in Figure 23.5.[2] The final potential method for detection of burn severity exploits thermal infrared imagery. Smoldering fires in the organic soil can burn for days in boreal forests. These smoldering fires may result in an elevated ground temperature, which may be detected on thermal infrared imagery.

When the information on the distribution of biomass (Fig. 23.4) and burn severity (Fig. 23.5) are combined, a spatial estimate on the amounts of carbon consumed during fires can easily be generated (Fig. 23.6[3]). Such a map is useful for several purposes. First, it can produce a much better estimate of the average amount of carbon released during a single fire event. In this case, a total of 30 t C ha^{-1} was released. In addition, it is easy to determine the average amount of carbon released in the different forest types.

One of the disadvantages of the above approach is the need for a certain amount of field data and high-resolution satellite imagery, which, is not always available. The greatest value of these types of studies, however, lies in their ability to provide more accurate information on the patterns of biomass burning, which in turn, can be used to provide inputs to models that use coarser-resolution satellite imagery.

Monitoring of Postfire Biogenic Emissions in Boreal Forests

Field and laboratory studies have shown that patterns of microbial respiration in the organic soils of boreal forests are dependent primarily on three factors: (1) the

[2]Figure 23.5 will be found in the color insert.
[3]Figure 23.6 will be found in the color insert.

composition and depth of the organic soil layers; (2) ground temperature; and (3) soil moisture. Studies by Bunnell and co-workers (1977) use a single-layer organic soil model to calculate seasonal patterns of soil respiration in Alaskan boreal forests (Schlentner and Van Cleve 1985; Bonan and Van Cleve 1992). Total soil respiration (R in g CO_2 m^{-2} h^{-1}) is estimated as

$$R = \frac{M}{a_1 + M} \frac{a_2}{a_2 + M} a_3 \, a_4^{(T-10)/10} \tag{23.2}$$

where M is the average forest floor moisture content (percentage by dry weight), T is the soil temperature (°C) at a depth of 15 cm, a_1 is the percentage soil water at half field capacity, a_2 is percentage water at half retentive capacity, a_3 is the theoretically optimal respiration rate at 10°C, and a_4 is the temperature Q_{10} value. Bonan and Van Cleve (1992) assume that 20% of R is due to microbial respiration and that R can be converted to a daily decay rate constant (k in g dry matter (DM) g DM^{-1} day^{-1}) as

$$k = \frac{0.2\,R}{\gamma_d \Delta L} \frac{0.61 \text{ g DM}}{1 \text{ g } CO_2} \frac{1 \text{ kg}}{1{,}000 \text{ g}} \frac{24 \text{ h}}{1 \text{ day}} \tag{23.3}$$

where γ_d is the bulk density of organic matter (55.3 kg DM m^{-3}) and ΔL is the forest floor thickness.

Field measurements have shown that although the model presented in Equations (23.2) and (23.3) works well for organic soils with a limited number of layers (e.g., those found in aspen and white spruce forests), the model does not accurately predict variations in soil respiration in black spruce forests, which has a very complex profile (Schentner and Van Cleve, 1985; O'Neill et al. 1997). However, O'Neill and co-workers (1997) and Richter and colleagues (this volume) showed that soil respiration in fire-disturbed black spruce and white spruce forests is sensitive to variations in ground temperature and moisture. It is likely that some form of Equation (23.2) can be developed to estimate soil respiration in burned forests.

To monitor seasonal patterns of soil respiration over extended areas, such a model will require spatially and temporally varying estimates of ground temperature and moisture, parameters that can be provided by using satellite remote sensing systems. Toward this end, efforts are now underway to develop approaches to use satellite remote sensing data to more accurately estimate patterns of postfire biogenic emissions in the forests of interior Alaska (French et al. 1996). Field and laboratory measurements of fluxes of carbon dioxide are being obtained to develop the coefficients required for Equations (23.2) and (23.3), with the eventual goal of using satellite data as a key input. This overall approach is outlined in Figure 23.7 and discussed below.

- *Empirically derived coefficients* (a_1, a_2, a_3, a_4, γ_d). These parameters are all based on preburn forest type and depth of the organic soil profile. Preburn forest type information can easily be derived from visible and near-infrared satellite

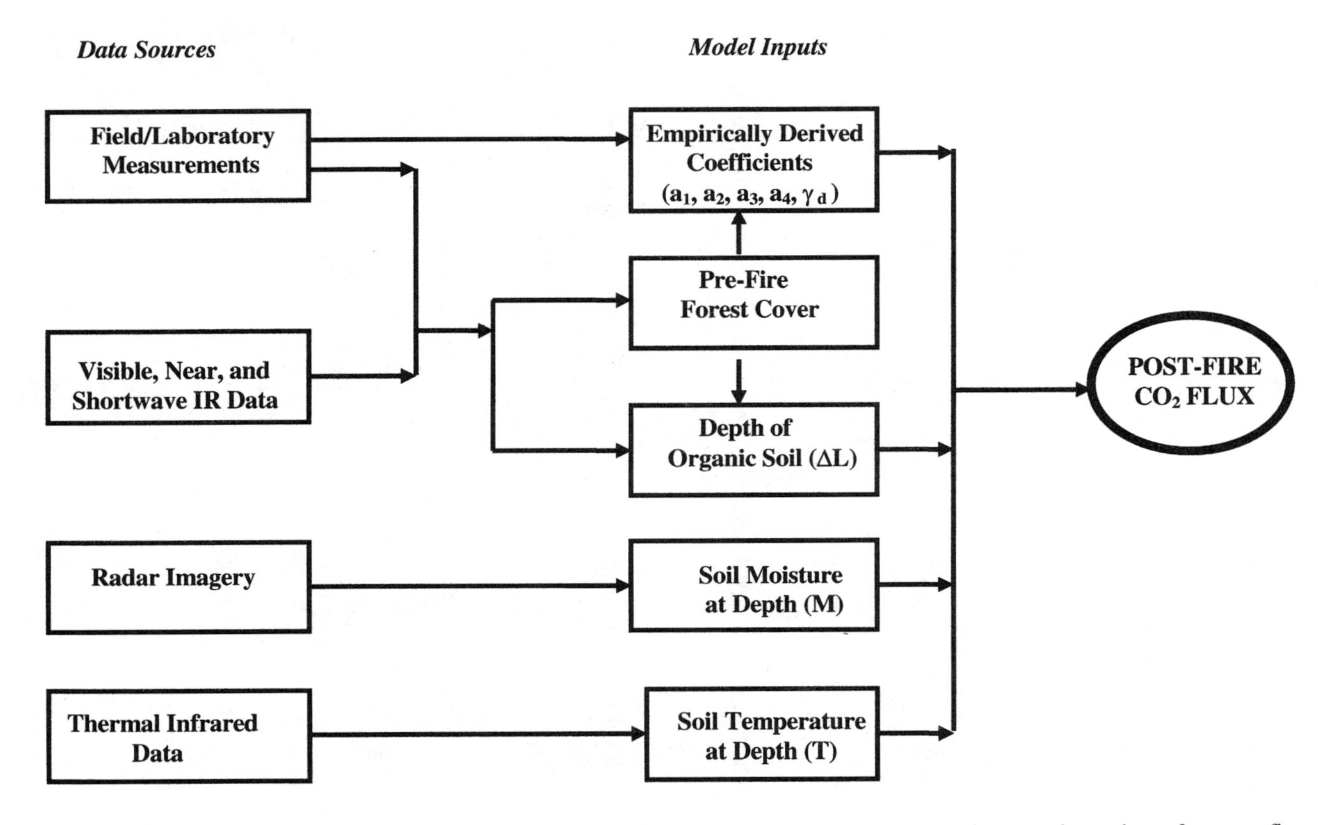

Figure 23.7. Approaches to use information derived from satellite remote sensing systems to estimate carbon release from postfire biogenic emissions.

imagery, whereas the depth of the organic soil can be inferred from field measurements and the forest-cover map. These coefficients are empirically derived or measured by using laboratory measurements.

- *Depth of organic soil profile (ΔL).* This information depends on preburn organic soil depth and the severity of the burning in the ground layer. This information can be obtained by using field measurements alone for the different forest-cover types or in combination with fire severity estimates derived from visible and infrared imagery.
- *Soil moisture at depth (M).* Imaging radar systems are sensitive to variations in soil moisture in the upper 5–10 cm of the ground surface in fire-disturbed boreal forests and provide a unique means to derive this parameter (Fig. 18.5).
- *Soil temperature at depth (T).* Thermal infrared systems can measure the depth of the soil surface in boreal forests. Although vegetation regrowth will mask surface temperature to a certain degree, most of the surface temperature signatures from fire- disturbed forests originate from the ground layer for the first 10–15 years after a fire. To estimate soil temperature at depth requires the development of models from field data that relates surface temperature to temperature at different depths.

Estimating Energy Exchange Between the Land Surface and Atmosphere in Fire-Disturbed Forests

A significant amount of research over the past several decades has been devoted toward developing approaches to using satellite imagery in quantifying the surface energy budget (Hall et al. 1992, 1996; Norman et al. 1995; Sellers et al. 1995). Such research is considered essential in global change studies because of the direct role the earth's vegetated surface plays in regulating the climate.

In terms of understanding the boreal forest carbon budget, the surface energy budget is critical because of the central role ground temperature plays in regulating biogenic releases of carbon dioxide and methane through microbial respiration in organic soils, as well as controlling patterns of succession. The role that fire plays in regulating ground temperature in the boreal forest has been well documented (Brown 1983; Viereck 1983; Kasischke et al., this volume, Chapter 12).

What is clearly lacking at this time is in-depth understanding of the mechanisms resulting in the dramatic increases in ground temperatures after fires in boreal forests. Understanding the surface energy budget requires balancing two sets of equations describing (1) the sources of radiation present at the earth's surface and (2) the fate of the energy that is absorbed at the earth's surface. The net radiation, R_n, at the earth's surface can be described (Norman et al. 1995) as

$$R_n = R_s(1 - \alpha) + (1 - \varepsilon)R_{\text{sky}} - \varepsilon\sigma T_{sh}^4 \tag{23.4}$$

where R_s is incoming solar radiation, α is the broadband surface albedo, ε is the broadband surface emissivity, σ is the Stefan Boltzman constant, R_{sky} is the sky thermal irradiance, and T_{sh} is the hemispherical infrared temperature of the surface.

The energy at the earth's surface is either absorbed or transported back to the atmosphere. This energy exchange process is described by Norman and associates (1995):

$$R_n = H + LE + G \tag{23.5}$$

where H is the sensible heat release by conductive heating of the atmosphere from the ground, LE is the latent heat release or loss via evapotranspiration from the vegetation and ground surface, and G is the ground heat conduction.

Although a fire significantly alters the energy budget in any vegetated landscapes, its effects are particularly pronounced in boreal forests underlain by permafrost. Prior to the fire, the surfaces of landscapes underlain by permafrost are losing energy to the atmosphere. After a fire, the land surface becomes a significant energy sink, as evidenced by the increasing ground temperatures.

Figure 11.2 illustrates the influence of fire on the ground temperature in a black spruce ecosystem in interior Alaska. Over the past several years, soil temperature profiles have been collected in late August at several black spruce stands that have been burned along with profiles from adjacent unburned stands. These data show that elevated temperature exists in stands as old as 25 years (where the soil temperature in the upper 1m of the mineral soil in the burned stand was an average 3.5°C warmer than in the unburned stand). These data also showed that the degree of warming depended on the depth of organic soil remaining after a fire. In one area where all the organic soil had been removed, the soil temperature in the burned stand was 6°C warmer than in an adjacent unburned stand 2 years after the fire. In another region of the same burn in an area where 15 cm of organic soil remained, the temperature in the burned stand was only 2°C warmer than in the unburned stand.

These measurements illustrate two important points. First, there has been a significant increase in the ground heat conduction in Equation (23.5), which means there has to be significant changes in the other terms in Eqations (23.4) and (23.5) caused by the fire. Second, the increased ground temperature is controlled by severity of burning of organic soil.

In terms of input parameters for Equations (23.4) and (23.5), approaches have been developed to use satellite sensors to monitor R_s and models and atmospheric sounding satellite data to estimate R_{sky} and T_{sh} (Hall et al. 1991b; Norman et al. 1995; Sellers et al. 1995). Approaches are being developed to use land-surface remote sensing instruments to α, ε, H, and LE, as discussed below.

- *Surface albedo (α).* The albedo of the earth's surface can be estimated from Landsat TM data (Starks et al. 1991). To estimate albedo requires either ground-based reflectance measurements to calibrate the satellite measurements or the development of absorption models to account for the effects of the atmosphere on the satellite reflectance values. Figure 23.8 illustrates the effect of fire on relative surface albedo in an Alaskan boreal forest and shows that the

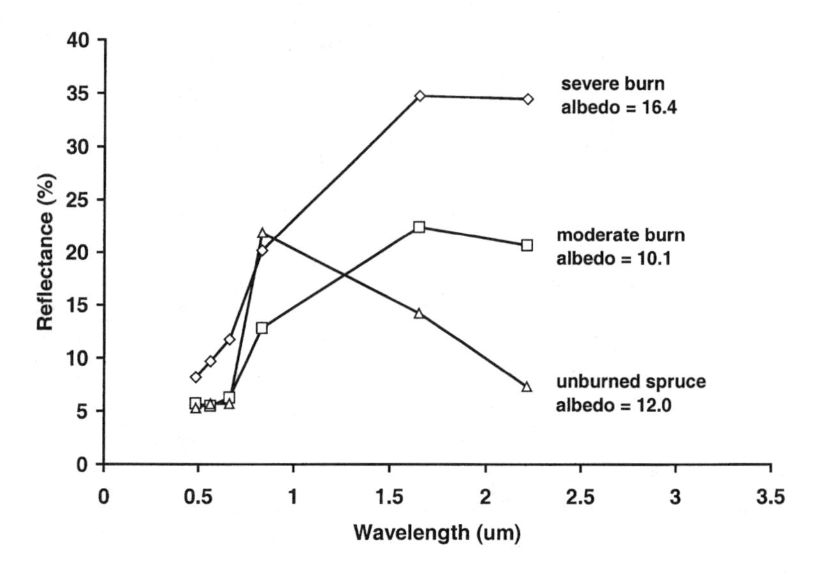

Figure 23.8. Effects of fire on reflectance and albedo derived from Landsat TM imagery.

change in albedo is dependent on severity of burning. In areas of severe burns, there is an increase in albedo, whereas in areas of moderate burning, there is a decrease in albedo. It is interesting to note that the strongest change in surface reflectance from burning occurs in the short-infrared region (wavelength > 1 μm) of the electromagnetic spectrum, where there is an overall increase in reflectance.

- *Surface emissivity (ε).* For surface energy balance studies, the emissivity of the earth's surface at longer wavelengths (8.0–14.0 μm) is important because black body emittance at lower temperatures is greater at longer wavelengths than at shorter wavelengths. Surface emissivity is dependent on land-cover type and, for vegetated surfaces, the amount of vegetation present. Vegetated surfaces have emissivities between 0.95 and 0.99, whereas bare soils have emissivities between 0.94 and 0.98. Approaches using satellite remote sensing data to estimate emissivity are based on land-cover categorization or variations in vegetation indices (Norman et al. 1995). Fires in boreal forests are thought to decrease the emissivity by 1 or 2%, which in turn, decreases the amount of energy emitted from the ground to the atmosphere.
- *Sensible heat flux (H).* Sensible heat flux is modeled by Norman and assoicates (1995) as

$$H = C_a(T_{\text{aero}} - T_a)/r_a \tag{23.6}$$

where C_a is the volumetric heat capacity of air, T_{aero} is the surface aerodynamic temperature (which is dependent on the vegetation kinetic temperature, soil kinetic temperature, and the aerodynamic resistances of the soil and canopy layers), T_a is the air temperature, and r_a is aerodynamic resistance (that is dependent on wind speed and the aerodynamic resistances of the vegetation and soil layers).

Most remote sensing studies have focused on relating the brightness temperature recorded by thermal infrared sensor (T_B) to T_{aero}. The first step involves estimating the infrared temperature (T_{IR}) by correcting for emissivity and sky emittance. This calculation is defined by Norman and associates (1995) as

$$T_B \approx [\varepsilon T_{IR}{}^n + (1 - \varepsilon)T_{sky}{}^n]^{1/n} \tag{23.7}$$

where $n = 4$. The thermal infrared temperature is then related to soil kinetic temperature (T_{soil}) and vegetation kinetic temperature (T_{veg}) (Norman et al. 1995):

$$T_{IR} = [f_v T_{veg}{}^n + (1 - f_v)T_{soil}{}^n]^{1/n} \tag{23.8}$$

where f_v is the fraction of the ground covered by vegetation. To estimate *H,* ground-based measurements are used to establish the relationships between vegetation type, percentage vegetation cover and the aerodynamic resistances, and the relationship between T_{veg} and T_{soil}. Thermal infrared satellite data are used to estimate T_{IR}. Finally, visible and near-infrared satellite data are used to determine vegetation-cover type as well as fraction of vegetation cover, f_v.

When a fire occurs on a forested landscape, it greatly simplifies the calculation of *H* because of the elimination of the vegetation canopy, thus eliminating or minimizing those terms associated with vegetation in Equations (23.6) to (23.8).

- *Latent heat flux (E).* Latent heat flux essentially represents the amount of energy dissipated from the earth's surface through evapotranspiration from the vegetation cover, which is defined by Norman and co-workers (1995) as

$$E = C_a(e_s - e_a)/[\gamma(r_a + r_s)] \tag{23.9}$$

where e_s is the saturation vapor pressure at the temperature of the surface, e_a is the saturation vapor pressure in the surface layer just above the surface, γ is the psychrometer constant, and r_s is the resistant of the surface to the diffusion of water. All the parameters are related to the conductance of water from the ground to the air via canopy conductance (e.g., transpiration).

One approach to use remote sensing data to estimate E has focused on relating the rates of transpiration to (1) vegetation-cover type, (2) the amount of photosynthetically active radiation present in a canopy, and (3) the simple ratio of the canopy reflectance (e.g., near-infrared reflectance divided by visible reflectance) (Hall et al. 1991b). Another approach uses the differences between surface temperatures collected during midday and the middle of the night to estimate thermal inertia, which, in turn, is related to *E* (Price 1980). In either case, the relationships between the remotely sensed data and surface processes are complex, and satisfactory results have yet to be demonstrated for a wide range of ecosystems. However, as with sensible heat exchange, the elimination of the vegetation canopy by fire significantly alters latent heat exchange in boreal forests. In this case, latent heat exchange is dramatically reduced, even eliminated, by the virtual removal of the vegetation canopy by fire.

Summary

In Chapters 17 and 18, different methods for using satellite remote sensing systems to monitor different surface characteristics of the boreal forest were reviewed. These systems are ideal tools for monitoring the effects of fire and recovery from fire in the boreal forest for several important reasons:

1. Fires result in dramatic changes in wide range of surface characteristics, including changes in surface reflectance, temperature, and moisture. These changes result in alterations to the signatures detected in all wavelength regions of the electromagnetic spectrum, including the visible, near-, shortwave, and thermal infrared, and microwave regions.
2. Fires affect large areas of forest in very remote regions, making it difficult (if not impossible) to completely survey the effects of fires by using ground-based measurements alone.
3. Because of temporal dynamics, it is difficult, if not impossible, to monitor fires and their effects by using ground-based monitoring approaches alone. In some years, there is relatively little fire activity, whereas in others, fires occur frequently. The repetitive nature of satellite data collection provides a means for efficient as well as cost-effective monitoring of these sporadic events.

Other recent developments have served to increase the demand for satellite data products from the boreal forest. First, landscape approaches to study ecological processes over the past two decades have led to a number of new modeling paradigms (as discussed in this chapter), which can effectively exploit the information available from satellite remote sensing systems. Second, new spatial analysis technologies, especially the rapid evolution of geographic information systems, have enabled the integration of data from remote sensing systems with other data sets. Third, the development of electronic data exchanges, such as the World Wide Web, has enabled a wider range of users to gain access to the large data sets required for landscape-scale analyses. Finally, the deployment of a large number of different remote sensing systems by different countries of the world has guaranteed a continuing supply of data for use by the scientific community. Given these developments, the use of satellite remote sensing data to monitor the effects of fires on the ecology of boreal forests will only continue to increase in the future.

References

Bonan, G.B., and K. Van Cleve. 1992. Soil temperature, nitrogen mineralization, and carbon source-sink relationships in boreal forests. *Can. J. For. Res.* 22:629–639.

Braswell, B.H., D.S. Schimel, E. Linder, and B. Moore III. 1997. The response of global terrestrial ecosystems to interannual temperature variability. *Science* 278:870–872.

Brown, R.J.E. 1983. The effects of fire on the permafrost ground thermal regime, pp. 97–110 in R.W. Wein and D.A. MacLean, eds. *The Role of Fire in Northern Circumpolar Ecosystems.* J. Wiley & Sons, New York.

Bunnell, F.L., D.E.N. Tait, P.W. Flanagan, and K. Van Cleve. 1977. Microbial respiration and substrate weight loss—I: a general model of the influences of abiotic variables. *Soil Biol. Biochem.* 9:33–40.

Cahoon, D.R., Jr., B.J. Stocks, J.S. Levine, W.R. Cofer III, and J.M. Pierson. 1994. Satellite analysis of the severe 1987 forest fires in northern China and southeastern Siberia. *J. Geophys. Res.* 99:18,627–18,638.

Cahoon, D.R., Jr., B.J. Stocks, J.S. Levine, W.R. Cofer III, and J.A. Barber. 1996. Monitoring the 1992 forest fires in the boreal ecosystem using NOAA AVHRR satellite imagery, pp. 795–802 in J.S. Levine, ed. *Biomass Burning and Climate Change,* Vol. 2: *Biomass Burning in South America, Southeast Asia, and Temperate and Boreal Ecosystems, and the Oil Fires of Kuwait.* MIT Press, Cambridge, MA.

Dye, D.G., and S.N. Goward. 1993. Photosynthetically active radiation absorbed by global land vegetation in October 1984. *Int. J. Remote Sens.* 14:3361–3364.

Foley, J.A., I.C. Prentice, N. Ramankutty, S. Levis, D. Pollard, S. Sitch, and A. Haxeltine. 1997. An integrated biosphere model of land surface processes, terrestrial carbon balance, and vegetation dynamics. *Global Biogeochem. Cycles* 10:603–628.

French, N.H.F., E.S. Kasischke, R.D. Johnson, L.L. Bourgeau-Chavez, A.L. Frick, and S.L. Ustin. 1996. Using multi-sensor satellite data to monitor carbon flux in Alaskan boreal forests, pp. 808–826 in J.S. Levine, ed. *Biomass Burning and Climate Change,* Vol. 2: *Biomass Burning in South America, Southeast Asia, and Temperate and Boreal Ecosystems, and the Oil Fires of Kuwait.* MIT Press, Cambridge, MA.

Goward, S.N., and K.F. Huemmrich. 1992. Vegetation canopy PAR absorptance and the normalized difference vegetation index: an assessment of using the SAIL model. *Remote Sens. Environ.* 39:119–140.

Hall, F., D. Botkin, D. Strebel, K. Woods, and S. Goetz. 1991a. Large-scale patterns of forest succession as determined by remote sensing. *Ecology* 72:628–640.

Hall, F.G., P.J. Sellers, D.E. Strebel, E.T. Kanemasu, R.D. Kelly, B.L. Blad, B.J. Markham, J.R. Wang, and F. Huemmrich. 1991b. Satellite remote sensing of surface energy and mass balance: results from FIFE. *Remote Sens. Environ.* 35:187–199.

Hall, F.G., K.F. Huemmrich, S.J. Goetz, P.J. Sellers, and J.E. Nickeson. 1992. Satellite remote sensing of the surface energy balance: success, failures, and unresolved issues in FIFE. *J. Geophys. Res.* 97:19,061–19,089.

Hall, F.G., P.J. Sellers, and D.L. Williams. 1996. Initial results from the Boreal Ecosystem-Atmosphere Experiment, BOREAS. *Silva Fennica* 30:109–121.

Johnson, R.D., and E.S. Kasischke. 1998. Automatic detection and classification of land cover characteristics using change vector analysis. *Int. J. Remote Sens.* 19:411–426.

Kasischke, E.S., N.H.F. French, L.L. Bourgeau-Chavez, and N.L. Christensen, Jr. 1995. Estimating release of carbon from 1990 and 1991 forest fires in Alaska. *J. Geophys. Res.* 100:2941–2951.

Myneni, R.B., C.D. Keeling, C.J. Tucker, G. Asrar, and R.R. Nemani. 1997. Increased plant growth in the northern high latitudes for 1981 to 1991. *Nature* 386:698–702.

Norman, J.M., M. Divarkarla, and N.S. Goel. 1995. Algorithms for extracting information from remote thermal-IR observations of the earth's surface. *Remote Sens. Environ.* 51:157–168.

Oechel, W.C., and K. Van Cleve. 1986. The role of bryophytes in nutrient cycling in the taiga, pp. 121–137 in K. Van Cleve, F.S. Chapin III, P.W. Flanagan, L.A. Viereck, and C.T. Dyrness, eds. *Forest Ecosystems in the Alaskan Taiga.* Springer-Verlag, New York.
O'Neill, K.P., E.S. Kasischke, D.D. Richter, and V. Krasovic. 1997. Effects of fire on temperature, moisture and CO_2 emissions from soils near Tok, Alaska, pp. 295–303 in I.K. Iskandar, E.A. Wright, J.K. Radke, B.S. Sharrett, P.H. Groenvelt, and L.D. Hinzman, eds. *Proceedings of the International Symposium on Physics, Chemistry, and Ecology of Seasonally Frozen Soils,* Fairbanks, Alaska, June 10–12, 1997. U.S. Army Cold Regions Research and Engineering Laboratory, Hanover, NH.
Price, J.C. 1980. The potential of remotely sensed thermal infrared data to infer surface soil moisture and evaporation. *Water Resources Res.* 16:787–795.
Prince, S.D. 1991. A model of regional primary production for use with coarse resolution satellite data. *Int. J. Remote Sens.* 12:1313–1330.
Prince, S.D., and S.N. Goward. 1996. Evaluation of the NOAA/NASA pathfinder AVHRR land data set for global primary production modeling. *Int. J. Remote Sens.* 17:217–221.
Running, S.W., and R.R. Nemani. 1988. Relating seasonal patterns of the AVHRR vegetation index to simulated photosynthesis and transpiration of forests in different climates. *Remote Sens. Environ.* 24:347–367.
Schlentner, R.E., and K. Van Cleve. 1985. Relationships between CO_2 evolution from soil, substrate temperature, and substrate moisture in four mature forest types in interior Alaska. *Can. J. For. Res.* 15:97–106.
Seiler, W., and P.J. Crutzen. 1980. Estimates of gross and net fluxes of carbon between the biosphere and atmosphere. *Clim. Change* 2:207–247.
Sellers, P.J., B.W. Meeson, F.G. Hall, G. Asrar, R.E. Murphy, R.A. Schiffer, F.P. Bretherton, R.E. Dickinson, R.G. Ellingson, C.B. Field, K.F. Huemmrich, C.O. Justice, J.M. Melack, N.T. Roulet, D.S. Schimel, and P.D. Try. 1995. Remote sensing of land surfaces for global change: models—algorithms—experiments. *Remote Sens. Environ.* 51:3–26.
Smith, T.M., H.H. Shugart, G.B. Bonan, and J.B. Smith. 1992. Modeling the potential response of vegetation to global climate change. *Adv. Ecol. Res.* 22:93–116.
Starks, P.J., J.M. Norman, B.L. Blad, E.A. Walter-Shea, and C.L. Walthall. 1991. Estimation of shortwave hemispherical reflectance (albedo) from bidirectionally reflected radiance data. *Remote Sens. Environ.* 38:123–134.
Thompson, M.V., J.T. Randerson, C.M. Malmstrom, and C.B. Field. 1997. Change in net primary production and heterotrophic respiration: how much is necessary to sustain the terrestrial carbon sink? *Global Biogeochem. Cycles* 10:711–726.
Tucker, C.J., C. Vanpraet, E. Boerwinkel, and A. Gaston. 1983. Satellite observation of total dry matter production in Senegalese Sahel. *Remote Sens. Environ.* 13:461–474.
Tucker, C.J., I.Y. Fung, C.D. Keeling, and R.H. Gammon. 1986. Relationship between atmospheric CO_2 and a satellite-derived vegetation index. *Nature* 319:195–199.
Viereck, L.A. 1983. The effects of fire in black spruce ecosystems in Alaska and northern Canada, pp. 201–220 in R.W. Wein and D.A. MacLean, eds. *The Role of Fire in Northern Circumpolar Ecosystems.* John Wiley & Sons, New York.

24. Influences of Fire and Climate Change on Patterns of Carbon Emissions in Boreal Peatlands

Leslie A. Morrissey, Gerald P. Livingston, and Steven C. Zoltai

Introduction

Projected changes in temperature and precipitation over the next several decades (Houghton et al. 1996) are expected to significantly enhance the release of gaseous carbon from northern peatlands. This enhancement is expected in response to increased rates of microbial decomposition acting on the vast carbon stores underlying these ecosystems (Oechel et al. 1993; Goulden et al. 1998) and to an increase in the frequency, extent, and intensity of peatland wildfires (Zoltai et al. 1998). Over much of the past 10,000 years, northern peatlands have collectively functioned as a globally important sink of atmospheric carbon dioxide (CO_2) (Mitsch and Gosselink 1986; Gorham 1991), that is, net primary productivity in these ecosystems has exceeded losses. As a result, nearly one-third of all soil organic matter on earth underlies northern peatlands in the form of partially decomposed organic matter (peat) (Gorham 1991). Evidence is building, however, to suggest that rates of carbon loss from these northern ecosystems due both to microbial decomposition (e.g., Livingston and Morrissey 1991; Carroll and Crill 1997) and fire (Levine et al. 1995) have increased dramatically, even exceeding net primary productivity in some areas (Oechel et al. 1993; Goulden et al. 1998). A dramatic shift in the ecological function of northern peatlands from that of a net carbon sink to a net carbon source could potentially enhance climatic change due to the resultant increased atmospheric loading of radiatively important gases such as CO_2 and methane (CH_4) (Fung et al. 1991).

In this chapter, we review the role and importance of climate and wildfires in northern peatlands with regard to carbon accumulation and emissions. This review builds on our recent analysis of the role of wildfires in North American peatlands (Zoltai et al. 1998), highlighting uncertainties and areas requiring further observation to improve understanding of carbon dynamics in these globally important ecosystems.

Peatland Environments

Peatlands are an important component of the boreal landscape, typically developing in low, poorly drained areas. Worldwide, peatlands occupy an estimated 3.37×10^6 km^2 (Zoltai and Martikainen 1996). Although relatively productive, the frequently waterlogged and oxygen-poor substrates in these ecosystems greatly limit rates of microbial decomposition, and thus excess dead organic material accumulates as peat. An estimated 397 (Zoltai and Martikainen 1996) to 455 Gt (Gorham 1991) of organic carbon presently underlies northern peatlands, exceeding the carbon stores of any other terrestrial biome on earth.

Peatlands are distinguished from other northern wetlands by the thickness of their peat (Zoltai and Vitt 1995). By definition, peat must contain more than 75% organic matter on a dry-weight basis (Andrejko et al. 1983). Peat thickness varies in response to regional differences in climate and hydrology. In Canada, only those areas having organic soil layers 40 cm thick or more are defined as peatlands (Canada Soil Survey Committee 1978), whereas in the United States peatlands are defined as having between 20 and 30 cm of organic soil (Heinselman 1963).

Peatlands differ greatly in structure and function, having developed under individual climate and hydrological (water quality and quantity) regimes. Climate and hydrology, in turn, influence the vegetation and soils communities that become established, net ecosystem productivity, pathways and rates of decomposition, and susceptibility of the peatlands to wildfire. Peat accumulation rates, as well as the composition and consistency of the peat, varies regionally and between peatland types.

Peatlands are often grouped into functionally similar classes that broadly reflect their inherent environmental controls, vegetation communities, and net productivity (Zoltai 1988). Three peatland classes are common throughout the boreal region: bogs, fens, and swamps (Zoltai and Vitt 1995). In North America, fens are the most extensive, occupying about 46% of the peatland area, followed by bogs underlain by permafrost (31%), permafrost-free bogs (15%), and forested swamps (8%) (Fig. 24.1; Zoltai et al. 1998). Boreal peatlands occupy about 1.14×10^6 km^2 in North America (Zoltai et al. 1998).

Bogs are characteristically nutrient-poor and dependent on precipitation for both their water and nutrient input. Bogs typically develop in areas with closed drainage; thus, there is little to no nutrient input from adjacent mineral soils. The water table is usually near or above the surface immediately following snowmelt but typically falls well below the surface throughout most of the summer. Sub-

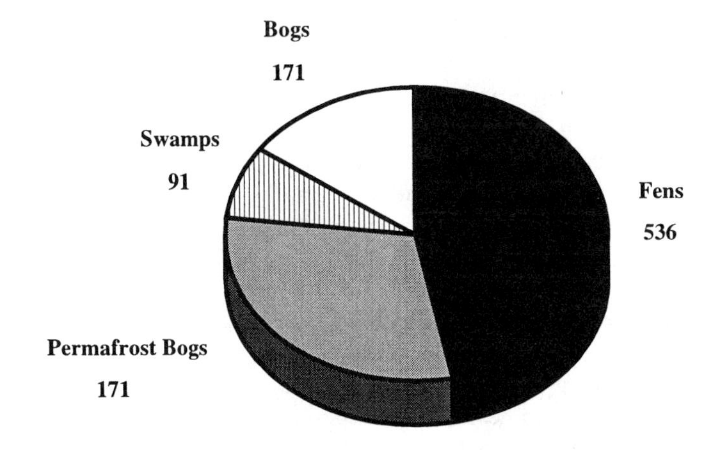

Figure 24.1. Areal extent of North American peatlands (area in 10^3 km^2).

strate conditions are often highly acidic. Primary productivity and rates of decomposition are low. Vegetation is represented by small forbes, low woody shrubs, and in particular, sphagnum mosses. At sites where the water table is often well below the surface, bogs usually support open-canopied stunted conifers (e.g., black spruce and tamarack). In the northern reaches, underlying permafrost often results in the development of raised bogs because the increased volume of the soil water on freezing causes the uplifting of the peat surface well above the water table. Within these elevated well-rained areas, primary productivity and decomposition may often nearly balance and peat formation ceases.

By contrast, fens typically develop in areas with a water table at or near the surface throughout much of the thaw period. Because much of the water originates in surrounding mineral soils, they are often enriched with dissolved elements. As a result, fens are more nutrient-rich and productive than bogs. Substrate conditions are only slightly acidic. Species diversity is relatively low, often dominated by one or two species, depending on the location of the water table. Moss and sedge communities typically dominate in areas where the water table is near or above the surface; tall shrub and tree communities develop when the depth to the water table falls 20 cm or more below the surface.

Swamps are peatlands with a relatively strong water flow through the peat, although water levels often fluctuate seasonally. Swamps in general are nutrient-rich and highly productive. Rates of decomposition vary seasonally but, on an annual basis, may nearly balance primary productivity; thus, peat formation is often minimal. The seasonal drop in the water table to depths of 30 cm or more often allows the development of forests. In northern swamps, closed-canopy conifer forests (e.g., black spruce) dominate; in southern reaches, white cedar or deciduous trees are more common. Mosses, if present, are only a minor component of the vegetation community.

Wildfires in Peatlands

Wildfires are also an important component of boreal ecosystems (Wein and MacLean 1983), affecting both the immediate release of carbon into the atmosphere as well as decadal-scale changes in the structure, hydrological, and biogeochemical dynamics. Fires not only directly affect the area burned but also exert indirect effects when dissolved or particulate materials are deposited as a result of the burn (Levine et al. 1990; Schmalzer et al. 1991). Wildfires thus influence carbon emissions to the atmosphere both directly through combustion and indirectly through their impact on productivity and carbon sequestration in peatlands. Although more common in upland forests, wildfires often spread into adjacent peatlands, consuming all or part of the surface vegetation (live and dead) and, depending on the kind of peatland and soil moisture conditions, the upper layers of the peat itself. Annually, wildfires in boreal ecosystems consume an estimated 6.4×10^3 km^2 or 0.6% of the areal extent of North American peatlands (Zoltai et al. 1998).

The importance and role of fire differs greatly between peatland types (Warner et al. 1991; Kuhry 1994; Zoltai et al. 1998). Given favorable fuel continuity and weather, fire can sweep across the surface of almost any wetland (Foster and Glaser 1986; Jasieniuk and Johnson 1982; Wein et al. 1987). As with peatland formation, regional differences in the frequency, extent, and intensity of fires in northern peatlands are closely linked to climatic and hydrological controls on the depth of the water table relative to the surface. Specifically, the position of the water table directly influences the moisture content and thus the susceptibility of both surface organic materials (live and dead) and near-surface peats to ignite and sustain combustion (Christjakov et al. 1983).

Estimation Methods

Our analysis of the role and importance of peatland wildfires is based on a variety of published studies on North American ecosystems. Only a limited number of observations regarding peatland fires are available. Annual carbon emissions were calculated separately for above- and below-ground combustion for each peatland class (permafrost bogs, nonpermafrost bogs, forested swamps, fens) based on estimates of available fuel, the fraction consumed, and emission ratios of the respective combustion by-products. Estimates of available fuel, fraction consumed, and fire periodicity were derived from direct and published observations of peat moisture content, air photo analyses, and charcoal analyses in peat profiles. Emission ratios of combustion by-products reported in the literature were assigned each peatland class in proportion to estimated fuel types (herbaceous, woody, peat) and combustion modes (flaming, smoldering, and mixed flaming and smoldering). Carbon emissions were derived by weighting above- and below-ground carbon losses due to fire by the assigned emission ratios. Details are provided in the work of Zoltai and colleagues (1998).

Areal Extent of Peatland Wildfires

Peatlands having water tables well below the surface throughout much of the growing season (permafrost bogs and forested swamps) are the most susceptible to burning. Surface fires consume about 1% of all permafrost bogs and 0.9% of forested swamps annually. It is estimated that 6,420 km^2 of peatlands burns annually in the North American boreal region. About 54% of the total occurs in permafrost peatlands, and 24% in fens, particularly in well-drained treed fens or in herbaceous fens following senescence or under drought conditions (Fig. 24.2). Despite their large areal representation, however, the relatively high water tables characteristic of fens clearly reduces their susceptibility to fire; only 0.3% of the total area of fens is burned annually.

Wildfires consume not only the aboveground standing biomass but also can burn into the underlying peat if moisture conditions allow. The susceptibility of peats to ignite and sustain combustion, however, is generally very low. North American peat fires consume only 1,200 km^2 or 0.1 % of the continental peatland area annually. In addition, moisture conditions typically limit peat fires to the upper 10–20 cm of the organic soil horizon (Dyrness and Norum 1983). Only under extremely dry conditions or if the water table is artificially lowered can peat fires penetrate to depths of several tens of centimeters (deep peat fires). Most peat fires occur in permafrost bogs, accounting for 84% of the total peat burn area (Fig. 24.2).

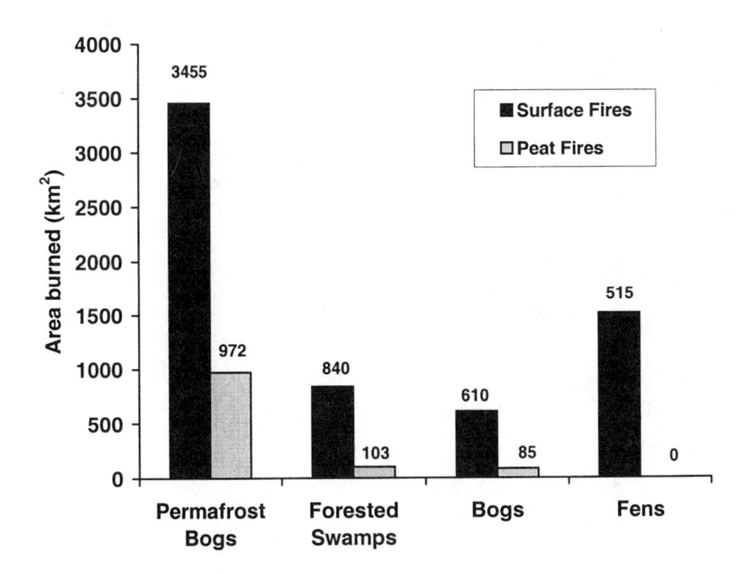

Figure 24.2. Areal extent of North American peatlands consumed annually due to surface and peat fires (km^2).

An analysis was conducted on the moisture contents of peat profiles from 151 sites across Canada and the northern United States (Zoltai et al. 1998). This analysis revealed that 66–75% of the peats underlying raised permafrost peatlands, 38% of those underlying conifer swamps, and 36% of those underlying nonpermafrost bogs are susceptible to combustion in an average year. By contrast, peat moisture levels in fens were consistently above the combustible limit, suggesting peat fires in fens are rare.

Analysis of the number and thickness of charcoal layers in peatland profiles also confirms that historically peat fires have occurred primarily in bogs, particularly in the subhumid continental and subarctic regions (Warner et al. 1991; Zoltai 1993; Kuhry 1994; Zoltai et al. 1998). Peat profiles from humid coastal regions either do not contain charcoal remains (Warner et al. 1991) or suggest that fires are typically small, patchy, or confined to the surface biomass only (Foster and Glaser 1986; Wein et al. 1987).

Carbon Emissions

Wildfires in North American peatlands contribute an estimated 9.6 Tg C yr^{-1} to the atmosphere on a long-term average (Zoltai et al. 1998). Fire in bogs underlain by permafrost and forested swamps are responsible for nearly 94% of all carbon emissions due to peatland fires (Fig. 24.3), with permafrost peatlands alone contributing nearly two-thirds of the total. Fires in forested swamps, however, are the largest aboveground emissions source, accounting for 70% of all surface emissions. Incomplete combustion within the near-surface peats is estimated to ac-

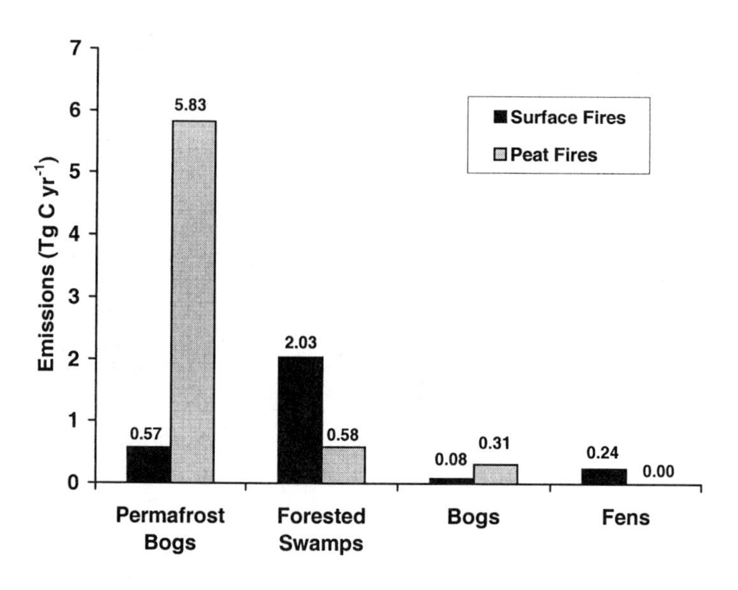

Figure 24.3. Estimated total annual carbon emissions due to wildfires in North American peatlands (Tg C yr^{-1}).

count for at least 70% (6.7 Tg C) of the total annual carbon emissions during wildfires in North American peatlands. The areal extent of peat fires, however, is only 18% that of surface fires. Carbon emissions due to surface combustion in forested swamps were 3.5-fold greater than estimated emissions from underlying peat combustion. By contrast, emissions from below-ground combustion in permafrost and nonpermafrost bogs ranged from three to ten times higher than from aboveground fires.

Averaged over their individual aboveground burn areas, annual carbon emissions from wildfires in North American fens, bogs, permafrost bogs, and forested swamps are estimated to be 1.6, 6.5, 18.5, and 31.1 t C m^{-2}, respectively (Zoltai et al. 1998). Overall, North American peatlands are estimated to release an average of 15 t C m^{-2} yr^{-1} during a wildfire. This estimate is well within the emissions estimates reported for fires in boreal forests, which range from 11.3 t C m^{-2} (Stocks 1991; Cahoon et al. 1994) to 24.5 t C m^{-2} (Kasischke et al. 1995). A direct comparison is difficult, however, as the boreal forest estimates do not differentiate forested peatland and nonpeatland contributions. Given that the carbon released on a per-unit area basis during fire from peatlands and upland forests is comparable, the significant difference in their global contributions appears to be due to differences in their respective areas susceptible to burning.

Global Estimates

If we assume that North American peatlands (1.14×10^6 km^2) are representative of boreal peatlands globally and the worldwide extent of peatlands is 3.37×10^6 km^2 (Zoltai and Martikainen 1996), then total carbon emissions from all northern peatlands due to biomass burning are estimated to be about 29 Tg C yr^{-1}. Annual carbon emissions due to boreal peatland wildfires is thus about the same as the contribution from burning of peat for fuel (26 Tg C yr^{-1}; Gorham 1991) and approximately one-fifth the total contribution of wildfires in boreal and temperate ecosystems worldwide (130 Tg C yr^{-1}; Levine et al. 1995).

Combustion By-Products

Combustion by-products in natural ecosystems typically include CO_2, carbon monoxide (CO), CH_4, and nonmethane hydrocarbons (NMHC) in various proportions that vary with fuel type, moisture content, and the mode of combustion (e.g., flaming or smoldering). Smoldering combustion is characteristic of peat fires due to the moist and often oxygen-limiting conditions in the peat profile. Flaming combustion, representing nearly complete oxidation, is more typical of surface fires in herbaceous and shrubby fens.

Assumed emission ratios corresponding to different combustion modes for each peatland type were estimated at 11.9, 1.19, and 1.07% for CO, CH_4, and NMHC relative to CO_2 (Zoltai et al. 1998; Fig. 24.4). In the absence of observations, it was assumed that CO_2 represented 90% of the total estimated carbon released during biomass burning (Levine et al. 1993). On this basis, it was estimated that 25.2 Tg C is released annually as CO_2 from northern peatlands world-

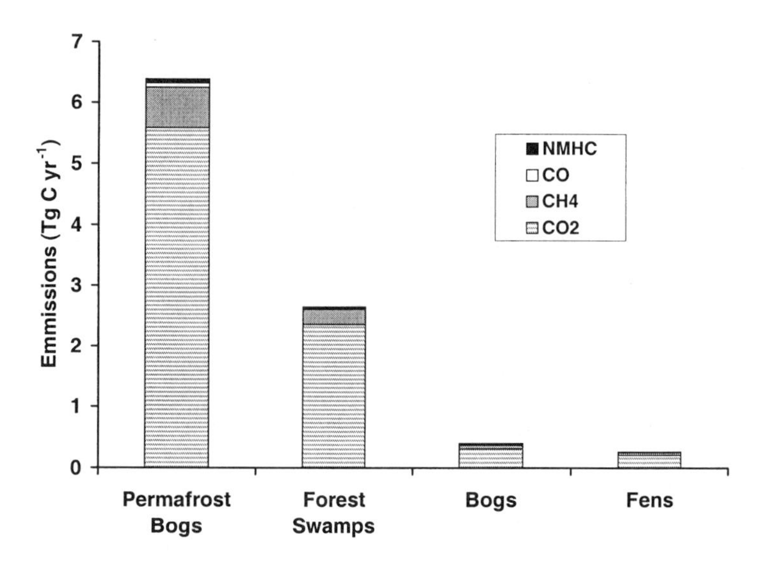

Figure 24.4. Estimated annual emissions of CO_2, CO, CH_4, and NMHC from North American peatlands (Tg C yr^{-1}).

wide due to wildfires. An additional 3.0, 0.3, and 0.27 Tg C yr^{-1} is also released as CO, CH_4, and NMHC due to combustion. These results clearly suggest that on a global basis wildfires in boreal peatlands are a significant source of atmospheric carbon. Although the uncertainties are high, peatland fires may account for upward of 50% of the total CO emissions (5.4 Tg C yr^{-1}) and 40% or more of the total CH_4 emissions (0.7 Tg C yr^{-1}) from all boreal wildfires (Hao and Ward 1993). Total carbon emissions from peatland fires, however, are still relatively small in comparison to biogenic emissions due to microbial decomposition and soil respiration. For example, CH_4 emissions due to boreal peatland fires account for only about 1.5% of the net 19.5 Tg CH_4 contributed annually by microbial activity in northern wetlands (Bartlett and Harriss 1993).

Wildfires in northern peatlands influence carbon emissions to the atmosphere not only directly through biomass burning but also indirectly through its impacts on primary productivity and microbial decomposition processes. Limited observations suggest that postburn soil microbial emissions both in peatlands burned and peatlands receiving dissolved or suspended materials from a burned area may actually far exceed direct losses. Lack of observations on wildfires in boreal ecosystems, however, prohibits estimating emissions from peatlands following fire with any certainty at this time.

Limited observations suggest that postburn soil microbial emission rates may far exceed preburn emission rates. The areal extent and time period over which the accelerated rates apply, however, is unknown. Dixon and Krankina (1993), Levine and colleagues (1993), and Mellilo and co-workers (1988) report a two- to eightfold increase in CO_2 emissions from boreal forest soils, and Levine and associates (1990) report a two-fold increase from wetlands following fire due to increased soil microbial activity. It is also known that fire can significantly change ecosystem structure as well as function. For example, removal of the insulating moss layer by fire can lead to increased soil temperatures and increased thaw depths in permafrost regions. Increased rates of microbial decomposition, in response to the warmer soil temperatures, may then lead to severalfold higher emission rates of CO_2 and CH_4 (Peterson and Billings 1975; Livingston and Morrissey 1991).

Increased nutrient availability due to the distribution of ash, dissolved materials, or sediments following fire may also enhance microbial and plant productivity in peatlands receiving these materials. Experimental application of ash following burning in a Canadian wetland increased CH_4 emissions eightfold (Hogg et al. 1992). Microbial emissions from all boreal peatlands fertilized as a direct result of fire in adjacent upland areas may thus approach or exceed, if expressed over a substantial area and time period, the instantaneous emissions from burning peatlands.

Impacts of Climate Change

Projected climatic changes over the next century are expected to significantly affect carbon dynamics in northern peatlands, particularly with regard to carbon storage and release to and from the atmosphere (Tans et al. 1990; Shugart et al. 1992; Houghton et al. 1996). Any climatic change that results in a drying of the surface peats or a lowering of the water table will almost certainly affect the present peat formation/depletion balance (Gorham 1991). This, in turn, will lead to increased depletion of the vast carbon stores underlying these ecosystems and add to the already increasing atmospheric burden of greenhouse gases such as CO_2 and CH_4 (Wein 1989; Livingston and Morrissey 1991; Oechel et al. 1993; Houghton et al. 1996; Carroll and Crill 1997). Increased soil temperatures are expected to result in increased carbon loss due to increased thaw depths, extended thaw seasons, and increased rates of microbial activity (Livingston and Morrissey 1991; Oechel et al. 1993; Carroll and Crill 1997).

Reduced soil moisture in northern peatlands (Houghton et al. 1996) is expected to result in a shift from anaerobic to aerobic microbial decomposition in the well-drained layers and thus to significantly higher rates of decomposition even without a change in soil temperature. In addition, drier surface peats are expected to increase fire frequency, intensity, and the areal extent of fires in boreal peatlands as well as increase the probability of shallow peat fires (Simard et al. 1985; Flannigan and Harrington 1988; Stocks 1990). How climate change will be regionally expressed will thus determine the geographic distribution of burns, fire frequency, areal extent of burns, and depth to which peats burn (King and Neilson 1992).

An increase in the frequency, extent, or intensity of peat fires is expected to significantly enhance gaseous carbon emissions and decomposition in peatlands.

By contrast, the present balance between peat accumulation and decomposition may remain fairly stable under a wetter climate regime that results in infrequent surface fires. For example, Kuhry (1994) demonstrated through analysis of charcoal layers in peat profiles that fire frequency was negatively correlated with rates of carbon accumulation and peat height in bogs. Carbon sequestration in bogs with less than one fire per millennium were more than double the rate in bogs subject to three to four fires over the same period. The greatest increase in fire susceptibility is expected to occur in the continental and subarctic regions as opposed to areas influenced by northern or maritime climates, where only small changes in water table levels are expected (Zoltai et al. 1998).

Regional differences in peatland type and successional changes in the vegetation community either as a result of fire or in response to climate change are expected to significantly affect net peat accumulation/depletion balances in peatlands (Laine et al. 1994). The reduction or loss in primary productivity resulting from destruction of peatland vegetation during a burn is generally short-lived. Most of the carbon lost directly as a result of burning is quickly reaccumulated in plant biomass through vegetation regrowth (Houghton et al. 1996). Most peatland vegetation communities recover on time scales of a few years to a few decades even under different climatic regimes (Houghton 1991; Zoltai et al. 1998). Thus the temporary decrease in the rate of peat accumulation over a period of a few decades after a fire is relatively insignificant. Only if the fire were to cause a long-term change in the environment (e.g., by initiating thermokarst degradation of permafrost) would the loss in carbon accumulation become important.

Forestry studies demonstrate that lowering the water table in previously waterlogged peatlands stimulates carbon loss and greatly decreases peat accumulation but generally tends to increase carbon accumulation in living biomass (Zoltai and Martikainen 1996). Laine and co-workers (1997) and Sakovets and Germanova (1992) reported four- to sevenfold increases in carbon storage in the living biomass of forested peatlands, despite net carbon losses from the underlying peat. At least on time scales of a few decades, the net ecosystem carbon balance in forested peatlands in response to drier soils may result in an increase in carbon storage. By contrast, the draining and conversion of northern peatlands for agricultural use has resulted in extensive net carbon losses because of increased peat decomposition (Nyk-nen et al. 1995).

Climate models also predict a longer growing season and more extreme climate events in northern ecosystems. Simulation studies of fire occurrence in eastern Canada based on anticipated climate changes suggest longer fire seasons of greater severity, particularly in late summer (Street 1989). In similar studies, Wotton and Flannigan (1993) and Flannigan and Van Wagner (1991) suggest that the length of the fire season in boreal ecosystems could increase by up to 30% in response to anticipated climatic changes and that more than 40% increases in fire frequency and area burned could be realized. The observed increase in the incidence and severity of fires across Canada and Russia in association with the drier and warmer weather conditions of the 1980s and 1990s may provide evidence for these projections (Krankina 1992; Auclair and Andrasko 1993). Based on these findings, the impact of longer fire seasons alone could nearly double carbon emissions from northern peatlands. Peatlands are also sensitive to extreme climate events such as heavy spring flooding and summer drought (Gorham 1991). Potential changes in precipitation patterns, particularly in the continental interiors, could affect the amplitude, periodicity, and frequency of extreme events such as wildfire. Droughts also will make peatlands far more susceptible to fire (Gorham 1994).

Uncertainties

Much uncertainty underlies our current understanding of the role and importance of wildfires in northern peatlands (e.g., Payette et al. 1989; Timoney and Wein 1991). The estimates presented here represent only a first approximation. The uncertainties in these estimates arise largely due to limited observations. An understanding of combustion and ecosystem processes that control trace gas release during and following fire is required to gain insight into the potential short- and long-term response of boreal peatlands to wildfire. Programmatic attention to these areas both nationally and internationally offers the possibility of improved fire management and improved understanding of carbon dynamics in northern ecosystems.

Uncertainties in estimating the amount of biomass consumed in peatland fires arise from uncertainties in estimates of the areal extent of peatlands, quantities of available standing live and dead fuel material, fraction of surface materials consumed, areal extent and depth of peat fires, and their fire return intervals. Estimates of the areal extent of northern peatlands are derived from a variety of sources at different scales, for different purposes, and under different classification approaches (Zoltai 1988). Estimates of living aboveground biomass vary greatly, ranging from 1.2 t ha^{-1} dry weight for nonforested bogs to more than 150.0 t ha^{-1} for conifer swamps. In addition, biomass estimates vary regionally along both temperature and moisture gradients. Observations of the fraction of surface peat biomass consumed in peatland fires are very limited and highly variable with regard to both fuel type and moisture conditions (Seiler and Crutzen 1980; Hao et al. 1990; Hao and Ward 1993). Estimates of surface biomass consumed range from 35 to 37% (Kasischke et al. 1995) to 50 to 90% (Forestry Canada Fire Danger Group 1992) for forested peatlands and up to 100% for graminoid peatlands (Cofer et al. 1990). Estimates of peat biomass consumed are even more limited (Kuhry 1994; Zoltai et al. 1998).

Little information is available on the length of fire cycles in various kinds of peatlands, although some observations (Jasieniuk and Johnson 1982) suggest that fires occur in uplands about twice as frequently as in peatlands. Based on observations in the literature (e.g., Rowe 1983), air photos, and fire maps, we estimated that the fire return period for surface fires in North American peatlands was between 75 and 1,000 years, although this varies regionally and by peatland type. The fire return period for shallow peat fires was estimated at between 250 and 1,000 years (Zoltai et al. 1998).

Uncertainties regarding carbon release to the atmosphere during and following fire also are large. Little is known regarding either the relative emission rates of the carbon species released during a burn or of the magnitude and significance of postburn biogenic emissions either from burned peatlands or from peatlands fertilized by fire in adjacent uplands. Emissions data are needed, which will allow improved characterization of fire-related emissions regionally, by peatland type, and by above- and below-ground combustion. Additional observations are needed to characterize the role of postburn carbon dynamics in peatlands and to understand the role and significance of fire-related fertilization and the areal extent and time scales over which they apply.

Conclusions

Over the past many millennia, northern peatlands have functioned as a globally important sink of atmospheric carbon, accumulating excess dead organic material as peat. Anticipated increases in rates of decomposition and the role and importance of wildfires in northern peatlands over the next 50–100 years in response to ongoing and projected climatic change could result in both a dramatic decrease in carbon accumulation as peat and an increase in the release of carbon into the atmosphere. These responses would not only alter carbon dynamics in northern ecosystems but also potentially enhance climatic change due to the resultant increase in the atmospheric concentrations of CO_2 and CH_4.

Any climatic change that would result in a lowering of the water table and thus in a reduction in soil moisture content of near-surface peats will have the largest impact on the peat accumulation/depletion balance in northern peatlands. Reduced surface moisture levels could lead to a shift from aerobic to anaerobic decomposition, greatly increasing rates of decomposition and carbon loss. Decreased surface moisture may also significantly increase the frequency, areal extent, and intensity of wildfires in northern peatlands and, in particular, increase the probability of peat fires. The resultant emissions due to biomass combustion would increase proportionately, making peatlands a considerably larger source of greenhouse gases.

Wildfire is an important process in boreal ecosystems, influencing ecosystem structure as well as carbon and nutrient dynamics. Wildfires from upland areas often spread into adjacent low-lying peatlands, given suitable weather and fuel conditions. Based on observations in North America, it is estimated that about 19,000 km^2, or 0.6% of the areal extent of northern peatlands worldwide, is burned annually on average, releasing 29 Tg of C into the atmosphere. This represents approximately one-fifth of the total contribution of wildfires in boreal and temperate ecosystems (Levine et al. 1995). Peatlands in which the water table seasonally falls well below the surface (e.g., raised bogs underlain by permafrost or forested swamps) are most susceptible to burning, although surface fires are also common in fens following senescence or under very dry conditions. Nearly 94% of all carbon emissions due to wildfires in North American peatlands are from fire in continental subarctic permafrost bogs (66%) and forested swamps (27%). Incomplete combustion within the near-surface peat (0–20 cm depth) is estimated to account for at least 70% of the total carbon emissions, even though the areal extent of peat fires is only 18% that of surface fires. Fires confined only to the aboveground biomass are a relatively small source of atmospheric carbon. Normalized over the surface area of North American peatlands burned, direct carbon emissions from wildfires are estimated to be approximately 15.0 t C m^{-2} under present climatic conditions. Fire return periods in peatlands estimated from charcoal layers in peat profiles range from 75 to 1,000 years but are highly regionally variable. Continental areas are far more subject to fire than extreme northern or coastal regions.

Biomass combustion in northern peatlands constitutes a significant source of carbon emissions to the atmosphere. Atmospheric loading of CO_2 due to wildfires in boreal peatlands is estimated at 25.2 Tg C yr^{-1}. Emissions of CO, CH_4, and NMHCs are estimated at 3.0, 0.3, and 0.27 Tg C yr^{-1}, respectively. These estimates of direct carbon emissions to the atmosphere due to wildfires in boreal peatlands suggest that they are currently a globally significant but relatively small source in contrast to either emissions from wildfires in the adjacent and areally extensive upland forests or emissions from microbial decomposition in peatlands. Anticipated changes in the frequency and intensity of peatland fires given projected climatic changes, however, could dramatically increase the importance of these peatlands as a source of carbon to the atmosphere.

Fire effects not only the structure, productivity, and biogeochemical dynamics of the area burned but also those of an often much larger area located downstream or downwind of the burned area. Limited observations suggest postburn microbial emissions in these areas may be severalfold greater than preburn rates. The areal extent, magnitude, duration, and thus importance of these fertilization events constitute key areas for future research.

Net ecosystem carbon balances in view of climate change will depend greatly on regional and successional influences on primary productivity. Changes in water table depths, and thus in peat accumulation/depletion balances, are least expected in extreme northern and coastal regions. Limited observations also suggest that forested peatlands may continue to function as net carbon sinks (i.e., primary productivity would exceed carbon losses due to peat decomposition at least on time scales of a few decades) but through carbon storage in living biomass as opposed to peat. Reduced soil moisture conditions in nonforested peatlands, however, will almost certainly result in dramatic peat losses due both to increased rates of microbial decomposition and to wildfire. Nonforested peatlands under anticipated climatic change may thus function largely as net sources of atmospheric carbon.

References

Andrejko, M.J., L. Fiene, and A.D. Cohen. 1983. Comparison of ashing techniques for determination of the inorganic content of peats, pp. 5–20 in P.M. Jarrett, ed. *Testing of Peats and Organic Soils.* ASTM Special Publication 820. American Society for Testing and Materials, Philadelphia, PA.

Auclair, A.N.D., and K.J. Andrasko. 1993. Net CO_2 flux in temperate and boreal forests in response to climate change this century: a case study, in K.M. Peterson, ed. *Proceedings of the NATO Advanced Workshop on the Biological Implications of Global Climate Change,* October 1991, Clemson, SC.

Bartlett, K.B,. and R.C. Harriss. 1993. Review and assessment of methane emissions from wetlands. *Chemosphere* 26:261–320.

Cahoon, D.R., B.J. Stocks, J.S. Levine, W.R. Cofer, and J.M. Pierson. 1994. Satellite analysis of the severe 1987 forest fires in northern China and southeastern Siberia. *J. Geophys. Res.* 99:18,627–18,638.

Canada Soil Survey Committee. 1978. *The Canadian System of Soil Classification.* Publication 1646. Canada Deptartment of Agriculture, Research Branch, Ottawa, Canada.

Carroll, P., and P. Crill 1997. Carbon balance of a temperate poor fen. *Global Biogeochem. Cycles* 11:349–356.

Christjakov, V.I., A.I. Kuprijanov, V.V. Gorshkov, and E.S. Artsybashev. 1983. Measures for fire prevention on peat deposits, pp. 259–271 in R.W. Wein and D.A. MacLean, eds. *The Role of Fire in Northern Circumpolar Ecosystems.* John Wiley & Sons, Chichester, UK.

Cofer, W.R., J.S. Levine, E.L. Winstead, P.J. LeBel, A.M. Koller, and C.R. Hinkle. 1990. Trace gas emissions from burning Florida wetlands. *J. Geophys. Res.* 95:1865–1870.

Dixon, R.K., and O.N. Krankina. 1993. Forest fires in Russia: carbon dioxide emissions to the atmosphere. *Can. J. For. Res.* 23:700–705.

Dyrness, C.T., and R.A. Norum. 1983. The effects of experimental fires on black spruce floors in interior Alaska. *Can. J. For. Res.*13:879–893.

Flannigan, M.D., and J.B. Harrington. 1988. A study of the relation of meteorological variables to monthly provincial area burned by wildfire in Canada. *J. Clim. Appl. Meteorol.* 27:441–452.

Flannigan, M.D., and C.E. Van Wagner. 1991. Climate change and wildfire in Canada. *Can. J. For. Res.* 21:66–72.

Forestry Canada Fire Danger Group. 1992. *Development and Structure of the Canadian Forest Fire Behavior Prediction System.* Information Report ST-X-3. Forestry Canada, Ottawa, Canada.

Foster, D.R., and P.H. Glaser. 1986. The raised bogs of south-eastern Labrador, Canada: classification, distribution, vegetation and recent dynamics. *J. Ecol.* 74:47–71.

Fung, I., J. John, J. Lerner, E. Matthews, M. Prather, L.P. Steele, and P.J. Fraser. 1991. Three-dimensional model synthesis of the global methane cycle. *J. Geophys. Res.* 96:13,033–13,066.

Gorham, E. 1991. Northern peatlands: role in the carbon cycle and probable responses to climatic warming. *Ecol. Appl.* 1:182–195.

Gorham, E. 1994. The future of research in Canadian peatlands: a brief survey with particular reference to global change. *Wetlands* 14:206–215.

Goulden, M.L., S.C. Wofsy, J.W. Harden, S.E. Trumbore, P.M. Crill, S.T. Gower, T. Fries, B.C. Daube, S.M. Fan, D.J. Sutton, A. Bazzaz, and J. W. Munger. 1998. Sensitivity of boreal forest carbon balance to soil thaw. *Science* 279:214–217.

Hao, W.M., and D.E. Ward. 1993. Methane production from global biomass burning. *J. Geophys. Res.* 98:20,657–20,661.

Hao, W.M., M.H. Liu, and P.J. Crutzen. 1990. Estimates of annual and regional releases of CO_2 and other trace gases to the atmosphere from fires in the tropics, based on FAO statistics for the period 1975–1980, pp. 440–462 in J.G. Goldammer, ed. *Fire in the Tropical Biota.* Ecological Studies 84. Springer-Verlag, New York.

Heinselman, M.L. 1963. Forest sites, bog processes, and peatland types in the Glacial Lake Agassiz region, Minnesota. *Ecol. Mon.* 33:327–374.

Hogg, E.H., V.J. Lieffers, and R.W. Wein. 1992. Potential carbon losses from peat profiles: effects of temperature, drought cycles, and fire. *Ecol. Appl.* 2:298–306.

Houghton, J.T., M. Filho, B.A. Callander, N. Harriss, A. Kattenberg, and K. Maskell, eds. 1996. *Climate Change 1995: The Science of Climate Change.* Intergovernmental Panel on Climate Change. Cambridge University Press, Cambridge, UK.

Houghton, R.A. 1991. Releases of carbon to the atmosphere from degradation of forests in Asia. *Can. J. For. Res.* 21:132–142.

Jasieniuk, M.A., and E.A. Johnson. 1982. Peatland vegetation organization and dynamics in the western subarctic, Northwest Territories, Canada. *Can. J. Bot.* 60:2581–2593.

Kasischke, E.S., N.H.F. French, L.L. Bourgeau-Chavez, and N.L. Christensen, Jr. 1995. Estimating release of carbon from 1990 and 1991 forest fires in Alaska. *J. Geophys. Res.* 100:2941–2951.

King, G.A., and R.P. Neilson. 1992. The transient response of vegetation to climate change: a potential source of CO_2 to the atmosphere. *Water Air Soil Pollut.* 64:365–383.

Krankina, O.N. 1992. Contribution of forest fires to the carbon flux of the USSR, pp. 179–186 in T. Kolchugina and T. Vinson, eds. *Proceedings of a Workshop on Carbon Cycling in Boreal Forests and Subarctic Ecosystems,* September 9–12, 1991, Corvallis, OR. Oregon State University Press, Corvallis, OR.

Kuhry, P. 1994. The role of fire in the development of *Sphagnum*-dominated peatlands in western boreal Canada. *J. Ecol.* 82:899–910.

Laine, J., K. Minkkinen, A. Puhalainen, and S. Jauhiainen. 1994. Effect of forest drainage on the carbon balance of peatland ecosystems, pp. 303–308 in M. Kanninen and P. Heikinheimo, eds. *The Finnish Research Programme on Climate Change. Second Progress Report.* Publications of the Academy of Finland 1/94. Painastuskeskus, Helsinki, Finland.

Laine, J., K. Minkkinen, J. Sinisalo, I. Savolainen, and P.J. Martikainen. 1997. Greenhouse impact of a mire after drainage for forestry, pp. 437–448 in C.C.Trenttin, ed. *Ecology and Management of Northern Wetlands,* August 1994, Traverse City, MI. CRC Lewis, Boca Raton, FL.

Levine, J.S., W.R. Cofer, D.I. Sebacher, R.P. Rhinehart, E.L. Winstead, S. Sebacher, C.R. Hinkle, P.A. Schmalzer, and A.M. Koller. 1990. The effects of fire on biogenic emissions of methane and nitric oxide from wetlands. *J. Geophys. Res.* 95:1853–1864.

Levine, J.S., W.R. Cofer, and J.P. Pinto. 1993. Biomass burning, pp. 299–313 in M.A.K. Khalil, ed. *Atmospheric Methane: Sources, Sinks, and Role in Global Change.* Springer-Verlag, Berlin.

Levine, J.S., W.R. Cofer, D.R. Cahoon, and E.L. Winstead. 1995. Biomass burning: a driver for global change. *Environ. Sci. Technol.* 29:120–125.

Livingston, G.P., and L.A. Morrissey. 1991. Methane emissions from Alaska Arctic tundra in response to climatic warming, pp. 372–377 in G. Weller, C. Wilson, and B. Severin, eds. *Proceedings of the International Conference on the Role of Polar Regions in Global Change,* June 1990, Fairbanks, AK. Geophysical Institute, University of Alaska, Fairbanks, AK.

Mellilo, J.M., J.R. Fruci, R.A. Houghton, B. Moore, and D.L. Skole. 1988. Land-use change in the Soviet Union between 1850 and 1980: causes of net CO_2 release to the atmosphere. *Tellus* 40:116–128.

Mitsch, M.J., and J.G. Gosselink. 1986. *Wetlands.* Van Nostrand Reinhold, New York.

Nyk-nen, H., J. Alm, K. Lang, J. Silvola, and P.J. Martikainen. 1995. Emissions of CH_4, N_2O, and CO_2 from a virgin fen and a fen drained for grassland in Finland. *J. Biogeog.* 22:1149–1155.

Oechel, W.C., S.J. Hastings, G. Vourlitis, M. Jenkins, G. Reichers, and N. Grulke. 1993.

Recent change of Arctic tundra ecosystems from a net carbon dioxide sink to a source. *Nature* 361:520–523.

Payette, S., C. Morneau, L. Sirois, and M. Desponts. 1989. Recent fire history of the northern Quebec biomes. *Ecology* 70:656–673.

Peterson, K.M., and D.W. Billings. 1975. Carbon dioxide flux from tundra soils and vegetation as related to temperature at Barrow, Alaska. *Am. Mid. Nat.* 94:88–98.

Rowe, J.S. 1983. Concepts of fire effects on plant individuals and species, pp. 135–154 in R.W. Wein and D.A. MacLean, eds. *The Role of Fire in Northern Circumpolar Ecosystems.* John Wiley & Sons, Chichester, UK.

Sakovets, V.V., and N.I. Germanova. 1992. Changes in the carbon balance of forested mires in Karelia due to drainage. *Suo* 43:249–252.

Schmalzer, P.A., C.R. Hinkle, and A.M. Killer. 1991. Changes in marsh soils for six months after a fire, pp. 272–286 in J.S. Levine, ed. *Global Biomass Burning: Atmospheric, Climatic, and Biospheric Implications.* MIT Press, Cambridge, MA.

Seiler, W., and P.J. Crutzen. 1980. Estimates of gross and net fluxes of carbon between the biosphere and the atmosphere from biomass burning. *Clim. Change* 2:207–247.

Shugart, H.H., R. Leemans, and G.B. Bonan, eds. 1992. *A Systems Analysis of Global Boreal Forests.* Cambridge University Press, Cambridge, UK.

Simard, A.J., D.A. Haines, and W.A. Main. 1985. Relations between El Nino/southern oscillation anomalies and wildland fire activity in the United States. *Agric. For. Meteorol.* 36:93–104.

Stocks, B.J. 1990. Global warming and the forest fire business in Canada, pp. 223–229 in G. Wall, ed. *Proceedings of a Symposium on Impacts of Climate Change and Variability on the Great Plains.* September 11–13, 1990, Calgary, Alberta, Canada. University of Waterloo, Waterloo, Ontario, Canada.

Stocks, B.J. 1991. The extent and impact of forest fires in north circumpolar countries, pp. 197–202 in J.S. Levine, ed. *Global Biomass Burning: Atmospheric, Climatic, and Biospheric Implications.* MIT Press, Cambridge, MA.

Street, R.B. 1989. Climate change and forest fires in Ontario, pp. 177–182 in D.C. MacIver, H. Auld, and R. Whitewood, eds. *Proceedings of the 10th Conference on Fire and Forest Meteorology.* Forestry Canada, Ottawa, Ontario, Canada.

Tans, P.P., I.Y. Fung, and T. Takahashi. 1990. Observational constraint on the global atmospheric CO_2 budget. *Science* 247:1431–1438.

Timoney, K.P., and R.W. Wein. 1991. The areal pattern of burned vegetation in the subarctic region of northwestern Canada. *Arctic* 44:223–230.

Warner, B.G., K. Tolonen, and M. Tolonen. 1991. A postglacial history of vegetation and bog formation at Point Escuminac, New Brunswick. *Can. J. Earth Sci.* 28:1572–1582.

Wein, R.W. 1989. Climate change and wildfire in northern coniferous forests: climate change scenarios, p. 169 in H. Auld, R. Whitewood, and. D.C. MacIver, eds. *Proceedings of the 10th Conference on Fire and Forest Meteorology.* Forestry Canada, Ottawa, Ontario, Canada.

Wein, R.W., and D.A. MacLean. 1983. An overview of fires in northern ecosystems, pp. 1–18 in R.W. Wein and D.A. MacLean, eds. *The Role of Fire in Northern Ecosystems.* John Wiley & Sons, New York.

Wein, R.W., M.P. Burzynski, B.A. Sreenivasa, and K. Tolonen. 1987. Bog profile evidence of fire and vegetation dynamics since 3000 years BP in the Acadian Forest. *Can. J. Bot.* 65:1180–1186.

Wotton, B.M., and M.D. Flannigan. 1993. Length of the fire season in a changing climate. *For. Chron.* 69:187–192.

Zoltai, S.C. 1988. Wetland environments and classification, pp. 1–26 in National Wetlands Working Group, ed. *Wetlands of Canada.* Ecological Land Classification Series 24. Sustainable Development Branch, Environment Canada, Ottawa, Ontario, and Polyscience Publications, Montreal, Quebec, Canada.

Zoltai, S.C. 1993. Cyclic development of permafrost in the peatlands of northwestern Alberta, Canada. *Arct. Alp. Res.* 25:240–246.

Zoltai, S.C., and P.J. Martikainen. 1996. The role of forested peatlands in the global carbon cycle, pp. 47–58 in M.J. Apps and D.T. Price, eds. *Forest Ecosystems, Forest Management and the Global Carbon Cycle.* NATO Advanced Science Institutes Series, Vol. I40. Springer-Verlag, Heidelberg, Germany.

Zoltai, S.C., and D.H. Vitt. 1995. Canadian wetlands: environmental gradients and classification. *Vegetatio* 118:131–137.

Zoltai, S.C., L.A. Morrissey, G.P. Livingston, and W.J. de Groot. 1998. Effects of fires on carbon cycling in North American boreal peatlands. *Environ. Rev.* 6:13–24.

25. Effects of Climate Change and Fire on Carbon Storage in North American Boreal Forests

Eric S. Kasischke

Introduction

There is little doubt in the scientific community that if the current patterns of climate warming continue, there will be dramatic changes in the distribution of vegetation and forest cover throughout the boreal region. Most research has concentrated on predicting how rises in air temperature will (1) cause shifts in vegetation cover and the carbon balance in the terrestrial biome (Smith and Shugart 1993) or (2) change rates of soil respiration (Townsend et al. 1992; Randerson et al. 1996; Thompson et al. 1997). Recently, more sophisticated biogeochemical cycling models have been developed that actually model the processes responsible for carbon storage in vegetation to assess changes in the terrestrial carbon budget over large areas (Haxeltine and Prentice 1996; McGuire et al. 1997; Peng et al. 1998). None of these approaches takes into consideration the effects of fire on the longer-term patterns of carbon storage in the boreal forest.

Kasischke and associates (1995) introduced a technique that allows exploration of how variations in rates of net primary productivity, soil respiration, and the fire cycle resulting from climate warming interact to influence the carbon balance of the boreal forest. This study showed that increases in fire frequency, when combined with increases in soil respiration, result in much lower levels of carbon being stored in the boreal forest, despite significant increases in net primary productivity. This approach, however, was carried out over the entire boreal forest region and therefore did not address the complex distribution of forest types that occurs throughout this biome.

This chapter presents the results of a sensitivity study that was carried out for the North American boreal forest region. The basic approach of Kasischke and associates (1995) was used, but the boreal forest region was divided into nine unique ecozones (Fig. 14.1) and specific forest-cover types were defined for each ecozone (Table 14.2). The fire return interval for each ecozone was determined by using the large-fire database for North America (Chapter 15; Table 14.3). The average carbon levels for each ecozone were determined by using the databases described in Chapter 14 (Table 14.3).

Methods

The model developed by Kasischke and colleagues (1995) was used to examine the overall effects of climate change on processes or factors that control carbon storage in the boreal forests, including (1) fire return interval, (2) net primary productivity, (3) soil respiration, and (4) aerial extent of different forest/vegetation-cover types. The approach used by Kasischke and co-workers (1995) calculates the average carbon present in a given forest type by integrating two curves: (1) the stand age distribution based on fire return interval, $p(t)$; and (2) the pattern of carbon density as a function of stand age.

Van Wagner (1978) showed that where fire is the dominant disturbance factor, the stand age distribution in boreal forests can be determined by the fire frequency (f), which is the reciprocal of the fire return interval. Using a Weibull distribution, the stand age distribution can be calculated as

$$p(t) = (f/\Gamma(1/c + 1)]) \exp(-(ft)^c) \qquad (25.1)$$

where t is the stand age in years, Γ is the gamma function, and c is a shape parameter dependent on the flammability of the forest type. For the stand replacement fires that occur throughout the North American boreal forest, stand age is usually equivalent to time since the last fire. If forest flammability is independent of stand age, then $c = 1$, and Equation (25.1) reduces to the negative exponential distribution. If flammability increases with stand age, then $c > 1$. Finally, if flammability decreases with stand age, then $c < 1$. Yarie (1981) showed that for the boreal forest types found in central Alaska, c ranged between 0.9 and 2.4. For this analysis, it was assumed that $c = 1.5$. Figure 25.1 presents stand age distribution curves based on the Weibull distribution for two different fire return intervals. These curves illustrate the fundamental nature of the relationship between fire and stand age—the shorter the fire return interval, the larger the number of younger stands.

Two sets of curves were used to describe the carbon level as a function of time after fire [($B(t)$)]. One set of curves was developed for black spruce forests found on less productive sites that have a slower rate of aboveground accumulation and a

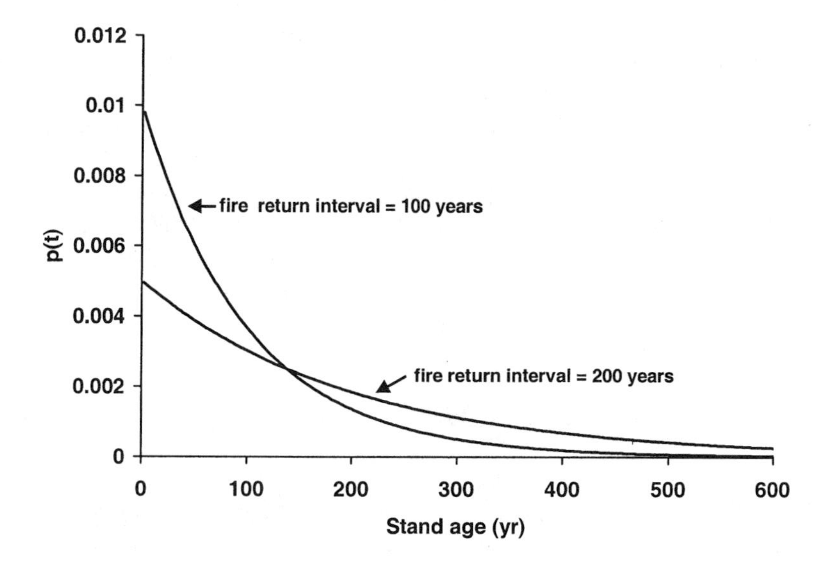

Figure 25.1. Effects of fire return interval on the stand age distribution in boreal forests based on a Weibull distribution.

faster rate of below-ground accumulation (Fig. 25.2a). A second set of curves was used for forest types found on more productive sites (e.g., pine, white spruce, fir, and deciduous species) that have a faster rate of aboveground accumulation and a slower rate of below-ground accumulation (Fig. 25.2b). The ground-layer carbon curves presented in Figure 25.2a and b differ from those used in previous studies (Kasischke et al. 1995; Kasischke 1996). These previous studies used curves that represented the carbon present in both the mineral and organic soils, whereas the present study only considered carbon present in the upper 30 cm of the soil profile.

The average level of carbon (C_a) can be estimated as

$$C_a = \int_t p(t)B(t) \tag{25.2}$$

By varying the fire return interval in Equation (25.1), it is possible to estimate the average carbon present in a region as a function of fire return interval. Figure 25.3 presents this relationship for the baseline carbon accumulation curves presented in Figure 25.2a. These curves illustrate the fundamental relationship between carbon storage and the fire return interval. When all other factors are held constant (e.g., net primary productivity, soil respiration, and fire severity), as the fire return interval decreases, the amount of stored carbon decreases.

In the sensitivity analyses performed for this study, the forest/vegetation covers were distributed in the different ecozones according to the fractions presented by Lowe and associates (1996) and summarized in Table 14.2. An overall fire return interval was assumed for each ecozone that was 30% higher than those presented in Table 14.3, based on the observations that (1) the length of the data record used

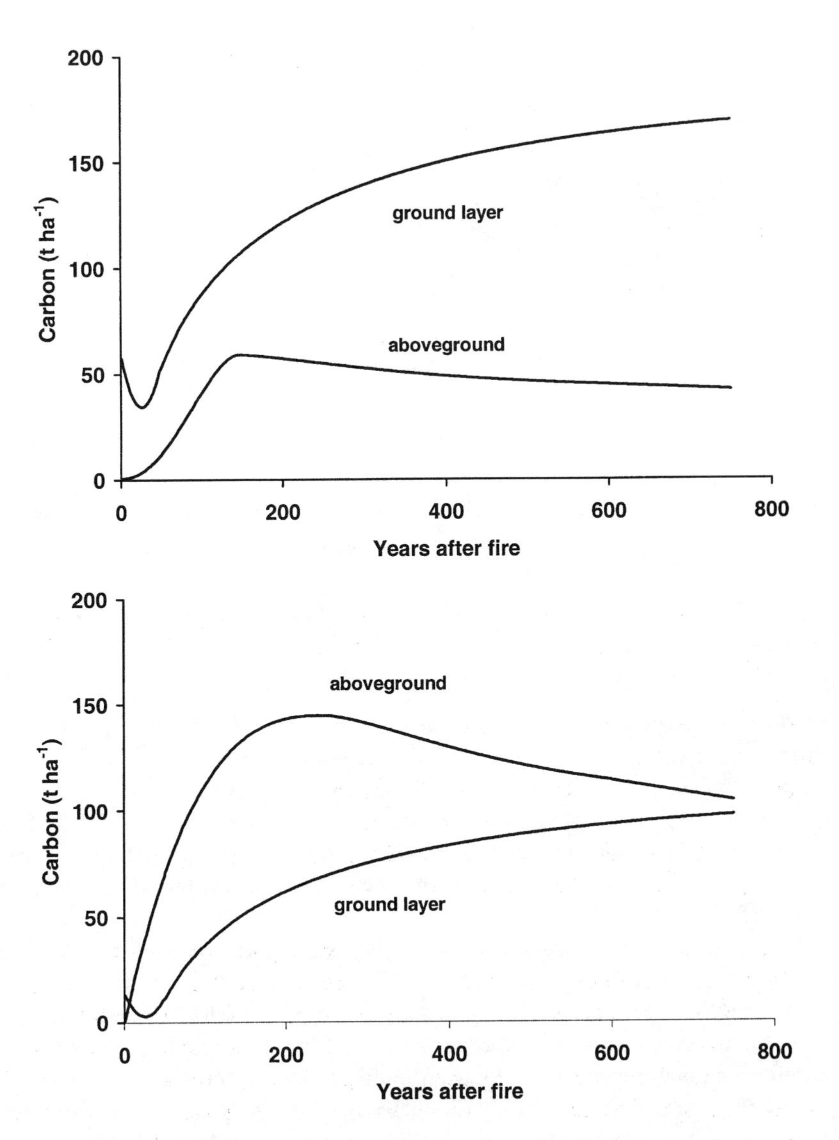

Figure 25.2. Carbon levels as a function of years after a fire for (a) nonproductive forest types and (b) productive forest types.

to estimate these values was too short to produce an accurate estimate and (2) the longer data record for the state of Alaska showed that the 15-year data record underestimated the fire return interval for this region by about 30% (see Chapter 14). Because of the short fire return interval for the Western Boreal ecozone, we increased the annual area burned by only 10% because of the low fire return interval in this region. We then assigned an area burned to the different forest types in each ecozone, assuming that the nonforest areas had the highest fire

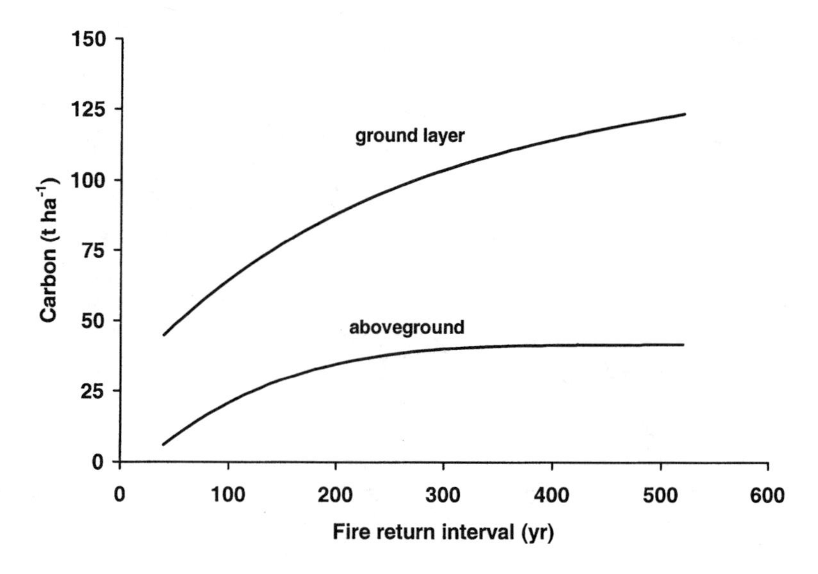

Figure 25.3. Effects of variations in fire return interval on average carbon storage for the nonproductive forest type.

return interval and the unproductive black spruce forests had the lowest fire return intervals (Table 25.1). The areas burned were distributed such that the fire return interval for the entire ecozone matched that presented in Table 25.1.

The overall carbon levels that were used are summarized in Table 25.1 for each ecozone. Again, different carbon levels were assigned to the different forest types in each ecozone such that the weighted averages equaled the values listed in Table 25.1.

The next step in the analysis was to apply a multiplicative weighting factor to the biomass accumulation curve (Fig. 25.2a and b) such that when these curves were integrated with the stand age distribution calculated for the ecozone/forest type, the average carbon level for that ecozone/forest type resulted. The resulting carbon accumulation curves were then multiplied by a factor to account for the specified increases in net primary productivity or soil respiration. The fire return interval for each forest type was then altered, assuming a 25% and 50% increase in the annual area burned. This increase is based on model predictions that the frequency of burning in the boreal forest will increase if global warming occurs (Chapter 20).

It was assumed that global warming would result in an overall increase in the levels of aboveground carbon through increases in net primary productivity and a decrease in ground-layer carbon because of increases in heterotrophic (soil) respiration. Several different scenarios were developed. Scenario I was based on the general assumptions made by Kasischke and co-workers (1995). Scenario II was based on recent research on the effects of climate warming on net primary productivity and soil respiration in the Canadian boreal forest (Peng and Apps 1998;

Peng et al. 1998). Scenario III was similar to scenario I except that it used a single value for average aboveground and ground-layer carbon and fire return interval. Scenario IV was similar to Scenario II except that it again used a single value for average aboveground and ground-layer carbon and fire return interval.

Scenario I

Scenario I used assumptions that were similar to those used by Kasischke and colleagues (1995). Net primary production was assumed to be 25% higher for all forest types, and a factor of 2 to 3 higher for the nonforest type. It was assumed that the increases in soil respiration were sufficient to result in the ground-layer carbon levels being 20% lower in all forest classes and 10% lower in the nonforest class. It was assumed that climate warming would also increase the levels of ground-layer carbon consumption in the nonforest and black spruce forest type by 10 t ha^{-1}, except in the nonforest type of the boreal cordillera ecozone and the black spruce and nonforest types in the west taiga shield ecozone. Finally, it was assumed that there would be a change in the aerial extent of the different forest/vegetation classes as a result of climate warming. Overall, it was assumed that there would be a decrease in the nonforest and unproductive black spruce areas and an increase in the unproductive pine and productive forest-type areas (Table 25.1).

Four different cases were evaluated:

- *Case a.* Changes in net primary productivity, soil respiration, and ground biomass consumption during fires.
- *Case b.* Changes in net primary productivity, soil respiration, ground biomass consumption during fires, and the area of each forest-cover type.
- *Case c.* Changes in net primary productivity, soil respiration, ground biomass consumption during fires, the area of each forest-cover type, and a 25% increase in annual area burned.
- *Case d.* Changes in net primary productivity, soil respiration, ground biomass consumption during fires, the area of each forest-cover type, and a 50% increase in annual area burned.

Scenario II

A recent study by Peng and Apps (1998) used a set of models developed by Peng and colleagues (1998) to assess the effects of projected climate changes on net primary production and soil respiration in three different boreal forest types along a north-to-south transect in central Canada. This study predicted that both the net primary productivity and soil respiration in these forests would increase by 40% under the $2 \times CO_2$ climate. Under this scenario, these increases were assumed for all cases described for scenario I. However, no increase was assumed in the levels of ground-layer carbon consumption during fires.

Table 25.1. Summary of Baseline Parameters and Assumptions Used in the Sensitivity Analyses

	Baseline fraction coverage	Climate change fraction coverage	Fire return interval (yr) baseline	Average carbon (t ha^{-1})	
				Aboveground	Belowground
Boreal Cordillera					
Nonforest	0.37	0.30	748	2	25
Productive spruce-fir-pine	0.30	0.36	303	77	45
Unproductive pine	0.10	0.17	242	44	50.5
Unproductive black spruce	0.23	0.17	165	20.2	110
Ecozone average			303	32.9	53.2
Taiga Plains					
Nonforest	0.12	0.10	350	2	215
Productive spruce-pine	0.30	0.35	195	50	70
Unproductive pine	0.12	0.15	136	30	70
Unproductive black spruce	0.46	0.40	104	18.5	153
Ecozone average			139	27.3	125.9
West Taiga Shield					
Nonforest	0.41	0.30	392	2	40
Productive spruce-pine	0.06	0.12	126	52	40
Unproductive pine	0.16	0.28	88	32	40
Unproductive black spruce	0.37	0.30	77	17	61
Ecozone average			122	15.3	47.8
East Taiga Shield					
Nonforest	0.41	0.30	611	1	100
Productive spruce-pine-fir	0.15	0.12	335	34	50
Unproductive pine	0.13	0.28	221	23.5	40
Unproductive black spruce	0.31	0.30	133	9.4	92.6
Ecozone average			248	11.5	82.3

West Boreal Shield					
Nonforest	0.07	0.05	251	2	118
Productive spruce-pine-fir	0.60	0.65	127	42	64
Unproductive pine	0.17	0.17	47	17	54
Unproductive black spruce	0.17	0.13	46	12	109.5
Ecozone average			82	30.1	73.6
East Boreal Shield					
Nonforest	0.07	0.05	988	2	150
Productive spruce-pine-fir	0.70	0.65	910	46	100
Unproductive pine	0.07	0.17	401	19	91
Unproductive black spruce	0.16	0.13	354	14	141
Ecozone average			682	35.9	109.5
Hudson Plains					
Nonforest	0.67	0.60	819	24	125
Productive spruce-pine	0.04	0.08	293	80	50
Unproductive black spruce	0.29	0.32	177	34	112
Ecozone average			386	29.1	118.1
Boreal Plains					
Nonforest	0.23	0.20	519	2	124
Productive deciduous-spruce-pine	0.52	0.58	153	56	60
Unproductive pine	0.05	0.06	97	25	62
Unproductive black spruce	0.20	0.16	71	20	150.5
Ecozone average			139	34.8	92.9
Alaska					
Nonforest	0.30	0.20	500	1	100
Productive white spruce	0.10	0.15	208	70	40
Unproductive white spruce	0.12	0.25	125	48	30
Unproductive black spruce	0.48	0.40	93	20.1	110
Ecozone average			139	22.7	90.4

Scenario III

Under scenarios III and IV, the effects of climate change and changes to the fire regime were assessed by using average values for the entire North American boreal forest. An average aboveground carbon level of 27.4 t ha^{-1}, an average below-ground carbon level of 88.4 t ha^{-1}, and a fire return interval of 173 years were assumed. It was also assumed that global warming would result in: (1) a 25% increase in net primary productivity; (2) a 15% increase in soil respiration; and (3) a net increase of 5 t ha^{-1} in consumption of ground-layer biomass during fires. Finally, no changes in the distribution of the different forest types. Therefore, three cases were evaluated (cases a, c, and d).

Scenario IV

The assumptions for scenario IV were identical to those used in scenario III, except that it was assumed that the net primary production and soil respiration both increased by 40%.

Results

Table 25.2 presents the model-estimated changes in the baseline carbon estimates for the entire North American boreal forest region using the ecozone approach. These results are consistent with previous studies (Kasischke et al. 1995; Kasischke 1996) in that they show there will be (1) increases in aboveground carbon because of the increased rates of net primary production, (2) decreases in below-ground carbon because of increases in soil respiration, and (3) an overall decrease of stored carbon because of the dominant influence of ground-layer carbon loss combined with a shortened fire return interval. Depending on the assumptions used, the model predicts there will be a 6.4–23.1 Gt loss of stored carbon, or a loss of 10–36%. Although there is an overall gain of 0.3–9.1 Gt gain of carbon in aboveground biomass, there is a predicted 10.2–25.5 Gt loss of carbon in the ground layer.

The changes in carbon levels predicted by using averages for the entire North American boreal forest (scenarios III and IV) tend to be greater than those predicted by using values for individual ecozones (scenarios I and II). It is thought that using more refined geographic divisions presents a more accurate assessment of the effects of future climate changes on boreal forest carbon storage than does using average values for the entire region. Finally, on a fractional basis, the carbon losses predicted for the North American boreal forest zone are considerably greater than those estimated by Kasischke and co-workers (1995) for the entire boreal forest region.

Discussion

The formulation of the modeling approach by Kasischke and associates (1995) used in the analyses presented in this chapter actuallly began in the fall of 1992.

Table 25.2. Modeled Changes in Total Carbon Storage levels in the North American Boreal Forest Region Based on Calculations for Individual Ecozones Using Scenarios I and II

Case[a]		a				b		c				d			
Scenario[b]	Baseline[c]	I	II	III	IV	I	II	I	II	III	IV	I	II	III	IV
Total carbon (Gt)															
Aboveground	15.2	19.3	22.0	19.0	21.6	21.2	24.4	19.5	22.2	17.2	19.5	18.1	20.8	15.5	17.6
Below-ground	49.2	38.3	32.2	38.9	29.4	36.2	30.0	32.7	26.9	34.0	26.0	31.0	24.7	30.4	23.7
Total	64.4	57.6	54.2	57.9	51.0	57.4	54.4	52.2	48.1	51.2	45.5	49.1	45.5	45.9	41.3
Change in total carbon (Gt)															
Aboveground		4.1	6.7	3.8	6.4	6.0	9.1	4.2	7.2	2.0	4.3	2.9	5.6	0.3	2.4
Below-ground		−10.9	−17.0	−10.2	−19.7	−12.9	−19.2	−16.5	−22.3	−15.2	−23.2	−18.2	−24.5	−18.8	−25.5
Total		−6.8	−10.3	−6.4	−13.3	−6.9	−10.1	−12.3	−15.1	−13.2	−18.9	−15.3	−18.9	−18.5	−23.1
Fraction of change in total carbon															
Aboveground	1.25	1.26	1.44	1.25	1.42	1.39	1.60	1.28	1.47	1.13	1.28	1.19	1.37	1.02	1.15
Below-ground	0.77	0.78	0.65	0.79	0.60	0.74	0.61	0.67	0.55	0.69	0.53	0.63	0.50	0.62	0.48
Total	0.86	0.89	0.84	0.90	0.79	0.89	0.84	0.81	0.77	0.79	0.71	0.76	0.71	0.71	0.64

[a]Case a, changes in net primary productivity and soil respiration. Case b, changes in net primary productivity, soil respiration, and the area of each forest-cover type. Case c, changes in net primary productivity, soil respiration, the area of each forest-cover type, and a 25% increase in annual area burned. Case d, changes in net primary productivity, soil respiration, the area of each forest-cover type, and a 50% increase in annual area burned.
[b]Scenarios I and III assumed lower increases in net primary productivity and soil respiration than scenarios II and IV. Scenarios I and II were based on assessments for individual ecozones, whereas scenarios III and IV were for average values over the entire North American boreal forest.
[c]Baseline values are total carbon for the entire North American boreal forest and fraction of change estimated by Kasischke and colleagues (1995) using the assumptions of case d.

The impetus behind this initial modeling study actually came from the scientific community studying the influence of biomass burning on the global carbon budget. At that time, this community thought that compared with other biomes (e.g., savannas and tropical forests), fire in the boreal forest played a relatively minor role in regulating the atmospheric concentration of carbon-based greenhouse gases because of two factors (Seiler and Crutzen 1980): (1) the relatively small areas burned by fires in the boreal forests (estimated to be about 1 million ha yr^{-1}); and (2) the perception that fire in the boreal forest represented no net transfer to the atmosphere because the carbon released during a fire was subsequently taken out of the atmosphere through the regrowth of plants and trees (e.g., that the boreal forest is in a steady state).

The purpose of the original modeling study was twofold: (1) to demonstrate the controlling role fire plays in carbon cycling in the boreal forest region; and (2) that under a climate warming scenario, boreal forests would not remain in a steady-state condition, and large amounts of carbon would be transferred to the atmosphere through a variety of pathways.

The original model developed by Kasischke and associates (1995) was developed not on direct observations or experiments but on a fundamental understanding of the role of fire in the boreal forest and hypotheses on how climate will interact with these processes to control longer-term carbon storage in this region. These hypotheses led to the research in the Alaska boreal forest discussed in this book. In addition, many of these hypotheses are supported by recent research conducted in other regions of the boreal forest. This recent research has shown the following hypotheses to be true:

- *Hypothesis 1. Increasing atmospheric temperatures in the boreal forest region will result in increased fire activity.* Since the initial study of Kasischke and colleagues (1995) nearly a full decade has passed. During this time period, analysis of climatic data has shown that there has clearly been a warming trend in the boreal forest region since the mid-1960s (Fig. 1.1). This warming trend correlates to increased fire activity in the boreal forest region during the same time period (Fig. 1.2; Chapter 15). In addition, more refined modeling approaches show that increases in air temperature in the future should lead to increased fire activity in the boreal region (Chapter 20).
- *Hypothesis 2. Increased fire severity will result in increased soil temperatures in boreal forests.* This hypothesis has been directly confirmed through field measurements in the Alaskan boreal forest (Chapter 12), which showed that higher levels of organic soil consumption during fire resulted in a greater warming of the mineral soil profile. This observation is also consistent with modeling studies (Chapter 11).
- *Hypothesis 3. The increased soil temperatures following fires result in increased rates of soil respiration in boreal forest.* Although this hypothesis was partially addressed through previous experiments that experimentally manipulated the soil temperatures in Alaskan boreal forests, during the past several years, field measurements have demonstrated elevated soil-respiration rates in fire-disturbed boreal forests (Chapter 11).

- *Hypothesis 4. Increased fire severity will result in higher levels of carbon being released through consumption of large amounts of organic soil.* One of the key uncertainties in estimating the levels of carbon release from fires in boreal forest are the amounts of ground-layer carbon consumed. Until recently, the prevailing consensus was that fires consumed relatively little organic soil and released about 6 t C ha^{-1}. Measurements in the Alaskan boreal forests have clearly shown that much higher levels of carbon are released during fires, especially in areas underlain by deep organic soil profiles. Data presented in Chapter 10 show that in black spruce forests, up to 75 t C ha^{-1} can be released during fires. These studies show that the potential for increased consumption of ground-layer carbon as a result of more severe fires certainly exists.

In summary, the sensitivity study presented in this chapter supports the overall theme of this book (i.e., that fires in the boreal forest play a central role in the exchange of carbon between this biome and the atmosphere). Through a series of processes, the continuing rise in atmospheric temperatures projected for this region will act to release much of the carbon stored in the deep organic soils present in this region. Under the worst-case scenario presented in this chapter, the upper organic-soil layers of the North American boreal forest will lose more than 50% of their stored carbon if the projected rises in atmospheric temperature occur. Because much of this carbon took centuries to millennia to accumulate, this represents a net loss of carbon to the atmosphere. In effect, fire represents a positive feedback to the atmosphere because the gases released from fires will serve to increase the rates of global warming.

One important aspect that the sensitivity model presented in this chapter cannot address is the rate at which the change in the boreal forest carbon will occur. Such an estimate can only be made by using more sophisticated models that account for the effects of climate warming and fire on the rates of soil respiration and soil carbon accumulation, as well as their effects on the patterns of forest cover and net primary production. The development of these higher-order models is only now just beginning.

References

Haxeltine, A., and I.C. Prentice. 1996. BIOME3: an equilibrium terrestrial biosphere model based on ecophysiological constraints, resource availability, and competition among plant functional types. *Global Biogeochem. Cycles* 10:693–709.

Kasischke, E.S. 1996. Fire, climate change, and carbon cycling in Alaskan boreal forests, pp. 827–833 in J.S. Levine, ed. *Biomass Burning and Climate Change,* Vol. 2: *Biomass Burning in South America, Southeast Asia, and Temperate and Boreal Ecosystems, and the Oil Fires of Kuwait.* MIT Press, Cambridge, MA.

Kasischke, E.S., N.L. Christensen, Jr., and B.J. Stocks. 1995. Fire, global warming and the mass balance of carbon in boreal forests. *Ecol. Appl.* 5:437–451.

Lowe, J.J., K. Power, and M.W. Marsan. 1996. *Canada's Forest Inventory 1991: Summary by Terrestrial Ecozones and Ecoregions.* Information Report BC-X-364E. Pacific Forestry Centre, Canadian Forest Service, Victoria, BC, Canada.

McGuire, A.D., J.M. Melillo, D.W. Kicklighter, Y. Pan, X. Xiao, J. Helfrich, B. Moore III, C.J. Vorosmarty, and A.L. Schloss. 1997. Equilibrium responses of global net primary production and carbon storage to doubled atmospheric carbon dioxide: sensitivity to changes in vegetation nitrogen concentration. *Global Biogeochem. Cycles* 11:173–189.

Peng, C., and M.J. Apps. 1998. Simulating carbon dynamics along the boreal forest transect case study (BFTCS) in central Canada. 2. Sensitivity to climate change. *Global Biogeochem. Cycles* 12:393–402.

Peng, C., M.J. Apps, D.T. Price, I.A. Nalder, and D.H. Halliwell. 1998. Simulating carbon dynamics along the boreal forest transect case study (BFTCS) in central Canada. 1. Model testing. *Global Biogeochem. Cycles* 12:381–392.

Randerson, J.T., M.V. Thompson, C.M. Malmstrom, C.B. Field, and I.Y. Fung. 1996. Substrate limitations for heterotrophs: implications for models that estimate the seasonal cycle of atmospheric CO_2. *Global Biogeochem. Cycles* 10:585–602.

Seiler, W., and P.J. Crutzen. 1980. Estimates of gross and net fluxes of carbon between the biosphere and atmosphere from biomass burning. *Clim. Change* 2:207–247.

Smith, T.M., and H.H. Shugart. 1993. The transient response of carbon storage to a perturbed climate. *Nature* 361:563–566.

Thompson, M.V., J.T. Randerson, C.M. Malmstrom, and C.B. Field. 1997. Change in net primary production and heterotrophic respiration: how much is necessary to sustain the terrestrial carbon sink? *Global Biogeochem. Cycles* 10:711–726.

Townsend, A.R., P.M. Vitousek, and E.A. Holland. 1992. Tropical soils could dominate the short-term carbon cycle feedbacks to increased global temperatures. *Clim. Change* 22:293–303.

Van Wagner, C.E. 1978. Age-class distribution and the forest fire cycle. *Can. J. For. Res.* 8:220–227.

Yarie, J. 1981. Forest fire cycles and life tables: a case study from interior Alaska. *Can. J. For. Res.* 11:554–562.

Index

A

Abies balsemea, 69, 115, 120–121, 126–127, 160, 262, 263, 268
Abies lasiocarpa, 77, 115–116, 121–122, 124, 127, 260, 268
Abies sibirica, 248
Abies spp., 133, 159, 247, 248
Aboveground biomass/carbon, 19–20, 33–34, 39, 178–179, 186–188, 240–255, 266–270, 290, 292–299, 337–339, 379–380, 411, 427, 433, 442–443, 446–447, 449
  sampling, 178–179
Acer negundo, 116
Acer saccharum, 116, 263
Acetonitrile (CH_3CN), 37
Advanced Very High Resolution Radiometer, *see* AVHRR
Acrylonitrile (CH_2CHCN), 37
Alaska, 24, 259, 264–265, 277–279, 280–283, 322–323, 337–338, 368–369, 408–411
  Alaska Range, 24, 199, 223, 260, 264, 285
  Anchorage, 90, 278
  Barrow, 12, 13
  Big Lake, 90
  Bonanza Creek, 341
  Brooks Range, 24, 112, 115, 199, 224, 260, 264, 285
  Copper Plateau, 259–260
  Copper River, 278
  Delta Junction, 201, 209
  Dry Creek, 175–176
  Fairbanks, 217, 260, 278, 341
  Fort Greely, 175–176, 217, 224–225, 227
  Gerstle River, 326, 340
  Hajdukovich Creek, 174–176, 217–218, 223–225, 340
  Interior Highlands, 259
  Kuskokwim Mountains, 260
  North Slope, 14
  Nulato Hills, 260
  Old Crow Basis, 259, 265
  Old Crow Flats, 259, 265
  Porcupine Creek, 175–176
  Porcupine River, 224
  Prudhoe Bay, 279
  Rosie Creek, 175–176
  Tanana River, 174, 175, 199, 216–217, 223, 224, 232
  Tetlin, 175–176, 204
  Tok, 26, 174–176, 186–194, 199, 201–205, 209, 217, 220, 220
  Yukon Flats, 259, 265

Alaska (*cont.*)
  Yukon River, 199, 273
Albedo, 25, 44, 109, 113–114, 152, 207, 215, 416–418
Alnus spp., 52, 122, 126, 262, 265
Ammonia (NH_3), 37, 38
Ammonium (NH_4^+), 43–44, 45
Arctostaphylos spp., 215
Ash
  from biomass burning, 37, 45, 52, 57, 198, 208–209, 431
  from volcanoes, 164, 260
Atmospheric temperature increase, 1–2, 7, 14, 24, 26–27, 44, 50, 74, 82, 85–90, 93–94, 155, 160, 163, 195, 220–221, 227, 232, 364, 368–371 378, 408, 423, 440, 450–451
AVHRR (Advanced Very High Resolution Radiometer), 312–316
  mapping of fire boundaries, 34, 140, 321–324, 343
  mapping of forest cover, 319–321, 342
  monitoring of seasonal vegetation patterns, 319, 327, 407–408

B

Belarus, 56
Betula alleghaniensis, 116, 263
Betula ermanii, 248
Betula nana, 254, 261
Betula papyrifera, 77, 115, 121, 123, 126, 127, 159, 163, 214, 219, 260–261, 264–265, 407
Betula platyphylla, 54
Betula spp., 133, 159, 247, 248
Biomass burning, 10, 22, 31–37, 40, 43–45, 60, 82, 108, 119, 137, 173–194, 220–221, 227, 233, 279, 289, 320, 326, 343, 378–379, 384–386, 394–395, 411, 413, 429–430, 450
Bogs, 78, 113, 125–126, 132, 262–263, 265, 292, 304, 392, 395, 424–429, 431–433, 435
Bromine, 42–43

C

^{14}C analysis, 209, 214
C/N ratio, 108, 202
Calcium, 208
Canada, 1, 15, 66–69, 91, 112–113, 153–154, 157–159, 166, 241, 259, 266, 275–277, 279–280, 283–286, 320–321, 339, 357–358, 364, 369–374, 381, 428
  Alberta, 66, 68, 115, 117, 153, 163–165, 261, 264, 280
  British Columbia, 68, 68, 79, 115, 117, 163, 260–261, 264, 280, 362
  Caribou Range, 79
  Chinchaga River, 117
  Great Slave Lake, 79
  Hay River, 79
  Hudson Bay, 69, 77, 112, 112, 159, 259, 262–263
  Inuvik, 163
  James Bay, 69, 263
  Labrador, 66, 116, 159, 262, 280
  Liard River, 78
  MacKenzie River, 77–78, 115
  Manitoba, 14, 66, 163, 201, 261–264, 280, 374, 400
  New Brunswick, 115
  Newfoundland, 66, 111, 115, 263, 280
  Northwest Territories, 60, 66, 67, 68, 77–82, 93, 115, 117, 261, 280, 324–326
  Nova Scotia, 66, 159
  Ontario, 44, 66, 69–77, 113, 117, 161, 167, 262–263, 280, 320, 360, 370
  Prince Edward Island, 66
  Quebec, 66, 67, 115, 156, 160–161, 167, 262–263, 280
  Red River, 215
  St. Lawrence Region, 152, 263
  Saskatchewan, 66, 163–165, 261–262, 280, 400
  Saskatchewan, River, 275
  Slave River, 78, 79
  Yukon Territory, 66, 67, 68, 77, 113, 115, 260, 280
Carbon cycle, 3, 10–14, 19–22, 24–25, 103–109, 173, 212, 239, 327–328, 342–343, 347–353, 377–378, 406, 450
Carbon dioxide (CO_2)
  atmospheric concentration, 1–2, 7–13, 25, 31, 85, 221, 347, 407, 434–435
  emissions from fires, 25, 27, 35–36, 38, 39, 173–195, 197, 200, 211, 305–306, 382–386, 409–413, 428–431, 434–435
  oceanic uptake, 10
  soil emissions, *see* Soil, respiration
  terrestrial exchange, 11, 173, 239, 255, 327, 343, 344, 348–353, 364, 369, 423

Carbon emissions, *see* Carbon dioxide
Carbon monoxide (CO), 31–32, 35, 37–39, 40–41, 43, 45, 429–430
Carbon pool (reservoir), 1, 10, 49–50, 103–106, 108, 152, 202, 247, 254–255, 289, 328, 348, 353, 377, 399
Carbonyl sulfide (COS), 37
Ceratodon prupures, 199
Charcoal, 36
  in lake sediments, 51, 106, 156–157, 159, 160, 161–166, 276
  in soil, 51, 208–209, 426, 428, 432, 437
China, 34, 53–54, 57, 92, 111, 323–324, 401
Chlorine, 42
Chosenia arbutifolia, 248
Cladonia sp., 38
Climate change, 1, 3, 14, 20, 25, 27, 44–45, 50, 67–68, 75, 83, 85–86, 88–89, 92–94, 104, 152, 155–159, 161, 162, 165, 167, 168, 221, 232, 279, 285, 343, 347, 349, 357, 364–365, 368–371, 374, 378, 408, 423, 431–433, 440–441, 445, 448, 450–451
Climate warming, *see* Atmospheric temperature increase
Coarse woody debris, 296, 299
Controlled burns, *see* Fire experiments
Cyanogen (NCCN), 37

D

Daily severity rating (DSR), 371, 373–374
Disease, *see* Insect, disease and pathogens
Drought code, 358
Duff, 38–39, 87–88, 90, 96, 122, 358–359
Duff moisture code (DMC), 358

E

Earth hummocks, 182, 199, 220, 222–223
Earth Resources Satellite (ERS), 331–333
  estimating aboveground biomass, 337–339
  estimating soil moisture, 339–341
  monitoring fire-disturbed forests, 339–341
  monitoring temperature-related phenomena, 341–342
Ecoclimatic regions of Canada, 259
Ecological models, 389–403
  boreal landscape models, 397–402
  carbon fluxes, 391, 394–395
  disturbance processes, 393, 397
  individual-based models, 400
  landscape dynamics, 392–393, 397–402
  regional models, 401–402
  role-based models, 400–401
  scaling processes, 390–392, 396
  succession, 14, 49, 72, 92, 104–109, 115, 121–125, 127, 134–135, 145, 159, 199, 214–233, 265, 305, 353, 378, 392–393, 395, 401–402, 408–411, 432, 435
Ecoregions of Alaska, 259
Ecotone
  boreal-mixed forest, 113
  boreal-tundra, 22, 113, 259–260
  steppe-forest, 63, 250
  taiga-tundra, 134
Ecozones of North America
  Alaska Boreal Interior, 259, 264–265, 268–269, 286, 380, 382–384, 445, 447
  Boreal Cordillerra, 259–261, 268–269, 286, 380, 383–384, 446
  Boreal Plains, 259, 263–264, 268–269, 286, 380, 383–384, 447
  East Boreal Shield, 259, 263, 268–269, 272, 286, 380, 382–383, 447
  East Taiga Shield, 259, 262, 268–269, 286, 380, 383–384
  Hudson Plains, 259, 261, 263–264, 266, 268–269, 286, 380, 383, 447
  Prairie, 267
  Taiga Cordillera, 259
  Taiga Plains, 77, 259, 261, 266, 268–269, 286, 380, 383–384, 446
  West Boreal Shield, 259, 262, 268–269, 272, 282, 286, 380, 383, 447
  West Taiga Shield, 77, 259, 261–262, 268–269, 286, 380, 383–384, 445–446
Ecozones (ecoregions) of Russia
  forest steppe, 249–250, 253
  forest tundra, sparse taiga, and meadow forests, 132, 133, 134, 137, 141, 144, 145, 247, 249, 253, 290, 293, 294, 297, 298, 302, 304, 307
  middle taiga, 133, 141, 144, 249, 253–2544, 290, 293, 294, 295, 297, 298, 302, 307
  mixed forests, deciduous forests, and

Ecozones (ecoregions) of Russia (*cont.*)
forest steppe, 133, 141, 290, 293, 294, 297, 298, 302, 304, 307
northern taiga, 133, 134, 137, 141, 144, 145, 253–254, 290, 249, 293, 294, 295, 297, 298, 302, 307
southern taiga, 133, 136, 141, 144, 250, 252–254, 249, 290, 293, 294, 295, 297, 298, 300, 302, 307
steppe, semidesert, and desert, 132, 133, 141, 249, 253, 290, 293, 294, 302, 307
subarid, 249
subarctic and tundra, 133, 141, 249, 253, 290, 293, 297, 302, 307
sub-boreal, 249
Energy exchange, between land and atmosphere, 49, 60, 104, 107–109, 114, 203–208, 215–216, 221–224, 231–237, 337, 378, 416–420
Estonia, 52, 56
Eurasia, 3, 15, 23, 49–61, 111, 134–135, 137–142, 237, 239–240, 245–252, 255, 292, 295–296

F

Feathermoss (*Pleurozium,* spp.), 115, 123, 125, 175
Fens, 118, 265, 424–429, 431, 435
Fibric soil, 22, 177–178, 179–186, 199–202, 208–209, 231, 290
Fine fuel moisture code (FFMC), 358, 360–361
Finland, 50, 51–52, 55, 111
Fire, 1, 10, 11, 12, 14, 15, 20, 22–25, 33–35, 37–39, 49–50, 52–56, 63, 103–104, 108–109, 133–134, 173–174, 197–198, 255. *See also* Biomass burning
adaptations by vegetation, 22–23, 120–127, 153, 167
area, 2, 23–24, 32–34, 50, 52–56, 67, 69, 73–74, 78–79, 85, 137–141, 269, 275, 277, 280, 283–286, 293, 323, 426–427. *See also* Fire, size
behavior, 36, 44, 53, 70, 87–88, 90, 91, 96 117–120, 135, 156, 173, 326, 358–359, 363, 368–374, 393
characteristics, 23, 33–35, 36, 44, 69, 116–120, 134–137, 142–145, 153–156, 284–286
cycle, 116, 121, 126, 134, 136, 140, 263, 269, 272, 303, 433, 440, 441–444
danger, *see* Fire, hazards; Forest Fire Danger Rating System
database, *see* Large-fire database
distribution, 3, 23, 73, 284–286
ecology, 120–127, 132–134, 151, 197–198, 214–233, 391–392
emissions, *see* greenhouse gas emissions from fires
experiments, 37–40, 60, 87, 120, 136, 173, 195, 233, 300
frequency, 14, 23–24, 27, 35, 44, 55, 60, 96, 120, 125, 126, 136, 139, 142–143, 145, 151, 154, 195, 229–230, 232, 262, 292, 295, 364, 397, 440, 431, 441
hazards, 53, 54, 55, 57, 59, 60, 68, 70, 81, 87–88, 90, 91–92, 93, 95, 97, 123, 126, 134, 166, 275, 358, 360, 362, 365, 369–374
history, 51–56, 60, 61, 67, 69, 72–73, 78–80, 139, 151–168, 274–276, 279
ignition, 22, 66, 88, 89, 95, 120, 125, 142, 151, 154, 156, 160, 166, 349, 357–359
human, 33, 54, 55, 69–70, 87, 139, 275, 276, 324, 357, 360–362, 364, 370
lightning, 23, 26, 33, 51, 54, 55, 57, 70, 79, 85, 87, 92, 117, 120, 126, 138, 139, 161, 223, 265, 275, 277, 279, 324, 357, 362–363, 370
intensity, 35, 79, 96, 118–119, 136, 151, 192–193, 195, 221, 223–225, 227–231, 232, 295, 370–372
interval, 117
management, 25–27, 55, 56–59, 61, 66–84, 85–97, 276–279
detection, 23, 57, 59, 61, 67–68, 70, 83, 87, 89, 97, 276–277, 326, 359, 362, 363
fuel manipulation, 87
mitigation, 25–27, 72–74, 76–77, 82, 85–97
prescribed fire, 51, 52, 57, 59, 61, 63, 67, 75, 77, 87–88, 91–92, 97, 139, 143, 304
prevention, 67, 87, 97
protection, 57, 58–59, 67, 70–71, 74, 76, 83, 97, 142, 277
suppression, 25–27, 57, 59, 67–68, 70–71, 74, 83, 88–97, 277, 326
of wildlife, 27, 75, 81, 92–93, 94, 326

mapping, 279–284, 321–326
monitoring with satellites, 321–327, 339–340
occurrence, 2–3, 33, 72–73, 138–142, 240, 357–365
occurrence models, 357–365
peatland, 24, 423–424, 426–428
policy, 25–27, 56–59, 70, 80–82, 83–84, 85–87, 94–97
prediction, 70, 76, 277
records, 279–284
regime, 1–2, 14, 27, 50, 66, 67, 68, 75–76, 82, 116, 135–136, 137–142, 151, 153, 155, 159–166, 369–371, 397, 400, 403
return interval, *see* Fire cycle
season, 14, 26, 50, 53–55, 57, 59, 70, 75, 78–79, 83, 90, 95, 96, 116–117, 120, 143, 151, 166, 265, 281, 323, 324, 361, 364, 370, 372, 374, 378, 432–433
severity, *see* Fire, intensity
size, 23–24, 72–73, 79, 90, 91, 117, 145, 151, 160, 277, 280, 321, 323, 325–326. *See also* Fire, area
spread rate, 44, 96, 118–120, 126, 136, 358–359, 363
types
crown fire, 38, 39, 44, 49, 60, 96, 116–120, 121–124, 126, 135–138, 141, 143, 153, 197, 290, 293, 299, 304, 305, 306, 359
flaming fire, 90, 118–119
ground fires, 117–118, 121, 135–138, 141–143, 173, 197, 199, 290, 292, 300–301, 304, 305–306, 427–428, 431, 433
on-ground fires, *see* surface fires
smoldering fire, 24, 36, 37–39, 118, 136, 223, 358–359, 362–363, 413, 426, 429
surface fires, 38, 39, 50, 116–119, 123–124, 134, 135–138, 140–144, 153, 290, 292, 300–301, 304–306, 359, 427, 427, 432, 433, 435
weather, 14, 75, 81, 153, 160, 279, 365, 370, 372
Fire Behavior Prediction (FBP) system, 120, 123, 124, 363
Fire incidence probability (FIP), 152
Fire Weather Index (FWI), 154–155, 221, 341, 358, 365, 371
First Nations people, 67, 77–78
Forest combustibles (FCs), 135–137, 289–292, 295–297, 299, 305. *See* also Coarse woody debris
Forest Fire Danger Rating System, 120, 257
Forest fund, (FF), 59, 132, 137–141, 239, 244, 289, 303, 305
Forest inventory method, 241–243
Fagus grandifolia, 159
Fraxinus nigra, 116, 263–264
Fraxinus pennsylvanica, 116

G
General circulation model (GCM), 1, 14, 154–155, 220, 364–365, 369–372, 374, 399
Geobotanical method, 240–241
Geographic information system (GIS), 60–61, 269, 274, 279, 285, 317, 325, 371, 379, 381, 383, 420
Germany, 51, 56
Global warming, *see* Atmospheric temperature increase
Great Lakes, 157, 162
Great Lakes–St. Lawrence region, 69, 72, 152, 161
Greenhouse effect, 7, 15
Greenhouse gas, 7, 9. *See also* Carbon dioxide; Carbon monoxide; Methane; Nonmethane hyrdocarbons
emissions from fires, 24, 31–33, 35–40, 74, 82
reducing emissions, 85, 88, 90, 92, 94–97
Gross primary production (GPP), 350, 377–378

H
Heterotrophic respiration, *see* Soil respiration
Holocene, 157–159, 163, 166, 232
Humic soil, 52, 115, 118, 177–178, 182–186, 199–203, 208–209, 220, 228, 231 290, 304–305
Hydrogen cyanide (HCN), 37
Hydroxyl radical (OH), 40–41, 43, 44–45
Hylocomium splendens, 200

I
Imaging radar
basic characteristics, 332–333
monitoring of soil moisture, 343
monitoring of wetland ecosystems, 342
principles of monitoring vegetated landscapes, 333–336

Insect, disease and pathogens, 20, 22, 26, 50, 53, 93, 111, 126–127, 160, 255, 296, 300, 328, 333, 364, 369, 390–391, 400,

J

Japanese Earth Resources Satellite (JERS), 333, 343
 estimating aboveground biomass, 337–338

K

Kyoto Protocol, 2, 15, 368, 403

L

Landsat, 312–314, 316–317
 mapping of fire severity, 324–328, 386
 mapping of forest cover, 320, 337, 342, 386, 409–411
 mapping of surface albedo, 417–418
Large-fire database (LFDB), 279, 283–286
Larix gmelinii, 53
Larix laricina, 115, 125, 219, 260–265
Larix sibirica, 54
Larix spp. 50, 133, 145, 245, 248
Latent heat flux, 419–420
Latvia, 52, 56
Ledum palustre, 261
Lichens, 23–25, 78, 106, 108, 112, 113, 115, 123, 124, 125, 126, 137, 200, 223, 244, 261, 262, 264, 294, 300, 326, 339, 357, 379, 391, 394
Lightning detection systems, 87, 90
Lithuania, 52, 56
Litter, 22–25, 38–38, 88, 90, 103, 105, 108, 118, 122–123, 135–137, 143, 177, 180–183, 199, 223, 240–241, 244, 250, 253, 260, 290, 295–297, 300–301, 304–305, 326, 379
Little Ice Age, 159, 161–165

M

Magnesium, 208
Mauna Loa Observatory, 7–8, 12, 13
Medieval Warm Period, 159, 162
Mesic soil, 177–178
Methane (CH_4), 31–32, 35, 37–39, 40–41, 42, 45, 343, 364, 416, 423, 429–431
Methyl bromine (CH_3Br), 32, 42–43
Methyl chloride (CH_3Cl), 31–32, 38
Minnesota, 153, 162, 167
Models of, *See also* Ecological models
 aboveground biomass, 189–191, 240–242
 biogenic flux, 290–293, 301–303, 414–416
 biomass burning, 32, 188–195, 379–382, 409–413
 boreal forest carbon cycling, 348–352
 carbon emissions from biomass burning, 191–194, 289–290, 292–301, 409–413
 carbon level as a function of fire frequency, 442
 changes in carbon storage, 440–451
 emission ratios, 35, 429
 emission factors, 35
 fire intensity, 118, 191–194
 fire occurrence, 360–363
 fire regimes, 369–372
 fraction of carbon consumed, 189–192, 299–301, 381–382
 ground-layer carbon, 189–191
 net biome production, 348
 radar scattering from vegetated surfaces, 334–336
 soil heat diffusion, 203, 206–208
 soil decomposition, 302–304
 stand age distribution, 441
 surface energy exchange, 416–420
Mongolia, 54–55, 57–58
Monthly severity rating (MSR), 371–373
Mortmass, 105, 240–242, 249, 250, 295
Moss, 23–25, 38, 55, 78, 104, 106, 108, 112, 114, 115, 119, 123, 124–125, 126, 137, 175, 177, 180–183, 197–198, 199–201, 208–210, 223, 231, 244, 261, 264, 265, 292, 294, 299, 300, 326, 339, 357, 379, 394, 408, 425, 431
Muskeg, 48, 113, 262–264

N

Native people, 66, 71, 74, 93–94, 275, 278. *See also* First Nations people
Net ecosystem production (NEP), 200, 353, 377, 385, 397, 424
Net primary production (NPP), 22, 103–104, 106, 198, 209, 328, 337, 341, 349–350, 353, 378, 385, 407–408, 423, 440–442, 444–445, 448–449, 451
Nitrate (NO_3^-), 37, 43–44
Nitric oxide (NO), 31–32, 38, 42, 44, 45

Nitrogen, 35, 36, 43–44, 45, 108, 178, 200, 202, 389, 399
Nitrogen dioxide (NO_2), 42, 43
Nitrous oxide (N_2O), 32, 37, 44, 45
Nonmethane hydrocarbons (NMHC), 31, 35, 37–38, 429–430
Normalized difference vegetation index (NDVI), 315, 319–320, 322, 408
North America, ecozones, *see* Ecozones of North America
North Korea, 401
Norway, 50, 56, 111

O
On-ground vegetation combustibles, 295–297
Ozone (O_3), 32, 38, 41, 42–43

P
Pathogens, *see* Insect, disease and pathogens
Peat fire, *see* Ground fires
Peatland, 19–20, 22, 24, 27, 50, 55, 163, 223, 239, 240, 242, 244, 245, 246, 250–251, 253, 264, 290, 353, 392, 423–435
Permafrost, 21, 22, 24–25, 50, 90, 104, 106, 109, 111, 112, 113–114, 115, 121, 125, 137, 143, 147, 173, 185, 186, 189, 198, 199, 203–206, 214–215, 220, 223–224, 231, 250, 261, 263, 265, 295, 301, 304, 306, 370, 381, 392, 395, 413, 417, 424–425, 428–429, 431–432, 435
Phosphorous, 36, 208
Photosynthesis, 10, 12, 21, 31, 44, 104, 327, 348, 351, 377, 390–392, 394
Phytomass, *see* Aboveground biomass
Picea, glauca, 69, 77, 114, 116, 120, 122–124, 126–127, 174–177, 180–184, 187–195, 214–215, 218–219, 224, 228, 232, 260–265, 268, 408
Picea mariana, 69, 77, 114, 116, 120–126, 164, 174–177, 180–195, 214–228, 230–232, 260–265, 268, 408–411, 412–414, 425
Picea obovata, 248
Picea rubens, 116, 263
Picea spp., 133, 160, 161, 166, 245, 247, 248
Pinus banksiana, 69, 77, 115, 120–124, 162, 262–264, 268
Pinus contorta, 77, 115–116, 120–124, 261, 268
Pinus koraiensis, 133, 248
Pinus pumila, 245–247, 254
Pinus resinosa, 69, 116, 162, 262, 263
Pinus spp., 50, 133, 145, 160, 161, 163, 164, 247
Pinus siberica, 81, 245, 247, 248
Pinus strobus, 69, 116, 162, 262, 263
Pinus sylvestris, 37, 53, 54, 245, 247, 248
Plant detritus, *see* Litter, Mortmass
Pleurozium, schreberi, 200
Pleurozium spp., 115, 123, 125, 175
Pohlia nutans, 199
Poland, 56
Pollen analysis, 157, 159, 161–164, 166, 168, 276
Polytrichum juniperinum, 199
Populus balsamifera 78, 115–116, 163, 214, 219, 260–265
Populus spp., 133, 247
Populus tremula, 248
Populus tremuloides, 78, 115–116, 120–123, 127, 164, 166, 174–175, 180–184, 186–195, 197–202, 215, 219–220, 222–226, 260–265, 408–411
Populus trichocarpa, 116
Potassium, 36, 208
Prairie, 113, 162, 264, 275
Propionitrile (CH_3CH_2CN), 37

Q
Quercus macrocarpa, 116
Quercus mongolica, 248
Quercus, spp., 133, 161, 162 247

R
Radioactive contamination, 56–57, 61, 63
Rangifer tarandus, 81
Russia (including Former Soviet Union), 1–2, 15, 20, 34, 49–53, 56, 58–61, 109, 112, 132–143, 145, 237–238, 239–255, 274, 286, 289, 292–307, 321, 368–369, 371–372, 378, 401, 432. *See also* Ecozones of Russia
  Altai, 246, 251
  Amur Oblast, 246, 249
  Amur River, 34, 50, 140, 144
  Arkhangelsk, 60
  Azerbaijan, 52
  Belarus, 56
  Bor Island, 37–40
  Chita, 246, 249

Russia (*cont.*)
Far East, 59, 137, 140, 142, 144–145, 239, 245–246, 250, 299–300, 302, 304
Irkutsk, 246, 249
Kamchatka, 246, 249
Karelian Republic, 252
Kazakhstan, 52, 56–57, 299
Kemerov, 246, 249
Khabarovsk, 61, 246, 249
Komi Republic, 302
Krasnoyarsk, 61, 244, 246, 249, 299
Kyrgystan, 52
Lake Baikal, 300
Lena River, 34
Magadan, 246, 249
Moscow, 61
Novosibirsk, 246, 249
Omsk, 249
Primorskye, 246, 249
St. Petersburg, 61
Sakhalin Island/Oblast, 139, 144, 246, 249, 300
Siberia, 26, 37, 53, 59–60, 135–137, 139–140, 142–143, 145, 239, 245–246, 250, 292, 299–300, 304, 323–324, 368
Tajikistan, 52
Tomsk, 139, 246, 249
Tyumen, 139, 245–246, 249
Ural Mountains, 300
Uzbekistan, 52
Yakutia, 244–245, 249–250

S

Salix spp., 122, 126, 174–175, 177–179, 182, 184, 187–194, 261–262, 265
Savannas, 11, 31, 33–34, 36, 39, 161, 162, 350, 400
Sensible heat flux, 418–419
SIR-C/X-SAR, 333, 336, 339
Smoke
aerosols, 32, 57, 323
impact of, 62, 71–72, 74, 87, 275
plume (column), 44
sampling, 36, 38–39
Soil
bulk density, 178–181
burning, 24, 87, 90, 109, 121, 122, 123, 125, 135, 136, 137, 140, 142, 143, 174, 175–178, 179–186, 188–193, 221, 223–225, 413, 290, 304–306, 326–327, 379–382, 393, 413, 427–431, 433–435, 444–445, 448, 451
carbon, 20, 22, 34, 49, 50, 103–106, 177–178, 179–186, 190, 200–201, 242, 244, 250–252, 269, 270, 379–381
carbon percentage, 178–181
chemistry, 202, 208–209
decomposition, 12, 14, 25, 108, 197–203, 209–211, 302–305, 394
emissions of carbon, 200, 209–211, 305–306
emissions of nitrogen, 45
formations, 112, 261–264
moisture conditions, 22, 24–25, 87, 104, 107–109, 112–113, 117, 122, 124–125, 134, 189–194, 214–216, 220, 227–233, 339–341, 343, 414–416, 428, 431–432, 434
organic, 22, 34, 52, 177–178
pH, 200, 202, 208–209
respiration, 12, 14, 25, 45, 103–104, 121, 413–416, 197–203, 209–211, 290–292, 301–303, 305, 307, 353, 390, 394, 413–416, 430–434, 440–441, 444–445, 448, 450–451
sampling, 175–178, 200
temperature, 24–25, 104, 107–109, 111, 113–115, 121, 122, 125, 134, 189–194, 203–208, 214–216, 220, 221–224, 227–233, 414–417, 431–434, 450–451
South Pole, 12–13
SPOT, 313–314, 316–317
Sphagnum, spp., 115, 124–126, 137, 175, 200, 295, 300, 425
Steppe, 53–55, 63, 135, 153
Sulfate (SO_4^-), 37
Sulfur, 36, 37, 38
Sulfur dioxide (SO_2), 37
Surface emissivity, 418
Surface energy budget, models, 416–420
Swamps, 55, 261–263, 395, 424–429, 431, 433, 435
Sweden, 50, 56–57, 111

T

Taiga, 22, 55, 113, 139–140, 142, 143, 198, 250, 259–261, 304
Temperate forests, 11, 15, 19–20, 33, 36–37, 53, 55, 56, 111, 113, 114, 241, 370, 377

Thuja occidentalis, 116, 122, 125, 263–264, 425
Tilia amurensis, 248
Tree mortality, 22, 26–27, 107, 119, 124, 136, 137, 143–145, 160, 167, 289, 290, 370, 400,
Tropical forests, 11, 15, 19–20, 33–36, 315, 350
Tsuga canadensis, 116, 159, 263
Tundra, 14, 19–20, 22, 24, 69, 111, 113, 132, 133, 134, 135, 140, 141, 142, 153, 158, 160, 163, 260, 261, 262, 290, 293, 295, 320, 342, 343, 369, 370

U
Ulmus americana 116, 263–264
Ulmus, spp., 133

V
Vegetation zone, *see* Ecotone, Ecozones of North America

W
Wisconsin, 162, 167

Y
Yellowstone National Park, 26, 165

Ecological Studies

Volumes published since 1993

Volume 98
Plankton Regulation Dynamics (1993)
N. Walz (Ed.)

Volume 99
Biodiversity and Ecosystem Function (1993)
E.-D. Schulze and H.A. Mooney (Eds.)

Volume 100
Ecophysiology of Photosynthesis (1994)
E.-D. Schulze and M.M. Caldwell (Eds.)

Volume 101
Effects of Land-Use Change on Atmospheric CO_2 Concentrations: South and South East Asia as a Case Study (1993)
V.H. Dale (Ed.)

Volume 102
Coral Reef Ecology (1993)
Y.I. Sorokin (Ed.)

Volume 103
Rocky Shores: Exploitation in Chile and South Africa (1993)
W.R. Siegfried (Ed.)

Volume 104
Long-Term Experiments With Acid Rain in Norwegian Forest Ecosystems (1993)
G. Abrahamsen et al. (Eds.)

Volume 105
Microbial Ecology of Lake Plußsee (1993)
J. Overbeck and R.J. Chrost (Eds.)

Volume 106
Minimum Animal Populations (1994)
H. Remmert (Ed.)

Volume 107
The Role of Fire in Mediterranean-Type Ecosystems (1994)
J.M. Moreno and W.C. Oechel (Eds.)

Volume 108
Ecology and Biogeography of Mediterranean Ecosystems in Chile, California, and Australia (1994)
M.T.K. Arroyo, P.H. Zedler, and M.D. Fox (Eds.)

Volume 109
Mediterranean Type Ecosystems: The Function of Biodiversity (1994)
G.W. Davis and D.M. Richardson (Eds.)

Volume 110
Tropical Montane Cloud Forests (1994)
L.S. Hamilton, J.O. Juvik, and F.N. Scatena (Eds.)

Volume 111
Peatland Forestry: Ecology and Principles (1995)
E. Paavilainen and J. Päivänen

Volume 112
Tropical Forests: Management and Ecology (1995)
A.E. Lugo and C. Lowe (Eds.)

Volume 113
Arctic and Alpine Biodiversity: Patterns, Causes and Ecosystem Consequences (1995)
F.S. Chapin III and C. Körner (Eds.)

Volume 114
Crassulacean Acid Metabolism: Biochemistry, Ecophysiology and Evolution (1995)
K. Winter and J.A.C. Smith (Eds.)

Volume 115
Islands: Biological Diversity and Ecosystem Function (1995)
P.M. Vitousek, H. Andersen, and L. Loope (Eds.)

Volume 116
High-Latitude Rainforests and Associate Ecosystems of the West Coast of the Americas: Climate, Hydrology, Ecology and Conservation (1995)
R.G. Lawford, P.B. Alaback, and E.R. Fuentes (Eds.)

Volume 117
Anticipated Effects of a Changing Global Environment on Mediterranean-Type Ecosystems (1995)
J.M. Moreno and W.C. Oechel (Eds.)

Volume 118
Impact of Air Pollutants on Southern Pine Forests (1995)
S. Fox and R.A. Mickler (Eds.)

Volume 119
Freshwaters of Alaska: Ecological Synthesis (1997)
A.M. Milner and M.W. Oswood (Eds.)

Volume 120
Landscape Function and Disturbance in Arctic Tundra (1996)
J.F. Reynolds and J.D. Tenhunen (Eds.)

Volume 121
Biodiversity and Savanna Ecosystem Processes: A Global Perspective (1996)
O.T. Solbrig, E. Medina, and J.F. Silva (Eds.)

Volume 122
Biodiversity and Ecosystem Processes in Tropical Forests (1996)
G.H. Orians, R. Dirzo, and J.H. Cushman (Eds.)

Volume 123
Marine Benthic Vegetation: Recent Changes and the Effects of Eutrophication (1996)
W. Schramm and P.H. Nienhuis (Eds.)

Volume 124
Global Change and Arctic Terrestrial Ecosystems (1996)
W.C. Oechel (Ed.)

Volume 125
Ecology and Conservation of Great Plains Vertebrates (1997)
F.L. Knopf and F.B. Samson (Eds.)

Volume 126
The Central Amazon Floodplain: Ecology of a Pulsing System (1997)
W.J. Junk (Ed.)

Volume 127
Forest Design and Ozone: A Comparison of Controlled Chamber and Field Experiments (1997)
H. Sanderman, A.R. Wellburn, and R.L. Heath (Eds.)

Volume 128
The Productivity and Sustainability of Southern Forest Ecosystems in a Changing Environment (1998)
R.A. Mickler and S. Fox (Eds.)

Volume 129
Pelagic Nutrient Cycles (1997)
T. Andersen

Volume 130
Vertical Food Web Interactions (1997)
K. Dettner, G. Bauer, and W. Völkl (Eds.)

Volume 131
The Structuring Role of Submerged Macrophytes in Lakes (1998)
E. Jeppesen, M. Søndergaard, M. Søndergaard, and K. Christoffersen (Eds.)

Volume 132
Vegetation of the Tropical Pacific Islands (1998)
D. Mueller-Dombois and F.R. Fosberg

Volume 133
Aquatic Humic Substances (1998)
D.O. Hessen and L.J. Tranvik

Volume 134
Oxidant Air Pollution Impacts in the Montane Forests of Southern California: A Case Study of the San Bernardino Mountains (1999)
P.R. Miller and J.R. McBride (Eds.)

Volume 135
Predation in Vertebrate Communities: The Bialowieza Primeval Forest as a Case Study (1998)
B. Jedrzejewska and W. Jedrzejewski

Volume 136
Landscape Disturbance and Biodiversity in Mediterranean-Type Ecosystems (1998)
P.W. Rundel, G. Montenegro, and F.M. Jaksic (Eds.)

Volume 137
Ecology of Mediterranean Evergreen Oak Forests (1999)
F. Roda, J. Retana, C.A. Gracia, and J. Bellot (Eds.)

Volume 138
Fire, Climate Change, and Carbon Cycling in the Boreal Forest (2000)
E.S. Kasischke and B.J. Stocks (Eds.)

Volume 139
Responses of Northern U.S. Forests to Environmental Change (2000)
R. Mickler, R.A. Birdsey, and J. Hom (Eds.)